High Content Screening

METHODS IN MOLECULAR BIOLOGY™

John M. Walker, SERIES EDITOR

383. **Cancer Genomics and Proteomics:** *Methods and Protocols,* edited by *Paul B. Fisher, 2007*
382. **Microarrays, Second Edition:** *Volume 2, Applications and Data Analysis,* edited by *Jang B. Rampal, 2007*
381. **Microarrays, Second Edition:** *Volume 1, Synthesis Methods,* edited by *Jang B. Rampal, 2007*
380. **Immunological Tolerance:** *Methods and Protocols,* edited by *Paul J. Fairchild, 2007*
379. **Glycovirology Protocols,** edited by *Richard J. Sugrue, 2007*
378. **Monoclonal Antibodies:** *Methods and Protocols,* edited by *Maher Albitar, 2007*
377. **Microarray Data Analysis:** *Methods and Applications,* edited by *Michael J. Korenberg, 2007*
376. **Linkage Disequilibrium and Association Mapping:** *Analysis and Application,* edited by *Andrew R. Collins, 2007*
375. **In Vitro Transcription and Translation Protocols:** *Second Edition,* edited by *Guido Grandi, 2007*
374. **Quantum Dots:** *Methods and Protocols,* edited by *Charles Z. Hotz and Marcel Bruchez, 2007*
373. **Pyrosequencing® Protocols,** edited by *Sharon Marsh, 2007*
372. **Mitochondrial Genomics and Proteomics Protocols,** edited by *Dario Leister and Johannes Herrmann, 2007*
371. **Biological Aging:** *Methods and Protocols,* edited by *Trygve O. Tollefsbol, 2007*
370. **Adhesion Protein Protocols,** *Second Edition,* edited by *Amanda S. Coutts, 2007*
369. **Electron Microscopy:** *Methods and Protocols, Second Edition,* edited by *John Kuo, 2007*
368. **Cryopreservation and Freeze-Drying Protocols,** *Second Edition,* edited by *John G. Day and Glyn Stacey, 2007*
367. **Mass Spectrometry Data Analysis in Proteomics,** edited by *Rune Matthiesen, 2007*
366. **Cardiac Gene Expression:** *Methods and Protocols,* edited by *Jun Zhang and Gregg Rokosh, 2007*
365. **Protein Phosphatase Protocols:** edited by *Greg Moorhead, 2007*
364. **Macromolecular Crystallography Protocols:** *Volume 2, Structure Determination,* edited by *Sylvie Doublié, 2007*
363. **Macromolecular Crystallography Protocols:** *Volume 1, Preparation and Crystallization of Macromolecules,* edited by *Sylvie Doublié, 2007*
362. **Circadian Rhythms:** *Methods and Protocols,* edited by *Ezio Rosato, 2007*
361. **Target Discovery and Validation Reviews and Protocols:** *Emerging Molecular Targets and Treatment Options, Volume 2,* edited by *Mouldy Sioud, 2007*
360. **Target Discovery and Validation Reviews and Protocols:** *Emerging Strategies for Targets and Biomarker Discovery, Volume 1,* edited by *Mouldy Sioud, 2007*
359. **Quantitative Proteomics by Mass Spectrometry,** edited by *Salvatore Sechi, 2007*
358. **Metabolomics:** *Methods and Protocols,* edited by *Wolfram Weckwerth, 2007*
357. **Cardiovascular Proteomics:** *Methods and Protocols,* edited by *Fernando Vivanco, 2006*
356. **High Content Screening:** *A Powerful Approach to Systems Cell Biology and Drug Discovery,* edited by *D. Lansing Taylor, Jeffrey Haskins, and Ken Guiliano, 2007*
355. **Plant Proteomics:** *Methods and Protocols,* edited by *Hervé Thiellement, Michel Zivy, Catherine Damerval, and Valerie Mechin, 2006*
354. **Plant–Pathogen Interactions:** *Methods and Protocols,* edited by *Pamela C. Ronald, 2006*
353. **DNA Analysis by Nonradioactive Probes:** *Methods and Protocols,* edited by *Elena Hilario and John. F. MacKay, 2006*
352. **Protein Engineering Protocols,** edited by *Katja M. Arndt and Kristian M. Müller, 2006*
351. ***C. elegans:*** *Methods and Applications,* edited by *Kevin Strange, 2006*
350. **Protein Folding Protocols,** edited by *Yawen Bai and Ruth Nussinov 2007*
349. **YAC Protocols,** *Second Edition,* edited by *Alasdair MacKenzie, 2006*
348. **Nuclear Transfer Protocols:** *Cell Reprogramming and Transgenesis,* edited by *Paul J. Verma and Alan Trounson, 2006*
347. **Glycobiology Protocols,** edited by *Inka Brockhausen-Schutzbach, 2006*
346. ***Dictyostelium discoideum* Protocols,** edited by *Ludwig Eichinger and Francisco Rivero, 2006*
345. **Diagnostic Bacteriology Protocols,** *Second Edition,* edited by *Louise O'Connor, 2006*
344. **Agrobacterium Protocols,** *Second Edition: Volume 2,* edited by *Kan Wang, 2006*
343. **Agrobacterium Protocols,** *Second Edition: Volume 1,* edited by *Kan Wang, 2006*
342. **MicroRNA Protocols,** edited by *Shao-Yao Ying, 2006*
341. **Cell–Cell Interactions:** *Methods and Protocols,* edited by *Sean P. Colgan, 2006*
340. **Protein Design:** *Methods and Applications,* edited by *Raphael Guerois and Manuela López de la Paz, 2006*
339. **Microchip Capillary Electrophoresis:** *Methods and Protocols,* edited by *Charles S. Henry, 2006*
338. **Gene Mapping, Discovery, and Expression:** *Methods and Protocols,* edited by *M. Bina, 2006*
337. **Ion Channels:** *Methods and Protocols,* edited by *James D. Stockand and Mark S. Shapiro, 2006*
336. **Clinical Applications of PCR,** *Second Edition,* edited by *Y. M. Dennis Lo, Rossa W. K. Chiu, and K. C. Allen Chan, 2006*
335. **Fluorescent Energy Transfer Nucleic Acid Probes:** *Designs and Protocols,* edited by *Vladimir V. Didenko, 2006*
334. **PRINS and *In Situ* PCR Protocols,** *Second Edition,* edited by *Franck Pellestor, 2006*
333. **Transplantation Immunology:** *Methods and Protocols,* edited by *Philip Hornick and Marlene Rose, 2006*
332. **Transmembrane Signaling Protocols,** *Second Edition,* edited by *Hydar Ali and Bodduluri Haribabu, 2006*
331. **Human Embryonic Stem Cell Protocols,** edited by *Kursad Turksen, 2006*
330. **Embryonic Stem Cell Protocols,** *Second Edition, Volume II: Differentiation Models,* edited by *Kursad Turksen, 2006*

METHODS IN MOLECULAR BIOLOGY™

High Content Screening

*A Powerful Approach to Systems Cell Biology
and Drug Discovery*

Edited by

D. Lansing Taylor
Cellumen, Inc., Pittsburgh, PA

Jeffrey R. Haskins
Cellomics, Inc., Pittsburgh, PA

Kenneth A. Giuliano
Cellumen, Inc., Pittsburgh, PA

HUMANA PRESS ✷ TOTOWA, NEW JERSEY

© 2007 Humana Press Inc.
999 Riverview Drive, Suite 208
Totowa, New Jersey 07512

www.humanapress.com

All rights reserved. No part of this book may be reproduced, stored in a retrieval system, or transmitted in any form or by any means, electronic, mechanical, photocopying, microfilming, recording, or otherwise without written permission from the Publisher. Methods in Molecular Biology™ is a trademark of The Humana Press Inc.

All papers, comments, opinions, conclusions, or recommendations are those of the author(s), and do not necesarily reflect the views of the publisher.

This publication is printed on acid-free paper. ∞
ANSI Z39.48-1984 (American Standards Institute)

Permanence of Paper for Printed Library Materials.
Cover illustration:

Production Editor: Jennifer Hackworth

Cover design by Patricia F. Cleary

Cover illustrations: Figures 2 and 4 from Chapter 16, "Microtextured Polydimethylsiloxane Substrates for Culturing Mesenchymal Stem Cell," by Erik T. K. Peterson and Ian Papautsky.

For additional copies, pricing for bulk purchases, and/or information about other Humana titles, contact Humana at the above address or at any of the following numbers: Tel.: 973-256-1699; Fax: 973-256-8341; E-mail: orders@humanapr.com; or visit our Website: www.humanapress.com

Photocopy Authorization Policy:
Authorization to photocopy items for internal or personal use, or the internal or personal use of specific clients, is granted by Humana Press Inc., provided that the base fee of US $30.00 per copy is paid directly to the Copyright Clearance Center at 222 Rosewood Drive, Danvers, MA 01923. For those organizations that have been granted a photocopy license from the CCC, a separate system of payment has been arranged and is acceptable to Humana Press Inc. The fee code for users of the Transactional Reporting Service is: [1-58829-731-4/07 $30.00].

Printed in Singapore. 10 9 8 7 6 5 4 3 2 1
1-59745-217-3(e-book)
ISSN 1064-3745

Library of Congress Cataloging-in-Publication Data

High content screening : a powerful approach to systems cell biology and
   drug discovery / edited by D. Lansing Taylor, Jeffrey R. Haskins,
   Kenneth A. Giuliano.
       p. ; cm. -- (Methods in molecular biology, ISSN 1064-3745 ; 356)
   Includes bibliographical references and index.
   ISBN 1-58829-731-4 (alk. paper)
   1. Biological systems--Research--Methodology. 2. Computational biology. 3. Drug development--Computer simulation.   I. Taylor, D. Lansing. II. Haskins, Jeffrey R.   III. Giuliano, Kenneth A. IV. Series: Methods in molecular biology (Clifton, N.J.) ; v. 356.
   [DNLM: 1. Systems Biology--methods. 2. Drug Design. 3. Image Processing, Computer-Assisted--methods.   W1 ME9616J v.356 2006 / QU 26.5 H638 2006]
   QH324.H49 2006
   615'.190072--dc22

2006009223

Part V. Assays and Applications of High Content Screening

25 Systems Biology in Cancer Research: *Genomics to Cellomics*
    **Jackie L. Stilwell, Yinghui Guan, Richard M. Neve,
    and Joe W. Gray** 353

26 Target Validation in Drug Discovery
    Robert A. Blake 367

27 High Content Screening as a Screening Tool in Drug Discovery
    Anthony Nichols 379

28 Discovery of Protein Kinase Phosphatase Inhibitors
    Andreas Vogt and John S. Lazo 389

29 High Content Translocation Assays for Pathway Profiling
    **Frosty Loechel, Sara Bjørn, Viggo Linde,
    Morten Præstegaard, and Len Pagliaro** 401

30 In Vitro Cytotoxicity Assessment
    Peter O'Brien and Jeffrey Haskins 415

31 Neurite Outgrowth in Retinal Ganglion Cell Culture
    John B. Kerrison and Donald J. Zack 427

Index 435

Contributors

MATTHEW J. ANSTETT • *Spotfire, Inc., Somerville, MA*
DANIEL BASSONI • *Vitra Bioscience, Inc., Mountain View, CA*
OREN BESKE • *Vitra Bioscience, Inc., Mountain View, CA*
SARA BJØRN • *BioImage A/S, Copenhagen, Denmark*
ROBERT A. BLAKE • *Exelixis, Inc., South San Francisco, CA*
ANDREJ BUGRIM • *GeneGo, St. Joseph, MI*
GRACE K. Y. CHAN • *The Neuroscience Research Centre, Merck Sharp and Dohme Research Laboratories, Essex, UK*
RAJESH V. CHAVLI • *Panomics, Inc, Fremont, CA*
MARK A. COLLINS • *Panomics, Inc., Pittsburgh, PA*
PETER G. CONRAD, II • *Genospectra, Inc., Fremont, CA*
WALLACE J. CZEKALSKI • *Cellomics, Inc., Pittsburgh, PA*
R. TERRY DUNLAY • *Sage Sciences, Inc., Albuquerque, NM*
SEAN EKINS • *GeneGo, Jerkintown, PA, and School of Pharmacy Department of Pharmaceutical Sciences, College Park, MD*
JOHN T. ELLIOTT • *National Institute of Standards and Technology, Gaithersburg, MD*
JAMES G. EVANS • *Whitehead·MIT BioImaging Center, Computational and Systems Biology Initiative, Massachusetts Institute of Technology, Cambridge, MA*
LEON S. GARFINKEL • *Hoffman-La Roche Inc., Global Research Infrastructure, Nutley, NJ*
RALPH J. GARIPPA • *Hoffmann-La Roche, Inc., Nutley, NJ*
RICHIK N. GHOSH • *Cellomics, Inc., Pittsburgh, PA*
KENNETH A. GIULIANO • *Cellumen, Inc., Pittsburgh, PA*
RICHARD S. GIVENS • *Department of Chemistry, University of Kansas, Lawrence, KS*
SIMON GOLDBARD • *Vitra Bioscience, Inc., Mountain View, CA*
ALBERT H. GOUGH • *Cellumen, Inc., Pittsburgh, PA*
JOE W. GRAY • *Life Sciences Division, Lawrence Berkeley National Laboratory, Berkeley, CA, and the Comprehensive Cancer Center, University of California, San Francisco, CA*
DALE E. GREENWALT • *Cambrex Bio Science Walkersville, Inc., Walkersville, MD*
YINGHUI GUAN • *Lawrence Berkeley National Laboratory, Berkeley, CA*
GEORGE T. HANSON • *Invitrogen Corporation, Madison, WI*
JEFFREY R. HASKINS • *Cellomics, Inc., Pittsburgh, PA*
RONALD P. HERZIG • *Upstate USA/Chemicon International, Charlottesville, VI*
ANN F. HOFFMAN • *Hoffmann-La Roche, Inc., Nutley, NJ*
JEFFREY T. HUNG • *Molecular Probes/Invitrogen, Eugene, OR*
MICHAEL J. IGNATIUS • *Molecular Probes/Invitrogen, Eugene, OR*
PAUL A. JOHNSTON • *University of Pittsburgh School of Medicine, Department of Pharmacology, Pittsburgh, PA*

JULIE E. KERBY • *The Neuroscience Research Centre, Merck Sharp and Dohme Research Laboratories, Essex, UK*
JOHN B. KERRISON • *Wilmer Eye Institute, Johns Hopkins Hospital, Baltimore, MD*
EUGENE KIRILLOV • *GeneGo, St. Joseph, MI*
JUERGEN A. KLENK • *Booz Allen Hamilton, Inc.,McLean, VA*
KURT J. LANGENBACH • *National Institute of Standards and Technology, Gaithersburg, MD*
OLEG LAPETS • *Cellomics, Inc., Pittsburgh, PA*
JOHN S. LAZO • *Department of Pharmacology, University of Pittsburgh Drug Discovery Institute, University of Pittsburgh, Pittsburgh, PA*
VIGGO LINDE • *BioImage A/S, Copenhagen, Denmark*
FROSTY LOECHEL • *NeuroSearch A/S, Ballerup, Denmark*
GEORGYI V. LOS • *Promega Corporation, Madison, WI*
MARNIE L. MACDONALD • *Odyssey Thera, Inc., San Ramon, CA*
THOMAS MACHLEIDT • *Invitrogen Corporation, Madison, WI*
DANIEL R. MARSHAK • *Cambrex Corporation, Baltimore, MD*
PAUL MATSUDAIRA • *Whitehead·MIT BioImaging Center, Computational and Systems Biology Initiative, Massachusetts Institute of Technology, Cambridge, MA*
DENNIS MCDANIEL • *National Institute of Standards and Technology, Gaithersburg, MD*
K. GREGORY MOORE • *Serologicals Corporation, Charlottesville, VA*
RICHARD M. NEVE • *Life Sciences Division, Lawrence Berkeley National Laboratory, Berkeley, CA, and the Comprehensive Cancer Center, University of California, San Francisco, CA*
ANTHONY NICHOLS • *Molecular Screening and Cell Pharmacology Department, Serono Pharmaceutical Research Institute, Geneva, Switzerland*
TATIANA NIKOLSKAYA • *GeneGo, St. Joseph, MI*
YURI NIKOLSKY • *GeneGo, St. Joseph, MI*
PETER O'BRIEN • *Safety Sciences Europe, Pfizer Global Research and Development, Sandwich, UK*
LEN PAGLIARO • *BioImage A/S, Copenhagen, Denmark*
ANNE L. PLANT • *National Institute of Standards and Technology, Gaithersburg, MD*
MORTEN PRÆSTEGAARD • *BioImage A/S, Copenhagen, Denmark*
GILLIAN R. RICHARDS • *The Neuroscience Research Centre, Merck Sharp and Dohme Research Laboratories, Essex, UK*
MATT ROBERS • *Invitrogen Corporation, Madison, WI*
PETER B. SIMPSON • *The Neuroscience Research Centre, Merck Sharp and Dohme Research Laboratories, Essex, UK*
WAYNE SPECKMANN, *Upstate USA, Charlottesville, VA*
JACKIE L. STILWELL • *Lawrence Berkeley National Laboratory, Berkeley, CA*
D. LANSING TAYLOR • *Cellumen, Inc., Pittsburgh, PA*
DONALD P. TAYLOR • *VIVISIMO, Inc., Pittsburgh, PA*
NICK THOMAS • *GE Healthcare, The Maynard Centre, Cardiff, UK*
ALESSANDRO TONA • *National Institute of Standards and Technology, Gaithersburg, MD, and Geo-centers, Inc. Newton, MA*

Contributors

ANDREAS VOGT • *Department of Pharmacology, University of Pittsburgh Drug Discovery Institute, University of Pittsburgh, Pittsburgh, PA*
ALAN S. WAGGONER • *Carnegie Mellon University, Pittsburgh, PA*
JOHN K. WESTWICK • *Odyssey Thera, Inc., San Ramon, CA*
KEITH WOOD • *Promega Corporation, Madison, WI*
DONALD J. ZACK • *Wilmer Eye Institute, Johns Hopkins Hospital, Baltimore, MD*

COMPANION CD

for *High Content Screening*

All of the electronic versions of illustrations in this book may be found on the Companion CD attached to the inside back cover. The image files are organized into folders by chapter number and are viewable in most web browsers. The number following "f" at the end of the file name identifies the corresponding figure in the text. The CD is compatible with both Mac and PC operating systems.

CHAPTER 1 FIGS. 2, 3, 7

CHAPTER 2 FIG. 2

CHAPTER 5 FIGS. 1, 6

CHAPTER 7 FIGS. 1, 3

CHAPTER 10 FIG. 1

CHAPTER 14 FIGS. 5–7

CHAPTER 16 FIGS. 1–4

CHAPTER 18 FIG. 2

CHAPTER 20 FIG. 2

CHAPTER 21 FIGS. 1–5

CHAPTER 23 FIGS. 1, 2, 7, 10, 13–16

CHAPTER 24 FIGS. 1–10

CHAPTER 25 FIGS. 2, 4

CHAPTER 28 FIGS. 1, 2

CHAPTER 29 FIG. 1

CHAPTER 30 FIG. 2

I

INTRODUCTION TO HIGH CONTENT SCREENING

1

Past, Present, and Future of High Content Screening and the Field of Cellomics

D. Lansing Taylor

Summary

High content screening (HCS) was created in 1996 to offer a new platform that could be used to permit relatively high-throughput screening of cells, in which each cell in an array would be analyzed at a subcellular resolution using multicolored, fluorescence-based reagents for both specificity and sensitivity. We developed HCS with the perspective of the history of the development of the automated DNA sequencers that revolutionized the field of genomics. Furthermore, HCS was based on a history of important developments in modern cytology. HCS integrates the instrumentation, application software, reagents, sample preparation, and informatics/bioinformatics required to rapidly flow from producing data, generating information, and ultimately creating new cellular knowledge. The HCS platform is beginning to have an important impact on early drug discovery, basic research in systems cell biology, and is expected to play a role in personalized medicine.

Key Words: Bioinformatics; cellome; cellomics; fluorescence; high content screening; informatics; light microscopy; multiplexed fluorescence; reagents; systems biology; systems cell biology.

1. Introduction

My cofounders and I formed Cellomics, Inc. in 1996 to create a platform technology that would permit large-scale screening of cells, with subcellular spatial resolution, using multiplexed fluorescence in arrays of cells on either microplates or other substrates such as chips *(1)*. In the years since the introduction of high content screening (HCS), there has been a growing acceptance of the technology in both drug discovery and basic biomedical research markets and numerous companies are now offering various components of a complete platform. This chapter is designed to give insights into how HCS was conceived and implemented in the past, the present state of evolution of the technology, and what the future holds for this rapidly emerging field.

There have been a variety of definitions of terms related to HCS over the last few years since we first introduced the field. The following definitions are based on the early perspectives of the cofounders and early employees of Cellomics and are relevant to today's use of the technology.

1.1. Definitions

Cellome: The complete complement of all cell types in an organism and their constituent molecules.

Cellomics: The study of the dynamic functions of cells and their constituent molecules.

High content screening: Platform and methods, including instruments, biological application software, reagents, assays, and informatics software used to automatically screen and analyze arrays of cells to define the temporal and spatial activities and functions of cells, and their constituents, on a cell-by-cell basis, including subcellular features.

HCS assays: The integration of the optimal biological application software with the optimal fluorescence-based reagents and protocols used to extract the type of cellular data defined by a particular experiment on the desired cell type(s).

Multiplexed HCS assays: HCS assays in which multiple parameters are not only measured within single cells using multiple reagents and morphometrics, but relationships between the parameter values are calculated, analyzed, and interpreted on a cell-by-cell basis. It is also possible to make a population average of any or all of the parameters measured on a cell-by-cell basis for some analyses.

Systems cell biology: The understanding of how the integration of the complex biochemical and molecular processes, occurring in time and space, are responsible for cell functions and the complex behavioral responses of cells to natural environmental changes or experimental treatments. The integration can occur within one or more cell types incorporated into an assay and involve panels of multiplexed HCS assays extracting up to hundreds of cellular measurements.

1.2. Why Do We Need HCS?

HCS was developed to meet the needs of research scientists in both basic biomedical research and early drug discovery. In basic biomedical research, the human genome project has identified approx 20,000–25,000 human genes that code for proteins. This coding portion of the human genome represents only about 2–4% of the total genome; the remaining 96–98% was originally assigned "junk" DNA status by many scientists, as a byproduct of evolution *(3,4)*. Over the last few years there has been a drive to define the pathways formed by interactions of the proteins encoded by the coding portion of the genome. These interactions, or the protein "interactome," are believed to bring about specific cell functions. It has become evident that the cell consists of many interacting pathways that are highly regulated. Hundreds of interacting proteins involving a variety of post-translational modifications create a complex network of activity that is only partially defined and understood *(5)*.

As if the complexity of the protein interactions and regulation was not difficult enough, the "new" genomics of noncoding RNA (ncRNA) has recently caused researchers to pursue the apparent parallel world of ncRNA in regulating cell functions. It has been established that about 50% of the human genome is transcribed into RNA, whereas only a small fraction codes for proteins. Recent studies have demonstrated that a growing number of ncRNA exhibit functions like proteins in regulating gene expression and even developmental changes *(6)*. Now it is clear that the proteome and ncRNA species must be investigated together in order to more fully understand cell functions and their regulation. This increases the value and importance of HCS, especially with multiplexed assays to define the functions of proteins and ncRNAs, as well as other cellular constituents, within the context of the living cell system.

Early drug discovery steps traditionally used primarily homogeneous, solution assays containing the protein targets, because they were relatively simple "mix and read" assays. This approach was compatible with the prevailing view in the 1990s that stressed fast measurements on a growing list of targets with large numbers of compounds. The chosen metric was how fast a plate could be read on a plate reader. However, the implementation of ultrahigh throughput screening did not have the desired impact on the number of investigative new drugs generated by this approach. In fact, the productivity of the whole pharmaceutical industry has decreased over the last couple of decades leading industry leaders to make changes in the process of drug discovery *(7)*.

An alternative approach of generating deep, functional information based on screens using cells with temporal and spatial information was suggested, based on HCS *(1)*. Now the metric is how much time is required to make a good decision on whether to continue pursuing a compound.

HCS, especially with multiplexed assays, should play an increasingly important role in drug discovery. In fact, HCS has opened the opportunity to perform drug discovery, not just on a preselected target, but to screen for compounds, singly or combined, that impact cellular functions such as cell cycle, cell motility, apoptosis, and so on *(8,9)*.

1.3. The Cell: First Level of "Systems Biology"

It is a fact that using cells in the early drug discovery process is more complicated and expensive than performing homogeneous protein assays in screens. However, cells offer the first level of the complexity that living systems exhibit and results using cells are more meaningful than those obtained from isolated proteins. In addition, cell-based assays are less complex and expensive than using whole organisms. There is great potential in performing "systems cell biology" screens on the optimal cells (validated cell line or primary cells) and then apply the systems cell biology information as a bridge to higher order systems biology studies. Today, most HCS is performed on two-dimensional (2D) arrays of cell lines. However, more complex cell-based assays can be performed on 2D and 3D cocultures of different primary cell types using tissue-engineering approaches to create functional arrays of tissue models. Information and knowledge gained at this next level of complexity can then be related to higher order systems biology studies.

The amount and quality of information and knowledge that can be obtained by cell-based discovery far outweighs the higher upfront costs in early drug discovery. In addition, the real costs in drug discovery increase as a compound continues down the pipeline. Better, deeper information early should become the new standard. High throughput HCS using multiplexed HCS assays with advanced reagents and informatics will play a major role in this paradigm shift. Success at this level will increase the need and demand for sophisticated systems biology databases that will be populated by mining the literature and the information derived from systems cell biology screens.

1.4. The Concept and Development of the Field of "Cellomics" Mirrored the Developments in the Field of Genomics

The field of genomics was driven by the need to sequence the genomes of organisms in order to understand the complexity and regulation of life processes starting with the DNA "blueprint" of life. Manual DNA sequencing by gel electrophoresis, "reading" the ladder patterns that defined the sequence and then manually entering sequences into spreadsheets was a major development in biotechnology and became a well established method by the late 1970s. Fundamental knowledge about selected genes and genome organization was created by the manual processes involved in this early approach to DNA sequencing. However, the human genome project demanded that automated instrumentation, with the optimal reagents and informatics/bioinformatics software tools be developed to permit the human genome to be defined in a reasonable period of time and cost. In the early 1990s, Applied Biosystems (Foster City, CA), as well as others, developed "complete" solutions to automatically prepare the DNA samples, fluorescently tag the four nucleotides, run the gels, read the ladders, and then read-out the sequence into searchable databases. Bioinformatics tools rapidly evolved to identify genes in the growing genome sequences *(10,11)*.

The field of cellomics was driven by the need to define the functions of genes and the proteins that they encoded. It was apparent by the mid-1990s that knowing the human genome was the start, not the end of the biological challenge for basic research and drug discovery. Light microscopy, especially digital imaging fluorescence microscopy on living cells was chosen as the best approach to defining the functions of genes and proteins *(1)*. Human interactive, imaging methods were pretty well developed by the 1980s and fundamental information about the temporal and spatial dynamics of cells and their constituents was being published by a growing academic community *(12–18)*. However, the human interactive imaging tools in the absence of automated imaging methods and informatics tools to archive, mine, and display complex imaging data made the process of studying cells time-consuming and complicated. Similar to the field of genomics, there was a need for the development of an automated system to acquire, process, analyze, display, and mine massive amounts of cellular data derived from arrays of cells treated in various ways. In 1997, Cellomics, Inc. offered a "complete solution" with the introduction of the ArrayScan platform that consists of the instrument, biological application software, reagents for a specific assay, and the first generation informatics for cell analyses. This was the starting point for the large-scale investigation of the function of genes and the proteins they encoded, as well as other cellular constituents *(1)*.

2. Past: Origins of HCS
2.1. Important Milestones in the Modern Era of Cytology and Cytometry

The development of HCS is rooted in the rich history of developments in cytometry going back more than 50 yr in the "modern" era of cytology. **Figure 1** depicts this authors view of the major advances that occurred over the last 50 yr that led up to the development of HCS. There have been many important developments over this period of time and not all of the important ones are depicted here. The development of immunofluorescence microscopy by Coons and Kaplan *(19)* was the first critical step in the modern era of fluorescence-based cytometry. For the first time, the specificity of labeling with antibodies was coupled to the sensitivity of fluorescence detection. Interestingly, a major advance in cytometry occurred in 1957 with the discovery of confocal scanning microscopy by Minsky more than 30 yr before optimal fluorescence dyes and imaging technologies made the method practical for fluorescence microscopy *(20)*. The early stages of the modern era of cytology also includes the development of the inverted fluorescence microscope *(21)* and dichroic filters for epifluorescence microscopy *(22)*, both of which created a system that produced the light throughput and signal/noise required for the practical use of fluorescence microscopy as a standard tool *(14)*.

The next phase in the modern era of cytometry consists of the development of fluorescence-based flow cytometry *(23–26)*, a method that blossomed with the development of specific antibodies to a range of cell surface molecules important in the immune responses and the use of multicolored fluorescent dyes to multiplex the measurements. Also in the late 1960s was the development of image intensifiers, imaging detectors that could record images of very low fluorescence signals in biological samples *(27,28)*.

The late 1960s and into the 1970s was a period of rapid developments in instrumentation *(14,29,30)* and imaging software *(31–33)*. Fluorescence-based reagents also emerged as a critical component of the detection systems (**Fig. 2**). In particular, Alan Waggoner created the modern field of fluorescence-based physiological indicators with the development of a series of membrane potential sensitive dyes *(34)* and Haugland *(35)* developed and/or commercialized a wide range of physiological indicator dyes. In addition, fluorescent analog cytochemistry, the original tool to measure the activities of specifically labeled proteins in living cells in time and space, was demonstrated *(36,37)*. However, the production of fluorescent analogs was a time-consuming process including protein purification, labeling, testing function in vitro and microinjecting into living cells *(37,38)*. Wide-spread use of this technology would require another technical development in the 1990s.

The 1980s were characterized by major developments in video microscopy to enhance the contrast and detection limits in both transmitted light and fluorescence *(12–15,39)*, and ratio imaging microscopy to quantify cellular physiological changes *(40,41)*. Ratio imaging was initially applied to pH *(40,41)* and then free calcium ion concentration *(42)*, but ultimately local protein concentrations and activation *(40,43)*, as well as cytoplasmic structure and rotational diffusion of proteins *(44)* (**Fig. 3**). The use of solid-state detectors improved the performance of imaging methods *(45)*, and the first practical use of laser scanning confocal fluorescence microscopy allowed 3D imaging of thicker biological samples *(46)*.

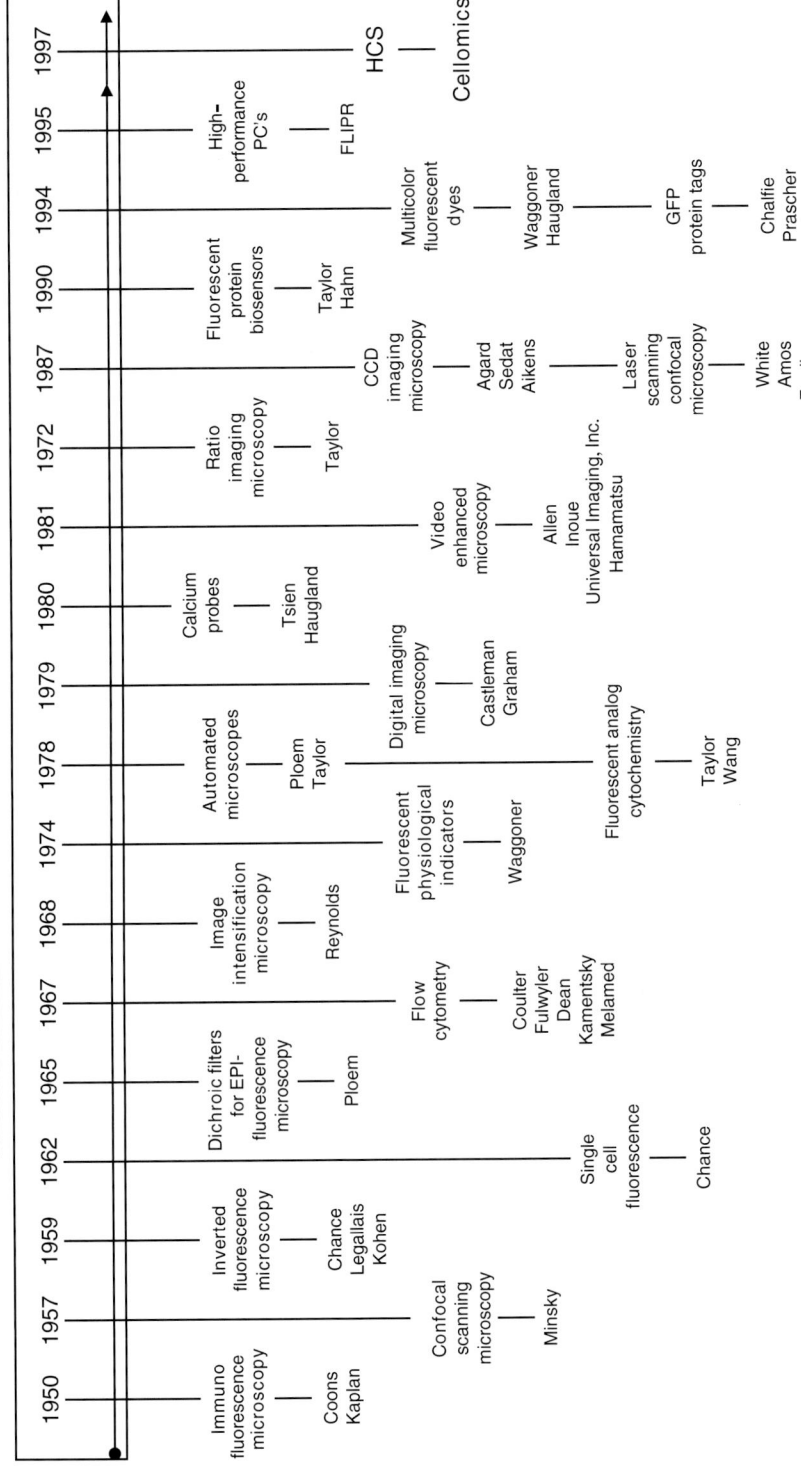

Fig. 1. Time line of the key developments that have occurred since 1950, the beginning of the modern era of cytometry. High content screening was based on a variety of important contributions that led to the development of HCS in 1997.

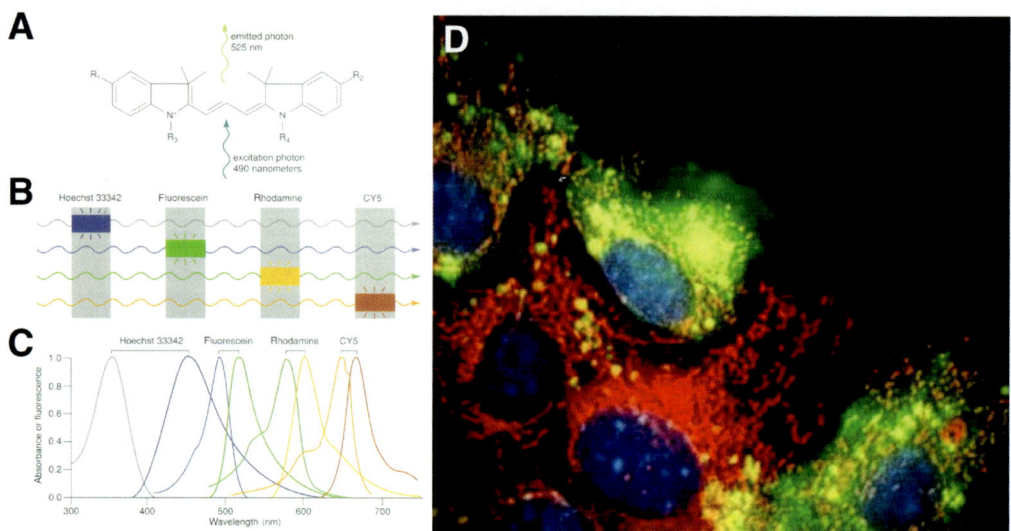

Fig. 2. Multicolor fluorescence imaging has been a key to the early stages of HCS. (**A**) Fluorescent molecules have a common motif in their structure that includes a pattern of alternating single and double bonds between carbon atoms, especially in ring structures. This example is a cyanine dye. (**B**) Multiple fluorescent dyes can be used in cells and can be distinguished with the use of optimal filter sets to produce multiparameter data sets or multiplexed measurements. (**C**) The goal is to assemble multiple fluorescent dyes in which the spectral overlaps are minimized so that filter sets can separate the distinct dyes. Recently, spectral deconvolution systems have been developed to extract the fluorescent spectrum of each dye rather that to separate them with filters. (**D**) An image of living mouse fibroblasts that have been labeled with a dye to label the DNA in nuclei (blue), a molecule that was endocytosed into endosomes (yellow), a dye to label and report the membrane potential of mitochondria (red) and labeled actin incorporated into the cells (green).

The early 1990s ushered in the era of advanced fluorescence-based reagents. Multiplexed imaging with water soluble, bright, stable, fluorescent dyes was optimized by the development of a range of cyanine dyes that emitted fluorescence from the blue to the near infrared portion of the light spectrum *(47)*. It was now possible to correlate multiple cellular parameters in the same cells *(48)*. The development of fluorescent analogs of proteins in the late 1970s *(36)*, led to the creation of fluorescent protein biosensors, reagents that reported biochemical or molecular changes, not just their distribution within the cell *(49–53)*. The original green fluorescent protein gene construct from the jelly fish (*Aequoria*), was also first used in a biological experiment *(54)*. Tsien et al. *(55)* subsequently optimized the properties through the selection of mutants making this the method of choice for creating fluorescent analogs of proteins.

My cofounders, early employees, and I were primarily influenced by three major types of fluorescence instrumentation used in cellular analyses in the creation of HCS (**Fig. 4**). First, flow cytometry created a platform in which multiplexed measurements of cells could be performed. Flow cytometry opened the door to fluorescence-based analyses and became a standard for cell-by-cell measurements in basic research and biotechnology *(23)*. Second, the fluorescence imaging plate reader (FLIPR; Molecular Devices, Inc., Sunnyvale, CA) the first whole plate reader designed for high throughput measurement of population averages of attached living cells, offered cell analyses as a powerful approach to drug discovery *(56)*. Finally, digital imaging microscopy created a tool that allowed multicolored and even multimodal microscopy to be used by biologists, not just biophysicists *(17)*.

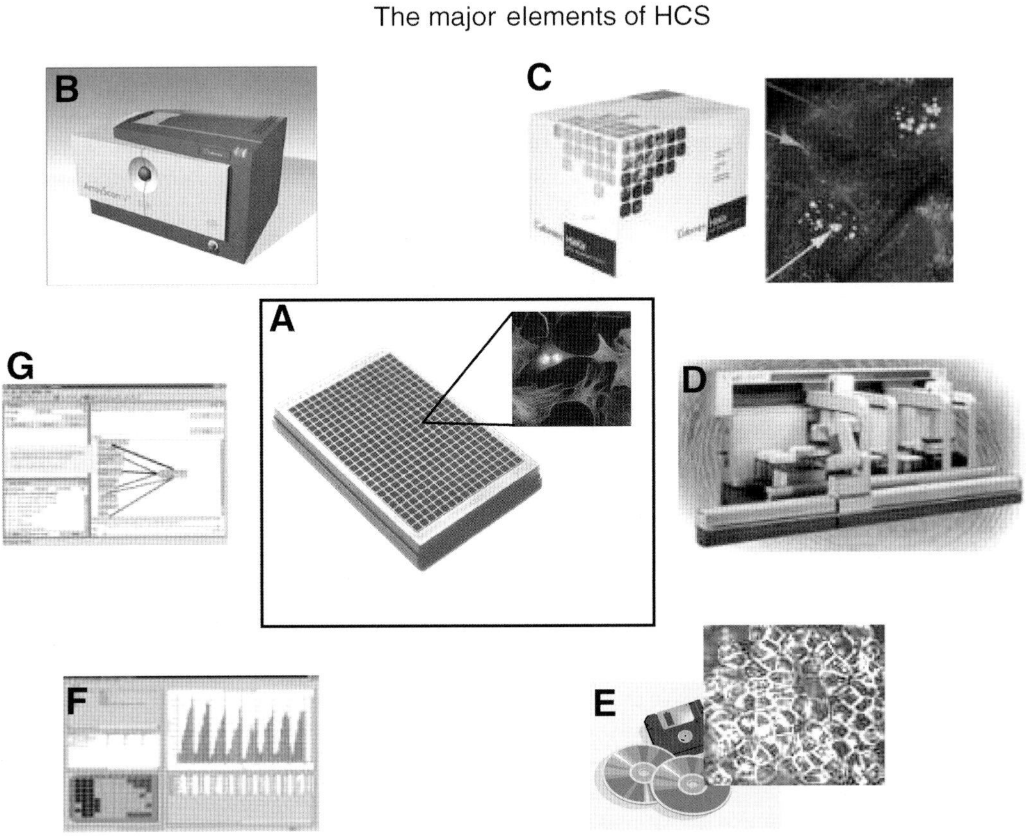

Past, Present, and Future of High Content Screening

The major elements of HCS

Fig. 6. HCS involves (**A**) the creation of arrays of cells, (**B**) instrumentation, (**C**) reagents, (**D**) sample preparation, (**E**) biological application software, (**F**) informatics software, and (**G**) cellular bioinformatics.

can be used, especially when the optimal cell parameters required for understanding the images are not specifically known. Both types of application software tools will continue to evolve (*see* Chapters 5 and 6).

HCS deals with large numbers of cells that are arrayed for combinatorial treatments, presently in microplates. In order to process large numbers of microplates and experimental treatments, it was critical to implement automated sample preparation for speed and reproducibility. The HCS instrumentation allows the reading of whole microplates in minutes, depending on the assay type, usually with the application software running on the instrument. However, it is currently possible to acquire images and then apply selected algorithms offline *(57)*.

Even a small screen can create hundreds of gigabytes to terabytes of data within a short period of time. Therefore, data storage (images, metadata, and numerical data), coupled with a variety of informatics software tools, were required to actually perform significant screens. Although presently limited in scope, cellular bioinformatics tools have also begun to evolve in order to help create knowledge from the information gleaned from the data using the application software and informatics software tools *(2,57–59)* (*see* Chapters 22–24).

3.3. Fixed End Point vs Live Cell HCS

All HCS assays begin with living cells that are treated with some combinatorial of manipulations including small molecules, biologicals, and/or physical treatments. Cells and plates that are fixed at some time-point after experimental treatment and then subsequently washed, labeled and read on an HCS instrument are called "fixed end-point assays." The sample preparation methods can automate all of these steps making it very fast and reproducible. However, the time domain of the biology is limited to a single time-point. Therefore, the investigator must either create a whole time-course by preparing multiple plates processed over time or initially define the half time of some cellular process of interest and set the time of fixation accordingly. The fixed end point approach can be a relatively high-throughput screening method *(57)* (*see* Chapters 2, 8, 13, and 26–28).

Live cell HCS is possible with the incorporation of on-board fluidics and an environmental chamber into the HCS reader. This can be accomplished with either a distinct instrument or by applying add-ons to a fixed end point reader *(2,57)*. Live cell assays can also be based on a single time-point with the use of fluorescent probes that are functional in living cells, or full kinetic measurements can be made, over time, starting before the experimental treatment and continuing over a defined period of time *(60)*. Kinetic assays are critically important in order to define the half-times of specific biological processes before setting up fixed end-point assays and/or for critically defining the complex temporal/spatial dynamics of cells and their processes.

4. Future Directions

HCS is still in its infancy. All of the elements of HCS depicted in **Fig. 6.** will evolve over time. Assays will become heavily multiplexed in which many cellular parameters will be measured in parallel in order to create a "systems cell biology" profile of cellular functions. The application software will be more powerful and will include elements of both directed algorithms and machine learning. A major direction will include the development of many classes of reagents that will "measure and manipulate" cellular constituents from DNA through all types of coding and ncRNA, proteins and metabolites. It will be possible to manipulate pathways and to measure the impact of the presence or absence of specific molecules on cell functions. A real systems cell biology profiling capability will emerge (*see also* Chapters 11, 12, 14, and 16–19).

The types of arrayed cells, the substrate structure, as well as the biology and chemistry of the environments will become more physiologically relevant. For example, biologically significant substrates using extracellular biochemistry will replace simple cells on plastic that now dominates the field. In addition, primary cells will be used to a greater extent based on improved methods for preparing and transporting these cells (*see* Chapter 9). Rather than investigating one cell type in 2D per well, the future will yield "tissue engineered" arrays of cells that have some tissue functions based on the optimal arrangement of specific types of cells (*see* Chapter 10). Further miniaturization will occur and cell chips will be engineered using a combination of nanotechnology and microfluidics *(1)*. Finally, it is not a distant vision to predict that cells will come stabilized (by freezing or freeze drying), prepackaged and containing a variety of biosensors ready for activation and screening.

The instrumentation will become more sophisticated. The next generation of instruments will be modular, allowing the end-user to define the specifications required for the desired applications. Options will include distinct types of light microscopy (**Table 1**) and multiple types of fluorescence measurements beyond intensity (**Table 2**). It is predicted that because the format for HCS will be miniaturized to chips, the instruments will be smaller and faster based on this transition. Finally, future generations of instruments will incorporate standards including intensity, spectral correction and size. These instrumentation standards will also be linked to standards incorporated into the cell array formats (*see* Chapters 4 and 7).

The informatics tools will become more sophisticated and automated to handle the massive cell data streams (*see* Chapter 20). These tools will include powerful data mining tools to extract patterns from multiplexed assay data sets, cell pathway tools to map the interactions of

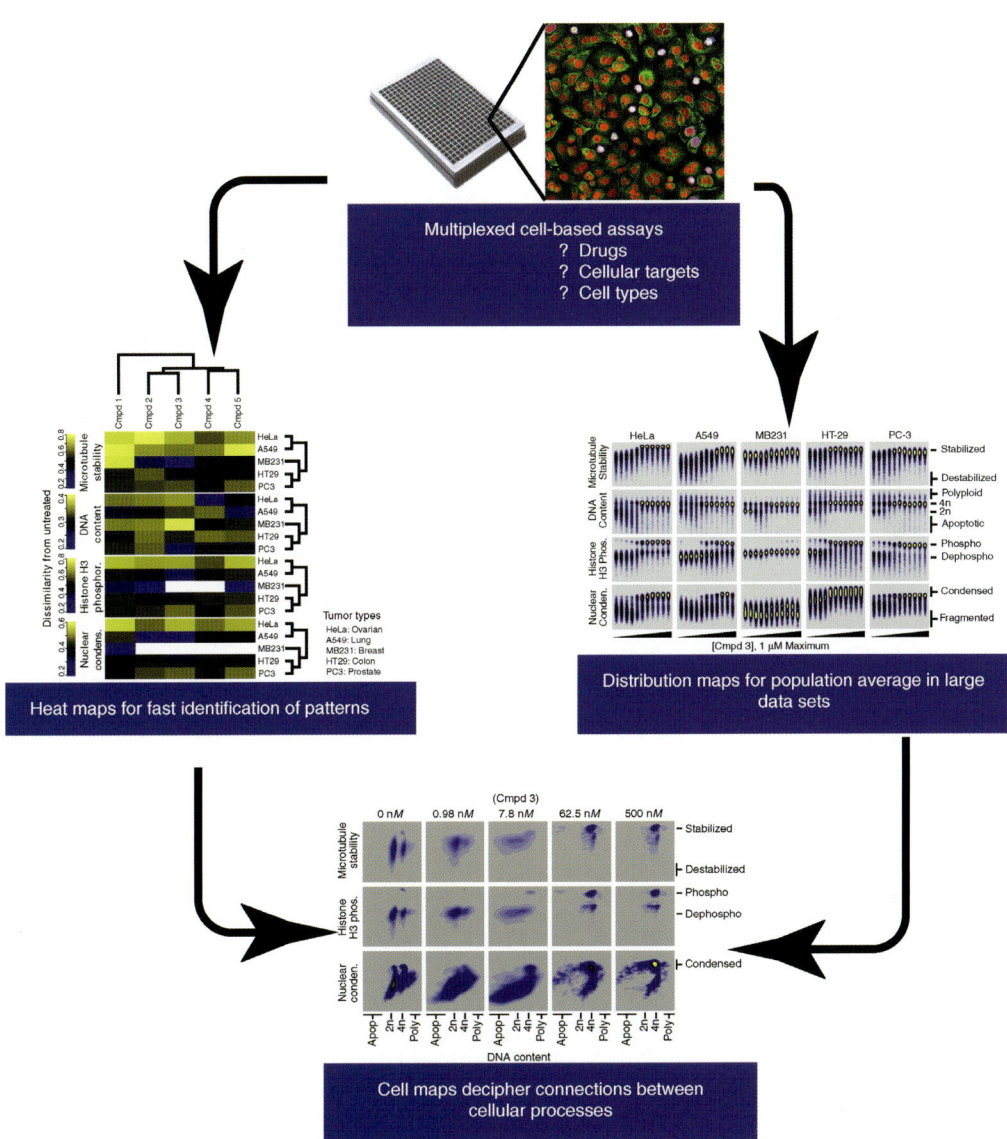

Fig. 7. Multiplexed HCS assays allow complex cell functions to be analyzed rapidly and in great detail. A complete systems cell biology knowledge building platform will include a continuum of software tools starting from the imaging algorithms to data archiving, mining, analysis, and cellular bioinformatics tools to rapidly traverse from collection of data, through the generation of information and the creation of knowledge *(2)*.

cell constituents, as well as visualization tools to rapidly detect important information for further exploration (**Fig. 7**) (*see* Chapters 22 and 24).

The entire work flow will become more integrated and continuous from growing cells, plating cells, incorporating reagents into cells, followed by acquiring the data with the instrumentation and application software, to the archiving, analysis, and visualization of data and information. In addition, the final step of interpretation of the information to create new knowledge will use cellular bioinformatics software. This integration will require more advanced software tools including advanced informatics and bioinformatics.

The field that was started 10 yr ago is only now beginning to blossom. It is my opinion that the greatest challenges and opportunities in HCS will involve the development and application of advanced reagents and informatics/bioinformatics tools. The next 10 yr will usher in tremendous opportunities for HCS-based discovery in basic biomedical research and drug discovery, but will also impact some industrial testing and personalized medicine. It is expected that in vitro toxicology will be the next major area of development that could ultimately produce a predictive tool *(61–64)* (*see* Chapter 30).

References

1. Giuliano, K. A., DeBiasio, R. L., Dunlay, R. T., et al. (1997) High content screening: a new approach to easing key bottlenecks in the drug discovery process. *J. Biomol. Screen* **2,** 249–259.
2. Taylor, D. L. and Giuliano, K. A. (2005) Multiplexed high content screening assays create a systems cell biology approach to drug discovery. *Drug Discov. Today Technol.* **2,** 149–154.
3. Mattick, J. S. (2003) Challenging the dogma: the hidden layer of non-protein-coding RNA's in complex organisms. *Bioassays* **25,** 930–939.
4. Gibbs, W. W. (2003) The unseen genome: gems among the junk. *Sci. Am.* **289,** 26–33.
5. Irish, J. M., Hovland, R., Krutzik, P. O., et al. (2004) Single cell profiling of potentiated phosphor-protein networks in cancer cells. *Cell* **118,** 217–228.
6. Poy, M. N., Eliasson, L., Krutzfeldt, J., et al. (2004) A pancreatic islet-specific microRNA regulates insulin secretion. *Nature* **432,** 226–230.
7. Posner, B. A. (2005) High-throughput screening-driven lead discovery: meeting the challenges of finding new therapeutics. Current Opinion. *Drug Discov. Dev.* **8(4),** 487–494.
8. Giuliano, K. A. (2003) High-content profiling of drug-drug interactions: cellular targets involved in the modulation of microtubule drug action by the antifungal ketoconazole. *J. Biomol. Screen.* **8,** 125–135.
9. Mitchison, T. J. (2005) Small-molecule screening and (profiling by using automated microscopy). *Chem. Bio. Chem.* **5,** 1–7.
10. Hunkapiller, T., Kaiser, R. J., Koop, B. F., and Hood, L. (1991) Large-scale and automated DNA sequence determination. *Science* **254,** 59–67.
11. Hood, L. and Galas, D. (2003) The digital code of DNA. *Nature* **421,** 444–448.
12. Taylor, D. L., Waggoner, A. S., Murphy, R. F., Lanni, F., and Birge, R. R. (eds.) (1986) *Applications of Fluorescence in the Biomedical Sciences.* Alan R. Liss, New York.
13. Taylor, D. L., Nederlof, M., Lanni, F., and Waggoner, A. S. (1992) The new vision of light microscopy. *Am. Scientist* **80,** 322–335.
14. Taylor, D. L. and Wang, Y. -L. (eds.) (1989) Fluorescence microscopy of living cells in culture. Parts A and B, in *Methods in Cell Biology*. Academic, New York, **29, 30.**
15. Inoue, S. and Spring, K. R. (1997) *Video Microscopy: The Fundamentals.* Plenum Press, New York.
16. Pawley, J. B. (ed.) (1995) *Handbook of Biological Confocal Microscopy.* Plenum Press, New York.
17. Farkas, D. L., Baxter, G., DeBiasio, R. L., et al. (1993) Multimode light microscopy and the dynamics of molecules, cells and tissues. *Annu. Rev. Physiol.* **55,** 785–817.
18. Denk, W., Strickler, J. H., and Webb, W. W. (1990) Two-photon laser scanning fluorescence microscopy. *Science* **248,** 73–76.
19. Coons, A. H. and Kaplan, M. M. (1950) Localization of antigen in tissue cells. II. Improvements in a method for the detection of antigen by means of fluorescent antibody. *J. Exper. Med.* **91,** 1–13.
20. Minsky, M. (1988) Memoir on inventing the confocal scanning microscope. *Scanning* **10,** 128–138.
21. Ploem, J. S., Tanke, H. J., Al, I., and Deedler, A. M. (1978) *Immunofluorescence and Related Staining Techniques,* (Knapp, W., Holubar, K., and Wick, G., eds.), Elsevier, Amsterdam.
22. Ploem, J. S. (1967) The use of a vertical illuminator with interchangeable dielectric mirrors for fluorescence microscopy with incident light. *Z. Wiss. Mikrosk.* **68,** 129–142.
23. Shapiro, H. M. (2003) *Practical Flow Cytometry*, Fourth ed. Wiley-Liss, New York.
24. Coulter, W. H. (1956) High speed automatic blood cell counter and cell size analyzer. *Proc. Natl. Electronics Conf.* **12,** 1034.

25. Fulwyler, M. J. (1965) Electronic separation of biological cells by volume. *Science* **150,** 910.
26. Kamentsky, L. A. and Melamed, M. R. (1969) Instrumentation for automated examinations of cellular specimens. *Proc. IEEE* **57,** 2007–2016.
27. Reynolds, G. T. (1972) Image intensification applied to biological problems. *Q. Rev. Biophys.* **5,** 295–347.
28. Reynolds, G. T. and Taylor, D. L. (1980) Image intensification applied to light microscopy. *Bioscience* **30,** 586–591.
29. Ploem, J. S. (1993) Fluorescence microscopy, in *Fluorescent and Luminescent Probes for Biological Activity,* (Mason, W. T., ed.), Academic, London, pp. 1–11.
30. Chance, B. (1962) Kinetics of enzyme reactions within single cells. *Ann. NY. Acad. Sci.* **97,** 431–448.
31. Ingram, M. and Preston, K., Jr. (1964) Automatic analysis of blood cells. *Scientific Amer.* **223,** 72.
32. Castleman, K. R. (1979) *Digital Image Processing.* Prentice-Hall, New Jersey.
33. Prewitt, J. M. S. and Mendelson, M. L. (1966) The analysis of cell images. *Ann. NY. Acad. Sci.* **128,** 1035.
34. Waggoner, A. S. (1979) Dye indicators of membrane potential. *Ann. Rev. Biophys. Bioeng.* **8,** 47–68.
35. Haugland, R. (1993) Intracellular ion indicators, in *Fluorescent and Luminescent Probes for Biological Activity* (Mason, W. T., ed.), Academic, London, pp. 34–43.
36. Taylor, D. L. and Wang, Y. -L. (1978) Molecular cytochemistry: incorporation of fluorescently labeled actin into cells. *Proc. Natl. Acad. Sci. USA* **75,** 857–861.
37. Taylor, D. L. and Wang, Y. -L. (1980) Fluorescently labeled molecules as probes of the structure and function of living cells. *Nature* **284,** 405–410.
38. Wang, Y. -L., Heiple, J. M., and Taylor, D. L. (1982) Fluorescent analog cytochemistry of contractile proteins. *Meth. Cell Biol.* **25(B),** 1–11.
39. Allen, R. D. (1985) New observations on cell architecture and dynamics by video-enhanced contrast optical microscopy. *Ann. Rev. Biophys. Chem.* **14,** 265–290.
40. Tanasugarn, L., McNeil, P., Reynolds, G., and Taylor, D. L. (1984) Microspectrofluorometry by digital image processing: measurement of cytoplasmic pH. *J. Cell Biol.* **98,** 717–724.
41. Bright, G. R., Fisher, G. W., Rogowska, J., and Taylor, D. L. (1987) Fluorescence ratio imaging microscopy: temporal and spatial measurements of cytoplasmic pH. *J. Cell Biol.* **104,** 1019–1033.
42. Williams, D. A., Fogarty, K. E., Tsien, R. Y., and Fay, F. S. (1985) Calcium gradients in single smooth muscle cells revealed by the digital imaging microscope using Fura-2. *Nature* **318,** 558–561.
43. Hahn, K. M., DeBiasio, R., and Taylor, D. L. (1992) Patterns of elevated free calcium and calmodulin activation in living cells. *Nature* **359,** 736–738.
44. Gough, A. and Taylor, D. L. (1993) Fluorescence anisotropy imaging microscopy maps calmodulin binding during cellular contraction and locomotion. *J. Cell Biol.* **121,** 1095–1107.
45. Aikens, R. S., Agard, D. A., and Sedat, J. W. (1989) Solid-state imagers for microscopy, in *Fluorescence Microscopy of Living Cells in Culture,* (Taylor, D. L. and Wang, Y. -L., eds.), Academic, New York, pp. 291–313.
46. White, J. G., Amos, W. B., and Fordham, M. (1987) An evaluation of confocal versus conventional imaging of biological structures by fluorescence light microscopy. *J. Cell Biol.* **105,** 41–48.
47. Waggoner, A. (1990) Fluorescent probes for cytometry, in *Flow Cytometry and Sorting,* (Melamed, M. R., Lindmo, T., and Mendelsohn, M. L., eds.), Wiley-Liss, Inc., New York, pp. 209–225.
48. DeBiasio, R., Bright, G. R., Ernst, L. A., Waggoner, A. S., and Taylor, D. L. (1987) Five-parameter fluorescence imaging: wound healing of living Swiss 3T3 cells. *J. Cell Biol.* **105,** 1613–1622.
49. Giuliano, K. A., Post, P. L., Hahn, K. M., and Taylor, D. L. (1995) Fluorescent protein biosensors: measurement of molecular dynamics in living cells. *Ann Rev Biophys. Biomol. Struct.* **24,** 405–434.
50. Giuliano, K. A. and Taylor, D. L. (1998) Fluorescent-protein biosensors: new tools for drug discovery. *Trends Biotech.* **16,** 135–140.
51. Giuliano, K. A., Chen, Y.-T., and Haskins, J. R. (2003) Positional biosensors: a new tool for high content screening. *Modern Drug Discov. (August),* 33–37.
52. Tsien, R. Y. (2005) Building and breeding molecules to spy on cells and tumors. *FEBS Lett.* **579,** 927–932.
53. Hahn, K. and Toutchkine, A. (2002) Live-cell fluorescent biosensors for activated signaling proteins. *Curr. Opin. Cell Biol.* **14,** 167–172.

54. Chalfie, M., Tu, Y., Euskirchen, G., Ward, W. W., and Prascher, D. C. (1994) Green fluorescent protein as a marker for gene expression. *Science* **263,** 802–805.
55. Heim, R. and Tsien, R. Y. (1996) Engineering green fluorescent protein for improved brightness, longer wavelengths and fluorescence resonance energy transfer. *Curr. Biol.* **6,** 178.
56. Schroeder, K. S. and Neagle, B. D. (1996) FLIPR: a new instrument for accurate, high throughput optical sectioning. *J. Biomol. Screen.* **1,** 75–80.
57. Giuliano, K. A., Haskins, J. R., and Taylor, D. L. (2003) Advances in high content screening for drug discovery. *ASSAY and Drug Dev. Tech.* **1,** 565–577.
58. Giuliano, K. A., Chen, Y.-T., and Taylor, D. L. (2004) Highcontent screening with siRNA optimizes a cell biological approach to drug discovery: defining the role of p53 activation in the cellular response to anticancer drugs. *J. Biomol. Screen.* **9,** 557–567.
59. Giuliano, K. A., Cheung, W. S., Curran, D. P., et al. (2005) Systems cell biology knowledge created from high content screening. *ASSAY and Drug Dev. Tech.* **3,** 501–514.
60. Abraham, V. C., Taylor, D. L., and Haskins, J. R. (2003) High content screening applied to large-scale cell biology. *Trends Biotech.* **22,** 15–22.
61. Taylor, D. L., DeBiasio, R., LaRocca, G., et al. (1994) Potential of machine-vision light microscopy in toxicologic pathology. *Toxicol. Pathol.* **22,** 145–159.
62. Haskins, J. R., Rowse, P., Rahbari, R., and de la Iglesia, F. A. (2001) Thiazolidinedione toxicity to isolated hepatocytes revealed by coherent multiprobe fluorescence microscopy and correlated with multiparameter flow cytometry of peripheral leukocytes. *Arch. Toxicol.* **75,** 425–438.
63. Abraham, V. C., Samson, B., Lapets, O., and Haskins, J. R. (2004) Automated classification of individual cellular responses across multiple targets. *Preclinica* **2,** 349–355.
64. Kolega, J. and Taylor, D. L. (1993) Gradients in the concentration and assembly of myosin II in living fibroblasts during locomotion and fiber transport. *Mol. Biol. Cell* **4,** 819–836.

2

A Pharmaceutical Company User's Perspective on the Potential of High Content Screening in Drug Discovery

Ann F. Hoffman and Ralph J. Garippa

Summary

It is early to fully reflect on the state of the art in high content screening (HCS), because it is still a relatively new approach in drug discovery. Although the development of the first microscopes are a century old and the first confocal microscope is only 20 yr old, the fluorescent probes used within HCS along with the combination of robotic automation and integrated software technologies are quite new. HCS will require a few more years to fully demonstrate its potential power in drug discovery. Within the last year, however, one has seen this ever-expanding field lure participants in from all areas of science, introducing newer versions of instruments and reagents such that the combined efforts result in platforms and tools that meet many organizational goals in multiple ways. The potential of HCS today lies in its versatility. HCS can be used for primary screening, basic research, target identification, biomarkers, cytotoxicity, and helping to predict clinical outcomes. HCS is being applied to stem cells, patient cells, primary hepatocytes, and immortalized cultured cells. We have noted for individual specialized assays, there are multiple solutions just as there are for those standardized universally accepted assays. Whether we have needed to query cellular processes under live conditions or wanted to follow kinetically the course of a compound's effects on particular cellular reactions, we have been hampered by only a few limitations. This chapter offers a glimpse inside the use of HCS in our drug discovery environment.

Key Words: High content screening; high throughput; GPCRs; cell-based assays; translocation; BacMam; imaging; arrestin; target identification; mitotic index.

1. Introduction

As high-throughput screening (HTS) has advanced from processing thousands of compounds to millions of compounds in screening campaigns, there has been significant progress in better understanding and profiling of the subsequent primary hits *(1)*. Significant scrutiny of physiochemical interactions and high standards for preliminary potency of these compounds is now followed with consecutive rounds of multidimensional optimization procedures, early structure activity relationships, and target selectivity evaluations. All of this occurs before the advancement stage in which medicinal chemistry determines how to modify these molecules or how to choose a completely different original core structure through *in silico* screening techniques. As one looks across multiple therapeutic areas, one possible option is to conduct early profiling using HCS technologies for both on-target and off-target liabilities determined by assessing functional cellular processes. Moving beyond the plate reader assays that are acquiring entire cell populations, these HCS assays (e.g., those assessing dead vs live responding cells) obtain further insight into the drug effects on each and every single cell *(2)*. Most of these HCS systems are set up to acquire multiplexed HCS data on various biological events that can be made simultaneously by simply performing sequential reads of different emission spectra corresponding to the varying fluorophore probes of interest.

One forward initiative is to use the complementary automation instruments that were initially installed in the biochemical HTS labs and to proceed with broad screening for cellular functions and phenotype elucidation. Envisioning a core or basic set of platform cellular assays for all targets classes might involve monitoring such compound effects as the consequences on cell cycle and whole cell or nuclear morphology changes. In any one of many disease or drug family-targeted disciplines, the HCS platform assays can be customized for specialized cell types, specialized cell effects, or revolve around a particular pathway that impinges on attractive drug targets. An example of this would be the G protein-coupled receptor (GPCR) targets in which receptor phosphorylation, receptor internalization, and the activation of cAMP pathways would all be simultaneously observed. With the acquired biological and chemistry indices, a broader picture emerges on how the mass of potential hits can be culled, clustered, or classified from the thousands to the selective hit molecules fulfilling the overall requirements of the biology along side the required drug-like properties necessary for new clinical candidate molecules.

Whether it is "blasting" the HTS hits against a battery of HCS assays or evaluating the latest new lead series in a single cell-based screen, a key for all drug targets is the knowledge of what effect on cellular homeostasis has occurred. This can represent both the desired effects and the liabilities (or the off-target effects). Although there have been long standing assays of cytotoxicity, HCS now offers a means to multiplex the biological results required to decide on pursuing one chemical series vs another chemical series. It also can be used to define both the range of effects and the magnitude of the events to address the "responder and nonresponder populations," to evaluate the variation among the cells and cell types that are related to particular disease states, or the progression toward those states. We have rolled-out a series of cell health assays as individual three-plexed cytotoxicity modules. These modules can be offered to the project team a la carte, depending on the degree of compound ranking which the project team wishes to conduct (3). Typically, HCS labs run retrospective analysis on failed compounds, which are those compounds that have demonstrated in vivo toxicity and are no longer viable as drug candidates. These compounds then can be used to define correlative HCS readouts, which in turn, can be used to build a database of predictive values. Once the correlative or predictive value of the HCS assay is recognized within an organization, project teams can begin to utilize the data for prospective analysis and ranking of lead compounds. As ever newer dyes and sensor probes are developed, they can be used to reveal more information in a single multifaceted cell health assay.

One up-and-coming paradigm shift is to combine the emerging multiple types of "differentiated" stem cells as the query cell type for HCS evaluations with the goal of approaching systems that mimic the compound's effects on the primary cells. This will be enabling to then focus efforts on cell specificity, pathway specificity, and particularly on developmental genes. By using the multiplicity of "combinatorial biology," the integrated HCS processes will operate like a magnifying glass on the general state of the cells with the focus on individual cells, which incidentally, is somewhat analogous to the goals of personalized medicine. The latter speaks to the importance of safety to the drug industry. With HCS and automated imaging, early safety can be assessed at every level in the development process from enzyme target, to the cells in which the target is localized, to the tissues and their surrounding and interacting cells, all the way to the effect on animal organs, and to whole animal studies.

Our laboratories have been redesigned over a 4-yr period since we have fully implemented HCS into the HTS department. This consists of having chosen a variety of flexible instrumentation for cell culture, cloning, expression as well as robotic automation compatible with many of the HCS instruments. We have not chosen to require that all of these instruments reside on fully robotically implemented platforms but have chosen a mixture of workstation and stand-alone options in addition to the fully automated systems. Our two Cellomics Arrayscans (Cellomics Inc., Pittsburgh, PA) fulfill most of our medium throughput assays regarding quantity as well as image analysis for most of the specialized cell biology that is requested. In these scenarios, the cell plates are prepared in the tissue culture lab from typical T225 flasks of passaged cells and protein expression of recombinant stable (or transient) transfections are monitored for cellular viability in order to maintain a standardized procedure for assessing cell health. The addition of the Guava Personal Cell Analysis System (Guava Technologies, Hayward, CA) to this laboratory has positively influenced our ability to quickly monitor cells on a daily basis for a preassessment concerning whether the HCS assay of the day should be commenced. By ascertaining the cell viability and the degree of protein expression on the cell surface, one is able to determine whether or not the cells are in a suitable condition for screening on a given day of experimentation. If the cell viability and/or the protein expression reach a level, which is below a predetermined quality set point, the experiment can be taken offline and reoptimized without waste of diluted library compounds. We have chosen to utilize the laser line scanning Acumen Explorer HTS (The Technology Partnership, Cambridge, UK) as one of our platforms for full library HCS/HTS screening of Transfluor assays. The "image analysis" on this system is user defined and based on feature characteristics such as multiple fluorescent wavelength, object width, and length as well as defining subpopulation analysis. An advantage of this platform is that 384 well plates might be read in HTS mode achieving plate acquisition and simultaneous analysis in less than 15 min per plate. In our HCS/HTS screening protocols, we have utilized Tomtec Quadra workstations (Tomtec, Hamden, CT) to complete compound addition to cell plates as well as for quenching and fixing protocols. Titertek's versatile Multidrop Microplate Dispensers and Microplate Washers (Titertek, Huntsville, AL) have become key units within the HCS lab.

Integral to the lab's effectiveness are the same work-flow solutions that have been successful to the biochemical screening in the past as well as the standardization of the processes that provide the final quantitative metrics of the projects. First and foremost is the flexibility that has been designed in allowing efficient use of resources to complete assay development and screening for projects requiring the testing of small numbers of compounds to those testing focused libraries and finally to those that range through the million compound screens. The work-flow solutions rely on the use of ID Business Solution's (IDBS) Activity Base software (Bridgewater, NJ) that can perform logical analysis of multiparameter data, as well as calculate z', and operate to document a plate-by-plate description of the assays *(4)*. Spotfire Decision Site software (Spotfire US, Somerville, MA) is used for analysis and visualization of quality hits and leads to further acquire a high level analysis of the full screening results *(5)*. The quality controls and operating procedures are similar in all extremes as the goals for readouts are to maintain consistent data no matter the plate geometry formats. These are the rules for statistically masking data points, making use of mean calculations vs averages and the assessment of frequency histograms in a multitiered approach.

2. Examples of Our Applications

2.1. GPCR Screening

Our first venture into using HCS for bonafide HTS was when our systems biology team approached us with a proposal designed to identify orphan GPCRs (oGPCRs), which are involved in appetite regulation, feeding behavior, and obesity. Their concept was to construct a genetic dendrogram of sequence homologies to known GPCRs involved in obesity, determine their expression profiles in discrete brain regions known to be involved in feeding behavior, and then to look for differential expression of those transcripts in tissue obtained through laser capture microdissection in diet-induced obese rodents. This approach resulted in a distilled list of candidate oGPCRs, which needed to be screened against the entire Roche compound collection (>700,000 compounds). We faced a dilemma in that we had no positive control compound or peptide for these receptors nor did we have any foreknowledge of their specific heterotrimeric G protein-coupled signaling pathway, details which would have been essential to develop a screening assay to identify agonist compounds. Our solution was to utilize the Transfluor technology in which one would track, through HCS, the movement of cytoplasmic diffuse green fluorescent protein (GFP)-labeled β-arrestin in U 2-OS cells (6). The assay principle utilizes the well-known universal desensitization mechanism of agonist-stimulated GPCRs (7). Basically, after agonist binding to a seven transmembrane receptor domain, the receptor undergoes a conformational change, which triggers the phosphorylation of cytoplasmic residues through G protein-related kinase. The receptors subsequently cluster into clathrin-coated pits on the cell surface and within minutes these receptors are internalized into cytoplasmic vesicles.

To date, we have successfully employed four different HCS reader systems to quantify the spatial redistribution patterns of GFP-β arrestin. Three of the HCS readers were used for four separate HCS/HTS oGPCR screening campaigns although one reader was used for follow-up analysis. The HCS/HTS readers are the Acumen Explorer, the Evotec Technologies Opera (Hamburg, Germany) and the INCell Analyzer 3000 (GE Healthcare, Franklin Lakes, NJ). The first of these instruments is a laser line scanning reader, the second is a Nipkow spinning disk confocal system and the third is a laser confocal slit system, respectively. In each case, we were able to read an entire 384 well plate in 11–12 min with two-channel color recording. Either a Hoechst 33258 or Draq5 dye (Biostatus Limited, Leicestershire, UK) was used for nuclear staining, along with the green fluorescent protein for receptor tracking. Fast autofocusing, together with robotic plate and liquid handling enabled us to have a daily throughput of 80–120 plates per day, or 25,000–40,000 compounds per day, well within the excepted range of a true HTS assay. As is shown, we analyzed a number of individual oGPCR clones on three of these aforementioned HCS readers (INCell Analyzer 3000) (**Fig. 1A**), Acumen Explorer HTS (**Fig. 1B**), and Cellomics Arrayscan (**Fig. 1C**). The results show clearly that each instrument would give an adequate signal to background ratio for screening for those active clones. Depending on one's particular needs for HTS, high definition screening or high content multiplexed screening, one can choose which of these platforms would be most appropriate for moving each type of HCS effort forward. In our experience, any of three different HCS readers were sufficient for HTS campaigns and we typically followed up on hit-to-lead activities with the fourth HCS reader, the Cellomics Arrayscan, although there is no inherent reason to switch HCS platforms as the programs move forward.

2.2. Choosing Lead Candidates

Occasionally, a chemist will come to us with the problem of, "I have dozens of active compounds in this lead series but I cannot discriminate them, based upon potency or binding affinity alone, as to which should be advanced for clinical candidacy." In this realm, HCS has shown,

---

Fig. 1. (**A**) *(Opposite page)* Quantification of receptor internalization in 10 stably transfected Transfluor oGPCR clones and one Transfluor clone of B_2-adrenergic receptor in U 2-OS cells as measured by Fgrains after acquisition and analysis on the A. INCell Analyzer 3000 using the Granularity Module. (**B**) Quantification of receptor internalization in 10 stably transfected Transfluor oGPCR clones and one Transfluor clone of B_2-adrenergic receptor as measured by Data Object No. after acquisition on the B. Acumen Explorer HT using a customized analysis program. (**C**) Quantification of receptor internalization in 10 stably transfected Transfluor oGPCR clones and one Transfluor clone of B_2-adrenergic receptor as measured in percentage Phase 3 after acquisition and analysis on the C. Cellomics Arrayscan 3.1 High Content Imaging Platform and its associated GPCR Bioapplication.

A Pharmaceutical Company User's Perspective

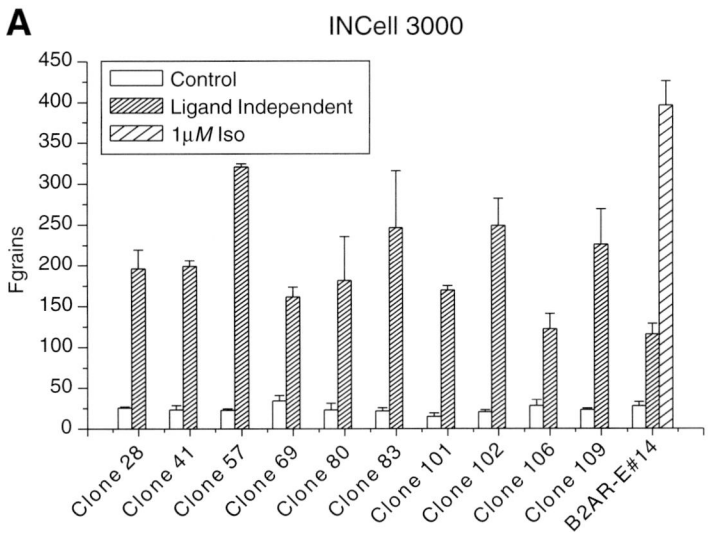

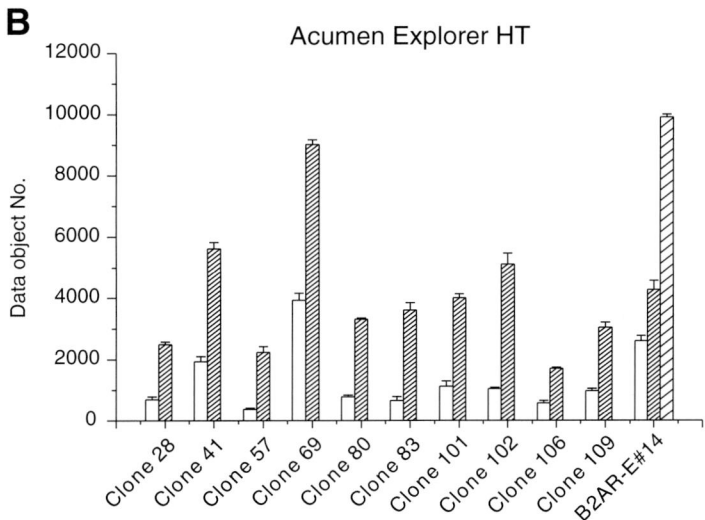

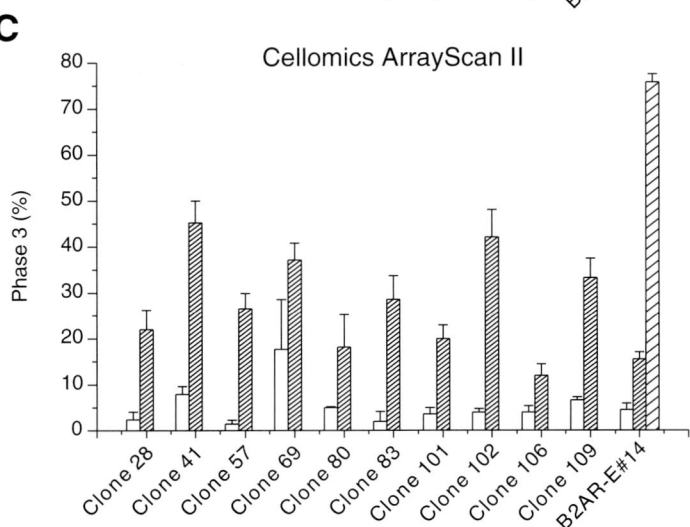

Fig. 1.

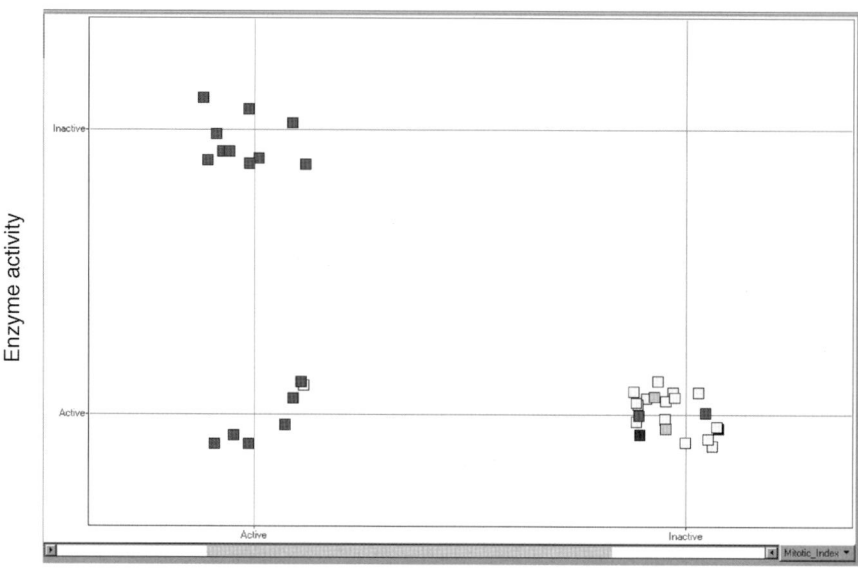

Fig. 2. Graphic analysis of mitotic index and enzymatic activities of a set of compounds. HCS was used to segregate potential chemistry classes with superior functional attributes. (Please *see* the companion CD for the color version of this figure.)

time and time again, to be an effective means to either cull out the compounds devoid of a strong functional impact in cells or, conversely, to allow the "cream" of the compounds to rise up to the top of the priority scale. The percentages of the compounds that "fall-out" vs "advance" depends on a multiplicity of factors, many of which can be accounted for by assaying the compound in a living cell background as opposed to a "test tube" assay of basic reagents/components and the target-of-interest (usually out-of-context). One such example is shown in **Fig. 2**, in which an early biochemical screen revealed compounds with a range of activities from high micromolar to low nanomolar activity. In order to discriminate the potentially most efficacious of the low nanomolar compounds, we employed a HCS mitotic index assay in HeLa cells. The resulting separation of the cell-based "active" from the "inactive" compounds was quite striking, allowing the chemist to move the program successfully forward using a more finely tuned structure–activity relationship, one which accounted for the functional effects of the compound in a living cell. A second example is shown in **Fig. 3**, in which a group of compounds were queried using two different noncell-based assays, one for electromobility (EMSA) shift assay and the other for an enzymatic kinase activity. When a HCS assay was deployed to evaluate the resulting active compounds, three of the compounds were shown to be inactive in cells. A third example is shown in **Fig. 4**, in which the correlation between a kinase activity using an IMAP reagent (bead assay using a fluorescent polarization readout, Molecular Devices, Sunnyvale, CA) was made to a nuclear transcription factor readout as measured using a translocation assay on the Cellomics Arrayscan . For this series of compounds, the correlation was highly (but not perfectly) associated. The conclusion was that the data from two different assay formats (biochemical and cell-based) agreed for a majority of compounds. As a result, the ability to screen and to eliminate dozens of compounds earlier in the drug discovery process became an effective way in which to move the projects forward. Lastly, a similar correlation was made between activity in a cellular nuclear translocation assay and the effect of those compounds on a three dimensional colony formation assay for an oncology project (**Fig. 5**). This information was interpreted to suggest that potent inhibitors in the HCS assay would be more effective in inhibiting tumor proliferation in the subsequent in vivo models.

A Pharmaceutical Company User's Perspective

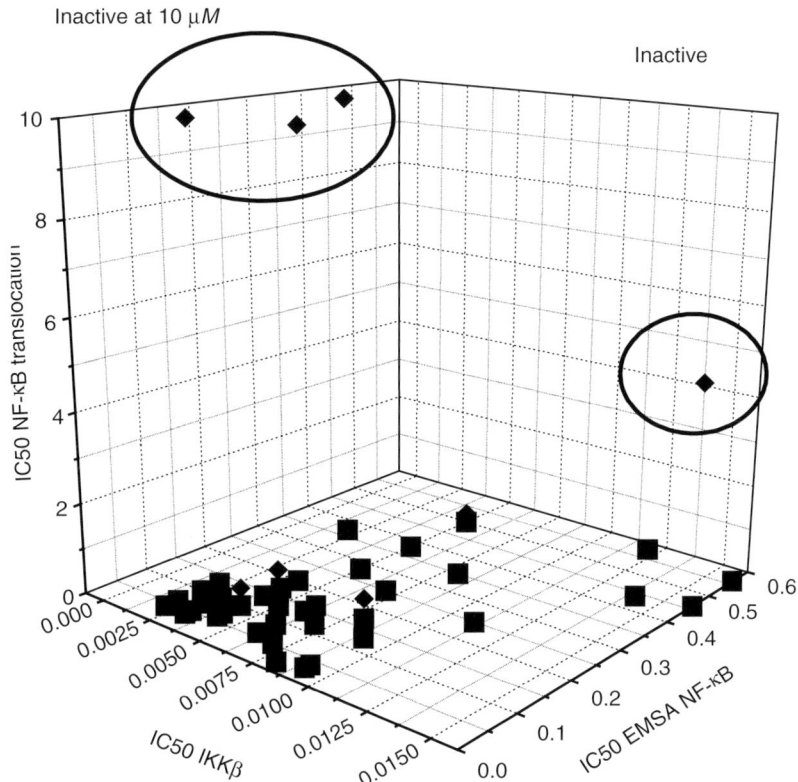

Fig. 3. Three-dimensional plot displaying the culling of inactive compounds using HCS after analysis via noncell-based assays. Enzymatic kinase activity and electromobility shift assays failed to sufficiently segregate the chemistries.

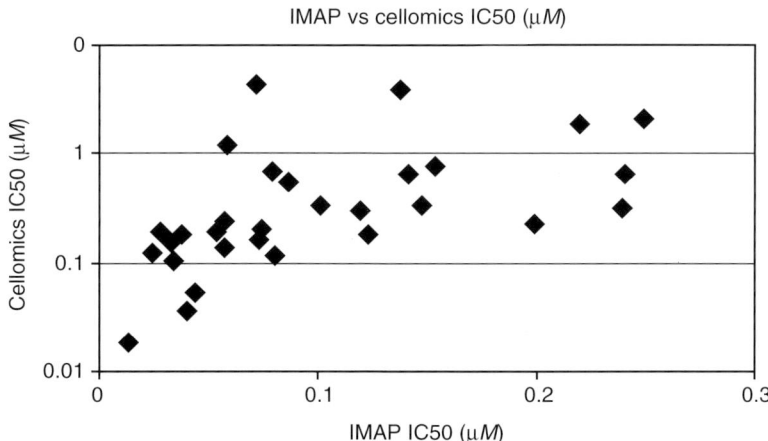

Fig. 4. Graphic correlation of Cellomics Nuclear Translocation HCS assay to in vitro fluorescence polarization assay in which the same target protein was queried with 30 compounds.

There have been other instances in which the Mitotic Index Bioapplication has assisted us. For example, in cases of oncology programs with defined molecular targets, the expectation would be that these targets have distinctly characterized "on-target" mechanism-based cell killing of the tumor cells. But how should one discriminate between generalized or "off target"

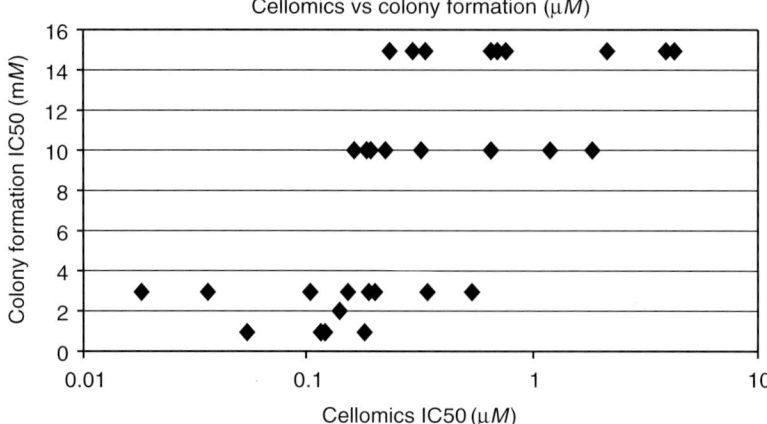

Fig. 5. Graph depicting the correlation between inhibition of a nuclear transcription factor and the inhibition of proliferation in a spheroid colony formation assay. HCS was utilized to separate chemistry classes possessing superior functional activities.

cell killing and the desired activity? An intriguing possibility would be to monitor the ability of the compound to cause a high percentage of the cells to become mitotically arrested whereas not giving evidence of cell rounding and de-adhesion. As seen in **Fig. 6A**, for the positive control compound nocodazole, the beginnings of mitotic arrest are closely associated with the concentration of the compound, which reduces the number of adherent cells in the well. However, for the experimental compounds (**Fig. 6B,C**), there is a clear 1.0–1.5 log window in which the cells exhibit mitotic block before the first evidences of cytotoxicity through cell de-adhesion. This type of dose-related effect for two end points, one target-based and desirable, whereas the other off-target based and undesirable, opens the possibility for finding oncology-related compounds with a greater therapeutic index (or margin of safety, translating into an effective tumoricidal concentration with minimal associated side effects in vivo).

2.3. Transcription Factors and Translocation

One of the most useful and versatile of all of the HCS assays to date has been the cytonuclear translocation module. The basics of this assay lie in the fact that one can discriminate and mask the nuclear border, the cell's periphery (plasma membrane), and the cytoplasm as the area lying in between them. This assay has been particularly effective for us in addressing several members of the nuclear transcription factor family. Simply by changing the primary antibody for a given transcription factor, the same basic algorithm template can be used again and again but for a different target. The assay itself can be configured to search for compounds, which either activate or inhibit the activation/translocation process. A variation of this assay deals with the tracking of an activation/translocation event from the cytoplasm to the plasma membrane, such as is the case with protein kinase C.

2.4. Angiogenesis Assay and Micronucleus Assay

The next two HCS assays highlight the importance of this medium to be able to eliminate subjectivity and bias from the investigator-reviewed image data but also to realize a great advantage in speed. The first of these is the tube formation/angiogenesis assay, in which one is able to quantify in great detail, the length and extent of anastamosed nascent vessels spreading on a two-dimensional substrate. As shown in **Fig. 7**, a dose-dependent difference in the degree of branching of the vascular network can be effectively quantified in an objective manner. This type of automation-assisted standardization minimizes technician-to-technician variation and also allows a greater

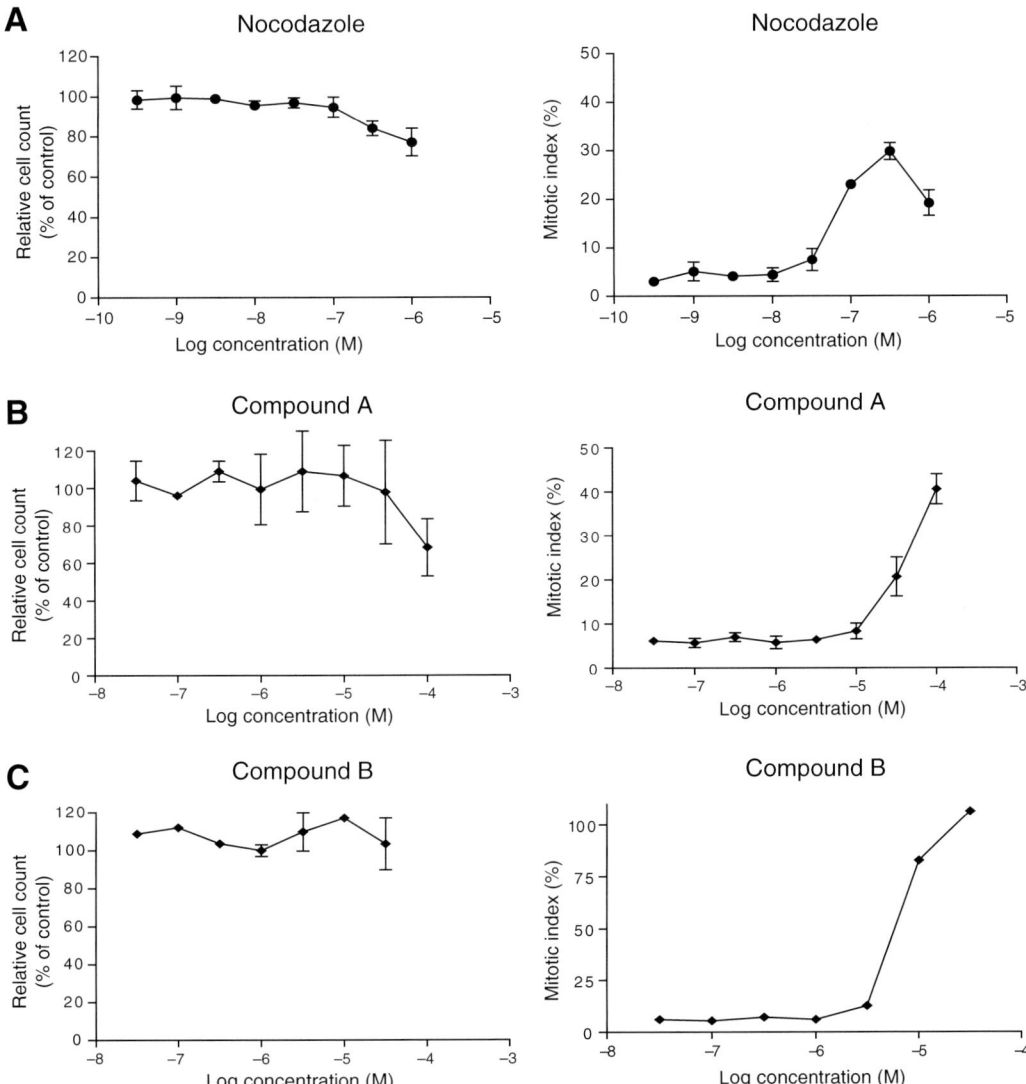

Fig. 6. (**A**) Paired graphs displaying the relative cell counts (left) and mitotic index in percent (right) compared with basal untreated control cells after an 18 h dose–response treatment of **A**. Nocodazole on HeLa cells. (**B**). Paired graphs displaying the relative cell counts (left) and mitotic index in percent (right) compared with basal untreated control cells after an 18-h dose–response treatment of **B**. Compound A on HeLa cells. (**C**). Paired graphs displaying the relative cell counts (left) and mitotic index in percent (right) compared with basal untreated control cells after an 18-h dose–response treatment of **C**. Compound B on HeLa cells.

level of rapid discrimination of specific perturbations in the vascular network. A second example of an HCS assay that has afforded drug investigators a speed advantage is the Micronucleus assay (MN). This test has been universally accepted as a standard for predicting genotoxic events (*8*). At its essence, the MN test quantifies the number of satellite nuclei seen in proximity of the nuclear envelope subsequent to compound exposure and a cytokinesis blocking procedure. The difference in throughput between the technician-curated MN test and the automation-assisted version is striking. It is estimated that one technician can effectively score 1000 cells/h of two

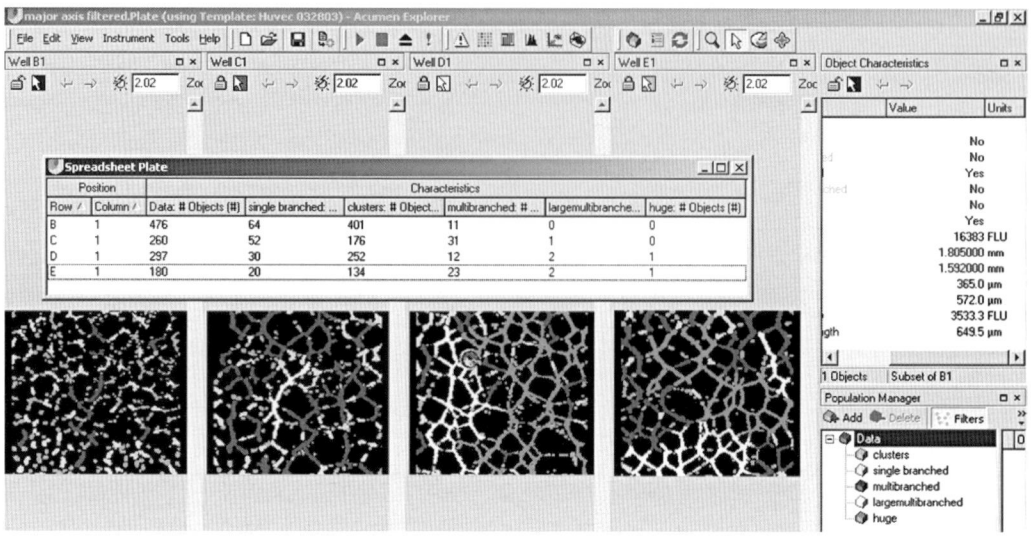

Fig. 7. Screen shot of Acumen Explorer HT software interface displaying the pseudoimages of four wells after treatment with untreated control (far left panel) and three ascending concentrations of a compound on HUVEC cells. Spreadsheet plate data depicts the subpopulation characteristics collected although the object characteristics are shown on right for well B1.

compounds per week, whereas one technician using an HCS module for MN can score 1000 cells/min or 50 compounds per week.

2.5. siRNA Profiling

The next key in advancing the future of HCS is in the development of its basic tools, the reagents. Reagents have been advancing almost as quickly as the next version of the HCS instruments have been introduced into the market and it is these newer reagents that are defining hardware specifications. The explosion of functional genomics from the previous generation in which one was deconvoluting DNA into proteins is now currently redefined by the use of small interfering RNAs (siRNA), which when combined with biological results, identifies the function of proteins within the context of the cell. By applying siRNA to modify the expression of selected genes, one is then able to analyze the specific cellular phenotypes and ultimately make correlations to similar phenotypes in response to specific drugs. In this manner, one uses HCS to examine the differential effects of siRNA and drug chemistries on the same biological functions. As can be seen for **Fig. 8A**, the quantitative assessment of gene knockdown via Taqman® (Roche Molecular Systems, Inc., Alameda, CA) analysis is a useful tool for evaluating the effectiveness of siRNA knockdown of a given target *(9)*. This data can then be further enhanced by introducing a HCS-based analysis, in this case Mitotic Index, for a functional correlate of the knockdown in a living cellular context (**Fig. 8B**). HCS can be used at this stage of target assessment to compare individual siRNAs or pools of siRNAs, based on their ability to elicit a particular cellular phenotype in cells. As is the case with various transfection procedures, the quality assessment measured by the effect of treatment on cell number (**Fig. 8C**), lends confidence to the interpretation that the gene knocked down using the specific siRNA had minimal interfering effect on the number of cells.

3. Looking at Improvements in HCS

3.1. Software Analysis Tools

What has lacked for many of the HCS applications, aside from those now considered standard (e.g., nuclear translocation, mitotic index, cell cycle, and granularity), is the flexibility to quantify a new "customized" assay and its image analysis programming. In the standard cases,

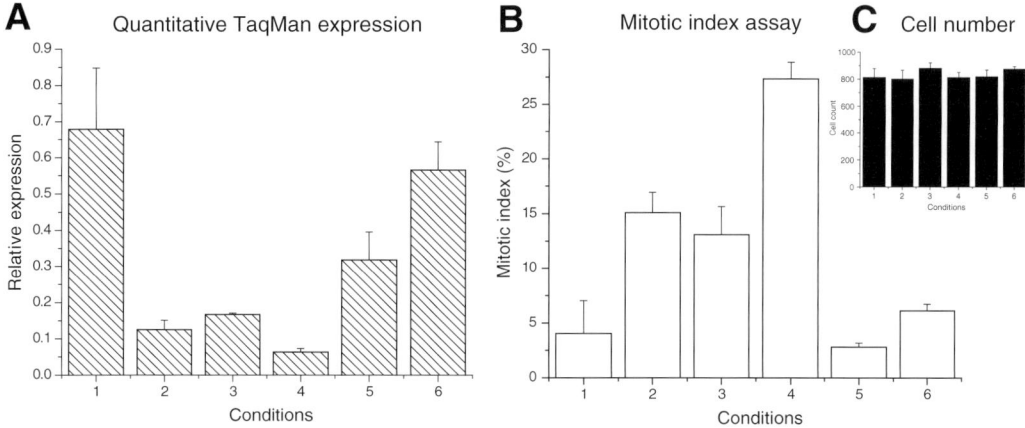

Fig. 8. (**A**). Histogram depicting the successful mRNA knockdown of a potential drug target involved in the cell cycle displaying the degree of mRNA suppression as quantified relative expression of GAPDH. Conditions 1–4 represent individual siRNA pools, condition 5 mock transfected controls, and condition 6 untreated controls. (**B**). Histogram showing the successful functional knockdown in an HCS mitotic arrest assay is inversely proportional to the degree of mRNA suppression. Conditions 1–4 represent individual siRNA pools, condition 5 mock transfected controls, and condition 6 untreated controls. (**C**). Histogram illustrating that under the conditions of the HCS assay the cell number for the individual conditions remained unchanged by transfection treatment and the siRNAs. Conditions 1–4 represent individual siRNA pools, condition 5 mock transfected controls, and condition 6 untreated controls.

there are several image analysis strategies to quantify examples of cellular translocation and other reactions. In these standard cases, most cytonuclear translocation applications can obtain sufficient confidence levels that the resultant data is good; however, often the performance is insufficient when using a diverse assortment of cell lines as is needed in real life assays. One method that is proliferating through the HCS community is the "open image format" allowing the analysis of any images from any instrument to be processed by third party software. Much of this comes from the unique position of the third party applications to apply very different image processing methods. These include processes using object-oriented filtering, those using machine learning (ML)-based on the reiterative process of applying known test sets to define the extremes and then using additional test sets for perfecting the analysis program. Even neural networks are becoming popular to define the HCS results *(10)*. The benefit to HCS is that the image processing becomes independent of the instrument and the HCS vendor. For those new applications that are highly unique or that have not been popularized in the HCS community, the third party vendors can offer the customization needed in the analysis (*see* Chapters 6 and 7). Hence, both image sets from stand-alone confocal microscopes as well as the automated HCS instruments become quite compatible.

With the multitude of choices that the experimentalist has regarding the defined feature systems and the personalized user defined algorithm choices, an effort to standardize assays from lab to lab has been discussed. This recent need to achieve standardization of assays throughout HCS laboratories might eventually result in more uniform data. This evokes the use of the third party image analysis options, particularly those with the ML-adapted solutions. In essence then, the new "student apprentice" in the laboratory is the ML-adapted algorithms enabled through the test sets. The use of this new partner also maintains an unbiased objectiveness to the analysis solutions, allowing the data to define itself as compared with biasing the data with the biological expectations of the outcome. It also facilitates the throughput of the HCS laboratory because it simplifies the tasks that are needed to be completed in the assay development process. In our experience the acquisition of those positive and negative control images of the assay clearly represents the diversity of the expected biology. Once defined and entered into one of the machine-learning processes, an evaluation is performed which maximally optimizes the quantification. Then this is integrated with a representative image series, which is referred to as the test image set. This process then allows for a reiteration and improvement of the image analysis ML algorithm. Sufficient optimization then occurs as measured by z′, manual inspection of the resultant data, and performance parameters such as timeliness of processing and robustness of the algorithm for significantly larger evaluation image sets *(11)*. The application of this new apprentice and the expertise he brings to the HCS analysis can then replace the trial and error customization of a new assay and certainly save considerable time in moving the assay into use.

3.2. Engineering Cells

Where are cell-based assays and their associated technologies situated in the drug discovery pipeline today and how are they going to improve the drug discovery process? The emerging new roles of what and how cell based assays are performed lends us clues to the future directions of drug discovery. Not only has the manipulation of recombinant-labeled cells over the past decades facilitated us to broadly use protein over expressed cell lines for querying enzymatic and protein protein binding, oncogene activation, and effects on various signaling cascades, the techniques involved have now been improved to carry multiple labels enabling duel and even triple labeling multiplexing. This has been achieved by creative experimentalists using cassette-based employment of Crelox internal ribosomal entry site and other systems as Gateway® (Invitrogen Life Technologies, Carlsbad, CA). As these cells can be queried to examine morphological changes via the imaging technologies of HCS, once again an improvement from a gross estimation to specific multiparameter quantification affords advances in drug discovery. In some cases, it is possible to use these recombinantly expressed cells in multiple assay formats to uncover information not initially apparent. The use of image processing affords that ability. Two examples include the definition of nuclear size where compounds that cause nuclear decrease might reflect adversely on the cells (permeabilization and toxicity) and second, in defining the adipocyte cell size and number in which correlations to drug efficacy have been made *(12)*.

Recent transfection reagent advancements involving recombinant baculoviruses that have been modified with mammalian expression cassettes have been used in our lab for primary screening *(13)*. These cassettes are referred to as "BacMams" (baculoviruses engineered to have mammalian promoter elements), and we have used them on occasion for hard-to-express protein in which we wish to follow transient protein expression. Our preliminary use was in a Tranfluor system in which a stably expressed oGPCR protein was unobtainable using conventional lipid-based transfection schemes, resulting in poor cell health even in early stable cell passages or with transient transfection. After multiple insertion techniques were evaluated to attempt to increase the protein expression, we turned to the BacMam system. This transient transfection system still required assay optimization to define the "window" of expression that overlapped with most robust receptor internalization response; however, the use of the BacMam system facilitated assay development and made it possible to run this problematic oGPCR target as a primary HCS/HTS screen. At 4 h post-transduction, the expressed protein could first be detected. However, by 24–30 h post-transduction, there were obvious signs of cytotoxicity. The cells could be easily transfected using this system by simply adding in an aliquot of the virus for 16–20 h and carrying out the assay at a point in which the transiently expressed oGPCR expression was high but cytotoxicity was low. Monitoring the protein expressed, as it was also Flag-tagged, with a standard antibody based measurement on the Guava PCA System. With BacMam expression, there is no viral replication in the transduced mammalian cells, thus providing a high degree of biosafety. Of late, our use of BacMam has significantly expanded, holding the potential that our target proteins of interest can be expressed functionally in primary hepatocytes, endothelial cells, or possibly even T cells.

3.3. The Next Phase

As we plan the implementation of the next HCS phase there is a realization that HCS has already become a "core competency" for many scientists doing basic research and drug discovery. This means that, aside from simply adding to the repertoire of new instrumentation, HCS has become a tool equally embraced by academia and drug screeners. The specialization once associated with this technology and automation-assisted platform instrumentation has been undergoing a metamorphosis, one that is widening its adoption within a greater audience possessing diverse scientific needs. Therefore, the future for us is to ensure HCS's rightful establishment as a core competency within the drug discovery process. By making use of its profiling abilities for early new lead chemistries regarding cytotoxicity screening, target selection, primary screens, and secondary assays, HCS will again and again prove its worth as a staple in the directed progression of a raw chemical into a well-honed new medicinal entity.

References

1. Abraham, V., Taylor, D. L., and Haskins, J. R. (2004) High content screening applied to large-scale cell biology. *Trends Biotechnol.* **22**, 5–22.
2. Vogt, A., Kalb, E. N., and Lazo, J. S. (2004) A scalable high-content cytotoxicity assay insensitive to changes in mitochondrial metabolic activity. *Oncol. Res.* **14**, 305–314.
3. Ainscow, E. (2004) Health check for cells. *Eur. Pharm. Rev.* **9**, 49–52.
4. Seeboth, P. and Hawkins, T. (2003) Implementing a commercial system for integrated discovery data management. *Sci. Comput. Instrum.* **8**, 20–27.
5. Ahlberg, C. (1999) Visual exploration of HTS databases: bridging the gap between chemistry and biology. *Drug Discov. Today* **4**, 370–376.
6. Oakley, R. H., Cowan, C. L., Hudson, C. C., and Loomis, C. R. (2006) Transfluor® provides a universal cell-based assay for screening G protein-coupled receptors, in *Handbook of Assay Development in Drug Discovery*, (Minor, L. K., ed.), CRC Press, Taylor & Francis Group, Boca Raton, FL, pp. 431–435.
7. Zhang, J., Ferguson, S. S. G., Barak, L. S., et al. (1997) Molecular mechanisms of G protein-coupled receptor signaling: role of G protein-coupled receptor kinases and arrestins in receptor desensitization and resensitization. *Receptors Channels* **5**, 193–199.
8. Fenech, M. (2005) In vitro micronucleus technique to predict chemosensitivity. *Methods Mol. Med.* **111**, 3–32.
9. Harborth, J., Elbashir, S. M., Vandenburgh, K., et al. (2003) Sequence, chemical, and structural variation of small interfereing RNAs and short hairpin RNAs and the effect on mammalian gene silencing. *Antisense Nucl. Acid Drug Dev.* **13**, 83–105.
10. Laghaee, A., Malcolm, C., Hallam, J., and Ghazal, P. (2005) Artificial intelligence and robotics in high throughput post-genomics. *Drug Discov. Today* **10**, 1253–1259.
11. Zhang, J., Chung, T. D. Y., and Oldenburg, K. R. (1999) A simple statistical parameter for use in evaluation and validation of high-throughput screening assays. *J. Biomol. Screen.* **4**, 67–73.
12. Chen, H. C. and Farese, R. V. Jr. (2002) Determination of adipocyte size by computer image analysis. *J. Lipid Res.* **43**, 986–989.
13. Ames, R., Fornwald, J., Nuthulaganti, P., et al. (2004) BacMam recombinant baculoviruses in G protein-coupled receptor drug discovery. *Receptors Channels* **10**, 99–107.

power can easily be scaled and applied in a distributed fashion to cope with increases in data acquisition, multiobjective parallel imaging platforms are currently being developed at the Whitehead MIT BioImaging Center to address the need for log orders of magnitude increases in sample throughput.

7.3. Data Processing, Analysis, and Management

Although hardware development remains firmly in the commercial sector, development of freely distributed image processing and analysis applications has long been available through the academic community. For small scale processing and analysis, the java-based ImageJ (formerly NIH Image) provides novice users the ability to segment and quantify cell features although this does not yet provide the necessary scalability to be effective in the HCS field. In contrast, the MatLab-based CellProfiler package (www.cellprofiler.org) was developed with large-scale analyses in mind and offers developers complete flexibility as an open source project.

Standard file formats in cellular imaging remain to be established, although many point to the success of the DICOM standard in medical imaging as a possible platform on which to build. Initatives such as the XML-based open microscopy environment (OME) and ExperiBase, although both still nascent, might become standards for general quantitative microscopy that also enable sharing of published HCS data *(18,19)*.

The increased sample throughput of HCS, compared with standard fluorescent microscopy, enables greater exploration of parameter space. Coupled with increasing resolution afforded by multiplexed and time-resolved assays, HCS is positioned as the predominant tool to used address the complexity of cell and tissue biology in the postgenomic era.

References

1. Lander, E. S., Linton, L. M., Birren, B., et al. (2001) Initial sequencing and analysis of the human genome. *Nature* **409,** 860–921.
2. Huh, W. K., Falvo, J. V., Gerke, L. C., et al. (2003) Global analysis of protein localization in budding yeast. *Nature* **425,** 686–691.
3. Huang, S. and O'Shea, E. K. (2005) A systematic high-throughput screen of a yeast deletion collection for mutants defective in PHO5 regulation. *Genetics* **169,** 1859–1871.
4. Harada, J. N., Bower, K. E., Orth, A. P., et al. (2005) Identification of novel mammalian growth regulatory factors by genome-scale quantitative image analysis. *Genome Res.* **15,** 1136–1144.
5. Perlman, Z. E., Slack, M. D., Feng, Y., Mitchison, T. J., Wu, L. F., and Altschuler, S. J. (2004) Multidimensional drug profiling by automated microscopy. *Science* **306,** 1194–1198.
6. Giuliano, K. A., Chen, Y. T., and Taylor, D. L. (2004) High content screening with siRNA optimizes a cell biological approach to drug discovery: Defining the role of p53 activation in the cellular response to anticancer drugs. *J. Biomol. Screen.* **9,** 557–568.
7. Evans, J. G., Correia, I., Krasavina, O., Watson, N., and Matsudaira, P. (2003) Macrophage podosomes assemble at the leading lamella by growth and fragmentation. *J. Cell Biol.* **161,** 697–705.
8. Linder, S., Hufner, K., Wintergerst, U., and Aepfelbacher, M. (2000) Microtubule-dependent formation of podosomal adhesion structures in primary human macrophages. *J. Cell Sci.* **113(Pt 23),** 4165–4176.
9. Blake, R. A., Broome, M. A., Liu, X., et al. (2000) SU6656, a selective src family kinase inhibitor, used to probe growth factor signaling. *Mol. Cell Biol.* **20,** 9018–9027.
10. Yarrow, J. C., Perlman, Z. E., Westwood, N. J., and Mitchison, T. J. (2004) A high-throughput cell migration assay using scratch wound healing, a comparison of image-based readout methods. *BMC Biotechnol.* **4,** 21.
11. Yarrow, J. C., Totsukawa, G., Charras, G. T., and Mitchison, T. J. (2005) Screening for cell migration inhibitors via automated microscopy reveals a Rho-kinase inhibitor. *Chem. Biol.* **12,** 385–395.
12. Ragan, T., Kim, K. H., Bahlmann, K., and So, P. T. (2004) Two-photon tissue cytometry. *Methods Cell Biol.* **75,** 23–39.
13. Ragan, T. M., Huang, H., and So, P. T. (2003) In vivo and ex vivo tissue applications of two-photon microscopy. *Methods Enzymol.* **361,** 481–505.

14. Ziauddin, J. and Sabatini, D. M. (2001) Microarrays of cells expressing defined cDNAs. *Nature* **411,** 107–110.
15. Voldman, J., Gray, M. L., Toner, M., and Schmidt, M. A. (2002) A microfabrication-based dynamic array cytometer. *Anal. Chem.* **74,** 3984–3990.
16. Gustafsson, M. G. (2000) Surpassing the lateral resolution limit by a factor of two using structured illumination microscopy. *J. Microsc.* **198(Pt 2),** 82–87.
17. Egner, A., and Hell, S. W. (2005) Fluorescence microscopy with super-resolved optical sections. *Trends Cell Biol.* **15,** 207–215.
18. Swedlow, J. R., Goldberg, I., Brauner, E., and Sorger, P. K. (2003) Informatics and quantitative analysis in biological imaging. *Science* **300,** 100–102.
19. Dewey, C. F., Jr., Downes, A., Chou, H., and Zhang, S. (2005) ExperiBase—an object model implementation for biology. *Proceedings of the Eighth Annual Bio-Ontologies Meeting*, June 24, 2005.

II

INSTRUMENTATION, BIOLOGICAL APPLICATION SOFTWARE, AND SAMPLE PREPARATION

4

Requirements, Features, and Performance of High Content Screening Platforms

Albert H. Gough and Paul A. Johnston

Summary

High content screening (HCS) platforms integrate fluorescence microscopy with image analysis algorithms and informatics to automate cell analysis. The initial applications of HCS to secondary screening in drug discovery have spread throughout the discovery pipeline, and now into the expanding research field of systems cell biology, in which new manipulation tools enable the use of large scale screens to understand cellular pathways, and cell functions. In this chapter we discuss the requirements for HCS and the systems that have been designed to meet these application needs. The number of HCS systems available in the market place, and the range of features available, has grown considerably in the past 2 yr. Of the two general optical designs, the confocal systems have dominated the high-throughput HCS market, whereas the more cost effective wide-field systems have dominated all other market segments, and have a much larger market share. The majority of available systems have been optimized for fixed cell applications; however, there is growing interest in live cell kinetic assays, and four systems have successfully penetrated this application area. The breadth of applications for these systems continues to expand, especially with the integration of new technologies. New applications, improved software, better data visualization tools, and new detection methods such as multispectral imaging and fluorescence lifetime are predicted to drive the development of future HCS platforms.

Key Words: Automation; confocal; drug discovery; fluorescence; high content screening; high-throughput; imaging; microscopy; systems cell biology.

1. Introduction

High content screening (HCS) is a major new advance in cell analysis technology introduced by Cellomics, Inc. (Pittsburgh, PA) in 1997. Fluorescence-labeling technologies were first combined with electronic imaging technology more than two decades ago and have been used extensively in basic research to study individual cells by light microscopy *(1–3)*. However, the innovation that drove the development of HCS, and what distinguishes HCS systems from the many confocal and wide-field microscopes, is the integration and automation of the entire analytical process. HCS platforms automate the capture and analysis of fluorescence images of millions of individual cells in tens of thousands of samples on a daily basis, and have made fluorescence-based cell analysis compatible with the needs of drug discovery *(4,5)*, and systems biology *(6)*. There are now more than 10 models of HCS imagers established in the market (*see* **Table 1**). Used in combination with appropriate probes, antibodies, fluorescent protein fusion partners, biosensors, environmentally sensitive probes, and stains. HCS systems can be applied

Table 1
Features of HCS Systems

System vendor	Optics	Autofocus	Optical sectioning	Cost ($$)	Other features
Single end point HCS					
ArrayScan® VTI (Cellomics)	WF	Software	Apotome	$$	Incubation[a]
Discovery-1 (Molecular Devices)	WF	Software Laser[a]	Software	$	Brightfield[a]
InCell 1000 (GE Healthcare)	WF	Software Laser		$$	Dispenser[a] Incubation[a]
Cell Lab IC 100 (Beckman Coulter)	WF	Software video		$$	
ImageExpress Micro (Molecular Devices)	WF	Software Laser[a,b]	Software	$	Brightfield[a]/ Phase contrast[a]
CellwoRx (Applied Precision)	WF	Software	Software	$	Uses cellomics bioapplications
Kinetic HCS					
KineticScan® (Cellomics)	WF	Software Laser	–	$$	Pipetor Incubation
ImageExpress 5000A (Molecular Devices)	WF	Software Laser	–	$$	Pipetor Incubation
Pathway HT (Becton-Dickinson)	WF CF	Software	Nipkow disk	$$	Pipetor Incubation
HT-HCS					
InCell 3000 (GE Healthcare)	CF	Laser	Confocal	$$$	Injector Incubation
Opera (Evotec)	CF	Laser	Disk scan confocal	$$$	–

The currently available HCS systems can be categorized by target application area, Single End Point HCS, Kinetic HCS, and HT-HCS. Features listed are limited to those features, which distinguish systems. All systems use either wide-field (WF) optics, confocal (CF) optics, or provide both. The approximate system costs are indicated as $ < $200K, $$ = $200K–$500K, or $$$ > $500K, but will depend on the specific configuration.
[a]Optional features.

to many drug target classes, can be configured for simultaneous multiple target readouts (multiplexing), and can provide information on distributions and cell morphology, in addition to many other fluorescence parameters. Image-based assays therefore provide multiparameter quantitative and qualitative information beyond the single parameter target data typical of most other assay formats, and thus are referred to as high "content" assays. In recent years there has been a growing trend in drug discovery toward the implementation of cell-based assays, in which the target is screened in a more physiological context than in biochemical assays of isolated targets *(7)*. Fluorescence microscopy, whether confocal or wide-field, is one of the most powerful tools that cell biologists can use to interrogate biomolecules and investigate the molecular mechanisms of the cell *(8)*. HCS platforms are therefore being deployed throughout the drug discovery process; target identification/target validation, primary screening and lead generation, hit characterization, lead optimization, toxicology, biomarker development, and diagnostic histopathology. Furthermore, these platforms are now spreading into the research markets for applications in high-throughput biology *(9–11)*.

The integration of new technologies in drug discovery by the pharmaceutical industry typically occurs in their high-throughput screening (HTS) groups. New technology integration is largely driven by a need to increase throughput and capacity, the ability to enable screens for a previously intractable drug target class, or to provide a novel method to rescreen high-priority targets in which the hits from other HTS formats fail to progress to quality leads. Dependent on the complexity of the assay, automated imaging platforms might only provide relatively modest throughput (10–30 K/d) when compared with other assay formats that are compatible with high (10–100 K/d) or ultrahigh (>100 K/d) throughput *(7)*. However, some high content assays are also high throughput *(12)*. Certainly imaging assays have provided a means to address many intracellular target classes (kinases, phosphatatses, proteases, and so on) that were previously challenging in cell-based HTS formats, and have provided novel methods to screen some target classes (e.g., G protein-coupled receptors [GPCRs]) that were well supported by other assay formats *(8,12)*. Owing to the relatively lower throughput and higher complexity of the multiparameter data generated, discovery scientist initially deployed HCS platforms for secondary and tertiary cell-based assays in hit characterization and lead optimization. However, as the HCS platforms and reagents have continued to mature, and as the adopters of the technology have gained more experience and understanding of its capabilities, HCS has been applied throughout the drug discovery pipeline.

The automated, multiplexed cell assays possible on HCS platforms are applicable not only in drug discovery, but also in basic biomedical research, especially in the emerging field of systems cell biology *(13–15)*. The analysis of complex cellular pathways and cell functions requires large scale experimental approaches that take advantage of powerful reagent tools such as siRNA *(5)* and controllable gene expression systems *(16)* for manipulating biological systems, in conjunction with HCS systems such as those described here. Although fluorescence imaging microscope systems are nearly ubiquitous in cell biology research labs, the applications have predominantly involved detailed, high resolution analysis of macromolecular structures, localization of cellular components, and measuring the dynamics of cellular functions, in a reductionist approach that has been used to identify and characterize many pathways, and molecular intermediates *(17)*. However, understanding the complex interplay between these molecules and pathways, and the emergent cellular functions, requires running larger numbers of multiplexed assays in a much more automated way than is possible on current research imaging systems. As a result, the application of HCS systems in the academic research market is expected to grow rapidly over the next few years.

The market for HCS technologies has grown to more than $100M and includes the instrumentation for automated image capture, reagents for sample preparation and staining, image/data analysis software, image/data visualization software, and image/data database applications to manage, and store the very large amount of data that these platforms can generate. In the past 2–3 yr the HCS instrument market has consolidated with many of the large HTS instrument vendors acquiring the smaller independent imaging platform manufacturers to provide their offerings in the HCS field. Amersham Biosciences (Piscataway, NJ) acquired Imaging Research Inc. (St. Catherines, Ontario, Canada) and Praelux Inc. (Lawrenceville, NJ) as the foundation of their In Cell Analyzer (INCA) 1000 and 3000 platforms, respectively, and then was acquired by GE Healthcare (Giles, UK). Molecular Devices (Sunnyvale, CA) acquired Universal Imaging (West Chester, PA) and Axon Instruments (Union City, CA) as the basis of their Discovery-1 and ImageXpress 5000 platforms, respectively. Beckman-Coulter (Fullerton, CA) acquired Q3DM (San Diego, CA) for their EIDAQ-100 platform, and Becton-Dickenson (Franklin Lakes, NJ) acquired Atto-Biosciences (Rockville, MD) for their Pathway-HT platform. Even Cellomics, Inc. the last of the independents and the company that created the HCS market, has recently been acquired by Fisher Scientific (Hampton, NH). However, even as this first round of consolidation is coming to a close, a new group of companies like Blueshift Biotechnologies (San Francisco, CA) and Applied Precision (Issaquah, WA) are entering the market with a new crop of novel HCS systems.

Although all of the HCS platforms were designed to meet common functional requirements discussed more in detail later, a diverse array of imaging technologies that have been integrated to meet these requirements. The systems discussed in this chapter will be limited to automated fluorescence imaging microscope systems designed for use in HCS. We will review the functional requirements that drive the development of HCS systems, and discuss the various solutions that have been used to achieve these requirements. It is important to keep in mind that the HCS imager is only a component (albeit an important component) of a larger assay platform that includes cells, reagents, image analysis applications, informatics, and automation (*see* related chapters in this volume).

The goals of this chapter are to:

1. Provide an introduction to the fundamental features required of all HCS systems.
2. Identify the key features that distinguish current HCS systems.
3. Introduce new technologies that are likely to, or should be integrated into HCS systems to provide expanded functionality.

2. System Requirements and Components

All HCS platforms require a process for the input and output of multiple microplates, mechanisms to position plates on a stage, the ability to position wells over the optics with precision and reproducibility, a method to capture quality images, image analysis applications (algorithms), data review tools, a mechanism for data management, and further analysis and visualization. The solutions offered by the many vendors of HCS systems vary in the degree to which all these components are integrated and provided as part of a complete screening system. Here, we will focus on the HCS imagers and software; discussion of the applications and informatics tools can be found elsewhere in this volume. The basic requirements for an HCS imager are the following:

1. Sufficient resolution and sensitivity to capture and analyze the cellular features of interest.
2. A field of view large enough to image multiple cells.
3. Spectral channels to distinguish multiple fluorescent labels.
4. Adequate speed to meet the needs of the planned screening volume.
5. Flexibility to address a wide range of assay requirements.

Unfortunately, these requirements cannot be addressed independently, but are interrelated, and therefore, the optimum design and configuration of an HCS system for a particular assay requires an understanding of the relationships between these requirements.

2.1. Biological Application Requirements

The basic cellular features of interest for HCS assays (at present) include subcellular localization and distribution; fluorescence intensity, and intensity ratios; texture within regions; cellular and subcellular morphometrics; and the total count of a particular feature, such as nuclei or cells. Numerical combinations of these features are also useful features, as are the correlations between features, for example, colocalization. HCS platforms must be designed to acquire images with sufficient contrast, resolution, and signal-to-noise ratio to allow image algorithms to extract these features of interest. These features span a wide range of sizes, from hundreds of nanometers, to hundreds of microns for colony features. A typical mammalian cell, attached and spread on the bottom of a microplate might be on the order of 20–50 µm across, with a nucleus of about 5–10 µm in diameter. One hundred of these cells in a confluent layer occupy an area of about 400 µm on a side, depending on the cell type, plating density, and amount of spreading. Assays vary widely in the maximum cell density that can be used. A cell-spreading assay naturally requires subconfluent cells, whereas receptor activation assays can often be done using highly confluent cells. The analysis of 100 to as many as 1000 or more cells, occupying an area as large as several square millimeters, might be needed to achieve an adequate level of statistical significance.

The area of a cell 30 µm in diameter is about 700 µm^2, or 700 pixels if sampled with 1 µm resolution. Except for very fine details, 700 pixels per cell provide adequate information on the spatial distribution of components within the cell. More resolution might make the edges of a

Features of HCS Platforms 49

Sophisticated software focus methods have the advantage of focusing on a feature of interest, ensuring sharp focus for that feature. The most sophisticated are designed to ignore dust and other contaminants when finding focus. Laser focus is very fast but the focus quality relies on a smooth substrate and a consistent position of the feature of interest relative to the substrate. Focus based on video signal processing is also very fast and focuses directly on the cells, although not necessarily the cellular feature of interest. All the focus methods should work well on cells, which are well spread on the substrate and therefore very thin, especially with magnification of ×10 or less, because of the large depth of field. Although laser and video-based focusing are certainly faster than software focus methods, the impact on speed is only significant for HT-HCS, in which upwards of 50,000 assays a day must be run.

2.7. Detectors

The ideal detector for quantitative fluorescence imaging has high sensitivity, broad spectral range from the near infrared to near ultraviolet wavelengths, high dynamic range, and a linear response. Scientific CCD cameras have all these features, and all of the systems listed in **Table 1** use a scientific grade CCD camera. Furthermore, many of the systems use the same camera or at least the same CCD detector in the camera. Of the systems, only the camera on the INCA 3000 is of a significantly different design, and that is principally because of the design of the system. Because the INCA 3000 is a line scan confocal system, it uses three line scan CCDs, one for each emission channel, rather than the two-dimensional CCD cameras used on the other systems. Some vendors offer alternative cameras with improved sensitivity and performance. For example, the new ImageExpress Micro (Molecular Devices) (**Table 1**) is available with either the CoolSnap ES (Photometrics, Tucson, AZ) or the higher sensitivity, lower noise, lower dark current CoolSnap HQ. The added cost of a higher performance camera will, in most cases, be offset by improved performance, and therefore is usually money well spent.

2.8. Summary

In summary, HCS systems provide flexibility to allow the optical configuration to be customized to address a wide range of specific assay requirements. The performance of an HCS system depends not only on the design, but on choosing the optimum configuration to meet the assay requirements. To take advantage of the flexibility in resolution, field of view, speed, and multiwavelength imaging, assay developers need to understand the relationships between configuration options, and determine the critical elements of the imaging requirements for each assay. For example, when running a Transfluor® (Xsira Pharmaceutical Inc., Research Triangle Park, NC) GPCR assay on the INCA 3000 platform, using Draq 5 to stain the nuclei of β-arrestin-GFP expressing cells avoids crosstalk and allows imaging both fluorophores with a single pass, collecting the emissions on separate red and green CCD cameras. By selecting glass bottom plates with the appropriate seeding density, only a single field of view will be required, and the width and length of the scan can be reduced to speed up image capture further. Binning 2×2 will also reduce scan times. Under these conditions, scan times of 8–10 min/384-well plate are attainable. However, scan times will increase significantly for other assays with different combinations of fluorophores in which more than one pass is required for excitation, or multiple channels of florescence and fields of view might need to be acquired. When comparing throughputs on different imaging platforms, it is good practice to run as many distinct biologies and combinations of fluorophores as possible, collecting data on instrument performance under a wide range of conditions, not just optimal conditions.

3. Imaging Software Solutions

A critical component of any automated imaging platform is the software, it serves to control and set up the instrument for image capture, but perhaps more importantly provides tools for image/data analysis, image/data visualization, and must integrate into a database application to manage and store the large volume of data generated *(22)*.

3.1. Acquisition and Control Software

During the development and optimization of image-based assays, a number of experimental parameters will be established that impact how the instrument must be set-up to capture the images. The number and types of fluorescent probes to be imaged will determine how many separate channels or passes will need to be collected, and which excitation and emission filters are required. The quality, reproducibility, and intensity of the fluorescent signal can affect whether auto-exposure or fixed exposure will be utilized, and the optical resolution required will determine, which objective will be employed. The cell seeding density and/or the relative frequency of the response being measured in the total population of cells will impact the number and/or the size of the image fields that need to be captured. For instruments with confocal capability, the appropriate focal offset must be selected based on a previous Z-stack analysis. Many of these instrument settings are either controlled by the software or have to be selected in the software, along with other experimental details such as the plate density, and number of wells to be imaged. It is important that the instrument setup procedure and software should be both intuitive and straightforward.

3.2. Image Analysis Software

There are multiple levels of image analysis/processing; at the pixel level, the object level, the semantic concept level, and the pattern and knowledge level *(22)*. Raw and filtered digital images are made up of pixels, or gray values captured by a CCD camera. The pixels are assigned to objects established through segmentation. Segmentation methods include marker-based, object-based, or contour-based segmentation. Information about the objects in an image is condensed into features such as intensity, color, shape, and texture. Objects and regions can be classified into user-defined categories based on these features or properties. A variety of object classification approaches are available; decision rules defined by the user as a set of boundary conditions, fuzzy decision rules that permit a gradual transition between classes, clustering methods that distinguish groups of similar objects based on a similarity measure defined on a set of features, and supervised learning methods such as neural networks or support vector machines *(22)*. Image analysis algorithms derive quantitative and qualitative measures of features such as object count, width, length, spatial distribution, motion, behavior over time, and feature ratios, which are calculated on a per cell basis and/or as a well (population) average (*see* Chapter 6).

Research fluorescence microscope imaging platforms typically utilize software packages such as Image-Pro (MediaCybernetics, Silver Spring, MD) or MetaMorph (Molecular Devices), for the acquisition, analysis/processing, archiving, and retrieval of raw and enhanced digital images. These software packages are designed for interactive image analysis and provide the high levels of flexibility and complexity that are required in a research environment. Most HCS platforms are operated by scientists that are not image processing experts, and therefore require more user friendly, or turnkey solutions to image analysis. More importantly, HCS is a production operation requiring methods that perform consistently and efficiently, and therefore requires applications that are designed and validated to be robust and efficient. Once again, research fluorescence microscopes and HCS systems can be used very effectively in combination to address a wide range of biological applications (for more information, *see* Chapter 1).

HCS platforms typically provide a number of canned algorithms to address specific biological applications (*see* **Table 3**). These automated image analysis algorithms produce a set of relevant features that might be tailored to specific biological applications *(22)*. The user identifies the objects and features to be extracted automatically from every image by adjusting algorithm parameters before running the analysis procedure. Although these canned algorithms limit the power of the image analysis application, too many options might be bewildering to the inexperienced user, or might significantly impede the assay development process. It is important therefore, that the

Table 3
Software Applications Available From HCS System Vendors

Category	Cellomics, Inc.	GE-Healthcare	Molecular devices	Becton-Dickenson	Beckman-Coulter
Translocation	Cytoplasm to nucleus translocation	Nuclear trafficking analysis module	Nuclear translocation	Nuclear translocation	Nuclear-cytoplasmic translocation
	Cytoplasm to cell membrane translocation	Plama membrane trafficking analysis module	Translocation		Invasion and motility
	Molecular translocation				
	Kinetic molecular translocation				
	Cell motility				
Cell cycle	Mitotic index	Cell cycle analysis module	Mitotic index		Cell cycle
	Cell cycle analysis		Cell cycle		Regulation of protein expression in the cell cycle
GPCR	Receptor internalization	Granularity analysis module	Transfluor®		GPCR validation report
	GPCR signaling				GPCR activation
	Spot detector				
Neurite outgrowth	Neurite outgrowth I	Neurite outgrowth	Neurite outgrowth		
	Extended neurite outgrowth				
Morphology	Cell spreading		Angiogenesis tube formation		
	Morphology explorer		Monopole detection		
Cell health	Live/dead		Live/dead	Live/dead	Proliferation and apoptosis
	Multiparameter necrosis		Cell health	Apoptosis	
	Multiparameter apoptosis		Cell proliferation	Mitochondrial dysfunction	
	Cell health profiling			Steatosis	
	Micronucleus detection				
General applications	Compartmental analysis	Object intensity analysis module	Granularity	Ion channels and ratio imaging	Aggregate formation
	Kinetic compartmental analysis	Bead analysis module	Count nuclei	cAMP	Transiently transfected cell populations
			Cell scoring		

automated image analysis software is intuitive and easy to navigate so that setting and validating the parameters for image segmentation, object identification/classification, and feature calculation are user friendly. The selection and optimization of the final image analysis parameters typically involves the use of a training set of images, most commonly the assay controls for the top and bottom of the signal window. Assay development might be done independently of the instrument, if the fully functional software is provided for offline use, such as the Cellomics, Inc. HCS toolbox. At the time of writing, the number of image analysis algorithms and the variety of biological applications addressed varied significantly between the HCS platforms, with the ArrayScan® Reader (Cellomics, Inc.) having the most extensive portfolio (**Table 3**). However, as users become more experienced or require methods for biological applications not adequately addressed by the available image analysis algorithms, it will become a requirement that image files can be exported to more powerful image analysis software to develop customized algorithms for analysis offline. Equally important, platforms will need to provide a process to import customized algorithms, and other software tools into their software. Another level of complexity is associated with the image analysis of multiplexed target readouts. Most of the basic canned image analysis algorithms require that separate target channels be analyzed independently, whereas many of the more advanced algorithms, such as Cellomics' Compartmental Analysis application, have been designed to handle multiplexed target readouts (*see* Chapter 5).

3.3. Data Mining

In addition to the image analysis algorithms, the automated imaging platform software must provide an integrated environment that supports visual data mining *(22)*. There are several levels that the software must achieve. First, when the instrument is acquiring images from a set of screening plates the software must be able to monitor and display the progress of the run in real time. This should include views to assess how the screening run is progressing at the plate level, plate views that illustrate which well is being imaged together with some representation, for example, a heat map, of the results for wells already imaged, and image views of the fields being captured for that well. These views serve to provide the operator a degree of confidence that the robotic plate loader is functioning, the instrument has been set up correctly and is acquiring quality images, the image analysis algorithm has been appropriately optimized, and that the data for the plate controls are consistent with expectations. Second, the software must provide a means to inspect and interpret the multiparameter analysis results in the context of the images, the raw data, the experimental conditions and procedures utilized. The user needs interactive software that allows them to assess the quality of the experiment and programmed analysis *(22)*. In addition to the specific target readout that the algorithm provides, the software should allow the user to mine the data to extract additional parameters such as morphology features, apparent cytotoxicity, or potential artifacts such as fluorescent compounds. The ability to toggle between the images (both fields and individual cells) and the data views is critical, and some software packages like that provided with the Cell Lab IC 100 platform (Beckman-Coulter), and others, allow the user to pick points or groups in a data viewer and pull back the individual cell images in a gallery. These tools enable the user to view the images from obvious outliers in the data set and can often provide a means to identify a plausible underlying cause, for example, cells that are in mitosis. Third, the software must provide tools to visualize, manipulate, and compare the multiparameter data and images, to help recognize high-level patterns and relationships that might assist interpretation of the data *(8,15,22)*. If the data visualization tools are limited, they should provide user friendly methods to extract and export multiparameter data, at both the cell and well levels, to more powerful external data visualization and analysis packages such as Spotfire® (Spotfire, Somerville, MA) or S-plus® (Insightful, Seattle, WA). If images are to be analyzed postacquisition, rather than in real time, additional fully functional software seats beyond that loaded on the instrument will be required. Additional copies of software might also be required to facilitate data sharing both on site and in a global organization. Although this might require more copies of the fully functional software, for some purposes simpler data viewer functionality might be adequate. The number of copies and/or cost of software seats that the vendor provides beyond that loaded on the instrument should be considered when selecting a platform to purchase (*see also* Chapter 23).

3.4. Data Storage and Management

Automated imaging platforms that are operated for HCS purposes generate large amounts of data and it is critical that these data be securely stored and effectively managed in a database *(22)*. In addition to the large raw image files, the data model also needs to capture and integrate the associated metadata together with the data generated by the image analysis/processing algorithm and information from corporate databases to provide an effective data-mining environment *(22)*. The metadata includes the nature of the samples, the experimental conditions, and the procedures used to acquire and analyze the images. The image-derived information includes the objects, features, classifications, and calculated data. In addition to the sheer amount of image data that needs to be managed and stored, the integration of these different data sources presents a significant challenge *(22)*. As the raw and derived data might be utilized to draw important conclusions about the actives, hits, and leads from drug discovery programs, these data must be archived, and stored in an unmodified form for scientific and regulatory reasons. The database should therefore provide user friendly and efficient methods to query and retrieve the images and data for review, and potentially for reanalysis. At the time of writing, only Cellomics and Molecular Devices were offering complete database solutions with their HCS platforms (*see also* Chapter 20).

4. Comparison of HCS Platforms

All of the HCS imagers available today are optical microscope systems principally designed for fluorescence imaging, but in some cases providing additional imaging modes. For the lab looking to add capability in HCS, as well as for the HCS veteran looking to add capacity, the array of choices for HCS systems today can be daunting. **Table 1** lists the established HCS readers at the time of writing. Although an extensive comparison of all the features and options for every system might help drive the choice, interpreting the matrix of features would be complicated. Instead the systems will be compared based on design features and target application areas.

4.1. Optical Designs: Confocal vs Wide-Field

The HCS readers can be broadly divided into two optical design types, confocal scanning or wide-field imaging. All confocal scanning systems work by illuminating the specimen in one or more small regions (spots or lines) and building up an image by scanning the illumination through the specimen while measuring the emission in synchrony with the scanning. Confocal HCS systems can be further divided based on illumination scan design, with systems available which use point scanning, line scanning, and multipoint scanning (e.g., spinning disk). In contrast, wide-field imaging systems illuminate a "large" area of the specimen, and directly image that area all at once. A direct comparison of the performance of wide-field and confocal microscope systems concluded that wide-field microscope systems perform better (have a higher signal-to-noise ratio) on thin specimens such as monolayers of cells, whereas confocal systems perform better on thick specimens such as tissue sections and multilayer cell preparations *(23)*. When imaging the microtubule cytoskeletal organization in *Toxoplasma gondii*, the same authors reported that the submicron, weakly fluorescent structures could not be reliably captured by point scanning laser confocal imaging, but were successfully imaged by wide-field microscopy *(24)*. It is difficult to draw a definitive conclusion from these limited comparison studies, especially given the number of different confocal systems and confocal designs available in the market place, but some useful observations can be made. Confocal systems have a definite advantage in rejecting background fluorescence from material outside the plane of focus, either because of the specimen being significantly thicker than the depth of field, or because of some fluorescent component in the surrounding media, such as excess label. However, on dim specimens, wide-field systems have an advantage in much longer integration times per pixel. For example, the integration time per pixel for a megapixel image acquired on a wide-field microscope will be a million times longer than the dwell time per pixel on a point scanned confocal image with comparable total acquisition time. Line scanning confocal systems and multipoint disk scanning systems are somewhat better in this regard, but still the dwell time per pixel is relatively short. Increasing the illumination intensity compensates for the short dwell time, but is ultimately limited by saturation of the fluorophore, and also results in undesirable photobleaching and phototoxicity *(25)*. Longer integration times can be used to improve the signal-to-noise of weaker fluorescence, but this compromises the throughput of the system.

Of the 11 HCS systems in **Table 1**, only three have confocal capability; the Pathway HT, the INCA 3000 and the Opera. The ArrayScan VTI offers a unique option, the Apotome (Carl Zeiss, Jena, Germany), which uses a grating illumination device to generate optical sections *(26)*. This device can easily be inserted when optical sectioning is needed and the cost is nominal compared with true confocal imaging systems. Although it does not reject background fluorescence like confocal systems, it does subtract it, and for occasional assays requiring measurement of a single cell layer in a clump or colony, isolating cells grown on feeder layers, or moderate solution background, the Apotome is certainly a cost effective solution.

Confocal HCS imagers are more complex, require more expensive light sources, and therefore are typically more expensive than wide-field HCS systems. Even though the majority of imaging applications perform well on wide-field HCS imagers, there is a perception that confocal capability is desirable, perhaps because many assays have their genesis on stand alone confocal microscope platforms. The decision on whether the HCS platform should be a wide-field or confocal imager should largely be driven by the scope and nature of the biologies that will need to be addressed. For assays on single layers of cells, confocal detection might not be necessary, and results in some loss in signal. For live cell imaging, the lower illumination intensities of wide-field imaging will cause less perturbation of the cells. For assays with a high solution background, confocal detection will be an advantage, but in most cases, solution background can easily be washed away. For thick, multilayer cell preparations, confocal imaging will certainly be an advantage. It is fair to say that both types of systems have a wide range of applicability in cell and tissue analysis.

4.2. HCS Systems Optimized for End-Point Assays

The HCS readers designed for fixed cell assays are generally the simplest in design, from both the hardware and software perspective. These readers can be easily integrated with a wide range of automated plate loaders to create a screening platform with a capacity to analyze on the order of 10–25,000 wells per day, depending on the assay complexity. All are of the wide-field optical design, owing to its well-deserved reputation for flexibility and high performance to cost ratio. These include the least expensive HCS systems, listed in **Table 1** for <$200K (cellWoRx, Discovery-1, and ImageExpress Micro) and three somewhat higher cost systems for >$200K but <$500K (**Table 1**). However, it should be noted that in addition to the listed instrument costs, the actual cost of any HCS system will depend on the options selected, the additional image analysis algorithms purchased, the number of software seats provided, and the purchase of hardware and software for a database solution, all of which contribute significantly to the bottom line.

The fixed end point cell reader is the largest category of HCS systems and by far the largest installed base. The available systems (*see* **Table 1**) consist of five relatively compact bench-top

Features of HCS Platforms

units, and one floor standing unit. The fifth generation ArrayScan VTI platform, the Discovery 1 and the INCA 1000 might be considered standards against which all other systems will be compared. The cellWoRx has been in production for less than a year, and the ImageExpress Micro is a recent launch.

The optical designs of five of these six imagers are essentially the same, including the illumination systems, lenses, and cameras, suggesting there will be very little difference in the imaging performance. The cellWoRx system from Applied Precision offers an innovative new optical design. The illumination system uses light guides to deliver the fluorescence excitation directly to the specimen in a "dark-field"-like arrangement. The principle advantage of this design is that there is very little backscatter, and it eliminates the need for a dichroic mirror. The result is fluorescence imaging with very low background. Although this technology is new to HCS, Applied Precision pioneered this illumination system for use in their arraywoRx® Biochip readers, which have outstanding performance.

The other differences between these systems are principally in a few specific features and the available options. The Discovery 1 and INCA 1000 both offer laser autofocus as a standard feature, whereas on the new ImageExpress Micro it is an option. The Cell Lab IC 100 has a unique and fast video autofocus system. Although fast hardware focusing is an essential feature on HT-HCS systems, for low to medium throughput applications, a well-implemented image autofocus algorithm might be more reliable, if not as fast. The two systems from Molecular Devices also offer a transmitted light option, although at present there are few validated algorithms available for this imaging mode. A few of the systems also offer an environmental control option, allowing these systems to be used for simple kinetic studies with manual or off system pipeting, as well as live cell end-point assays. Overall, the fixed cell HCS systems offer excellent performance and are the lowest cost systems.

4.3. Live Cell Kinetic Systems

The live cell kinetic HCS systems are significantly more complex. In addition to simply scanning a plate, the live cell systems require on board liquid handling for stimulus–response assays and an environmental control system. They also require sophisticated acquisition software with provision for scheduling various sequences of reads and liquid additions, scan sequences that accommodate a wide range of biological timing, and sophisticated software applications for analyzing the time-course of the response. The assay developer must be concerned with the status of the cells during the assay, whether the reagents impact the biological processes being assayed, and whether the system itself, through optical, or even mechanical stimulation, might affect the biological response. However, with careful system and assay design, these difficulties can be addressed and valuable information collected. Access to live cell kinetic capability can significantly enhance the assay development capability for fixed end point HCS assays. Although, live-cell systems could double as single time-point fixed cell imagers, the additional features, and cost might not be warranted unless a significant proportion of live cell assays will be run.

Of the 11 HCS systems in **Table 1**, four have been designed for live cell capability; the KineticScan® (Cellomics, Inc.), the Pathway HT (Becton-Dickinson), the ImageXpress 5000A (Molecular Diagnostics), and the INCA 3000 (GE Healthcare). The KineticScan pioneered the integration of a general purpose pipeting device with an HCS imager in order to provide flexibility for single cell kinetic analysis of many cells in many wells. The ImageXpress 5000A provides a similar configuration of pipetor and reader. The Pathway HT presents a variation on this concept, with a single channel pipet and a unique optical design in which the plates remains stationary, whereas the optical system moves to scan the plate. The goal of this design concept is to avoid movements, which could cause mechanical stimulation or motion of unattached cells. The Pathway HT also provides the ability to switch to a Nipkow disk confocal acquisition mode for assays that require sectioning, background rejection, or for three dimensional analysis during assay development. The INCA 3000 utilizes two peristaltic pumps for making additions to wells, but with a 7-mL dead volume and a limited flexibility in pipeting, the live cell imaging capability is severely restricted.

Because live cell kinetic assays are more challenging to implement, requiring interaction between liquid handling and imaging, and more complex algorithms, the selection of a system is better made on the basis of software, and supported applications rather than specific hardware features, the importance of which might be difficult to assess. Demonstration of a validated assay is always the best evidence that a system will provide all the required features, and that the vendor is ready and able to support customers' assay development needs.

4.4. HT-HCS

Achieving high throughput on an HCS system, which means routinely scanning more than 50,000 wells/d, is by no means an easy assignment. In fact, stopping to take images is one of the main impediments to fast scanning. Taking multiple images per well, whether they are of different wavelengths or fields, further slows the scanning process. Finally, taking time to focus, and mechanically switch components takes even more time. The INCA 3000 and Opera systems have been designed to minimize these time wasting operations and thereby quickly and efficiently scan a plate. However, there is a significant premium to be paid for this technology, and the solutions are not without their tradeoffs, as discussed later. These are the two most expensive HCS systems costing between $800K and $1M.

The INCA 3000 is a confocal line scanning reader, which projects a line of illumination into the specimen, and images the fluorescence emission simultaneously on three CCD line cameras. This system is capable of high resolution in X, Y, and Z. The multiwavelength line imaging allows for continuous scanning of the specimen from one end of the plate to the other; however, most often the system is used to image one field at a time. The speed of the INCA 3000 depends on the use of fluorophores with little or no crosstalk between channels, binning the camera, and an appropriate cell seeding density that allows imaging only a single field of view, and a reduction of scan length and width. If the use of a particular fluorophore requires taking a second image to avoid crosstalk, the scan time will essentially double, and the speed advantage will be lost. The INCA 3000 acquires and saves images, which can be analyzed on the fly, or postacquisition, using standard image processing routines to produce feature sets appropriate to the assay being run. The operating software was not designed to function as a slave to a robot, which makes the integration of automated plate handlers challenging. The INCA 3000 has an environmental chamber, two peristaltic pumps for pipeting, uses three independent lasers, and three CCD cameras, all of which contribute to largest footprint for all of the HCS platforms.

The Opera is also a confocal imaging system, that uses a high performance spinning disk with integral lenses to provide significantly higher sensitivity than that achieved by standard Nipkow disk confocal designs, like that used on the Pathway HT. Coupled with three (optionally four) area CCD cameras, the Opera provides very rapid confocal imaging. On the positive side, the Opera images a whole field at a time, so if a second image is required to avoid crosstalk, the extra acquisition time will increase the overall scan time by only a fraction. A choice of laser and arc lamp illumination provides more flexibility for compatibility with a wide range of fluorophores.

With careful selection of fluorescent labels, proper configuration of the scan parameters, and the right microplates, either of these systems can provide the throughput needed to scan more than a hundred 384-well plates in a day.

5. Installation and Operation Considerations

The integration of an HCS platform into the drug discovery process is a complicated and time-consuming process that will involve a significant investment by multiple components of

21. Keller, H. E. (1995) Objective lenses for confocal microscopy, in *The Handbook of Confocal Microscopy,* (Pawley, J. B., ed.), Plenum, New York, pp. 111–126.
22. Berlage, T. (2005) Analyzing and mining image databases. *Drug Discov. Today* **10,** 795–802.
23. Andrews, P. D., Harper, I. S., and Swedlow, J. R. (2002) To 5D and beyond: quantitative fluorescence microscopy in the postgenomic era. *Traffic* **3,** 29–36.
24. Swedlow, J. R., Hu, K., Andrews, P. D., Roos, D. S., and Murray, J. M. (2002) Measuring tubulin content in Toxoplasma gondii: a comparison of laser-scanning confocal and wide-field fluorescence microscopy. *Proc. Natl. Acad. Sci. USA* **99,** 2014–2019.
25. Waggoner, A. (1986) Fluorescence probes for analysis of cell structure, function and health by flow and imaging cytometry, in *Applications of Fluorescence in the Biomedical Sciences,* (Taylor, D., Waggoner, A., Murphy, R. F., Lanni, F., and Birge, R., eds.), Alan R. Liss, Inc., New York, pp. 3–28.
26. Lanni, F. and Wilson, T. (1999) Grating image systems for optical sectioning fluorescence microscopy of cells, tissues, and small organisms, in *Imaging Neurons: A Laboratory Manual,* (Konnerth, A., Lanni, F., and Yuste, R., eds.), Cold Spring Harbor Laboratory Press, Cold Spring Harbor, NY, pp. 8.1–8.9.
27. Zwier, J. M., Van Rooij, G. J., Hofstraat, J. W., and Brakenhoff, G. J. (2004) Image calibration in fluorescence microscopy. *J. Microsc.* **216,** 15–24.
28. Richardson, T. (2002) Test slide for microscopes and method for the production of such a slide, US, pat. no. 6,381,013.
29. Souchier, C., Brisson, C., Batteux, B., Robert-Nicoud, M., and Bryon, P. A. (2003) Data reproducibility in fluorescence image analysis. *Methods Cell Sci.* **25,** 195–200.
30. Michael, A. and Model, J. K. B. (2001) A standard for calibration and shading correction of a fluorescence microscope. *Cytometry* **44,** 309–316.
31. Gerena-Lopez, Y., Nolan, J., Wang, L., Gaigalas, A., Schwartz, A., and Fernandez-Repollet, E. (2004) Quantification of EGFP expression on Molt-4 T cells using calibration standards. *Cytometry A* **60,** 21–28.
32. Gratama, J. W., D'Hautcourt, J. L., Mandy, F., et al. (1998) Flow cytometric quantitation of immunofluorescence intensity: problems and perspectives. European working group on clinical cell analysis. *Cytometry* **33,** 166–178.
33. Purvis, N. and Stelzer, G. (1998) Multi-platform, multi-site instrumentation and reagent standardization. *Cytometry* **33,** 156–165.
34. Zucker, R. M. and Price, O. T. (2001) Statistical evaluation of confocal microscopy images. *Cytometry* **44,** 295–308.
35. Richardson, T. (2002) Test slide for microscopes and method for the production of such a slide. in *USPTO.* Northern Edge Associates, US, pat. no. 6,381,013.
36. Zimmermann, T., Rietdorf, J., and Pepperkok, R. (2003) Spectral imaging and its applications in live cell microscopy. *FEBS Lett.* **546,** 87–92.
37. Gough, A. H. and Taylor, D. L. (1993) Fluorescence anisotropy imaging microscopy maps calmodulin binding during cellular contraction and locomotion. *J. Cell Biol.* **121,** 1095–1107.
38. Wallrabe, H. and Periasamy, A. (2005) Imaging protein molecules using FRET and FLIM microscopy. *Curr. Opin. Biotechnol.* **16,** 19–27.
39. Suhling, K., French, P. M., and Phillips, D. (2005) Time-resolved fluorescence microscopy. *Photochem. Photobiol. Sci.* **4,** 13–22.
40. Booth, M. J. and Wilson, T. (2004) Low-cost, frequency-domain, fluorescence lifetime confocal microscopy. *J. Microsc.* **214,** 36–42.

5

Characteristics and Value of Directed Algorithms in High Content Screening

Richik N. Ghosh, Oleg Lapets, and Jeffrey R. Haskins

Summary

High content screening requires image processing algorithms that can accurately and robustly analyze large image numbers without requiring human intervention. Thus, a suite of algorithms that are directed by an understanding of the biology being studied was developed for the optimized automated acquisition and quantitation of cellular images. Two categories of directed algorithms were developed: Developer Tools for assay development and Specific Algorithms for turnkey screening of specific biological situations. The same basic sequence of analysis steps are used in these directed algorithms:

1. Primary object identification.
2. Measurement of primary object properties.
3. Identification and measurements of associated targets.
4. Analysis of raw measurements for specific biological problems.

The detailed application of these steps is guided by the biology being studied and the expected phenotypic changes. Most cell biological problems to be analyzed using high content screening can be categorized by either the phenotype of the problem or labeling pattern, or by a standard biological response behavior of the cells. This enables application of directed algorithms optimized for these categories. Examples of the use of directed algorithms for specific categories are discussed, as well as the detailed analysis steps for a specific directed algorithm.

Key Words: Directed algorithms; HCS; high content analysis; high content screening; image analysis; quantitative fluorescence imaging; translocation.

1. Introduction

Quantitative fluorescence microscopy, in which the images of the fluorescently labeled biological samples are quantitatively analyzed, takes microscopy from being a purely visual, descriptive, and subjective tool to one of much more power by offering an objective, quantitative dimension. Quantitative analysis of the cellular fluorescence can provide information on the spatial distribution, amounts and arrangements of the fluorescently labeled macromolecular targets, and the morphology of the cells. Fluorescence enables multiple fluorophores with different emission wavelengths in the visible spectrum to be used to simultaneously label and quantitatively monitor and correlate multiple targets. Cells can have a heterogeneous response; quantitation enables subpopulations of cells to be identified that exhibit the different categories of responses, leading to identification of more nuanced responses to changing biological conditions. The advantages of quantitative fluorescence microscopy in studying cells are further enhanced by the addition of automation. High content screening (HCS), or the automated acquisition of the fluorescently labeled cellular images and their subsequent automated analysis enables the quantitative assessment of large sample numbers leading the way for large scale biology. This automated approach requires image-processing algorithms that can accurately and robustly analyze large image numbers without the need for human intervention.

Over the past 20 yr, a growing number of image processing software packages have become available enabling users to write image analysis programs using core image processing functions (e.g., MetaMorph, ImagePro, and Axiovision). Each package allowed users to interactively determine a sequence of image processing steps to analyze their images, and then recorded these steps so that they could be reused for analysis. However, these recorded scripts could be used only for a limited set of similarly acquired images and biology. Such image processing and analysis approaches were not sufficient to meet the needs of analyzing large numbers of images or images acquired under differing imaging conditions in an accurate and robust manner.

The same key basic sequential analysis steps can be applied to most cell biological problems requiring the analysis of microscope images of fluorescently labeled cells (described later in this chapter). However, the specifics on how these steps are to be applied require an understanding of the biology that is being analyzed, and the expected phenotypic changes. Some of the changes that could be expected are changes in cell shape, movement of the cells, rearrangement of intracellular targets of interest, or an accumulation of particular macromolecules. We call this biological understanding of the problem being studied and the resulting informed expectation of the phenotypic changes as the "domain knowledge" of the biological system. Domain knowledge enabled us to develop directed image processing algorithms that could be optimized and tuned for specific types of biology, and report quantitative features appropriate for the biology. In other words, two options exist for analyzing a fluorescence microscopic image of cells: (1) the undirected approach in which domain knowledge is not needed, and everything in the image that is possible to measure is measured, and the measured features, which change between the positive and negative control cases are monitored and (2) the directed approach in which the domain knowledge is used to measure the properties of the cells in the image which are relevant for the biology, and also used to understand the constraints on the biological problem to facilitate and simplify the image processing. For example, an image-processing algorithm for the quantitation of neurite outgrowth would be guided by the domain knowledge of the morphology and arrangements of neurons and neurites, and would be optimized for this class of objects. Properties measured that are relevant for this type of biology include the number of neurites per neuron and the neurite length. A very different algorithm would be needed to quantify the translocation of a protein from the cytoplasm to the nucleus; in this case, domain knowledge would guide the algorithm to identify the nuclear and cytoplasmic regions of the cells, and then measure the intensity in these distinct cellular regions, as well as calculating the intensity ratios and differences between these regions as optimized metrics to quantify the translocation event.

Our strategy for developing image-processing software is to develop a suite of algorithmic software modules, also known as BioApplications, which carry out the automated acquisition and quantitation of cellular images in an optimized manner that is directed and guided by the domain knowledge of the biology being studied. Within the domain of particular biological situations, the relevant algorithm makes a wide range of measurements relevant to the problem being studied, and the algorithm needs only minor adjustments to deal with the particular biological situations it was being applied to or to images acquired under differing imaging conditions. For example, one algorithm is designed to simultaneously quantify and correlate the translocation of several proteins from the cytoplasm to the nucleus; to run this algorithm, the user only needs to adjust settings for intensity thresholds and object identification criteria to take into account the particular cell type, intensity staining level and imaging conditions under which the experiment was done. The basic sequence of steps identifying the nucleus, cytoplasm, the translocation of the different fluorescently labeled targets between these distinct cellular regions, and the correlation between the different targets analyzed, are all done automatically by the algorithm, without the need for the user to redefine the image processing steps each time the software is used.

In the next few sections of this chapter, we describe the strategy and thinking employed in developing these directed image-processing algorithms. We first describe the requirements needed for directed algorithms and the general analysis steps that are universal for HCS problems. We then describe categories of HCS problems defined by their biology or phenotypic response, and how this domain knowledge led to directed algorithms for their analysis. We end the chapter describing in more detail the specific analysis steps in one of our directed algorithms.

2. Categories of Directed Algorithms for HCS Assays: Developer Tools vs Specific Algorithms

While developing a comprehensive suite of directed algorithms, we found two distinct sets of capabilities that were required by users from their image processing algorithms. The first category of use and requirements was that the image-processing algorithms needed to be fairly general applications that could be easily configured and applied to a wide range of cell biological situations. These applications provided more of basic information about individual cells and cell populations, and were fairly flexible in their usage and application. Users having these requirements were typically involved in doing basic cell biological research, or were in drug discovery and were doing target identification, target validation, or screen development; all areas that required a flexible tool that could be easily configured and applied to give more of basic quantitative information about the state of cells. These users frequently had a sophisticated understanding of HCS technology, as well as its applications and potential, and required a tool that would allow them to develop their own approaches. The other category of use and requirements was that the image processing algorithms needed to be simple to use with rapid start-up time. Turnkey usage was the emphasis, such that a person with less sophisticated understanding of the details of HCS technology could still easily conduct HCS assays. In this case, the image analysis was targeted to very specific biological situations with their specific set of measurements, and any interpretive logic or additional analysis of the raw measurements pertinent to the biological situation was built in, adding to the application's "turnkey" design. Users in this category were often involved in primary screening efforts.

In our suite of directed image processing algorithms, we developed two categories of algorithms fitting the above sets of requirements: Developer Tools and Specific Algorithms. Developer Tools were designed as general purpose assay development tools, whereas Specific Algorithms were designed to be a turnkey solution. The user developing a HCS assay using the functionality offered by a Developer Tool can develop, customize, and validate the assay for their particular biological situation. Developer Tools have a broad biological range in which they can be configured and applied to a wide variety of cell biological situations. These algorithms make and report a large set of basic cellular domain measurements with their statistics, thus giving these tools a high degree of flexibility. Developer Tools were not designed to provide additional interpretive logic or analysis of the raw cellular measurements for particular cell biological situations; for these the Specific Algorithms are used.

Specific Algorithms were designed to be easily and rapidly applied for the screening of targets of a specific biology. They were designed to have rapid startup time, validated protocols, and specific assay classification features, all targeted at a specific biology with a specific set of assay measurements. There is flexibility, but this is within the targeted biological problem being solved. Classification of categories specific to the targeted biology are automatically provided, and logic to interpret results is integrated into the analysis.

Developer Tools are still directed algorithms in that they are still guided by the general biological context in which the measurements are made. However, domain knowledge is required for users to set the input parameters to focus these tools toward specifically analyzing a particular biology. In other words, both Developer Tools and Specific Algorithms are directed algorithms, with the Specific Algorithms being very biology driven and deeply analyzing its specific biology, whereas the Developer Tools can embrace many more biologies, but with a more shallow level of analysis. For example, a Developer Tool like Cellomics' Compartmental Analysis BioApplication is designed to identify several distinct regions within a cell, and then report properties related to the fluorescent target's intensities (e.g., differences, ratios) within these intracellular regions or compartments. Thus, domain knowledge directs the algorithm to identify the cell, the intracellular regions, and then make the measurements in the identified regions. In contrast, a Specific Algorithm like Cellomics' Cytoplasm to Nucleus Translocation BioApplication only identifies two intracellular regions, the cytoplasm and the nucleus, and makes intensity related measurements between these two regions. This too is a directed algorithm, but with a narrower scope, than the more broadly applicable Developer Tool.

3. Analysis Steps in Automated Directed Algorithms

In directed algorithms for HCS, whether a Developer Tool or Specific Algorithm, the same basic sequence of key analysis steps are employed. However, before applying any type of analysis, the user must identify the biological question he/she is trying to answer. Being able to articulate the question being answered enables deciding the objective of the analysis. Once this critical, but often overlooked step has been done, then the directed image analysis strategy can be devised. The steps and their sequence in the automated directed analysis approach are as follows:

Fig. 1. *(Opposite page)* Applications of directed algorithms to four different biological situations. The top row shows the raw fluorescence image, and the bottom row shows the overlays applied by the directed algorithm on automatically identifying key entities to analyze. First Column: Images of NF-κB translocation from the cytoplasm to the nucleus analyzed by Cellomics' Molecular Translocation BioApplication; specifically shown are the target channel (nuclear factor [NF-κB] labeling) in which the cells were stimulated with IL-1α to induce translocation from the cytoplasm to the nucleus. Note that in some cells translocation has not occurred, and the NF-κB is mainly in the cytoplasm. The color overlays show results of analysis by the Molecular Translocation BioApplication. The primary objects are the nuclei, which fit the criteria to be selected for analysis; these criteria were set such that binucleated cells were not analyzed, and thus the rejected objects have no overlay. The nuclear region is shown by a red overlay. The annular part of the cytoplasm that corresponds to the cytoplasm, and from where the cytoplasmic intensity was measured, is shown by a green overlay. Second Column: Image from the analysis of F-actin and whole cell morphology by Cellomics' Morphology Explorer BioApplication. The raw image shows the F-actin target, and the color overlays show the whole cell (yellow) and the F-actin fibers (red) that were identified and analyzed by the Morphology Explorer BioApplication. The primary object, the whole cell, was identified by a whole cell stain (not shown). Several of the cells were touching so advanced segmentation was used to separate and resolve the individual cells (yellow overlays). Third Column: Analysis of embryonic stem cell colonies grown on a feeder layer of mouse fibroblasts by Cellomics' Compartmental Analysis BioApplication. The raw image shows the cell nuclei labeled with Hoechst 33342. The primary object is the stem cell colony, but the image also has the smaller nuclei from the feeder layer cells. The overlays show identification and analysis by the Compartmental Analysis BioApplication. A size exclusion criteria was set to identify the large object as the primary object (stem cell colony) and retain it for analysis (blue overlay). Smaller objects were rejected from analysis (orange overlay), as they corresponded to the feeder layer cells' nuclei. Once the primary object was identified, the oct4 labeled nuclei in the target channel could be identified and counted, to calculate what percentage of the stem cells in the colony were pluripotent (not shown). Fourth Column: Analysis of angiogenic tube formation by Cellomics' Tube Formation BioApplication. The raw image of the connected angiogenic tube (primary object) is shown. The overlays show the results of the analysis by the Tube Formation BioApplication. The algorithm identifies the connected tube (primary object; blue overlay), and branch nodes (pink dots). Unconnected tube segments are also identified and shown by a light blue overlay, and debris rejected from analysis are shown by an orange overlay.

Table 1
Categories of Problems Encountered for HCS Analysis, and Some of the BioApplications Designed to Solve Them

Phenotypic classifications	Biological classifications
Intracellular intensity changes • All BioApplications *Counting number of objects* • All BioApplications	*Intracellular translocation (movement of intracellular objects)* • Target Activation, Compartmental Analysis, Cytoplasm to Nucleus Translocation, Molecular Translocation, GPCR Signaling, Cytoplasm to Membrane Translocation
Spot analysis • Target Activation, Compartmental Analysis, Spot Detector, Morphology Explorer, Cell Health Profiling, Multiparameter Cytotoxicity, Micronucleus, GPCR Signaling, Multiparameter Apoptosis, Mitotic Index	*Internalization and receptor activation* • Target Activation, Compartmental Analysis, Spot Detector, Morphology Explorer, GPCR Signaling, Receptor Internalization
Colocalization • Target Activation, Compartmental Analysis, Cytoplasm to Nucleus Translocation, Molecular Translocation, Cytoplasm to Membrane Translocation, Cell Viability	*Cell cycle* • Compartmental Analysis, Cell Cycle, Mitotic Index
Cell size and/or shape changes • Target Activation, Compartmental Analysis, Morphology Explorer, Cell Spreading, Cell Motility, Cell Health Profiling, Extended Neurite Outgrowth, Tube Formation	*Neurite outgrowth* • Morphology Explorer, Neurite Outgrowth, Extended Neurite Outgrowth, Tube Formation
Analysis of interconnected tubular objects • Morphology Explorer, Tube Formation *Cell movement (fixed end-point assay)* • Cell Motility	*Monitoring cell health and toxicity* • All BioApplications can be used to monitor cell health, but especially: Target Activation, Compartmental Analysis, Spot Detector, Morphology Explorer, Cell Health Profiling, Multiparameter Cytotoxicity, Micronucleus, Cell Cycle, Cytoplasm to Nucleus Translocation, Molecular Translocation, Cell Spreading, Cell Motility, Cell Viability, Multiparameter Apoptosis, Mitotic Index
and more...	

Measurement of such intracellular intensity changes does not require the spatial information obtained from a microscopy-based approach, and the analysis could be done using non-microscopy-based instruments such as a flow cytometer, in which the fluorescence intensity of each cell is recorded, or even a fluorescence plate reader, in which the intensity of the entire detected field is measured. In HCS assays in which more than one color is used and the target fluorophore has different spectra from the primary object, then the microscope's spatial resolving power is used to first identify and resolve the primary object before detecting the target intensity in it. This approach also enables a subpopulation analysis where one can see the distribution of the target intensity among the primary objects (i.e., cells).

4.1.2. Counting Number of Objects

Another category of problems that arises frequently in HCS involves counting individual objects. The primary objects are typically individual cells or their nuclei, and for this category of problems, the spatial resolving power of microscopy is needed to identify and resolve the individual objects.

The types of biology that fall in this category include cell proliferation assays, migration assays through Boyden chambers, to the simple quantitation of cells containing a particular fluorescent stain *(8–10)*. The key issue here is to identify the primary object, and knowledge of the underlying biology helps in this identification. For example, consider the different types of primary objects involved in the examples shown in **Fig. 1**: cell nucleus, whole cell, cell colony, and angiogenic tube respectively. Biological situations exist in which knowing the number of these primary objects in an image are of interest. However, different strategies (and different algorithms) would be required to identify and count these different types of objects.

4.1.3. Spot Analysis

Another standard type of problem encountered in HCS is one of analyzing spots. In this type of problem, the fluorescent label is sequestered into discreet, punctate objects in the cell (i.e., spots), and the changes in spot properties are what are needed to be quantified. Spot Analysis requires the additional spatial information obtained from a microscopy-based imaging approach. Properties of spots that are of interest include the number of spots, spot intensity, spot size, spot shape, and spot locations inside the cell. One extreme of a spot analysis situation is that a cell only contains one spot defined by the fluorescently labeled target, and the existence and intensity of the spot is the property of interest; this was the assay strategy in early articles in which HCS was used to assay for G protein-coupled receptor (GPCR) activation and internalization *(11–13)*. Another extreme case is that the cell has many spots, and the number of spots and their intensity are of interest; recent articles in which this was the situation being analyzed include biological problems consisting of receptor tyrosine kinase (RTK) activation and internalization, and β-arrestin redistribution on GPCR activation *(1,14)*. Yet another extreme situation is in which the existence of a particular type of spot is a rare occasion, and the property of interest is in identifying and counting the cells which has this spot; this is the basis of micronucleus assays, in which a micronucleus is a special type of spot labeled by the DNA binding dye and adjacent to the nucleus, and its occurrence is a rare event.

The typical strategy in spot analysis is to identify the primary object, and then in an area defined by the primary object (either colocalized with the primary object, or in a region associated or adjacent to it), the spot(s) are detected, and their properties are measured. Spot detection is often enhanced by spatial filters applied during the image processing whose dimensions are tuned to the typical dimensions of the spots in question. For example, in a cell, both the nucleus and its endosomes can be considered as spots; the difference is that the nucleus is a much larger "spot" than an endosome. A spatial filter optimized to enhance identification of either the nucleus or the endosome will enable the algorithm to better analyze the spot under question. The spatial filter size requires domain knowledge, and is usually controlled by a user specified input parameter before applying the algorithm.

4.1.4. Colocalization

Another common type of problem requiring spatial information is one of colocalization. In colocalization problems, a particular region, compartment or organelle of the cell, or a group of intracellular molecules, are identified by a fluorescent label specific for that entity. Then, the target of interest, is identified by a fluorescent probe with a different emission wavelength, and the amount of the target in the cellular region masked by the first probe is measured. A variation on the colocalization problem is that the target is not in the area defined by the primary object mask, but in an area associated with or adjacent to it.

A common colocalization type of problem encountered in HCS is the translocation of an activated transcription factor from the cytoplasm to the nucleus. This type of problem typically involves a fluorescent-labeled nucleus as the primary object, which identifies individual cells as well as the cytoplasmic and nuclear areas between which the transcription factor's translocation occurs. Various labeling strategies can identify the transcription factor, and immunofluorescence

Directed HCS Algorithms

is a common labeling option. Unstimulated cells have fluorescent cytoplasm, and nuclei relatively devoid of signal. The translocation event results in the nuclear signal increasing at the expense of the cytoplasmic signal. The analysis strategy for this problem involves measuring the transcription factor's signal colocalized with the nucleus, which has been identified by a nuclear specific fluorescent dye, and also simultaneously monitoring it in the cell's cytoplasm (**Figs. 1–6**) *(1,15–17)*. The cytoplasm is usually defined as an annular region outside the nuclear mask; even though the nuclear label does not cover the cytoplasmic region, its existence allows definition of the cytoplasmic region adjacent to it. Understanding that the target has moved from the cytoplasm to the nucleus, thus causing the opposing intensity changes in these two regions of the cell, enables a directed analysis strategy in which the two distinct cellular regions (cytoplasm and nucleus) are first accurately and optimally identified, and then the translocation event is most sensitively and robustly measured by either the difference or the ratio in intensities between these two distinct regions (**Figs. 1–6**). Not understanding the cell biological process being measured might result in the targeted identification of the cells' distinct cytoplasmic and nuclear regions not being made. Thus, although the transcription factor's intensity would be measured, the actual translocation event would not be as sensitively captured because the intensity difference or ratio between the two cellular regions would not be done. Thus, domain knowledge of this biology is needed to devise the appropriate measurements for the translocation process.

This is an example of a nuclear colocalization problem, but colocalization problems are not restricted only to nuclei; other compartments, including the whole cell or even a multicellular assemblage, can also be the primary object. Cellomics' Compartmental Analysis BioApplication was designed to identify multiple different regions of a cell based on the area covered by the primary object, but some of these regions do not necessarily overlap with the primary object, and might often be adjacent to it *(1)*.

4.1.5. Quantifying Cell Size and Shape Changes

A standard cellular response to stimuli or changing conditions is that either the whole cell, or cellular components (such as organelles or cytoskeleton) might change in size or shape. Size and shape changes are subsets of a larger problem of quantifying the spatial rearrangement of the detected entities. As HCS is based on microscopy, it is well suited to quantifying these sorts of morphological and size changes, and related changes in spatial arrangement *(2,6,7,18)*. The entity whose morphology changes are being reported is either the primary object (e.g., the whole cell) or components of the primary object. The key in this sort of analysis is to again first identify the primary object, and then to identify the targets and cellular entities associated with the primary object whose analysis is of interest. Once these have been identified, it is relatively straightforward to measure their different morphological and size features, and other spatial information.

In a cell biological context, spatial changes resulting from stimuli can really occur over three different spatial dimensional scales: subcellular morphology, whole cell morphology, and multicellular morphology. Often, responses to stimuli can occur over all three scales and are correlated. We, thus, designed our algorithms in this category to quantitatively report phenomena that change over the multiple spatial scales. Cellomics' Morphology Explorer BioApplication reports spatial rearrangements over all three spatial scales. Features reported for whole cell morphology include cell shape, dimensions, orientation, and extent, as well as quantitation of cellular outgrowths. Features reported at the subcellular level include the intracellular location and amounts of macromolecules or discrete objects, their intracellular arrangement including a range of different texture measurements, properties of intracellular spots or fibers, and the shape and dimensions of major intracellular compartments such as the nucleus or Golgi. Multicellular features reported include the shape and dimensions of multicellular assemblages such as colonies or multinucleated cells, the proximity and spacing of similar and dissimilar cells or colonies, and the number of cells in such assemblages.

An example of a problem in this category is to measure the effects of a drug that affects the cytoskeleton. Such compounds cause changes both in the whole cell morphology as well as in the structure and arrangement of the cytoskeleton. Thus, the strategy employed is to use a whole cell stain to identify the cell as the primary object, measure its morphology, and then to specifically identify and measure the cytoskeletal fibers amidst any background fluorescence (**Fig. 1**). In **Fig. 1**, several of the cells are touching; so advanced segmentation techniques were employed to identify the distinct areas covered by the individual cells, even if they were touching. Understanding the biological situation would allow one to recognize that segmentation to separate and resolve touching cells was needed in this particular biological situation, but would not be needed in other situations in which the object being analyzed was a multicellular assemblage. Identification of the cytoskeletal fibers first required recognition that the fibers, in an associated fluorescent channel, were contained within the area covered by the individual cell, and then to recognize the fibers among the fluorescent background. Knowledge that cytoskeletal fibers were the targets that had to be identified and analyzed directed the algorithm to selectively detect such structures; an undirected approach might not have detected the fibers as optimally as the directed approach.

Another example in this category is the analysis of cell colonies. **Figure 1** shows a colony of embryonic stem cells growing on top of a mouse fibroblast feeder layer; all cells' nuclei were labeled with Hoechst 33342, and the stem cell colonies are the primary objects. Directed analysis using Cellomics' Compartmental Analysis BioApplication was done in which the nuclear label was used to identify the colony. A size exclusion principle was applied to ignore small objects (i.e., mouse fibroblast nuclei), and to only analyze large objects, which were the stem cell colonies. Lack of domain knowledge might cause all cells to be analyzed, thus obscuring the actual object of interest, the stem cell colonies. Once the colonies were identified, the number of Hoechst 33342 labeled nuclei, identified as spots within the colony, were counted as a measure of colony size, and in additional channels, October 4 labeling allowed identification of the number of pluripotent cells in the colony.

4.1.6. Analysis of Interconnected Tubular Objects

For many of the problems encountered in HCS, the primary objects are self-contained entities such as cells, nuclei, and colonies. However, the object of interest could be an interconnected structure with dark areas (i.e., holes) within the fluorescent label. An extreme case of this is a tubular network, which is made of many cells, and the primary object is the entire multicellular tube. Examples in this category include angiogenic tube formation (**Fig. 1**), differentiated myoblasts forming myotubes, and tangled interconnected webs of neurites *(19,20)*.

In angiogenic tube formation, individual endothelial cells associate with each other to form a tubular, branched capillary network. These connected tubes are the primary objects. Analysis of tubes is a specialized case of morphological analysis in which once the primary object (i.e., the tube) has been identified, questions asked of it include its length, width, area, number of branch nodes, and spacing between branch nodes. Domain knowledge leads one to recognize that the individual cells are associated with each other in a linear, interconnected tubular network, so the directed algorithm applied has to trace and connect the cells making up this network in order to identify the entire primary object. If the domain knowledge of tubulogenesis was not there, the analysis strategy might miss the association of the cells into tubes, and thus might not report the tube properties of interest.

4.1.7. Cell Movement

Another cellular response that can be quantitatively analyzed by HCS assays is cell motility, and there are several ways that have been designed to assay this behavior by HCS. Because motility is a dynamic process, one way of analyzing it is to take a sequence of images over time of the motile cells, and then using the spatial coordinates over time, compute motility parameters *(6)*. However, HCS assay methods of quantifying cell motility have been developed in which the cells on the sample plates being imaged are fixed. One assay technique is to use modified

Boyden chamber inserts for microtiter plates, and to count the number of cells that migrated through the Boyden chamber membrane; this reduces the analysis problem to one of just counting the number of cells *(8,21)*.

An alternative fixed end-point assay technique for cell motility is based on the method of Albrecht-Buehler et al. *(22)*. In an adaptation of this method, cells move on a lawn of fluorescent beads clearing the beads as they move and leaving behind dark tracks; the track areas are proportional to the cells' mean-squared displacement movement over time, and is the primary object that is quantified *(21)*. In this case, the primary object is neither the fluorescently labeled entities (the lawn of fluorescent beads), nor the cells, but the dark tracks that mark the area of the cell's movement. This is akin to analyzing a structure in a negative image. Once the dark tracks or regions have been identified, their properties (such as their length and area) can be measured. The cells in the tracks can be detected by other dyes, and the number of cells per track can also be reported.

4.2. Biological Categories of HCS Problems

HCS problems can also be classified by the broad categories of their biological response. This classification facilitates identification of particular solution strategies. Many of these solutions involve recognition of a certain phenotypic response, which fits into one of the categories described in the previous section, thus, further facilitating the solution of the problem.

4.2.1. Intracellular Translocation

A common response of cells to stimuli in HCS assays is that the fluorescently labeled cellular target moves from one region of the cell to another. In this category of problem, understanding of the source and destination of the target's movement is necessary. Then the assay strategy involves identifying the source and destination regions, and quantifying the target's presence in them (e.g., intensity). Arithmetic operations (e.g., differences, ratios) between the amount of target in the two regions further facilitates quantitative assaying of the translocation event. One of the earliest, and most common examples of this cellular response subjected to HCS analysis was the translocation of a protein from the cytoplasm to the nucleus of the cell *(1,15–17)*.

For translocation events between distinct compartments or regions of the cell, algorithms such as Cellomics' Compartmental Analysis or Molecular Translocation BioApplications can be used to identify the regions, and then measure the translocation (as discussed in **Subheading 4.1.4.**). If one of the regions where the target needs to be detected is made up of a punctate pattern (i.e., spots), then a spot analysis approach, as previously discussed, could be used. Sometimes, the region might involve translocation to or from the plasma membrane, and this requires identification of the cell surface. One way to do this is to directly label the cell surface in a colocalization type of approach. An alternative is to recognize that the cellular image is a two-dimensional projection of the entire cell, and that if the label is not present on the cell's surface, then the area of plasma membrane over the nucleus might not have any label, giving the cells an area devoid of labeling where the nucleus. When the target translocates to the plasma membrane, the patch above the nucleus gets filled in and the dark region above the nucleus disappears. Thus, this allows an analysis strategy similar to the cytoplasm to nucleus translocation approach, because even though the translocation is between the cytoplasm and the cell membrane, the labeling pattern is the same as a cytoplasm to nucleus translocation situation, allowing the same quantitation technique to be used.

4.2.2. Internalization and Receptor Activation

One form of translocation is the internalization of a receptor from the cell surface. Internalization of signaling receptors, such as GPCRs and RTKs, occurs after the receptor's activation by an agonist. GPCRs internalize and then get recycled back to the plasma membrane as part of their desensitization and resensitization steps, whereas RTKs often get internalized as a part of their down-regulation. Nevertheless, the internalization of these signaling receptors indicates that their activation by an agonist has occurred, and, thus, can be used as a surrogate assay for receptor activation *(1,11–14)*. The internalized receptors in endosomes typically have a punctate appearance, and thus the problem becomes one of spot analysis.

4.2.3. Cell Cycle

As described earlier, HCS can be used for cell cycle analysis. Cell cycle analysis has been done in several different ways including: (1) analyzing the presence of a specific cell cycle indicator to determine whether the cell was in a particular cell cycle state (e.g., the presence of phosphorylated histone H3, detected by immunofluorescence, indicates that the cell is mitotic); (2) measuring the DNA content of the cells from a DNA binding dye to determine the cell's ploidy; and (3) to combine and correlate the cell ploidy measurements with other cell cycle associated proteins for a better indication of the cell's particular state *(3–5,9,23)*.

As discussed earlier, identification of nuclei is enabled by labeling with a DNA binding dye, and the DNA content is obtained from the total intensity of the dye in the nucleus. Population analysis of these intensities over all the nuclei defines the categories of different ploidy of cell cycle states (i.e., cells in G1, S, or G2), and then each nucleus can be assigned a state. Identifying and measuring nuclear intensity is straightforward, but additional analysis in identifying and assigning cell cycle states is needed. Several directed algorithms, including Cellomics' Cell Cycle BioApplication, have been introduced to provide this enhanced analysis. Measurement of other cell cycle targets usually involves measuring the intensity or spot analysis of the targets, and their correlations can either be built in to the analysis packages, or can be done offline with third party software.

4.2.4. Neurite Outgrowth

Often, particular cells such as dendritic cells and neurons have processes. The item of interest are these processes, and the questions asked of them include the number of them per cell, their length, and their degree of branchedness. We developed several different directed algorithms to quantify neurite outgrowth, and which can be applied to this specific type of morphological problem *(24–28)*.

We have implemented three approaches for quantifying neurite outgrowth types of problems. In the approach offered by Cellomics' Morphology Explorer BioApplication, the neuron is the primary object. Once the neuron has been identified, the algorithm identifies and traces any processes emanating from the cell body. It then reports size and shape properties of the cell body, as well as the number of processes emanating from it, and their lengths.

An alternative cell-based approach is employed by Cellomics' Extended Neurite Outgrowth BioApplication. This was designed to be used in mixed cell cultures in which not all of the cells might be neurons with neurites, or a neuron of the correct type. In this situation, the primary object label is still a neuronal stain, which also contains a valid nucleus. The presence of both the neuronal specific stain and a labeled valid nucleus identifies the primary objects as being valid neurons, and it is only on those that the algorithm specifically identifies and traces neurites.

A third approach is where neurite outgrowth is reported at the field level. This approach is often used when the situation involves an interconnected mat of overgrown neurites. In this case, because of the overgrown neurites, it is often difficult to match neurites to their cell bodies. One way to solve this is by using Cellomics' Neurite Outgrowth BioApplication, which is a fast, field-based approach in which accurate identification of the cell body for each neurite is not needed, and an overall measure of neurite outgrowth for the field suffices. The situation is also analogous to the phenotypes seen in a tube formation situation, and thus a similar analysis strategy can be applied. The analysis strategy is to treat the entire mat of tangled neurites and cell bodies as the primary object, and then report the length of the web of neurites, the number of branch or cross points, and the length of neurite segments between branch and cross points.

4.2.5. Monitoring Cell Health and Toxicity

An area of growing popularity for HCS is in the monitoring of cell health and cytotoxicity. Although any HCS assay that reports on the physiological state of cells could be used as a cell health or cytotoxicity assay, there have been some specific applications of HCS toward monitoring cytotoxicity and apoptosis *(2,6,7,18,29)*. These applications have combined and correlated measurements several of the phenotypic classes discussed such as nuclear size changes, with cell intensity changes from a membrane-impermeant dye, to spot analysis monitoring the changes in mitochondrial transmembrane potential *(2,6,7,18,29)*. Measurement and correlation of multiple different properties related to cell health measured simultaneously in the cell gives a better understanding of the cell's health, and also enables uncovering the sequence of events leading to a cell's death in response to a toxic insult. This class of assays is discussed in other chapters in this book.

5. Practical Utilization of Directed Algorithms to HCS Problems

Cytoplasm to nucleus translocation is one of the earliest HCS problems worked on, and one of the first directed algorithms released by Cellomics *(15)*. In this section, we describe it in detail, as a practical example of a directed algorithm. The cytoplasm to nucleus translocation algorithm measures translocation events by quantifying the relative distribution of target fluorescence intensities between the cytoplasm region and the nuclear region of a cell (**Fig. 2**). Measuring the intensity within nucleus and then comparing it with the intensity of cytoplasmic region provides a simple and very robust way to assess quantitatively the degree of cytoplasm to nucleus translocation. The issue remaining is how to specifically measure the intensities in these two regions.

Let's consider a real life example with the goal to characterize the activation and nuclear translocation of the transcription factor nuclear factor (NF)-κB in HeLa cells. HeLa cells plated in 96-well microplates were treated with two concentrations (Min/Max) of IL-1α, fixed and stained using the reagents and protocol in Cellomics' NFκB Activation HitKit® HCS Reagent Kit (**Fig. 3**). The algorithm takes advantage of two-color fluorescence labeling of the cells (**Fig. 4**). The image of the same field is acquired twice using different excitation-registration channels: one-nucleus channel (blue fluorescence), the other-target channel (green fluorescence). First the primary objects (nuclei) need to be identified. In a typical case a limited number of image processing procedures is needed. Those include background correction, smoothing and intensity thresholding. **Figure 5** shows an overlay of the object mask created as a result of applying of these procedures. In some cases additional steps like object segmentation, border touching object removal, and so on might be required.

After identification of the primary object, target identification is performed. It might consists of similar processing steps: background correction, smoothing and intensity thresholding and, if necessary, segmentation (**Fig. 6B**). The next step is to create measurement masks corresponding to the already defined regions of interest: nucleus and cytoplasm. We have found that the best way to accomplish this is to use the nucleus mask and modify it to derive masks that correspond to the areas (regions) of interest (**Fig. 6**). The mask that covers the nuclear region of interest (called CIRC), can be created by simple erosion of the nucleus mask (**Fig. 6A**). At the same time the mask that covers the cytoplasmic region of interest (called RING), can be created by dilation of the nucleus mask and applying logical XOR operation to the dilated and the original nucleus mask (**Fig. 6B**). To move the RING mask further away from the nuclear region, the original nucleus mask must be also dilated. The size of dilation must correspond to the distance in pixels the inner edge of RING to the nuclear region. Thus, the CIRC and RING masks, their size (width) and location relative to the nucleus edge, can be fully controlled through the combination of erosion/dilation applied to the nucleus mask. In some cases, when cells are elongated, the cytoplasmic region might become very narrow and as a result the RING mask can fall (extend) beyond the target and has to be confined to the target mask (**Fig. 6F**).

Table 2
Intensity Measurements (Arbitrary Units) Made From the Target Channel:
Before and After Translocation

	Avg. inten CIRC	Avg. inten RING	Avg. inten diff CIRC–RING	Avg. inten ratio CIRC/RING
Before	760	1200	–440	0.63
After	1840	640	1200	2.85

After CIRC and RING masks are created measurements of intensity can be made from a target by simply calculating the aggregate intensities of the target image under the CIRC and confined RING masks (**Fig. 6C,F**) and then computing the average pixel intensity within the CIRC and confined RING masks. Typically the average pixel intensities under the CIRC and RING masks, but not the aggregate intensities under the original nucleus mask and target mask is used to measure the degree of the translocation response.

Visually the process of cytoplasm to nucleus translocation can be described as an increase of transcription factor concentration within the nuclear region and its decrease with the cytoplasmic area around the nucleus. In image analysis terms the concentration of the labeled material (transcription factors) within a particular region is nothing but the average pixel intensity within that region. Thus, comparing the average pixel intensity of the cytoplasmic area around the nucleus (RING) with its average pixel intensity inside the nuclear region (CIRC) provides a simple way for quantitative description of cytoplasm to nucleus translocation process. As a metric for the translocation process, typically two measures are used: the difference or the ratio between average pixel intensity under CIRC and RING masks (**Table 2**). The use of each has its pros and cons. The difference is insensitive to the nonspecific background intensity, but produces different (inconsistent) results if the intensity of illumination source changes or the excitation illumination is spatially nonuniform. The ratio, vise versa, is insensitive to the intensity of illumination source change or to spatial nonuniformity in the excitation illumination but produces poor (inconsistent) results if nonspecific background intensity is not removed.

References

1. Ghosh, R. N., Grove, L., and Lapets, O. (2004) A quantitative cell-based high content screening assay for the egf receptor specific activation of mitogen-activated protein kinase. *Assay Drug Dev. Technol.* **2(5),** 473–481.
2. Abraham, V. C., Samson, B., Lapets, O., and Haskins, J. R. (2004) Automated classification of individual cellular responses across multiple targets. *Preclinica* **2(5),** 349–355.
3. Ghosh, R. N., Grove, L., and Lapets, O. (2003) A quantitative high content screening assay for retinoblastoma protein phosphorylation and the cell cycle, *9th Annual Meeting of the Society for Biomolecular Screening*, September 21–25, 2003, Portland, OR.
4. Ghosh, R. N. and Grove, L. (2003) Decoupling retinoblastoma protein dephosphorylation from mitosis. *Mol. Biol. Cell* **14,** 2306.
5. Grove, L. E. and Ghosh, R. N. (2005) Quantitative characterization of mitosis-blocked tetraploid cells using high content analysis. *Assay Drug Dev. Technol.* (in press).
6. Abraham, V. C., Taylor, D. L., and Haskins, J. R. (2004) High content screening applied to large-scale cell biology. *Trends Biotechnol.* **22(1),** 15–22.
7. Giuliano, K. A., Haskins, J. R., and Taylor, D. L. (2003) Advances in high content screening for drug discovery. *Assay Drug Dev. Technol.* **1(4),** 565–577.
8. Mastyugin, V., McWhinnie, E., Labow, M., and Buxton, F. (2004) A quantitative high-throughput endothelial cell migration assay. *J. Biomol. Screen.* **9(8),** 712–718.
9. Gasparri, F., Mariani, M., Sola, F., and Galvani, A. (2004) Quantification of the proliferation index of human dermal fibroblast cultures with the ArrayScan high content screening reader. *J. Biomol. Screen.* **9(3),** 232–243.

10. Li, Z., Yan, Y., Powers, E. A., et al. (2003) Identification of gap junction blockers using automated fluorescence microscopy imaging. *J. Biomol. Screen.* **8(5)**, 489–499.
11. Conway, B. R., Minor, L. K., Xu, J. Z., et al. (1999) Quantification of G-protein coupled receptor internalization using G-protein coupled receptor—green fluorescent protein conjugates with the ArrayScanTM high content screening system. *J. Biomol. Screen.* **4(2)**, 75–86.
12. Conway, B. R., Minor, L. K., Xu, J. Z., D'Andrea, M. R., Ghosh, R. N., and Demarest, K. T. (2001) Quantitative analysis of agonist-dependent parathyroid hormone receptor trafficking in whole cells using a functional green fluorescent protein conjugate. *J. Cell Physiol.* **189(3)**, 341–355.
13. Ghosh, R. N., Chen, Y. T., DeBiasio, R., et al. (2000) Cell-based, high content screen for receptor internalization, recycling and intracellular trafficking. *Biotechniques* **29(1)**, 170–175.
14. Ghosh, R. N., DeBiasio, R., Hudson, C. C., Pamer, E. R., Cowan, C. L., and Oakley, R. H. (2005) Quantitative cell-based high content screening for vasopressin receptor agonists using transfluor technology. *J. Biomol. Screen.* **10(5)**, 476–484.
15. Ding, G. J., Fischer, P. A., Boltz, R. C., et al. (1998) Characterization and quantitation of NF-kappaB nuclear translocation induced by interleukin-1 and tumor necrosis factor-alpha. Development and use of a high capacity fluorescence cytometric system. *J. Biol. Chem.* **273(44)**, 28,897–28,905.
16. Vakkila, J., Demarco, R. A., and Lotze, M. T. (2004) Imaging analysis of STAT1 and NF-kappaB translocation in dendritic cells at the single cell level. *J. Immunol. Methods* **294(1–2)**, 123–134.
17. Vogt, A., Adachi, T., Ducruet, A. P., et al. (2001) Spatial analysis of key signaling proteins by high content solid-phase cytometry in Hep3B cells treated with an inhibitor of Cdc25 dual-specificity phosphatases. *J. Biol. Chem.* **276(23)**, 20,544–20,550.
18. Lovborg, H., Nygren, P., and Larsson, R. (2004) Multiparametric evaluation of apoptosis: effects of standard cytotoxic agents and the cyanoguanidine CHS 828. *Mol. Cancer Ther.* **3(5)**, 521–526.
19. Grove, L. E., Lapets, O., Pamer, E., and Ghosh, R. N. (2004) An Automated High Content Screening Assay To Quantify Myotube Formation And Morphology, 10th Annual Meeting, The Society for Biomolecular Screening, September 11–15, 2004, Orlando, FL.
20. Grove, L. E., Lapets, O., and Ghosh, R. N. (2005) A quantitative, automated, high content screening assay for angiogenic tube formation, *11th Annual Meeting, The Society for Biomolecular Screening,* September 11–15, 2005, Geneva, Switzerland.
21. Richards, G. R., Miller, R. M., Leveridge, M., Kerby, J., and Simpson, P. B. (2004) Quantitative assays of chemotaxis and chemokinesis for human neural cells. *Assay Drug Dev. Technol.* **2(5)**, 465–472.
22. Albrecht-Buehler, G. (1977) The phagokinetic tracks of 3T3 cells. *Cell* **11(2)**, 395–404.
23. Giuliano, K. A., Chen, Y. T., and Taylor, D. L. (2004) High content screening with siRNA optimizes a cell biological approach to drug discovery: defining the role of P53 activation in the cellular response to anticancer drugs. *J. Biomol. Screen.* **9(7)**, 557–568.
24. Arden, S. R., Janardhan, P., DeBiasio, R., Arnold, B., and Ghosh, R. (2000) An automated quantitative high content screening assay for neurite outgrowth. *Mol. Biol. Cell* **11**, 127a.
25. Simpson, P. B., Bacha, J. J., Palfreyman, E. L., et al. (2001) Retinoic acid evoked-differentiation of neuroblastoma cells predominates over growth factor stimulation: an automated image capture and quantitation approach to neuritogenesis. *Anal. Biochem.* **298(2)**, 163–169.
26. Arden, S. R., Janardhan, P., DeBiasio, R., Arnold, B., and Ghosh, R. (2002) An automated quantitative high content screening assay for neurite outgrowth. *Chem. Today* **20**, 64–66.
27. Ogbonna, G., Weiss, M., Jane, E., Ramer, E., Arden, S., and Ghosh, R. (2003) Quantitation of neurite outgrowth and neuronal cell differentiation by high content screening methods, *9th Annual Meeting of The Society for Biomolecular Screening,* September 21–25, 2003, Portland, OR.
28. Debiasio, R. and Ghosh, R. N. (2004) The use of cryopreserved primary neurons in high content screening assays, *10th Annual Meeting, The Society for Biomolecular Screening,* September 11–15, 2004, Orlando, FL.
29. Tencza, S. B. and Sipe, M. A. (2004) Detection and classification of threat agents via high-content assays of mammalian cells. *J. Appl. Toxicol.* **24(5)**, 371–377.

6

Characteristics and Value of Machine Learning for Imaging in High Content Screening

Juergen A. Klenk

Summary

Requirements for a flexible image analysis package for high content screening (HCS) are discussed. An overview of tools and techniques for image analysis and machine learning is given. Machine learning for classification and segmentation, the two fundamental elements of image analysis, is discussed. Next generation image analysis packages for HCS are reviewed. Recommendations for the development of image analysis solutions for advanced assays are given.

Key Words: Classification; computer vision; high content screening (HCS); image analysis; machine learning; morphology operations; neural networks; segmentation; semantic networks; thresholding; training.

1. Introduction

There is one universal constant in the field of high content screening (HCS): change—rapid change to be precise! Researchers' requirements for image quantification change constantly as new technologies, instruments, fluorophores, and labeling reagents become available. This continuous change in requirements makes it difficult, if not impossible, for providers of image analysis packages to keep up with the latest needs of their clients. Combining elements of machine intelligence with computer vision based algorithms might remove this bottleneck.

Both computer vision and machine learning are huge fields, and more number of researches has been conducted in refining methods to approach human vision capabilities. The bad news is that despite all efforts there is still no single generic method, which adequately replicates human vision. The good news is that acceptable results can be achieved by optimizing methods for particular domains. Accordingly, we will focus on image analysis methods for biomedical, specifically cellular images. And because this book serves as a practical guide, we will aim for practical recommendations over technical breadth and depth.

2. Problem Description

2.1. HCS User Types Requiring Imaging

Image analysis can be viewed as an independent component that is used along the generic HCS workflow during the three phases as shown in **Fig. 1**, i.e., assay development, screening, and assay evaluation. The usage of image analysis packages varies significantly between these three phases. Assay development users consists of an assay development scientist, a technician, and possibly an image analysis expert who often doubles as technician. They usually have lower

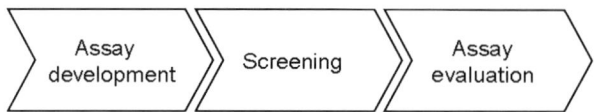

Fig. 1. Generic HCS workflow phases which include image analysis.

Fig. 2. Image analysis workflow.

throughput requirements as they use the image analysis software to validate their assays and to configure the corresponding image analysis protocol for screening.

Screening users consists of a screening scientist and a screening technician, with high throughput requirements to analyze the screens on- or near-line. They use the image analysis software for quality control and might perform minimal fine-tuning of the image analysis protocol.

During the assay evaluation phase a scientist uses data mining and visualization software such as Spotfire DecisionSite or ID Business Solutions (IDBS) ActivityBase to inspect the run data. The image analysis software comes into play by supporting simultaneous inspection of data and corresponding result images. Although image analysis plays a role in all three phases, it is predominantly during assay development that limited flexibility of existing image analysis packages presents a problem to the scientists. Thus, for the remainder of this article we will concentrate on how image analysis can be (and has been, in part) improved using machine learning techniques to optimally support the flexibility needed for rapid development of new assays. But first, let us review some terminology in image analysis and machine learning.

2.2. Image Analysis 101

Image analysis consists of several steps, starting from image loading to the extraction of the results. We specifically exclude the image acquisition part, and assume that knowledge about the camera, illumination, magnification, and so on is provided as metadata and available to the image analysis system. **Figure 2** illustrates the key steps of image analysis.

The processes of loading an image as well as extracting results and writing them to a database require a flexible input/output interface to handle data formats and other specificities of an existing HCS environment, a feature that every hardware-independent image analysis package should offer.

Image preprocessing predominantly deals with compensating acquisition artifacts. There is an abundance of preprocessing strategies. They include pixel brightness transformations (histogram-based equalization, contrast stretching) and local preprocessing (smoothing, convolution operations, noise reduction filtering), to name just a few of the more commonly used methods. Every comprehensive image analysis package should allow for a proper preprocessing of the raw image as this will dramatically improve analysis results. Preprocessing methods are well described in the literature *(1–5)* and can be viewed as independent of the subsequent image analysis—at least for the purpose of this article. Thus, we will assume throughout this article that images have already been preprocessed and artifacts have been removed as much as possible for optimal analysis results.

The two remaining elements of image analysis that we will concentrate on in this article are segmentation and classification (or recognition). Image segmentation is the process of partitioning an image into objects, and image classification is the process of naming these objects. Does

Machine Learning for HCS

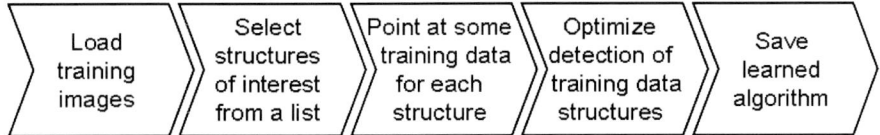

Fig. 4. Workflow of a machine learning based image analysis solution.

But now we have ended up with too many parameters to adjust manually. Imagine a nucleus detection algorithm with 100 independent dials to adjust. You would not know where to begin. And the effect of each dial is not always immediately apparent. There even seems to be crosstalk between the individual dials. A user would be simply overwhelmed by the complexity and we are only on our first structure, the nucleus.

This is where machine learning comes into play. It can assist the user and find the optimal algorithm for nucleus detection by automatically determining the best settings for all dials. As we now know, this takes some training data: the user must point to some examples of well-identified nuclei (and similarly for other structures of interest). From this the machine learning method can learn the settings, which optimally reproduce the training results. If the training data have been well selected (we shall come back to this later), then these settings should also work well on previously unseen test data.

2.5. Features of a Machine Learning-Based Imaging System

More generally, what is needed is a system that can learn to automatically detect structures of interest such as nuclei, cytoplasm, cell membrane, other subcellular structures, or entire cells or even cellular structures, independent of the assay type. It should be a relatively simple process to teach the system how to automatically and reliably detect these structures—ideally much in the same way, as we would educate a student—by pointing at them and naming them on training data. **Figure 4** provides an illustration of key elements of a machine learning-based image analysis solution. The key here is that the individual workflow components must be developed in a generic way so that they will work with (almost) any assay.

3. Tools and Techniques

3.1. Machine Learning for Classification

Using the terminology from the previous section, machine learning methods for classification must help us find an optimal classification scheme for the provided training data. Probably the best-known technique for classification is neural networks. This technique uses a supervised learning algorithm called the backpropagation algorithm. Providing input and output layers and a network of nodes (neurons) and weighted links (axons) in between, neural networks compute an output signal by propagating an input signal through the weighted network. The input nodes are connected to the objects' attributes, and the output nodes to the class types. The backpropagation algorithm changes the weights in the network until an optimal classification scheme for the training data is obtained. Usually this classification scheme provides a very good extrapolation to unseen test data. Furthermore, neural networks feature classification likelihood measures, which yield not just one classification result for each object, but a table with a list of classes it might belong to and the respective likelihoods.

Another technique for classification is known as nearest neighbor, with a learning algorithm based on clustering of the training data. There are many clustering strategies, most of which are geometrically intuitive. Their goal is to group the data points in attribute space which represent the training data objects into sets or clusters of data points which belong to the same class. This is achieved by drawing separation hyperplanes (or, more generally, hypersurfaces), and, thus, partitioning the attribute space in the best possible way so that each partition only contains data

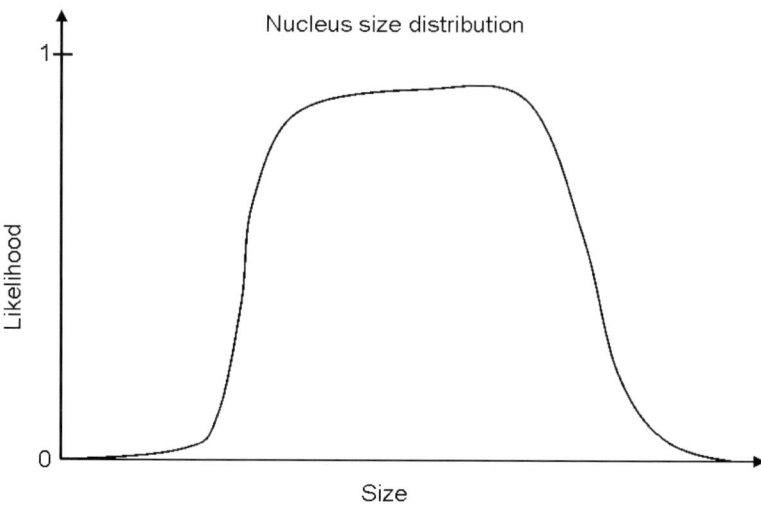

Fig. 5. Fuzzy logical expression of nucleus size.

points (objects) of one class. Methods such as principal component analysis might be employed to reduce the dimensionality of the problem by weeding out dependent attributes. Classification of training or test data objects is then achieved by choosing the class type of the nearest partition. Like neural networks, nearest neighbors also provides classification likelihoods, using a distance measure from partition centerpoints.

Decision trees are a very old technique for classification. A classification result is obtained by traversing a tree, with a decision being made at each branch point, until an end point is reached. Decisions are based on attribute values. The end points of a decision tree are class types. Algorithms have been described in the literature to optimize the decision points in order to improve the overall classification result. In practice, decision trees are limited to simply structured problems and are thus not too useful for our needs.

The last technique we discuss is a knowledge-based approach, also known as semantic networks. We touched on semantic networks already when we gave a knowledge-based description of a nucleus. Semantic networks are not completely independent of the previously discussed techniques. In fact, one can argue that the selection of input nodes for neural networks as well as the selection of attribute dimensions in nearest neighbors is best done by employing prior knowledge about the structures to be identified. However, a knowledge-based approach goes further in that it does not only select the relevant attributes, but also provides meaningful ranges for them. For nuclei, typical ranges for size, shape, color, texture, and so on are given. Mathematically this can be best expressed in terms of fuzzy logical, which provides a way to link attribute values to classification likelihoods. **Figure 5** shows a sample fuzzy logical expression for nucleus size. The basic idea of fuzzy logical is that shapes other than a step function are allowed as likelihood curves.

There are many ways to learn the attribute ranges from the training data. The simplest technology is the minimum to maximum approach, taking for each attribute the minimum and maximum value encountered in the training data and declaring the interval between minimum and maximum as the allowed range, with a classification probability of 1 (i.e., using a step function). This special case of a semantic network coincides with the nearest neighbors approach with mutually orthogonal separation hyperplanes. A more subtle technology employs Gaussian distributions, and yet more subtle technologies vary the shape of the curve using polynomial fitting strategies. It should also be noted that attributes can be connected with functions other than just Boolean, and allowing for even more flexibility in class modeling.

Machine Learning for HCS 89

As a rule of thumb, the more prior knowledge a method allows to be used, the more robust it is across the variety of images. This is especially true for images, which are as complex as the ones we are trying to analyze. Thus, the best techniques for classification in machine learning based, robust, generically applicable image analysis solution for HCS are neural or semantic networks.

3.2. Machine Learning for Segmentation

Once again, using our terminology, machine learning methods for segmentation must help us to find an optimal algorithm for segmentation of the provided training data. Before we look at examples, it is important to realize that there is a fundamental difference between classification and segmentation. Classification requires computing a classification result from attribute values, a process, which can be expressed in terms of inserting values into a mathematical function. Thus, improving the classification result by machine learning can be done by searching for an optimum for this mathematical function, a mathematical problem known by the name of variational calculus. Variational calculus is well understood and there are efficient algorithms, which solve the optimization problem (the backpropagation algorithm for neural networks being such an example). By contrast, measuring the quality of a segmentation algorithm requires executing it, and then comparing the produced (predicted) and actual segments. There is usually no closed expression in terms of a mathematical function by which a segmentation result can be computed. Thus, the problem of optimizing a segmentation algorithm can in most cases only be solved by an iterative, trial-and-error approach, which is usually much more computationally expensive. For this reason, machine learning for segmentation is not nearly as far developed as machine learning for classification. Nonetheless, there are some techniques, which apply to our problem.

Another important point to realize is that there are two different philosophies about segmentation. The first and conventional approach aims for a good segmentation result in one step, using a clever algorithm that does not depend on classification results. The second approach uses classification results to control the segmentation algorithm—this is the iterative strategy of segmentation and classification that we discussed earlier. This iterative approach requires some kind of bootstrapping: an initial segmentation must be performed to get a starting point. The segments of this bootstrapping step are often referred to as seed objects. The seeds then undergo an evolution and eventually some of them become the desired objects, whereas others might become helper objects (typically structures that are easily identified) that assist in detecting the desired objects. Because of their classification independence, the one-step segmentation algorithms are often used for the bootstrapping segmentation in the iterative approach.

The first and simplest one-step technique is segmentation by threshold. This is usually done on the basis of brightness, color, or texture, with a channel specific selection for color images. In our example on nucleus detection, all pixels with brightness greater than a given threshold in the Hoechst channel were grouped into objects, and these objects were then classified as nuclei. Because this technique depends on one parameter only, there are rather straightforward algorithms to automatically determine the threshold setting that produces the best segmentation results for the training data. One such algorithm works by lowering the threshold and thus increasing the object sizes until the average size of the produced (predicted) objects equals the average size of the actual objects. The resulting threshold is then the optimal threshold. If used as a bootstrapping algorithm, the threshold is usually set slightly less than the global maximum, so that for example the brightest 10% of all pixels are grouped into seed objects.

A related technique is searching for local extremes, again for brightness, color, or texture. This is often referred to as local thresholding. Every pixel that is only surrounded by darker pixels is a local brightness maximum. Each local maximum gets its own local threshold. All local thresholds are lowered by a certain percentage until again the average size of the produced objects equals the average size of the actual objects. Local thresholding can also be used to generate seeds for an iterative approach. Apart from outstanding brightness in their centers, another strategy to detect objects is to look for clearly visible boundaries. This is known as edge detection. Because edges partition an image into segments, edge detection is also a one-step segmentation technique.

One of the best-known edge detection techniques is called the Watershed Algorithm. The intuitive idea underlying this technique comes from geography: it is that of a landscape or topographic relief which is flooded by water, watersheds being the dividing lines of the domains of attraction of rain falling over the region. As a result, the landscape is partitioned into domains of attraction separated by watersheds. If the height of a pixel is given by its intensity, then the Watershed Algorithm tends to create objects, which are separated by a visible contrast. This technique usually works reasonably well to separate cells, which in turn can produce good seed objects for nucleus detection (note that this would involve a shrinking evolutionary strategy for nucleus detection). Parameters of the Watershed Algorithm that a machine learning method might adjust are the minimal size of the catchment basins, the stepping height during flooding, and what to do with plateaus. In addition, there are a number of distance measures that can be employed which might lead to different results.

The Hough Transform is another method to detect curves through near-extreme values. It is usually employed for straight lines and circle segments. For straight lines, the idea is to transform the pixel space into a so-called Hough space. Each point in the hough space identifies a unique line in the original space, and the intensity of that point is the accumulated brightness along its corresponding line. Extreme points in Hough space then identify visible lines in the image. The hough transform for circles is sometimes a good method to detect nuclei, especially if the nuclear membrane is well visible.

The next technique is our first iterative approach, and it is know as region growing. It is usually guided by additional local and global attributes, such as size, shape, smoothness, compactness, and neighboring relations criteria and so on, all of which can be captured in a knowledge-based description for the desired objects of interest. The algorithm starts out from a set of seed objects determined by some bootstrapping criterion, for example, finding local maxima, and then continues to grow the seeds by adding new objects or pixels for as long as their classification likelihood improves. During this process the objects and thus their attributes continuously change, which makes it very complicated to optimize the segmentation algorithm. Because the local stopping criterion for each object is reaching a maximum for its classification likelihood, the "learning" strategy must first extract optimal class descriptions from the training data. This can be done using the approach described for semantic networks in the subsection on machine learning for classification. Then all objects must be grown and continuously reclassified. The growing process for each object is individually stopped when its classification likelihood starts to decrease. A simple method to avoid stopping in a local maximum is to continue the growing process a little further it to see if the classification likelihood increases again.

The opposite of region growing is region shrinking, a process that might be employed when starting with seed objects which are larger than the desired objects. Region growing and region shrinking are specific forms of the more general morphology operations dilation and erosion. There are also boundary optimization techniques, known as opening and closing. Opening is the sequence of an erosion step followed by a dilation step, whereas closing is the sequence of a dilation step followed by an erosion step. They result in making a boundary rougher or smoother, respectively. Again, these processes can be guided by knowledge-based descriptions of the desired objects derived from training data. More generally, all of these morphology operations can also be combined with one another (e.g., grow first and then smoothen the boundary), and be tailored for specific needs, such as growing linear structures (for neurite outgrowth), or jumping gaps and building bridges to deal with incomplete structures.

This most general form of an evolutionary strategy often includes the use of helper classes. Helper classes might not have anything to do with real structures. Thus, they are not available from the training data, so that we have no *a priori* knowledge about their attributes. Optimizing the iterative algorithm then involves testing it with a variety of attribute settings for the helper classes. An example would be to determine the maximal bridge length for a linear growth strategy, the bridge being a helper class to close gaps in the linear structure.

Finding the best settings for the helper classes' attributes can be done in a brute force way by computing results for a large number of settings, or more intelligently by an interval halving method (assuming a certain linear relation between the settings and the results). It can also be done by asking the user to provide a first guess for the settings.

In conclusion, it should be noted that there is no single optimal method for segmentation and classification to reliably detect all structures. Rather, a good method will be tailored to the structure of interest it is supposed to detect. A good image analysis package should thus contain such tailored methods for most structures of interest, and provide an easy to use framework for new method development to extend it to new structures of interest.

3.3. Training Techniques

A good image analysis package should minimize the user interaction required for training and make the process as user-friendly as possible. The best results are achieved by wizards or guides, which help the user to prepare the training data. Let us first review some of the important steps in preparing the training data, and then look at the art of selecting good training data.

Preparing training data requires manual delineation of the structures of interest. Any drawing software could be used to do this, but such an approach would be far from practical. A training set requires a significant amount of images. On each image there are many structures of interest. The outlines of these structures are very difficult to trace with a drawing tool. All of this would result in an unacceptable amount of work for the user to prepare the training data. Instead, a point-and-click strategy should be employed to select objects of interest, with effective and easy-to-use tools to optimize the outlines. This approach of training data preparation requires prior segmentation so that objects can be offered for selection. But how can this segmentation are performed without having access to the training data in the first place?

We have, thus, encountered another chicken-and-egg-type problem, which we solve again by an evolutionary strategy. We start out with a structure of interest that is relatively easy to detect, such as a nucleus. Then we use some of our one-step segmentation strategies to get a rough outline of this structure. We let the user select which strategy worked best, and then provide him with morphology tools to optimize the outlines. The only morphology tools he needs are dilation, erosion, opening, and closing. This will allow him to fix up most nuclei very quickly. The nuclei, which do not come out nicely will simply be rejected. This is not a problem, however, because we do not have to outline all nuclei on every training image. Instead, a good set will suffice. Then we repeat this process for other structures of interest, using segmentation strategies, which can now be guided by the structures we already detected in the previous steps. For example, detecting cells is more easier if the algorithm can use the already detected nuclei as seeds. This process, if nicely packaged into a wizard, provides a speedy and user-friendly way to prepare training data.

Now we turn to the art of selecting "good" training data. This has a big impact on how well our "learned" algorithms can extrapolate results from training data to previously unseen test data. The point is to select a "representative training set." What does this mean for our case? First, it refers to the training set size, which depends on the number of classes, or structures of interest that we want to distinguish. For each class type the system must be presented with an adequate number of samples. Second, it refers to picking the training data such that the entire range of each class is covered. This means that all varieties of a particular structure of interest should be present in the training data, and ideally more than once. If samples are collected statistically, then both

---

Selection criteria for training data

• Pick a set of images which contains all structures of interest
• Ensure that the entire variety of each structure is covered

---

Fig. 6. Selection criteria for training data.

goals are achieved by just picking a large enough training set. There are statistical tools to compute how big the training set should be given the number of classes and their assumed distribution. But we are selecting the training data manually. Thus, a simple rule of thumb is that the training set needs to be characteristic of its usage in the real world. It is a very common mistake to pick only the best nuclei as training data, and then be surprised that only the best nuclei are detected, although most others are badly missed (*see* **Note 1**). All structures of interest vary significantly resulting from staining, biological, or other effects. A good selection includes a few samples from every variant of every class, thus covering the entire range of all classes reasonably well (*see* **Note 2**). This is summarized in **Fig. 6**.

4. Solutions

There are many image analysis packages for HCS on the market. For a good overview *see* **(6)**. If you are a user who is just entering the field of HCS, or a user who uses HCS only for existing, standardized assays, then you are probably best served by existing turnkey solutions. The best known and most widely used turnkey solution is Cellomics' (http://www.cellomics.com/) ArrayScan system (http://www.cellomics.com/content/menu/ArrayScan%AE/). Other frequently used turnkey systems are offered by GE Healthcare (http://www.gehealthcare.com/), BD Biosciences (http://www.bdbiosciences.com), Evotec (http://www.evotech-technologies.com/), and Molecular Devices (http://www.moleculardevices.com/). These systems allow running frequently used assays, and the corresponding image analysis can be done almost at the push of a button. In particular, no detailed knowledge about image analysis is needed to operate these systems.

In contrast, if you are a user who is developing new assays, you will need a system with open, flexible image analysis capabilities. A more flexible approach is to offer a package of modules instead of prebuilt solutions. A module comprises a configurable algorithm for the detection of a specific structure of interest, for example, the nucleus or the cytoplasm (*see* **Note 3**). Solutions can be assembled by combining the modules, which detect the desired structures into an image analysis workflow. Modules can also be configured to fine- tune analysis to specific assay conditions, and ideally this step is supported by machine learning techniques. GE Healthcare's IN Cell Developer Toolbox (http://www.gehealthcare.com/company/pressroom/releases/pr_release_10287.html), Definiens' (http://www.definiens.com/) Cellenger HCS (http://www.definiens.com/news/ releases/19_e_Bioimage.htm) (*see* **Note 4**), and Evotec's Acapella (http://www.evotec-technologies.com/opencms/export/et/products/life_science_software/software_products/acapella.html) are such modular systems that have just or will soon become available. Definiens' Cellenger HCS includes wizard guided module selection and machine learning supported configuration (*see* **Note 5**). Cellomics, Inc. has also developed a number of its BioApplications in a much more modular way (http://www.cellomics.com/content/menu/BioApplications/). CellProfiler (http://groups.csail.mit.edu/vision/cellprofiler/), an effort at Whitehead and the Massachusetts Institute of Technology (MIT) to develop open-source modules for image analysis, which plug into MATLAB (http://www.mathworks.com/) as well as the open microscopy environment (OME) (http://www.openmicroscopy.org/), might become a choice for the budget-oriented user. All of these next generation, modular image analysis packages provide a significant step forward for the field of HCS. But there will always be structures for which no corresponding module yet exists.

For this reason some of the above-mentioned providers are making available the very techniques and tools which they use to develop their modules. In most cases, their technique is based on some scripting language, public or proprietary, and they have a development environment to assist the programmer. Sometimes additional algorithms for segmentation and classification developed in C++ or Java can be plugged into the package. This is obviously the most flexible approach. But it comes at a price: someone will need to invest a significant amount of time to get acquainted with image analysis in general and with the package and its programing language in particular. This is actually already true for the next generation, modular packages mentioned in the previous paragraph, though to a lesser degree.

If you want to be flexible with your assay development, and you need similar flexibility in image analysis, you will need an image analysis expert in your team. Computer vision is still miles away from human vision. Compared with cars, we are still in the age in which you needed to carefully adjust several levers and crank a handle to start the car, and you needed a flag person to walk in front of the car to warn people, all of which required an expert to operate the machine. Likewise, there is no image analysis push-button solution for the kind of flexibility you are asking for. You will need an expert to run your image analysis. And it will take him at least 3–6 mo of training until he will have an impact. But if you accept this you will be able to support your cutting edge research with cutting edge image analysis.

There is no doubt that we will sometime have push-button image analysis modules for (almost) any cellular or subcellular structure. Getting there will be greatly supported by packages which provide a comprehensive set of machine learning based algorithms for classification and segmentation as previously described. Until then the 80/20 rule applies: push-button solutions only for those 80% which run assays that are in widespread use, the remaining 20% which spearhead the field of HCS will need to pick their package of choice and develop their corresponding image analysis solution themselves (*see* **Note 6**).

5. Notes

During his time at Definiens the author observed several projects to develop customized image analysis solutions for a variety of assays using Definiens' development environment, Cellenger Developer Studio (http://www.definiens.com/cellenger/files/cellenger.pdf). Based on this experience as well as customer feedback, we will now take a look at some effort estimates which should serve as a guideline for your own lab.

1. Another rule for modules is that absolute numbers are to be avoided, and relative expressions with respect to some calibration or metadata should be used instead. For instance, using absolute brightness or size of a structure makes no sense, when fluctuations must be expected between assay types, during image acquisition, or during incubation.
2. When developing a solution for a particular assay, it is best to start with a rather small yet reasonably representative training set, and to focus on the easiest structures first. For a reasonably difficult cellular assay this step takes in the order of 1–4 wk. Afterwards it is important to validate and refine the solution using a larger training set, a process than can take another 1–8 wk, depending on the desired accuracy. Do not forget to separate your data into training and test data beforehand. Test data must never serve for training purposes. If it does, it will immediately become training data as well, and new test data must be obtained. This might sometimes be necessary if the training data is found not to be representative.
3. Developing a module to generically identify a particular structure of interest, for example, the nucleus or the cell membrane, is significantly harder. Our research has shown that it makes sense to build modules from submodules, with each submodule being capable of analyzing a particular subgroup of the structure. For instance, it would be too hard to develop one single strategy to identify nuclei for all stains, because Hoechst only stains the nucleus, but DRAQ5 also stains the cytoplasm. Instead, a submodule for each stain should be developed, and the metadata, which contains information on the used stain triggers the appropriate submodule.
4. Cellenger Developer Studio is an integrated development environment, which uses a graphical programing paradigm to assemble image analysis solutions. It supports Semantic Networks (i.e., a knowledge-based

description of structures of interest) as well as a list of classification and segmentation algorithms. Machine learning techniques are not included, although the graphical programing language is flexible enough to allow users to include machine-learning elements in their programs.

5. For Cellenger's graphical programing language, just as for any other programing language, the average training time for a new user to become productive was in the order of 3–6 mo, although it would take an additional 6 mo for the user to really master the system to a degree that he could also crack hard problems. Training comprises learning the language (theory) as well as working through examples (practice). Programing experience is quite useful though not a must. A solid theoretical understanding is important early on, whereas later building a repertoire of recipes from practical experience is critical.

6. Our last remark returns us to the importance of machine learning for the development of generically applicable modules. We found that without some form of machine learning assistance, for instance to adjust values of a region growing segmentation step, or to optimize a classification step, it was not possible to develop a generically usable module, not even for a structure as simple as a nucleus. The variation of nuclei in assays is simply too big. We had to include a step in which knowledge, automatically gathered from the training data, was fed into the module to automatically fine-tune the evolution of structures to the specific assay in question.

Acknowledgments

The author thanks Prof. Gerd Binnig for stimulating and enjoyable years of research during which a new framework for human understanding on computers, called Cognition Networks *(7–9)*, could be developed. This article is based mostly on experience gained from applying Cognition Networks to image analysis for biomedical images. Most of the work was carried out at IBM Research and Definiens.

References

1. Gonzalez, R. C. and Woods, R. E. (eds.) (2003) *Digital Image Processing*, Prentice Hall, Englewood Cliffs, NJ.
2. Gonzalez, R. C., Eddins, S. L., and Woods, R. E. (eds.) (2003) *Digital Image Processing Using MATLAB*, Prentice Hall, Englewood Cliffs, NJ.
3. Parker, J. R. (ed.) (1996) *Algorithms for Image Processing and Computer Vision*, John Wiley and Sons Inc., New York, NY.
4. Myler, H. R. and Weeks, A. R. (eds.) (1993) *The Pocket Handbook of Image Processing Algorithms in C*, Prentice Hall, Englewood Cliffs, NJ.
5. Seul, M., O'Gorman, L., and Sammon, M. J. (eds.) (2000) *Practical Algorithms for Image Analysis*, Cambridge University Press, Cambridge, UK.
6. Comley, J. (2005) High content screening. *Drug Discov. World* Summer 2005, 31–53.
7. Klenk, J., Binnig, G., and Schmidt, G. (2000) Handling complexity with self-organizing fractal semantic networks. *Emergence* **2(4)**, 151–162.
8. Klenk, J., Binnig, G., and Bergan, E. (2001) Modeling knowledge and reasoning using locally active elements in semantic networks. In *Proceedings of ES2001, the Twenty-first SGES International Conference on Knowledge Based Systems and Applied Artificial Intelligence* (Bramer, M., Coenen, F., and Preece, A., eds., Springer, New York, NY.
9. Binnig, G., Baatz, M., Klenk, J., and Schmidt, G. (2002) Will machines start to think like humans? *Europhys. News* **33(2)**, 44–47.

7

Tools for Quantitative and Validated Measurements of Cells

Anne L. Plant, John T. Elliott, Alessandro Tona, Dennis McDaniel, and Kurt J. Langenbach

Summary

In this chapter, we describe the preparation of thin films of collagen that can serve as reference materials for assuring reproducible and predictable cell responses. Subtle differences in the molecular-scale characteristics of extracellular matrix proteins, including the supramolecular structure of type 1 collagen, can have tremendous influences on cell state and cell-signaling pathways; therefore the careful control and analysis of the culture surface is critical to assure a relevant and consistent response in cell-based assays. We also describe how cell-phenotypic parameters such as morphology, proliferation, and green fluorescent protein expression can be unambiguously quantified in adherent cells by automated fluorescence microscopy or high content screening. Careful consideration of protocols, and the use of fluorescent reference materials, are essential to assure day-to-day and instrument-to-instrument interoperability. The ability to collect quantitative data on large numbers of cells in homogeneous matrix environments allows assessment of the range of phenotypes that are reproducibly expressed in clonal cell populations. The inherent distribution of responses in a cell population will determine how many cells must be measured to reach an accurate determination of cellular response.

Key Words: Alkanethiol self-assembled monolayers; automated fluorescence microscopy; extracellular matrix protein; green fluorescent protein; type 1 collagen.

1. Introduction

The National Institute of Standards and Technology (NIST) is considered the nation's measurement institute. With more than a 100 yr history, NIST technology, measurements, and standards are geared to help US industry invent and manufacture superior products reliably, provide critical services, ensure a fair marketplace for consumers and businesses, and promote acceptance of US products in foreign markets. As industrial and national priorities have evolved, NIST has responded by expanding its portfolio into new strategical directions, such as biotechnology. The physical measurement capabilities and approaches that NIST is best known for are now routinely being applied to biological molecules such as DNA and proteins, and more recently, to cells (e.g., *see* http://www.cstl.nist.gov/biotech/Cell&TissueMeasurements/Main_Page.htm).

A major challenge facing the biotechnology and pharmaceutical industries today is the ability to quantitatively assess biomarker expression in living cells. Drug screening and toxicology relies increasingly on cell-based assays, with the hope that quantitative cell response will provide a more reliable predictor of clinical outcome than molecular-scale screening approaches. High content screening has permitted a high throughput approach to fluorescence-based imaging (*see* Chapter 1). However, the validity, reproducibility, interoperability, and interpretation of such data are currently less than perfect. Establishing an in vitro model system that accurately reflects in vivo response will likely remain a challenge for years to come. However, the development of improved matrices and quantitative methodologies will help the community make more reliable, reproducible, interoperable, and meaningful cell-based measurements.

Our group is developing cell metrology tools that will help minimize sources of variability in conducting quantitative cell biology measurements and facilitate interlaboratory data comparability. The components of this toolkit include thin films of extracellular matrix (ECM) proteins, which are highly reproducible and amenable to analysis by independent methods, and quantitative fluorescence microscopy protocols that allow accurate and reproducible quantification of cell response; these tools are discussed in this chapter. We are also developing lines of indicator cells that allow nearly real time, *in situ*, evaluation of growth conditions such as growth matrices, serum, and pharmaceuticals. These tools will facilitate applications such as diagnostic assays, drug screening, pharmaceutical development, development of biomaterials, and determination of culture conditions for stem cells. In addition, such reference materials and protocols will be essential for standardizing and validating data on cells and tissues that might be compiled into databases for use in modeling complex biological networks.

1.1. Thin Films of Collagen as Reference ECM

ECM proteins can induce diverse intracellular signals by providing both mechanical and chemical stimuli to cells (*1*). Owing to signal pathway crosstalk, these intracellular signals will mitigate and be affected by other sources of cell signals. Because of the importance of ECM cues in poising cell-signaling pathways, careful control of the ECM might be extremely important when studying cellular response to soluble factors such as pharmaceuticals. In addition, well-defined matrices provide the opportunity to understand the role of the matrix cues in cell response, which is difficult with serum protein-coated tissue culture polystyrene in which the identity and presentation of matrix molecules are difficult to know.

Collagen is the major structural protein in tissues. We have developed a reliable method for fabricating reproducible thin films of native type-1 and denatured type 1 collagen (*2*). Thin films of native collagen are determined by ellipsometry measurements to range from about 6 to 40 nm thick depending on the concentration of collagen in the solution from which they are assembled. The thin films are formed by allowing collagen in solution to self assemble onto gold-supported alkanethiol monolayers as depicted in **Fig. 1**. As shown in **Fig. 1**, at low concentrations of collagen in solution, the surface is primarily covered by a layer of very small collagen fibrils; and large fibrils grow out of the small fibrils (**Fig. 1C**). These large fibrils are about 200 nm in diameter, can be microns in length, and appear to be pinned at one end to small fibrils at the surface. A thin film made up of exclusively of short fibrils appears to present different mechanical signals to cells than do thin films containing large fibrils. This will be discussed in greater detail later.

Vascular smooth muscle cells (vSMC) and fibroblasts response to thin films of large collagen fibrils is analogous to their response to collagen gels with respect to cell morphology, proliferation rate, cytoskeleton organization, expression of the tenascin-C gene, ERK1/2 phosphorylation, and $\alpha_2 \beta_1$ integrin engagement (*3,4*). Unlike collagen gels, the thin films are rugged, highly reproducible, and homogeneous. Furthermore, they are optically superior to collagen gels because they scatter less light, and because they are homogeneous in thickness, cells in different fields are in approximately the same focal plane. These characteristics make them amenable to automated microscopy. These thin films provide a matrix environment that is identical from

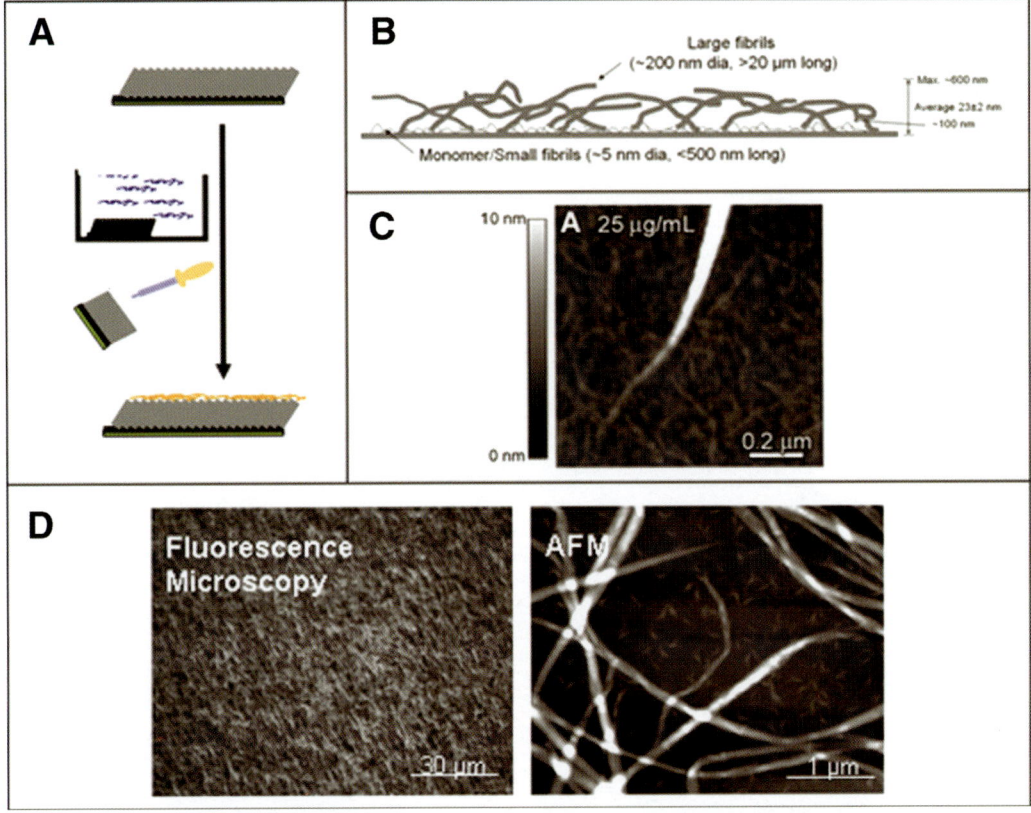

Fig. 1. Thin films of type 1 collagen. **(A)** Hydrophobic alkanethiol monolayers are self-assembled on translucent layers of gold on glass slides. Thin films of collagen are formed by placing the alkanethiol-coated slide into a solution of monomeric type 1 collagen at neutral pH and physiological ionic strength. After overnight at 37°C, the surfaces are rinsed, dried briefly under a stream of N_2, and replaced into buffer or medium. **(B)** Ellipsometry measurements indicate that thin films prepared from 300 µg/m collagen solutions have an apparent thickness of approx 23 nm according to an effective medium model. **(C)** Atomic force microscopy of collagen films prepared from a low concentration of collagen shows that a layer of very small fibrils underlay larger fibrils and that larger fibrils appear to be connected to, and grow out of, the smaller fibrils. The dimensions of the small fibrils are consistent with an aggregate of five collagen triple-helix monomers. **(D)** Atomic force microscopy (right) indicates that the large fibrils are on the order of 200 nm in diameter and tens of microns long. Fluorescence microscopy of fluorophore-labeled collagen thin films (left) indicates that the films are spatially homogeneous.

field-to-field and from experiment-to-experiment, and cells display a predictable and highly reproducible response on these films. Because these films can be characterized by a variety of surface analytical techniques such as ellipsometry, surface plasmon resonance, and scanning probe microscopies, one can verify the homogeneity and reproducibility of the cell culture surface independently from the cell response.

Thin films of collagen are excellent substrates for observing cell structure and cell-matrix interactions by light microscopy techniques. The thin film approach also allows us to systematically modify the ECM environment to observe that subtle differences in ECM protein conformation and supramolecular structure can alter cell response. For example, the surface free energy of the substrate onto which type 1 collagen adsorbs determines the supramolecular structure of

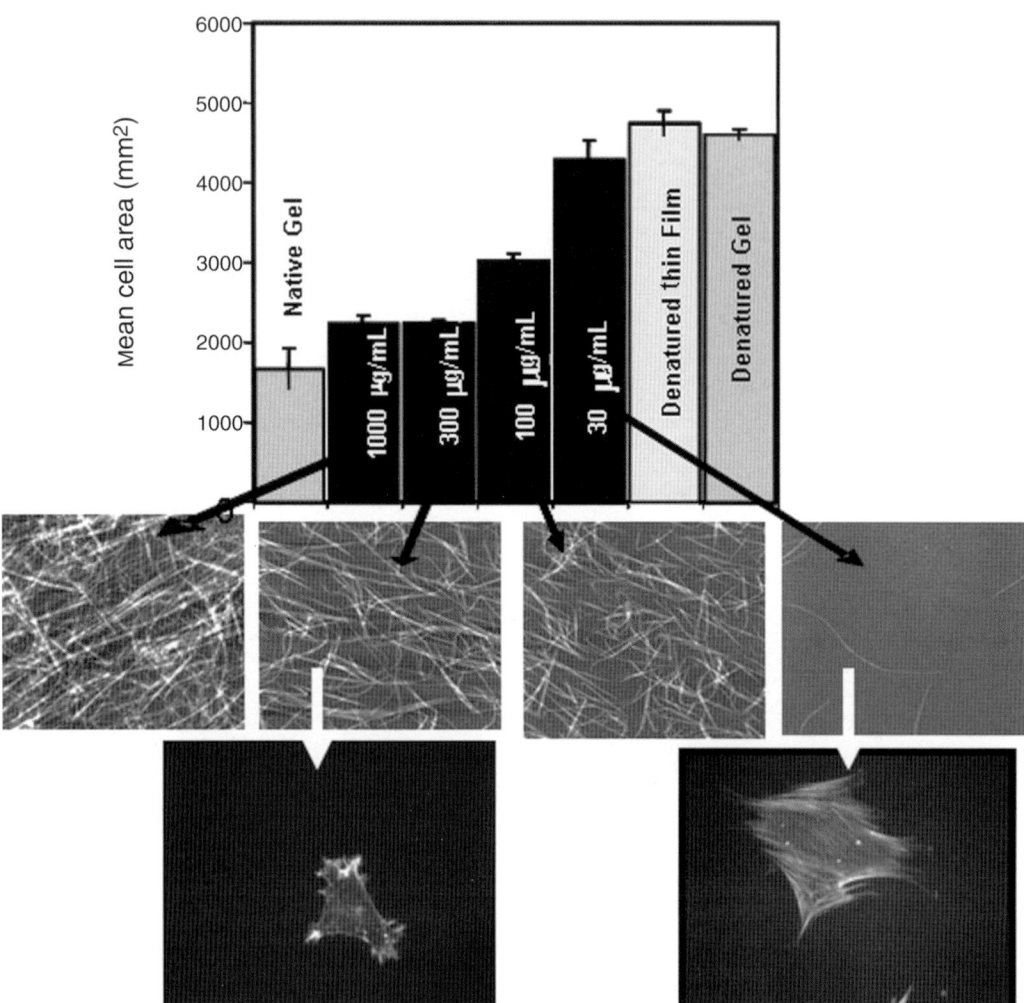

Fig. 2. Evidence for mechanical influence on cellular response. Thin films of collagen on alkanethiol monolayers prepared from higher concentrations of collagen in solution have a denser film of large collagen fibrils, and thin films prepared from low concentrations of collagen consist primarily of the very small fibrils of collagen that are in very close proximity to the alkanethiol monolayer. The cells that interact with large fibrils (left) do not spread significantly, and have a poorly organized cytoskeleton compared to the cells that interact primarily with small collagen fibrils, which are much more spread (left) and have well-organized stress fibers as seen by staining with phalloidin. Cells on the surfaces prepared from low concentrations of collagen also are more proliferative and demonstrate changes in gene expression compared to the cells on the more physiologically relevant collagen on the left. Other experiments have shown that cells interact with both surfaces identically through the β_1-integrin receptor.

adsorbed collagen. On hydrophilic surfaces such as glass, collagen adsorbs as a monomer, but on more hydrophobic surface collagen forms fibrillar structures (5). Cells respond very differently to monomeric collagen than to collagen fibrils. We have also demonstrated that on hydrophobic surfaces exposed to different concentrations of collagen in solution, thin films will result that have different densities of large collagen fibrils. The density of large fibrils has a remarkable effect on cell morphology, proliferation, and gene expression, even in the absence of any changes in chemistry or integrin recognition. **Figure 2** shows the obvious differences in the density of large collagen fibrils that form at alkanethiol surfaces as a function of differences in concentration of collagen in solution. These differences in collagen fibril density result in obvious differences in vSMC morphology and cytoskeletal organization. Further analysis has determined that the cells engage the same integrins on these surfaces, yet show distinct differences in gene expression and proliferation rate, in addition to differences in morphology. Ongoing work suggests that cells are responding to differences in the mechanical properties of large and small collagen fibrils.

1.2. Quantitative and Automated Microscopy

Quantitative measurements are essential for determining the results of a cell-based assay, and for verifying interoperability of cell-assay data collected on different days and in different laboratories (*see* **Note 1**). By making quantitative measurements, one can also assess the effect that time in culture or cryopreservation has on cell phenotype. Ultimately, quantitative interoperable data will be required for the development of the extensive cell-based bioinformatics databases and models that will make possible the understanding of signaling networks.

Figure 3 shows the novel use of Texas Red maleimide to provide a bright fluorescent stain to discern cell areas from noncell areas, which allows rapid cell identification, counting of cells, and morphology analysis on a cell-by-cell basis over large numbers of cells using automated microscopy *(6)*. The use of a nuclear stain allows discrimination of Texas-Red-labeled objects that are not cells, and determination of how many cells are present as a cluster *(2)*. The bright cell area defined by Texas Red staining can also be used as a "mask" for quantifying intracellular green fluorescent protein (GFP) and other intracellular fluorescent indicators. Robust protocols for quantitative automated fluorescence microscopy and imaging analysis facilitate measurements of statistically relevant numbers of cells in an unbiased fashion (*see* **Note 2**). Thin films of ECM aid automated microscopy because the optical properties of the substrate are homogeneous across the x–y plane. Thin films also tend to have reduced background scattering, making discrimination between cells and background greater. By automatically moving the stage to select fields, the experimenter is not responsible for locating "good" fields of cells. Many fields can be sampled, allowing for accumulation of data on large numbers of cells. By collecting data on a sufficient number of individual cells, the inherent variability of response within the population can be discerned.

1.3. Natural Variability Within Cell Populations

Even clonal populations of cells cultured on spatially homogeneous self-assembled matrices show a range of phenotypic responses. We have quantified the reproducible distributions of cell size, GFP expression, and phalloidin staining of filamentous actin in vSMC in contact with collagen films. This variability in response within the population is highly reproducible. **Figure 4** shows that vSMCs are less spread (i.e., assume a smaller area) on thin films of collagen with a large density of collagen fibrils, compared to vSMC on tissue culture polystyrene. Although the average areas for cells on the two surfaces are distinctly different, on each surface there is a range of cell areas. It is clear that there is overlap in the amount of spreading of cells on the two different matrices, i.e., for cells of intermediate size it cannot be distinguished based on the extent of spreading which surface they are growing on. Most cellular phenotypes will appear as a range of responses, even if the cell population arises from a single clone and the cells are all in a spatially homogeneous environment such as provided by a thin film of collagen fibrils. The average response is not necessarily an adequate representation of the cellular response especially when the distribution of responses is broad or non Gaussian. Analysis of the cumulative distributions provides a sensitive statistical treatment of the differences between distributions

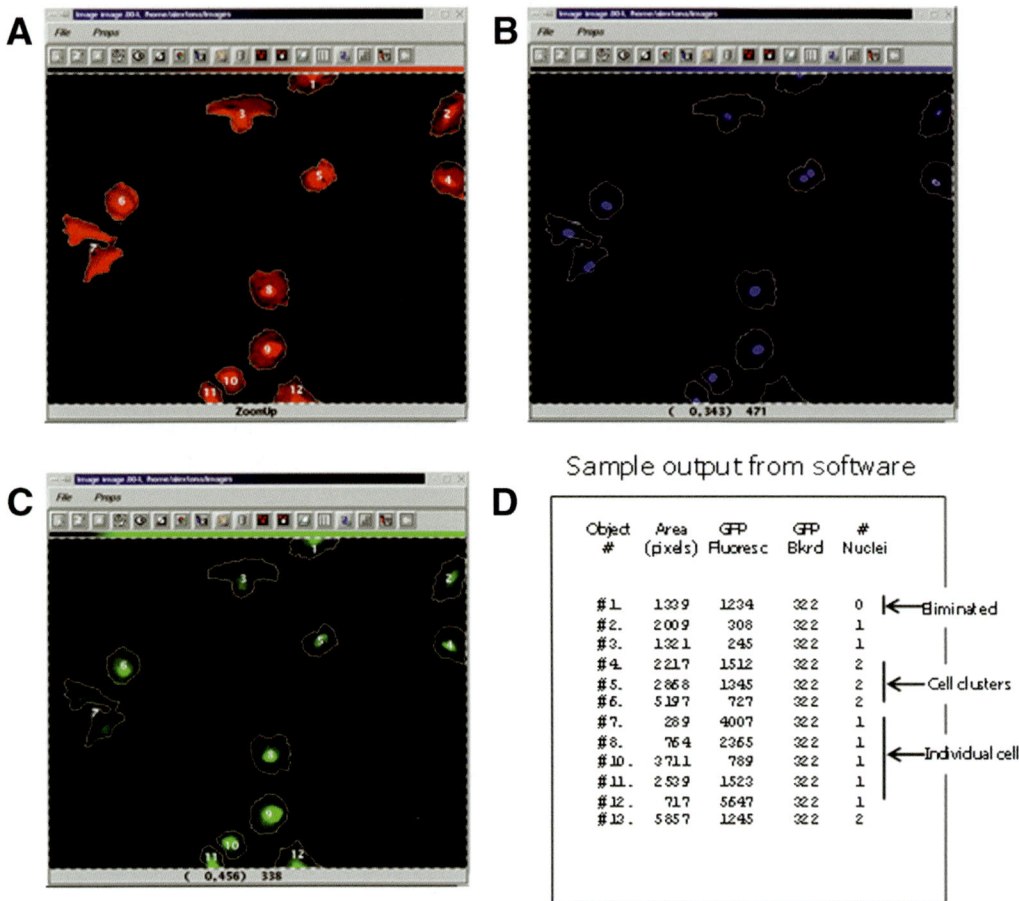

Fig. 3. Automated quantitative microscopy. (**A**) Staining with Texas Red maleimide allows good discrimination of cell edges from noncell background. The user sets a threshold pixel intensity value above which the pixel is recognized as part of an object. Each object is automatically detected and numbered. (**B**) Using another filter set, DAPI staining of cell nuclei is detected, allowing Texas Red-stained objects to be verified as cells or clusters of cells. (**C**) Using a third filter set, GFP inside of cells is detected allowing integration of the intensity of the pixels lying within the area identified by each cell object "mask" delineated in (**A**). (**D**) The presence or absence of nuclei in each object, the number of nuclei, the cell area (as determined from the number of Texas Red pixels per object), the integrated intensity of GFP (or other fluorophore) within the object mask, and the background fluorescence are automatically determined by the software routine along with other parameters.

(7,8). Quantifying and analyzing the statistics associated with the range of responses within the population also allows calculation of how many cells must be examined in order to achieve an accurate sampling of the population.

The collection of data from large numbers of cells (hundreds to thousands) provides the statistical assurance of the reproducibility of the observations. Although flow cytometry also allows collection of data on large numbers of cells, it does not permit examining cells adhered to matrix, eliminating the collection of data regarding cell morphology, and it does not provide the possibility of quantifying changes in cell response in real time. High content screening now allows these measurements in a high-throughput way, whereas human-interactive; imaging microscopy is still a valuable tool for lower throughput studies.

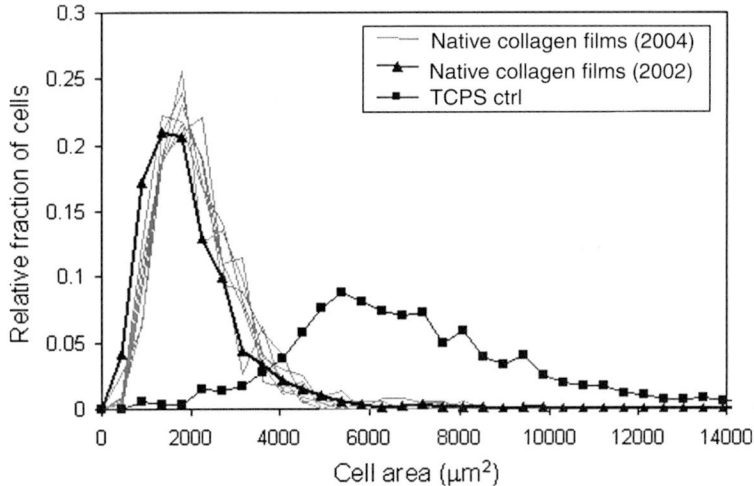

Fig. 4. Quantifying distributions of cellular responses. vSMCs (A10) respond to different kinds of collagen thin films in a large number of ways, including morphology. Cells on thin films of collagen with a high density of large collagen fibrils are minimally spread and assume a stellate morphology (gray lines). These cells spread to an average area of 2100 μm². The cells on fibrillar collagen are significantly less spread than the same cells cultured on tissue culture polystyrene (black, square). These averages, and the range of the size observed, are highly reproducible and consistent from experiment to experiment. For example, morphology distributions of A10 cells cultured on the same day on replicate thin films of fibrillar collagen are nearly identical (gray lines). These distributions are plotted with the A10 morphology distribution collected from thin films of fibrillar collagen prepared by identical methods collected 12 mo earlier (black, diamond).

2. Materials*

2.1. Preparation of Alkanethiol-Coated Supports and Thin Films of Collagen

1. Sodium dodecyl sulfate (Bio-Rad, Hercules, CA).
2. H_2SO_4 (Mallinckrodt Baker, Paris, KY).
3. Potassium persulfate (Sigma, St. Louis, MO).
4. Deionized H_2O, purified through a Milli-Q Academic system (Millipore Corp, Bedford, MA).
5. Particle-free polyester cloth (Texwipe TX1010, Fisher Scientific, Springfield, NJ).
6. Polished Si wafers (Silicon, Inc. Boise, ID).
7. Kimwipes (Kimberly-Clark, Roswell, GA).
8. 1-Hexadecanethiol (Aldrich, Milwaukee, WI), used as 0.5 mM solution in absolute ethanol.
9. Teflon squirt bottles (Nalgene Nunc, Rochester, NY).
10. Vitrogen, a sterile solution of purified, pepsin-solubilized bovine dermal collagen dissolved in 0.012 N HCl (Cohesion Technologies, Palo Alto, CA).
11. 10X Dulbecco's phosphate-buffered saline (DPBS, Gibco Invitrogen, Carlsbad, CA; when diluted to 1X concentration, contains: 0.1 g/L $CaCl_2$, 0.2 g/L KCl, 0.2 g/L KH_2PO_4, 0.1 g/L $MgCl_2 \cdot 6H_2O$, 8 g/L NaCl, and 2.16 g/L $Na_2HPO_4 \cdot 7H_2O$).
12. 0.1 M NaOH (Sigma), sterilized by passing through 0.2-μm filter.

2.2. Cell Culture

1. DPBS, Ca^{2+} and Mg^{2+} free (Gibco Invitrogen).
2. Trypsin-EDTA solution (Sigma).

---

*Indication of specific manufacturers and products is for clarity only and does not constitute endorsement by NIST.

3. Rat aortic vSMC line, A10 (ATCC, Manassas, VA).
4. Dulbecco's modified Eagle's medium (DMEM; Mediatech, Herndon, VA).
5. L-Glutamine, 100X (Gibco Invitrogen).
6. MEM nonessential aminoacid solution, 100X (Gibco Invitrogen).
7. Penicillin-streptomycin solution, 100X, (Gibco Invitrogen).
8. Fetal bovine serum (FBS; Gibco Invitrogen).
9. 8-well polystyrene plates (Nalgene Nunc).
10. Mouse embryonic fibroblast cells (NIH3T3; ATCC).
11. 25 cm^2, vented cap tissue culture flasks, for routine culture (Corning, Corning, NY).

2.3. Cell Fixation, Staining, and Fluorescence Microscopy

1. Hanks Balanced Salt Solution (HBSS; ICN Biomedicals, Costa Mesa, CA).
2. DPBS, without $CaCl_2$ (Sigma).
3. Texas Red-C_2-Maleimide (Molecular Probes Invitrogen).
4. Bovine Serum Albumin, (BSA, Fraction V; Sigma).
5. Triton X-100 (Sigma).
6. Glycerol (Research Organics, Cleveland, OH).
7. 16% Paraformaldehyde (Electron Microscopy Sciences, Hatfield, PA).
8. Ammonium chloride (Sigma).
9. 1,4-Diazabicyclo(2,2,2)octane (DABCO, Sigma).
10. 4′,6-Diamidino-2-phenylindole (DAPI, Sigma).
11. No. 1 coverglass slides (Nunc, Naperville, IL).
12. Fluorescein-isothiocyanate-phalloidin (Sigma).
13. 3-Malemido-benzoic acid-NHS ester (MBS), Sigma.
14. DMF (dimethylformamide, Sigma).
15. Microtubule stabilizing buffer (MTSB): 100 mM PIPES (piperazine-N,N'-bis[2-ethanesulfonic acid, Sigma), 1 mM ethylene glycol-bis(2-aminoethyl)-N,N,N',N'-tetraacetic acid (Fluka, Milwaukee, WI), 4% polyethylene glycol 8000 (Fluka), adjusted to pH 6.9.
16. Long Pass GG475 (Schott) Glass Filters (Edmund Optics, Barrington, NJ).
17. NIST SRM 1932 fluorescein solution (NIST, Gaithersburg, MD).
18. Optical filters: emission filters, D460/50, HQ525/50, S630/60, No. 84101 (multi bandpass); excitation filters, D360/40, HQ470/40, S555/25; Beam Splitters, BS No. 51019 (triple bandpass), No. 84100 (quadruple bandpass); (Chroma Technologies, Brattleboro VT).

3. Methods
3.1. Preparation of Alkanethiol-Coated Supports

1. Glass cover slips (No. 1, 22 × 22 mm^2) were washed in 1% (w/v) sodium dodecyl sulfate/water, rinsed extensively with deionized H_2O, and acid washed in fresh H_2SO_4 containing 10% potassium persulfate (30 min). The cover slips were rinsed extensively with deionized H_2O, transferred to acetone and dried on a particle free polyester cloth.
2. Polished Si wafers were cleaned by wiping them across an ethanol soaked Kimwipe and then a dry Kimwipe. Residual fiber and dust particles were removed under a stream of particle-filtered N_2.
3. Cover slips and wafers were coated with a layer of chromium (approx 1 nm) and a layer of gold (approx 7 nm) by magnetron sputtering using an Edwards (Wilmington, MA) 306 vacuum system equipped with two (approx 10 cm diameter) magnetron sputtering sources (one for chromium, the other for gold). The base vacuum level is typically better than 2×10^{-5} Pa (2×10^{-7} mbar). The argon pressure, 0.1 Pa (6×10^{-3} mbar), is kept constant during the sputtering process using an MKS Instruments (Andover, MA) Baratron butterfly valve. During sputtering the sample holder and samples are rotated over the gold and chromium targets (purity 99.99%) using a Superior Electric (Bristol, CT) SLO-SYN stepper motor system. A specially shaped mask is placed between the targets and the sample holder to ensure uniformity of film thickness over the whole sample surface. A DC power source applies 750 W for gold and 440 W for the chromium depositions. The substrate rotation time required to produce a given film thickness was determined with the help of X-ray and neutron reflectivity measurements on test samples. Atomic force

microscopy measurements on freshly prepared samples indicated that the gold surface roughness was <1 nm rms (measured over 25 µm^2).

4. The gold-coated wafers and cover slips were immersed in 1-hexadecanethiol (0.5 mM; Aldrich) in ethanol for at least 8 h before being rinsed with ethanol and dried with filtered N$_2$. Alkanethiol-coated samples could be stored under ethanol for at least 7 d without any loss in performance.

3.2. Preparation of Thin Films of Type 1 Collagen

1. To prepare thin films of native collagen, translucent alkanethiol-treated gold-coated cover slips or Si wafer pieces were placed into neutralized solutions of native type 1 collagen monomer in DPBS (4°C).
2. To prepare the neutralized collagen solution, the cold solution of Vitrogen (acid extracted collagen monomer) was mixed with cold concentrated (10X) DPBS, and then with cold 0.1 M NaOH in a ratio of 8/1/1 (v/v/v) to achieve physiological pH and ionic strength conditions under which collagen monomer readily polymerizes into fibrils at 37°C.
3. The cover slips were immersed in 3 mL of collagen solution in eight-well polystyrene plates and were then incubated overnight at 37°C to allow polymerization of the collagen.
4. The collagen-coated samples were removed, rinsed with a stream of DPBS, and then deionized water from sterile Teflon squirt bottles (Nalgene Nunc).
5. Once all loosely adhered gel was removed, the samples were dried briefly (for approx 30–60 s) under a stream of filtered N$_2$ and immediately placed back into a DPBS solution (*see* **Note 3**). The samples were stored in DPBS at 4°C until they were used.

3.3. Characterization of Thin Films of Type 1 Collagen

1. The thickness of the collagen thin films on Si was determined by spectroscopic ellipsometry (J. A. Woollam, Lincoln, NE, Model M-44) in air using a two-layer model. The optical constants of the first layer were determined empirically from ellipsometric data collected for a control sample of alkanethiol-treated, gold-coated Si wafer. The optical constants of the second layer were fixed to $n = 1.45$ and $k = 0$ to approximate the properties of the protein film and the average thickness of the second layer for each sample was determined using the manufacturer's fitting routine.
2. The collagen samples were imaged using an atomic force microscopy with a magnetically driven Si tip in intermittent contact mode (PicoScan; Molecular Imaging, Phoenix, AZ). The thin films of collagen were imaged in air before they were placed into the fluid cell and imaged under DPBS. The results indicated that rehydrated and dry samples were qualitatively the same topographically. Images were taken from several areas on each sample to ensure the homogeneity of surface features. Images were flattened and, in some cases, plane fitted with the PicoScan software to improve visualization, and analysis of the images.

3.4. Cell Culture and Preparation of Cells for Experiments

1. The rat aortic vSMC line, A10, was maintained in DMEM supplemented with nonessential amino acids, glutamine, penicillin (100 U/mL), streptomycin (100 µg/mL), and 10% (v/v) FBS, and maintained in a humidified 5% (v/v) CO$_2$ balanced-air atmosphere at 37°C.
2. Subconfluent cultures were switched to supplemented DMEM containing 2% (v/v) FBS 24 h before an experiment. The reduced serum concentration maximizes the extent of cell signaling that is because of the ECM and mimics conditions that have been typically used in characterizing the response of these cells to native and denatured collagen gels (*9*).
3. Cells were removed from tissue culture polystyrene flasks by trypsinization, washed with DMEM/2% FBS and plated in DMEM/2% FBS onto the collagen substrates at a density of 2000 cells/cm^2 except where indicated. The cell density was chosen to maximize the number of individual (nonclustered) cells on the substrate. Care was taken to ensure the seeding density was homogeneous over the surface of the substrates.
4. Cells were typically incubated at 37°C for 24 h except in the proliferation and integrin blocking experiments. All experimental incubations were performed with the thin films of collagen on gold-coated glass cover slips on the bottom of eight-well polystyrene plates.

3.5. Cell Fixation and Staining for Morphology Analysis

1. After incubation, samples were left in the wells of the polystyrene plates and the adhered cells were washed with warm HBSS, fixed in 4% (v/v) formaldehyde in DPBS (30 min) at room temperature, quenched in 0.25% (m/v) NH_4Cl in DPBS (15 min), and rinsed with DPBS.
2. Cells were permeabilized and stained (1 h) with Texas Red-C_2-maleimide (10 mg/mL in dimethylformamide stock, dissolved in DPBS (0.5 µg/mL) containing 0.1% (v/v) Triton X-100.
3. Cells were rinsed once with DPBS, DPBS containing 1% (m/v) BSA and DPBS. DPBS-glycerol (1/1 v/v) containing 0.25% (m/v) 1,4-DABCO to reduce photobleaching, and 1.5 µg/mL DAPI as a nuclear counterstain, was added to each well.
4. Samples were removed from the polystyrene wells and were placed upside down onto a drop of Tris-buffered saline (10 m*M* Tris, 140 m*M* NaCl, pH 8.5) containing 90% (v/v) glycerol, 0.25% DABCO and 1.5 µg/mL DAPI on No. 1 glass slides.
5. The samples were clamped to the slides with small alligator clips, rinsed extensively with distilled water, dried under a stream of air and sealed at the edges with nail polish. This procedure significantly reduces the optical artifacts because of the presence of excess glycerol or dried buffer salts on the cover slips. Throughout the fixation and staining procedure, cell samples were always kept immersed in solution.
6. For filamentous actin staining, cells were permeabilized with 0.1% (v/v) Triton X-100 in DPBS (5 min), rinsed with DPBS, blocked with DPBS containing 3% (m/v) BSA blocking solution (30 min), stained with fluorescein-isothiocyanate-phalloidin in blocking solution (200 n*M*, 1 h), and rinsed with blocking solution.
7. The phalloidin-stained samples were rinsed extensively with DPBS and mounted on slides in Tris buffered saline containing 90% (v/v) glycerol, 0.25% (m/v) DABCO and 0.05 µg/mL DAPI.
8. Cells were imaged with phase and fluorescence microscopy using the appropriate filter sets.

3.6. Cell Fixation and Staining for Quantifying Cellular GFP

1. Samples were removed from the incubator after 8 h, adhered cells were rinsed with warm HBSS supplemented with 10 m*M* HEPES and fixed for 24 h at room temperature in MTSB containing 100 µg/mL MBS as the crosslinker *(10)* (*see* **Note 4**).
2. Cells were permeabilized in 0.05% Triton X-100 in DPBS for 5 min, rinsed in DPBS, and incubated with DPBS containing Texas Red-C2-maleimide (1 ng/mL) as a general cell membrane stain and 0.05% DAPI as a nuclear counter stain.
3. After 2 h at RT, 1% BSA was added to quench the conjugation reaction and the samples were rinsed with DPBS.
4. The samples were mounted onto No. 1 glass slides with 9/1 glycerol/Tris (v:v), pH 8.0 and the cells were imaged with fluorescence microscopy in an automated mode.

3.7. Automated Microscopy for Quantitative Analysis

1. The fixed and stained cells were examined by phase contrast and fluorescence microscopy using a 10X objective on an inverted microscope (Zeiss Axiovert S100TV, Thornwood, NJ) with a 100 W mercury arc lamp for fluorescence excitation, a computer controlled stage (LEP, Hawthorne, NY), an excitation and an emission filter wheel (LEP, Hawthorne, NY), and a CCD camera (CoolSnap fx, Roper Scientific Photometrics, Tucson, AZ). Hardware operation, and image digitization and analysis were under software control (ISee Imaging, Cary, NC).
2. A modular software routine controlled automated movement of the stage, autofocusing, and collection of data from 50 to 100 independent fields (870×690 µm^2) of cells per sample. Autofocusing was performed on the Texas Red fluorescence of stained cells.
3. At each field cellular fluorescence from Texas Red, and then DAPI, was collected by automated switching of the appropriate excitation filters (S555/25 and D360/40, respectively) and passing the emitted light through a multipass beam splitter (No. 84100) and filter (No. 84101).
4. For measurements of GFP expressing cells, the multipass beamsplitter No. 51019 was used along with appropriate excitation and emission filters (D360/40 and D460/50 for DAPI, HQ470/40 and HQ525/50

for GFP, and S555/25 and S630/60 for Texas Red, respectively) (*see* **Note 5**). The excitation lamp intensity was checked every day using a 61 μM fluorescein solution (NIST SRM 1932) to allow for normalization of intensities with other data from other days.

5. Flat field correction was performed on all images. The flat fields were generated with a 475 nm long pass glass filter that exhibited broad fluorescence excitation and emission *(11)*, which was placed on a No. 1 glass slide on the stage.
6. Appropriate thresholding of image data allowed cell areas, as determined by cellular Texas Red fluorescence, to be accurately distinguished from the nonfluorescent-noncell areas (*see* **Notes 6** and **7**).
7. For quantitative analysis of intracellular fluorescence such as GFP, the areas of Texas Red fluorescence that were associated with cells served as a "mask" to delineate the pixels associated with cells in identical fields of images collected with other filter sets. In this way, even cells that were not producing much GFP could be quantified even though their low GFP fluorescence might have otherwise precluded observing them.
8. Non-GFP fluorescence background was determined by eroding the area of the cell (i.e., including pixels immediately beyond the area of the cell), and using adjacent but noncell pixels to define the background intensity (*see* **Note 8**). The number of nuclei, and therefore the number of cells, was determined from the corresponding images collected with the DAPI filter. The requirement for spatial correspondence of DAPI and Texas Red fluorescence ensures that only cell areas with nuclei are used during data analysis.
9. Each cell area, cell roundness, and intracellular GFP intensity, as well as the number of nuclei in each field, were determined with image analysis software (ISee Imaging) as previously described *(2,6)*. Data used to determine these parameters were collected from at least 200 cells on each sample.

4. Notes

1. *Intensity reference materials*: for standardizing fluorescence intensity measurements, a variety of reference materials can be used for checking lamp output. Fluorescent glass, solutions of high concentration of fluorophore, and fluorescent beads, for example, can all be used. Some materials prepared especially for this purpose can be purchased commercially. We have found that inexpensive semiconductor long pass glass filters that exhibit broad fluorescence excitation and emission have been useful for this purpose. (The glass filters in this series are fluorescent in different wavelength regions; coincidentally, 1 s integration times of GG475 at the wavelength regions of interest for DAPI, GFP, and Texas Red produced similar intensities.) The requirements for such a reference material include sufficient photostability, batch or solution reproducibility, appropriate intensity at desired wavelengths, and a linear fluorescence response to incident light. The use of this simple kind of intensity reference is only useful for providing relative information about the light source, and to verify that the instrument is functioning consistently; determining the transfer function for the collection optics is much more challenging.
2. *Image analysis software*: image analysis software packages are available from commercial and open source suppliers. One open source package is ImageJ, which is written in JAVA for platform-independent applications. Although any image analysis software package will have built in segmentation and data extraction techniques, it is likely that at least some algorithm development or testing will be required for particular applications. Development of robust algorithms for choosing appropriate threshold values or for reliable edge detection of cellular objects is still a challenge to be overcome. On our NIST web page we currently provide a set of image data for NIH3T3 cells that express GFP. The image set contains three fluorophores (Texas Red, DAPI, and GFP), and is provided for comparing edge detection and GFP quantification algorithms. With contributions from others in the community, we hope to add additional image sets from assays or staining procedures to create a comprehensive set of standard images to test algorithms development.
3. *Preparation of thin collagen films*: the volume, as well as the concentration of collagen monomer solution from which the thin films are made can effect the resulting structure of the collagen thin films. Brief drying with a stream of nitrogen or air after rinsing the thin films of collagen is required. If the films are not dried at all, many cells will remain round and die, possibly because the fibrils are not associated sufficiently with the surface. Long drying times (hours) will result in very stiff collagen fibrils

that appear to adhere to one another. Cells on stiff collagen fibrils respond to this mechanical stimulus by increased spreading, higher rates of proliferation, and changes in gene expression. Visualization of the collagen fibrils is possible with phase microscopy.

4. *Cell fixation and staining for quantifying cellular GFP*: we typically examine transfected cells with constructs of GFP linked to a promoter region of interest. As a result, we are measuring soluble GFP that is present in the cytoplasm of the reporter cells; the GFP is not fused to another protein. Fixing these GFP-expressing cells with formaldehyde in PBS results in the attenuation of the GFP signal by >50%. To prevent GFP signal reduction during fixation, we found that the use of MTSB *(10)* and the MBS crosslinker resulted in the preservation of greater than 90% of the initial GFP fluorescence signal with no loss in cell morphology.

5. *Bleed-through of fluorescence from other fluorophores*: careful control experiments must be performed when developing protocols for the use of more than one dye in quantitative analysis. Control samples that contain each dye alone, then in combination, must be examined under the conditions that will be ultimately used for data collection, including lamp intensity, filter sets, integration time, and so on. The conditions described above were selected for unambiguous quantification of GFP expression levels.

6. *Thresholding*: potential caveats associated with automated microscopy include assigning the threshold intensity that will be applied to images to discriminate between cell area and background. If the staining at the cell periphery is very bright relative to background, a range of threshold values can be applied that do not significantly affect the data *(6)*. However, under other circumstances, thresholding can be a source of inconsistency in datasets. Development of robust algorithms for choosing appropriate threshold values will a challenge to be overcome.

7. *Quantifying GFP*: the use of a general cell stain like Texas Red maleimide to provide a cell "mask," allows us to easily identify all cells and to quantify GFP even in those cells that do not express much GFP. In this way, we can quantify the difference between cells in conditions in which the associated promoter is very active, inactive, or of intermediate activity. It must be noted that we are describing a procedure for achieving only relative quantitative analysis; determining absolute concentration of a fluorophore by microscopy is by far a more challenging issue.

8. *Background determination using erosion*: erosion is a technique that can be used for determining the background fluorescence that should be subtracted from a cell image. It is particularly useful when the background varies widely across the field such that an average background value would not be appropriate for all cells in the field. A large standard deviation in the intensity of the eroded pixels can be the result of the presence debris or another cell, or an indication that the cell lies partially outside of the frame; the standard deviation thus can be a flag that precludes the use of that background determination.

References

1. Geiger, B., Bershadasky, A., Pankov, R., and Yamada, K. M. (2001) Transmembrane extracellular matrix-cytoskeleton crosstalk. *Mol. Cell Biol.* **2,** 793–805.
2. Elliott, J. T., Jones, P. L., Woodward, J. T., Tona, A., and Plant, A. L. (2002) Morphology of smooth muscle cells on thin films of collagen. *Langmuir* **19,** 1506–1514.
3. Elliott, J. T., Langenbach, K. J., Tona, A., Jones, P. L., and Plant, A. L. (2005) Vascular smooth muscle cell response on thin films of collagen. *Matrix Biol.* **24,** 489–502.
4. Langenbach, K. J., Elliott, J. T., Tona, A., and Plant, A. L. (2005) Evaluating the correlation between fibroblast morphology and promoter activity on thin films of extracellular matrix proteins. *BMC-Biotechnology* (in press).
5. Elliott, et al. The effect of surface chemistry on the formation of thin films of native fibrillar collagen. (in preparation).
6. Elliott, J. T., Tona, A., and Plant, A. L. (2002) Comparison of reagents for shape analysis of fixed cells by automated fluorescence microscopy. *Cytometry* **52A,** 90–100.
7. Perlman, Z. E., Slack, M. D., Feng, Y., Mitchison, T. J., Wu, L. F., and Altschuler, S. J. (2004) Multidimensional drug profiling by automated microscopy. *Science* **306(5699),** 1194–1198.
8. Giuliano, K. A., Chen, Y. -T., and Taylor, D. L. (2004) High content screening with siRNA optimizes a cell biological approach to drug discovery: defining the role of p53 activation in the cellular response to anti-cancer drugs. *J. Biomol. Screen.* **9,** 557–567.

9. Jones, P. L., Crack, J., and Rabinovitch, M. (1997) Regulation of tenascin-C, a vascular smooth muscle cell survival factor that interacts with the alpha v beta 3 integrin to promote epidermal growth factor receptor phosphorylation and growth. *J. Cell Biol.* **139,** 279–293.
10. Safiejko-Mroczka, B. and Bell, P. B., Jr. (1996) Bifunctional protein cross-linking reagents improve labeling of cytoskeletal proteins for qualitative and quantitative fluorescence microscopy. *J. Histochem. Cytochem.* **44,** 641–656.
11. Wyszecki, G. and Stiles, W. S. (1967) *Color Science: Concepts and Methods, Quantitative Data and Formula,* Wiley, New York, NY.

8

Automated Cell Plating and Sample Treatments for Fixed Cells in High Content Assays

Gillian R. Richards, Julie E. Kerby, Grace K. Y. Chan, and Peter B. Simpson

Summary
Robust and reliable methods for the manipulation of neural cell lines, by passaging, plating, dye labeling, imaging, fixation, and immunocytochemistry, are required to enable consistent, reproducible screens to be performed. We describe herein procedures and processes we have established to maximize the level of consistency of cell plating, fixation, and dye or antibody labeling, to ensure that assays which we are running on a routine basis remain consistent across long periods of time. These procedures involve a variety of fully or semiautomated steps, using high-quality commercially available liquid handling and dispensing technology.

Key Words: Cell plating; chemokinesis; fluorescence imaging; neurospheres; neurite outgrowth.

1. Introduction

Much attention in the high content screening field has been applied to optimizing the quality of image analysis to maximize value and information *(1,2)*. Equally important; however, is ensuring a high standard, and a high level of consistency, in the preparation of cells for analysis. This is particularly important for screening assays, which run on a week-by-week basis, whereby variability between plates and across weeks will critically affect the utility of an assay and the interpretability of the results. We have established procedures and processes to maximize the level of consistency of cell plating, fixation, and dye or antibody labeling, to ensure that high content assays which run on a routine basis remain reliable across long periods of time. These procedures involve fully and semiautomated steps, and we describe the high-quality commercially available liquid handling and dispensing technology we have used to implement these approaches.

2. Materials

2.1. Automated Cell Plating

1. Clonetics human neural progenitor cells.
2. Plating media (*see* **Subheading 2.2.1., step 5**).
3. 50- and 15-mL conical tubes.
4. Phosphate-buffered saline (PBS) Ca/Mg free.
5. Accutase.
6. Hanks' buffered saline solution (HBSS).
7. Trypan blue.

2.2. Sample Treatments for Fixed Cells

2.2.1. An Automated Chemokinesis Assay

1. Sterile PBS and PBS pH 7.4.
2. Blacksided clear bottomed 96-well plates of cells.
3. Cellomics® (Pittsburgh, PA) Cell Motility Hit Kit.
4. 5% Paraformaldehyde in PBS: stored at 4°C.
5. Sterile plating medium: Dulbecco's modified Eagle's medium (DMEM)/F12 3/1, transferrin 50 µg/mL, insulin 5 µg/mL, progesterone 20 nM, putrescine 100 µM, T3 30 nM, and selenium 30 nM.
6. Positive control and test compounds.
7. Bovine serum albumin (BSA).
8. Normal goat serum (NGS).
9. Triton X-100.
10. Monoclonal anti β-tubulin antibody (Sigma, St. Louis, MO).
11. Alexa-594 goat antimouse secondary antibody (Molecular Probes, Eugene, OR).
12. 2X Blocking buffer prepared freshly and stored at 4°C until required: PBS containing 0.2% Triton X-100, 10% NGS, and 2% BSA.
13. 2X Primary antibody solution prepared freshly and stored at 4°C until required: PBS containing 10% NGS, 2% BSA, and 1/500 primary antibody.
14. 2X Secondary antibody solution prepared freshly and stored at 4°C until required. (This solution is light sensitive.) PBS containing 10% NGS, 2% BSA, and 1/1000 secondary antibody, and 60 µM Hoechst 33342.
15. Cellomics Arrayscan II and Cellomics Cell Motility Algorithm.

2.2.2. Compound Incubation and Plate Processing of a Neurite Outgrowth Assay

1. Sterile PBS and PBS pH 7.4.
2. Black-sided clear-bottomed 96-well plates of cells.
3. Positive control and test compounds.
4. BSA.
5. NGS.
6. Triton X-100.
7. Ice-cold methanol.
8. Primary antibodies as appropriate to assay.
9. Secondary antibodies as appropriate to assay.
10. Hoechst 33342.
11. 2X Blocking buffer.
12. 2X Primary antibody solution.
13. 2X Secondary antibody solution.
14. Automated liquid handling and plate washing.
15. High content screening plate imager.

2.2.3. Fully Automated Single Cell Kinetic and Immunocytochemical Assay

1. KHB (Krebs-Henseleit-bicarbonate buffer) pH 7.4.
2. PBS pH 7.4.
3. Black-sided clear-bottomed 96-well plates of cells.
4. Test compounds.
5. Fluo 3-AM.
6. Pluronic F-127.
7. 2X Fluo3-AM dye solution (This solution is light sensitive.) KHB containing 8 µM Fluo 3-AM and 0.16% Pluronic F-127.
8. BSA.
9. NGS.
10. Triton X-100.
11. Ice-cold methanol.
12. Primary antibodies as appropriate to assay.

13. Secondary antibodies as appropriate to assay.
14. Hoechst 33342.
15. 2X Blocking buffer.
16. 2X Primary antibody solution.
17. 2X Secondary antibody solution.
18. Automated liquid handling, plate processing, and imaging on integrated robotic platform.

3. Methods
3.1. Automated Cell Plating

Clonetics human neural progenitor cells (Cambrex, NJ), growing as neurospheres, were treated with enzymes to create a single cell suspension, counted, and diluted to a predetermined number, and plated into 96-well plates using a Multidrop (Thermo Electron Corporation, Waltham, MA).

3.1.1. Generation of a Single Cell Suspension

1. Spheres were transferred into a 50-mL conical tube with a minimal volume of media, allowed to settle, the liquid removed, and the spheres washed twice with 50 mL PBS (*see* **Note 1**).
2. Following removal of the last wash of PBS, 5 mL of one-fifth dilution of Accutase in HBSS was added and the tube placed in a vigorously shaking water bath at 30°C for 30 min (*see* **Note 2**).
3. Cells were triturated using a 1000 µL pipet tip until most clumps were dispersed (10–15 times), 40 mL PBS added and the cell suspension centrifuged at 150 rcf for 5 min.
4. Supernatant was removed, 2 mL plating media added, and the cells triturated again.
5. The suspension was then transferred to a 15 mL conical tube, 6 mL plating media added, and centrifuged at 10 rcf for 1 min.
6. The supernatant was collected and the pellet discarded (*see* **Note 3**).
7. The cell suspension was counted using the Trypan blue exclusion method on a Cedex cell counter (Innovartis) (*see* **Note 4**) and diluted appropriately.

3.1.2. Plating (see **Note 5**)

1. To ensure sterility of the cultures, a multidrop was located in a biological safety cabinet and sprayed with 70% ethanol, with the tubing primed with 70% ethanol followed by sterile PBS.
2. The cell suspension was primed through the tubing until air bubbles were eliminated and then each plate was processed in turn.
3. The cell suspension was mixed by swirling between each plate. When the pipeting was complete, the plates were transferred to a cell culture incubator (humidified at 37°C with 5% CO_2) (*see* **Note 6**).

3.2. Sample Treatments for Fixed Cells
3.2.1. An Automated Chemokinesis Assay

1. Cell suspensions for this assay were plated into 96-well plates precoated with blue fluorescent beads from the Cell Motility HitKit. During incubation, moving cells displace or phagocytose the fluorescent beads leaving dark "tracks."
2. At an appropriate time-point, cells were fixed and fluorescently labeled, and the tracks quantified using a Cellomics Arrayscan II (Pittsburgh, PA). The assay described here was designed to screen for compounds, which increase the motility of human neural precursor cells (Clonetics®, Cambrex, East Rutherford, NJ) *(3)*, but could equally be applied to investigate inhibitors of cell motility, by adding a known stimulant to the plating medium.

3.2.1.1. Assay Plate Preparation

1. 96-well black-sided microtiter poly-D-lysine coated plates (Biocoat™, BD Biosciences, Oxford, UK) were coated with laminin 1 µg/mL (*see* **Note 7**).
2. For each 96-well plate, one vial of blue fluorescent beads from the Cellomics Cell Motility Hit Kit was resuspended by vortexing for 30 s then centrifuged for 1 min at 20,000g.
3. Supernatant was removed, and 0.5 mL PBS was added, before vortexing and centrifugation. This washing step was then repeated.

4. The resulting bead suspension was added to 7.5 mL PBS and vortexed for 1 min (*see* **Note 8**).
5. 75 µL of bead suspension was then added to each well of the laminin-coated 96-well plate using a multichannel pipetor.
6. The plate was incubated for 1 h at 37°C and 5% CO_2, then washed very gently five times using sterile PBS and a multichannel pipetor, leaving a residual volume of 100 µL per well.
7. Plates were wrapped in foil and stored at 4°C until required.

3.2.1.2. CHEMOKINESIS ASSAY

1. Test compounds were prepared at 2X[final] in plating medium containing 0.2% serum, and 50 µL added to the test wells in the bead-coated plates. 5% serum was used as a positive control.
2. Assay plates were prewarmed to 37°C and 2000 cells/well in 50 µL of plating medium were added using a Multidrop (*see* **Note 9**).
3. Plates were incubated for 18 h at 37°C with 5% CO_2.
4. All the following incubations were performed at room temperature, in the dark, to prevent photobleaching of the beads.
5. Cells were fixed by adding 100 µL/well of prewarmed 5% paraformaldehyde for 10 min at room temperature without removing the medium. Addition and washing steps were performed with a multichannel pipet and a Thermo Wellwash (Thermo Electron Corporation, Waltham, MA) plate washer.
6. Plates were washed three times with 200 µL/well PBS with a residual volume of 50 µL/well.
7. 50 µL of 2X blocking buffer was added to each well and incubated for 1 h.
8. 100 µL/well of 2X primary antibody solution was added without washing, and incubated for 1 h.
9. Plates were washed three times with 200 µL/well PBS with a residual volume of 50 µL/well.
10. 50 µL of 2X secondary antibody solution was added and incubated for 1 h.
11. Plates were then washed three times with 200 µL/well PBS and a residual volume of 200 µL/well, then sealed with thin adhesive plate seals (*see* **Note 10**).
12. Chemokinesis was quantified in an automated manner using a Cellomics ArrayScan II and Cellomics Cell Motility algorithm, acquiring nine fields per well with a ×5 objective. The algorithm outputs, which were found to be most useful for analysis were the average track area and the average number of tracks per field.

3.2.2. Compound Incubation and Plate Processing of a Neurite Outgrowth Assay

The methods described here detail the procedures used to perform a screen for promoters of neurite outgrowth in human neural precursor cells, although the general methodology and principles could be adapted to a range of fixed endpoint high content screening assays *(4,5)*. The equipment used for processing and reading plates were a Multimek (Beckman Coulter, Fullerton, CA), a PlateTrak equipped for 96-channel pipeting and plate washing (Perkin Elmer, Boston, MA) and a Cellomics Arrayscan II; a range of suitable alternatives are available.

3.2.2.1. COMPOUND ADDITION

1. Test compounds and positive and negative controls were prepared in deep 96-well microplates such that subsequent addition of 200 µL of sterile PBS on the Multimek would produce a solution with 10X final concentration. For this assay the positive control was 100 ng/mL platelet-derived growth factor (PDGF) final concentration, sterile PBS was used as the negative control and test compounds were screened at 1 µ*M*.
2. 96-well black-sided microtiter poly-D-lysine coated plates (Biocoat) were coated with 1 µg/mL laminin.
3. Cells were plated at 17,500 cells/well in 200 µL of plating medium using a Multidrop (*see* **Subheading 2.2.1.** and **Note 11**).
4. Following incubation for 1 h at 37°C with 5% CO_2, a Multimek program was used to dilute the compounds in 200 µL of sterile PBS and to immediately add 22 µL to a cell plate to produce the correct final desired concentrations (*see* **Note 12**).
5. Cell plates were incubated for 48 h at 37°C with 5% CO_2. The assay was typically performed on batches of 20 cell plates.

Automated Cell Plating

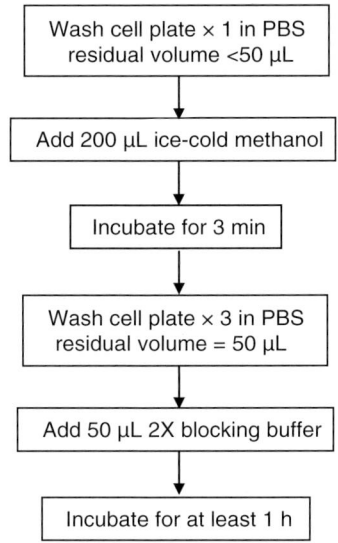

Fig. 1. A flow diagram representing the steps of the PlateTrak program used for fixing and blocking cells.

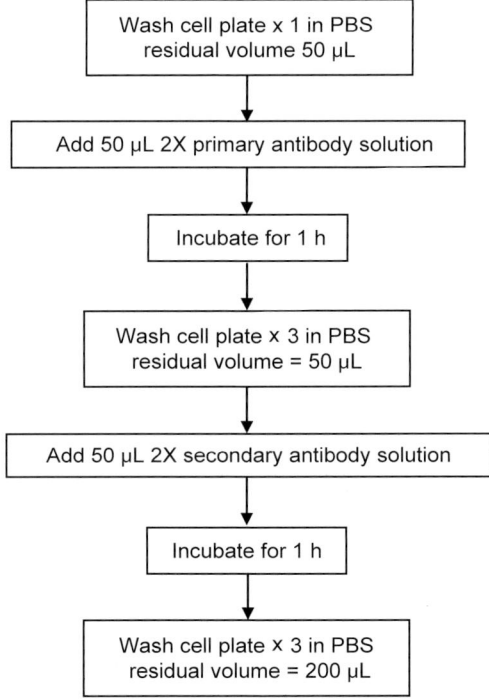

Fig. 2. A flow diagram representing the steps of the PlateTrak program used for applying and incubating primary and secondary antibodies with cells.

3.2.2.2. Cell Fixing and Labeling

1. Cell plates were fixed and antibody-labeled using a PlateTrak configured with two plate stackers, a 96-well plate washer, a 96-channel dispense head, and an autoreplenish reservoir. The cell washer reservoir was filled with PBS. A bottle of methanol standing in a bucket of ice was connected to the autoreplenish reservoir such that ice-cold methanol was continuously recycled to the reservoir during cell fixing.

2. For convenience, plate processing utilized two PlateTrak programs: one to fix cell plates and to add blocking buffer (*see* **Fig. 1**), and another to perform antibody labeling (*see* **Fig. 2** and **Note 13**).
3. Following this PlateTrak processing, cell plates were sealed using thin adhesive seals, wrapped in foil and stored at 4°C until they were read.
4. Antibodies were used at 1/1000 final concentration and typically included polyclonal or monoclonal anti-β-III tubulin (Covance), monoclonal antiGFAP (Sigma), Alexa-488 conjugated goat anti-mouse or anti-rabbit, and Alexa-594 conjugated goat anti-mouse or anti-rabbit anti-bodies (Molecular Probes) (*see* **Note 14**).

3.2.2.3. Cell Plate Imaging

1. Neurite outgrowth was quantified using a Cellomics Arrayscan II and Cellomics Neurite Outgrowth algorithm.
2. A Twister (Caliper Life Sciences, Mountain View, CA) attached to the ArrayScan allowed batched of up to 20 plates to be read automatically.
3. Satisfactory results were achieved by adapting with the Cellomics neurite outgrowth algorithm using a ×5 objective, with nine fields imaged per well. The algorithm outputs, which were found to be most useful were Neurite Outgrowth Index, and several indicators of neurite length and branching.
4. Plate data was accepted as valid if the positive control had produced at least a twofold increase in neurite outgrowth, length, and branching.
5. Compound results were normalized to positive control values to allow for inter-experiment variability, with active compounds defined as those producing increases more than 35% of control in all three parameters.
6. The images from wells with active compounds were viewed to confirm that labeling and image analysis algorithms had been applied correctly.
7. The antibody labeling produced using this automated methodology was consistent to such a degree, which identical image capture exposure times were used across an approx 8 mo screening period.

3.2.3. Fully Automated Single Cell Kinetic and Immunocytochemical Assay

The methods described here detail the procedures used to perform a kinetic signaling assay to measure dynamic changes in intracellular free calcium in individual human neural precursor cells in response to compound treatment. This was combined with posthoc immunocytochemistry to separately characterize and compare with the kinetic responses from specific subsets of cells, for example, precursor, glial, and neural cells within a mixed neural cell population *(6–8)*. The robotic platform used to perform this assay consists of a CRS Catalyst 5 robotic arm (Thermo, San Jose, CA) integrated with an Atto PathwayHT confocal imager (now marketed as BD PathwayHT Bioimager) *(6,7)* fitted with a one channel pipettor head for online compound addition (BD Biosciences), a PlateTrak equipped for 96-channel pipeting and plate washing as described in **Subheading 3.2.2.** and a Heraeus Cytomat CO_2 incubator (Kendro, Bishop's Stortford, UK). The robot protocol is outlined in a flow diagram (*see* **Fig. 3**). This general methodology can be adapted to a range of combined kinetic and immunocytochemical assays in high-content single cell imaging.

3.2.3.1. Cell and Compound Plates Preparation

1. 96-well black-sided microtiter poly-D-lysine coated plates (Biocoat, BD Biosciences) were coated with 1 μg/mL laminin.
2. Cells were plated at 17,500 cells/well in 200 μL of medium using a Multidrop and incubated at 37°C with 5% CO_2.
3. A 2.5 μL of test compounds at 2 m*M* concentration was plated into 96-well microplates such that subsequent addition of 97.5 μL of KHB buffer on the Multimek (Beckman Coulter) would produce a solution at 5X final. For this assay, the positive control was 10 μ*M* methacholine, KHB buffer was used as the negative control and test compounds were assayed at 10 μ*M* (final).
4. Compound plates were lidded after dilution until ready for use.

3.2.3.2. Dye Loading, Cell Fixing, and Labeling

1. Cell plates were washed and loaded with Fluo 3-AM for live cell kinetic imaging, then fixed and antibody labeled using the PlateTrak component of the integrated system (*see* **Note 15**).

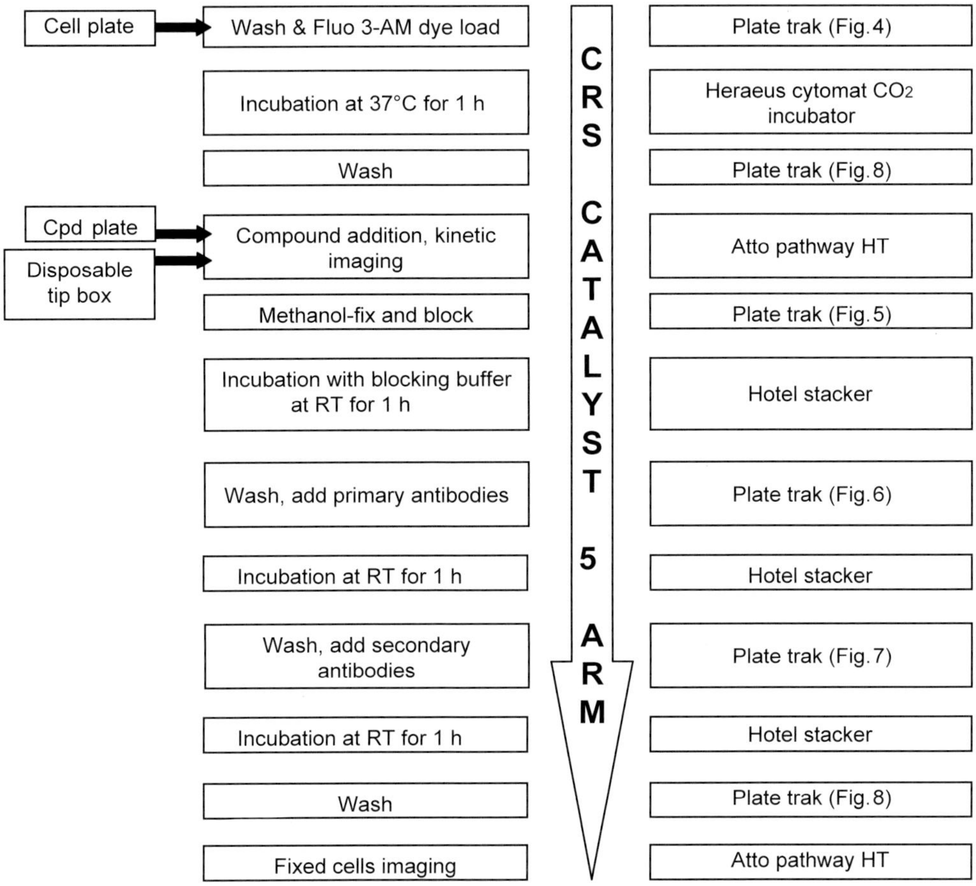

Automated Cell Plating

Fig. 3. Robot protocol for kinetic and endpoint combined assays on PathwayHT.

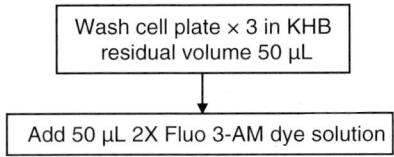

Fig. 4. Dye-loading protocol for live cell kinetic imaging on PathwayHT.

2. The PlateTrak was setup with the cell washer reservoir filled with KHB and the tip wash reservoir filled with dH_2O.
3. 2X Fluo 3-AM dye solution was placed in a foil-wrapped bird-feeder bottle connected to an autoreplenish reservoir (*see* **Note 16**).
4. A bottle of methanol standing in a bucket of ice was connected to another autoreplenish reservoir as described in (**Subheading 3.2.2.**) such that ice-cold methanol was continuously recycled to the reservoir during cell fixing.
5. Separate PlateTrak programs were executed within the robot protocol for dye loading, cell fixing, and blocking, primary and secondary antibody labeling, and cell washing (*see* **Figs. 4–8**, respectively).

3.2.3.3. COMPOUND ADDITION AND KINETIC IMAGING

1. Each cell plate was imaged on the BD (formerly Atto) PathwayHT one well at a time.
2. Single cells were automatically identified and marked as an individual region of interest (ROI) by intensity threshold (*see* **Note 17**).

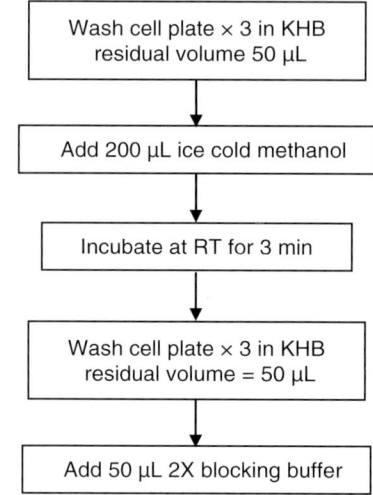

Fig. 5. Flow diagram representing the steps of the PlateTrak program used for fixing and blocking cells.

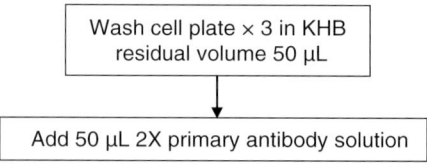

Fig. 6. Primary antibody-labeling procedure.

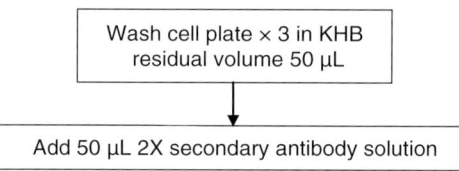

Fig. 7. Secondary antibody-labeling procedure.

Wash cell plate × 3 in KHB residual volume 100 μL

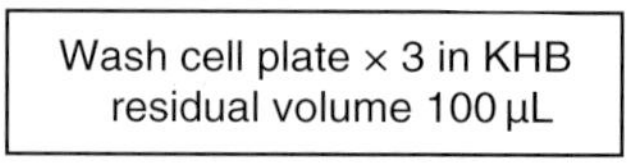

Fig. 8. Cell washing step.

3. Fluorescence in each ROI was then tracked throughout the experiment. Following 10 s of basal read, 25 μL of compounds from the 5X (final) compound plate was added to the well with 100 μL residual volume by the Atto PathwayHT one channel pipetor head, after which images were acquired for a further 60 s to capture the acute response. To ensure quality of data, the methacholine positive control peak and plateau response on each plate must fall within the normal expected amplitude range.

3.2.3.4. POSTHOC FIXED CELL IMAGING

1. After kinetic imaging on the BD PathwayHT, the cell plate was moved from the imaging system manually or by robotic arm and fixed and antibody labeled on the PlateTrak.
2. Monoclonal antiGFAP (Sigma), monoclonal antiTuj1 (Covance, Princeton, NJ) and monoclonal antinestin (Chemicon, Temecula, CA) primary antibodies were used at 1/1000, 1/500 and 1/500 (final), respectively.

3. Alexa-488 conjugated goat antimouse (Molecular Probes) secondary antibody was also used at 1/500 (final).
4. An image was then acquired sequentially in both Hoechst (blue) and Alexa-488 (green) channels of each well on the BD PathwayHT (*see* **Note 18**).
5. These two images were then combined to produce a composite image offline, which represented the group of cells that corresponded to the ROIs initially imaged during the kinetic experiment. This allowed users to identify, for example, the GFAP+ or Tuj1+ or nestin+ cells/ROIs, and extract their kinetic responses from which of the whole population.
6. Each well that was imaged resulted in output of a text file containing the raw fluorescence intensity data for each ROI at each time-point.
7. For analysis, text files from the kinetic experiment were exported into an inhouse Excel template, which normalized each data point to fluorescence intensity change over basal (Microsoft, Seattle, WA).
8. The normalized data for the ROIs representing GFAP+ or Tuj1+ or nestin+ cells were then exported to Graphpad Prism in which the peak response, time to reach peak response and area under curve were calculated and compared between each cell type for each compound treatment (Graphpad Software, San Diego, CA).
9. With the use of automation, assay steps such as cell washing, dye loading, and antibody labeling were standardized. This helped to reduce experimental variability and improve consistency of plate handling.
10. Liquid handling equipment, such as the Multidrop, Multimek, and PlateTrak are tested regularly to make sure their dispense volume and CVs fall within the manufacturer's specifications.

4. Notes

1. The contamination of cell cultures by *Mycoplasma* can remain undetected and yet cause effects on cell growth, cytokine production, DNA/RNA synthesis and so on. *Mycoplasma* can be detected using proprietary kits, and infected cell cultures treated with antibiotics or discarded.
2. It is preferable, wherever possible, to avoid the used of proteolytic enzymes when generating cells for an assay. There are a number of nonenzymatic cell dissociation solutions, which are very effective for cell lines growing as monolayers.
3. Sieves of 100 or 70 μM are an alternative to the final centrifugation step (Beckton Dickinson).
4. Hemocytometers are the traditional method of cell counting and could be used but may introduce more variation. Cell counters such as the Cedex provide a viable cell count per mL and a measure of whether the cell suspension was single cell.
5. It is important to minimize well-to-well and plate-to-plate variation. This can be achieved by steps as simple as using an eight-channel pipet rather than a single channel, and ensuring adequate mixing of the cell suspension. For larger plate numbers we use a Multidrop or similar liquid handler, these in turn can be semiautomated with Twister arms, or plate preparation can be performed on fully automated using systems such as SelecT® (The Automation Partnership, UK). SelecT is a Stäubli arm-controlled cell culture system, which can generate flasks and assay plates using an integrated Cedex cell counter to determine viability and seeding density. We use SelecT for the cell culture of robust, adherent cells, which form the majority of our cell lines: it is able to generate a very consistent, high quality cell output and, in our hands, has proven to be very reliable and capable of working 24/7. The advantages of using an automated system include the consistency of product, the ability to plates cells every day of the week and at specific time points in the day. SelecT is best suited to adherent cell lines such as 293, CHO, or SHSY5Y but it can be used for suspension cultures which can be passaged by dilution. Whatever liquid handling equipment is chosen to generate plates it is very important to ensure it is calibrated regularly: we check the calibration of our Multidrops weekly and change the tubing monthly. In addition once a week we clean the tubing by priming, and leaving for 1 h in a bleach solution to remove protein buildup (trypsin is an acceptable alternative) then washing with copious water.
6. Some cell types tend to settle towards the edges of the plate leading to an uneven density. This is particularly prevalent if the incubator has a vibration. If this occurs, a simple fix is to leave the plates for 1 h on the lab bench after pipeting before placing in the incubator.
7. Laminin was determined to be optimum substrate for this cell type in separate experiments. For alternative cell types the appropriate substrate would need to be investigated.

8. It is possible to prepare the bead suspension at half of the manufacturer's recommended concentration to reduce costs; however, this increases the fragility of the bead layer and the likelihood of assay failure.
9. Cell viability was assessed in parallel experiments, by preparing identical test compounds in standard 96-well laminin coated plates and plating cells at a variety of densities. Following 18 h incubation at 37°C with 5% CO_2 a staining solution comprising 30 µM Hoechst 33342, 5 µM calcein-AM, and 2 µM propidium iodide (all from Molecular Probes) was added to the plates, to indicate nuclei, live cells, and dead cells, respectively. Following a further 30 min incubation cells were imaged and quantified using a Cellomics Arrayscan II and Cellomics Cell Viability algorithm, acquiring nine fields per well with a ×10 objective. With the human neural precursor cells, 2000 cells/well was determined to be a suitable density to use for minimizing contact-inhibition of chemokinesis while maintaining cell viability and convenient image-analysis.
10. Methanol fixing was found to be unsuitable for this assay, as cells were prone to detaching. PFA fixing times were varied between 10 min and 1 h, with no difference in antibody labeling intensity. All solution addition and washing steps were performed manually with a multichannel pipet and a Thermo Wellwash plate washer. These steps could be automated to increase the throughput of the assay, but for sensitive cell types like neural cells, care is required in defining the pipeting and washing heights and speeds. Alexa-594 conjugated phalloidin was investigated as a cytoskeletal marker, but the labelling intensity was much less than using primary, and secondary antibodies as described.
11. The plating medium and cell density are critical factors, which must be determined empirically before performing an assay of this nature. The plating medium has to sustain viable cells, which are capable of responding to neuritogenic stimuli, without itself producing neurite outgrowth. The cell density had to be sufficient to promote cell viability but sparse enough to allow individual neurites to be delineated at the image analysis stage of the assay.
12. Pipeting heights and speeds need to be determined such that compound plates are thoroughly mixed and accurately pipeted to the cell plates. Compound addition to the cell plates, and subsequent mixing, needs to be performed very gently at a sufficient height within the well so that cell viability is not compromised. Cell plate lids were removed for as short a time as possible and sterility during the 48 h incubation was not found to be an issue.
13. The cell washer parameters need to be determined such that adequate washing was performed without disturbing the cells. Similarly, reagent addition needs to be performed gently to avoid disturbing the cells. Suboptimal parameters can sometimes be compensated for when imaging the plates, for example acquiring images from the edges of the wells if wash-off has occurred at the center. The secondary antibody solution is light sensitive and so was either added onto the PlateTrak deck just before it was required, and/or the whole PlateTrak protocol was run in subdued lighting. To reduce evaporation from the top plate in a stacker, and to allow antibody incubations to occur in the dark, an extra plate with blacked-out wells was added to each stacker but excluded from the number of plates entered into the software.
14. Antibody specificity and concentrations were determined in separate experiments. Antibodies which can be used at low concentrations are more suitable to this methodology, because of the relatively large volumes of reagents, which have to be prepared for automated liquid handling. Combining mono and polyclonal antibodies in the same assay allowed the effects of compounds on subpopulations of cells to be studied simultaneously, although the algorithm employed for analysis in this instance meant that each plate had to be read twice to achieve this.
15. The cell washer parameters need to be determined such that adequate washing is performed without disturbing the cells and the correct residual volumes are left in the wells. Pipeting heights and speeds also need to be determined for adequate transfer of reagents. Methanol's liquid property dictates that it required a substantial postaspirate air gap to avoid leakage during transfer. As the dye and secondary antibody solution are light sensitive, the assay was run in subdued lighting.
16. To save reagents in a small assay with only a few cell plates, the ×2 Fluo 3-AM dye and antibody solutions can be added from a 96-well microplate instead of a bird-feeder bottle and reservoirs. Pipeting heights would need to be changed accordingly.
17. Cell density need to be determined to allow adequate segmentation of the cells by the BD PathwayHT software and avoid clusters of cells being identified as one ROI.
18. Antibody concentrations were determined in separate experiments. In this assay, each well was only labeled for one cell type marker, thus three wells received the same compound treatment in the kinetic

experiment and then proceeded to be labeled with different antibodies. However, one could perform dual or triple labeling by using a combination of mono and polyclonal antibodies.

Acknowledgments

We gratefully acknowledge the involvement of other members of the Molecular & Cellular Neurosciences Dept in the development of these protocols, notably Alison Smith, Fran Parry, Matthew Leveridge and Rachel Millard.

References

1. Ramm, P. (2005) Image-based screening: a technology in transition. *Curr. Opin. Biotechnol.* **16,** 41–48.
2. Simpson, P. B. (2005) Getting a handle on neuronal behaviour in culture. *Eur. Pharm. Rev.* **10,** 56–63.
3. Richards, G. R., Millard, R. M., Leveridge, M., Kerby, J. E., and Simpson, P. B. (2004) Quantitative assays of chemotaxis and chemokinesis for human neural cells. *Assay Drug Dev. Technol.* **2,** 465–472.
4. Simpson, P. B., Bacha, J. I., Palfreyman, E. L., Woollacott, A. J., McKernan, R. M., and Kerby, J. (2001) Retinoic acid-evoked differentiation of neuroblastoma cells predominates over growth factor stimulation: an automated image capture and quantitation approach to neuritogenesis. *Anal. Biochem.* **298,** 163–169.
5. Ramos Villullas, I., Smith, A. L., Heavens, R. P., and Simpson, P. B. (2003) Characterisation of a sphingosine 1-phosphate activated Ca^{2+} signaling pathway in human neuroblastoma cells. *J. Neurosci. Res.* **73,** 215–226.
6. Zemanová, L., Schenk, A., Valler, M. J., Nienhaus, G. U., and Heilker, R. (2003) Confocal optics microscopy for biochemical and cellular high-throughput screening. *Drug Discov. Today* **8,** 1085–1093.
7. Chan, G. K. Y., Richards, G. R., Peters, M., and Simpson, P. B. (2005) High content kinetic assays of neuronal signalling implemented in BD PathwayHT *Assay Drug Dev. Technol.* **3,** 623–636.
8. Richards, G. R., Smith, A. J., Cuddon, P., et al. The JAK3 inhibitor WHI-P154 prevents PDGF-evoked process outgrowth in human neural precursor cells. *J. Neurochem.*, in press.

9

Differentiating Primary Human Cells in Rapid-Throughput Discovery Applications

Daniel R. Marshak and Dale E. Greenwalt

Summary

Primary cultures of human cells provide an increasingly important alternative to using virally transformed or otherwise immortalized cell lines or to using cloned cell lines derived from human or animal tumors. Advances in primary cell culture techniques, media formulations, and other reagents have enabled routine culture of primary cells derived from human tissues for biomedical research and drug discovery approaches such as high content screening. That primary cells retain the phenotypic characteristics of the original tissue is one main advantage over immortalized cell lines. However, securing reliable supplies of primary cells on a large scale has been problematic. Here, two primary differentiating cell types, preadipocytes and preosteoclasts, are described to illustrate the utility of commercially produced cell systems in discovery research and rapid-throughput applications.

Key Words: Differentiating cell systems; gene discovery; in vitro toxicology; preadipocytes; preosteoclasts; primary cell cultures; target validation.

1. Introduction

Primary cultures of human cells provide an increasingly important alternative to using virally transformed or otherwise immortalized cell lines or to using cloned cell lines derived from human or animal tumors. Advances in primary cell culture techniques, media formulations, and other reagents have enabled routine culture of primary cells derived from human tissues.

The use of primary cell cultures isolated from human tissues has several distinct advantages over immortalized cell lines. Primary human cell cultures retain many of the phenotypic characteristics of the tissue of origin, including normal physiological functions, and, therefore, can be highly relevant models for gene discovery, target validation, drug testing, and in vitro toxicology. The biological relevance is a key advantage in functional and toxicological studies, because immortalized cell lines might have compromised mechanisms of apoptosis, cell cycle checkpoint controls, or altered proliferative responses *(1–4)*.

The availability of primary cells of consistent viability and performance allows the widespread use of cell systems in research. Key to this is the acquisition of human tissues under proper legal and ethical oversight. It is often difficult and time-consuming for researchers to build the relationships and infrastructure necessary to have tissues available routinely. Access to human tissues is typically sporadic, and the expense of shipping, processing and characterizing the tissues, and ensuing cells can be prohibitive. In addition, primary cells have limited life span, reaching senescence within a finite number of cell divisions *(5–7)*. Therefore, reliable supplies of primary cells serve an important research need. In addition, utilizing multiple donors for samples of tissue to generate independent isolates of primary cells allows researchers to demonstrate consistent responses among individuals *(8)*.

Among the most useful primary cell types are progenitors that are not terminally differentiated in function or morphology *(4,9)*. Such differentiating cell systems often allow for significant expansion of the numbers of cells under mitogenic conditions, followed by functional differentiation under controlled conditions. This process adds another dimension to discovery research using primary cell cultures, allowing researchers to study factors that promote or inhibit differentiation of particular cell types. Two particular primary differentiating human cell types, preadipocytes *(8–11)* and preosteoclasts *(12,13)*, will be described to illustrate the utility of such cell systems in discovery research and rapid-throughput applications.

2. Materials

1. Laminar flow hood.
2. Tissue culture incubator, 5% CO_2, w vacuum aspiration.
3. Cryopreserved primary human preadipocytes (*see* **Note 1**).
4. Cryopreserved primary human osteoclast precursors (*see* **Note 2**).
5. Sterile pipets.
6. Sterile 50-mL polypropylene conical tubes with caps.
7. Cell culture media (*see* **Notes 3–5**).
8. Fetal bovine serum (FBS) (*see* **Note 6**).
9. Insulin (E. Lilly, Indianapolis; IN).
10. Macrophage colony-stimulating factor (M-CSF) (R&D Systems, Minneapolis; MN).
11. Soluble receptor activator of nuclear factor-κβ (RANK) ligand (Chemicon International; Temecula, CA).
12. Dexamethasone (Sigma; St. Louis, MO).
13. 3-Isobutyl-1-methylxanthine (IBMX [Sigma]).
14. Indomethacin (Sigma).
15. Trypan blue.
16. Glutamine.
17. Penicillin.
18. Streptomycin.
19. Adipored (Cambrex; Walkersville, MD; *see* **Note 7**).
20. Time-resolved fluorescence-capable plate reader with injector.
21. 96-well cell culture plates (black walls with clear bottoms).
22. OsteoLyse-collagen assay kit (Cambrex; *see* **Note 8**).
23. Phosphate-buffered saline (PBS).

3. Methods

The methods outlined in this section describe:

1. The culture and differentiation of primary human preadipocytes and their use in an assay of intracellular triglyceride accumulation.
2. The culture and differentiation of primary human osteoclast precursors and their use in an assay of in vitro bone resorption.

3.1. Culture of Primary Human Preadipocytes

1. Prepare Preadipocyte growth medium by adding FBS, L-glutamine, penicillin and streptomycin to Preadipocyte basal medium (The final concentrations of the supplements will be 10%, 2 m*M*, 100 U/mL, and 100 µg/mL, respectively). The medium should be warmed to 37°C before use.
2. Remove a cryovial of cells from liquid nitrogen storage and thaw rapidly in a 37°C water bath. Decontaminate the external surfaces of the cryovial of cells with 70% v/v ethanol or isopropanol.
3. Using a micropipet, gently add the thawed cell suspension to 50 mL of Preadipocyte growth medium.

4. Rinse the cryovial with medium and add the rinse to the cell suspension.
5. Centrifuge at 300g for 10 min at room temperature.
6. When washing the cells, do not attempt to remove too much of the wash. Leave a minimum of 1 mL of wash at the bottom of the tube. If the final cell count is low, some of the pellet might have been removed with the wash.
7. Add 2 or 3 mL of preadipocyte growth medium to the remaining 1 mL of wash and resuspend the pellet of cells. Dilute 20 µL of the cell suspension in 20 µL of 0.4% Trypan blue, do a cell count and determine percentage viability. Recovery should be greater than 90%.
8. Resuspend the preadipocytes at 100,000/mL in preadipocyte growth medium and plate the preadipocytes at 10,000 cells/well of a 96-well cell culture plate in 0.1 mL of preadipocyte growth medium (*see* **Note 9**).
9. Incubate at 37°C, 5% CO_2 and 90% humidity for 24 h. At this point, the preadipocytes should be confluent. Optimal differentiation of preadipocytes is obtained when the cells are confluent before to treatment with differentiation agents.
10. Prepare "2X" Adipocyte differentiation medium by the addition of insulin, dexamethasone, indomethacin, and IBMX to 100 mL of preadipocyte growth medium. After addition of the differentiation medium to the cells, the final concentrations of the differentiation agents, having been diluted 1/1, will be 10 µg/mL insulin, 1 µM dexamethasone, 50 µM indomethacin, and 200 µM isobutyl-methylxanthine.
11. Induce the preadipocytes to begin differentiating into adipocytes with the addition of 0.1 mL of "2X" Adipocyte differentiation medium to each well.
12. If the cells are to be treated with a series of test samples, set up several 24-well dilution plates with the appropriate volume of adipocyte differentiation medium/well and make the required serial dilutions of the test samples. Add 0.1 mL of each different concentration of test sample to wells of the preseeded confluent preadipocytes. Each assay should be done in triplicate.
13. "Control" wells should be setup, which contain **(A)** no added test sample, **(B)** "solvent only" treatments if the test samples were dissolved in solvents such as dimethyl sulfoxide (DMSO), ethanol, and so on and **(C)** 100 µL of preadipocyte growth medium instead of differentiation medium.
14. Culture the cells in a tissue culture incubator in 5% CO_2 and 90% humidity at 37°C. No further additions or medium changes are required for a period of 10 d (*see* **Note 10**).

3.1.1. Assay of Intracellular Triglyceride Accumulation

1. For maximum throughput, the assay will require a fluorimeter capable of reading multiwell plates. The fluorimeter should ideally be equipped with an injector capable of delivering 1 or 5 µL of reagent to each well of 384- and 96-well plates, respectively. The fluorimeter should also have the ability to mix or shake the plate for 1 s immediately on reagent addition.
2. Following the instructions for the fluorimeter to be used, load the injector with Adipored reagent. Program the instrument to inject 5-µL of Adipored reagent/well (96-well plate) and shake/mix the plate immediately after each injection.
3. Remove the tissue culture plate containing differentiated adipocytes from the incubator and cool to room temperature.
4. The culture supernatant should be removed and each well carefully rinsed with 200 µL of PBS. Be extremely careful not to remove cells from the wells during this rinse step (*see* **Note 11**).
5. Fill each well with 200 µL of room temperature PBS.
6. Place the assay plate in the fluorimeter and initiate the Adipored reagent addition program. Make sure that the plate's lid/cover has been removed (*see* **Note 12**).
7. On completion of reagent addition, the assay plate should be incubated at room temperature for 10–15 min. The plate can be removed from the fluorimeter or remain in the instrument during the incubation period.
8. After 10 min, place the plate in the fluorimeter and measure the fluorescence with excitation at 485 nm and emission at 572 nm.
9. The actual readout in relative fluorescence units will vary with the fluorimeter used. However, after 10 d of differentiation, the ratio of relative fluorescence units (differentiated cells) to relative fluorescence units (undifferentiated cells) should exceed 20 (**Fig. 1**).

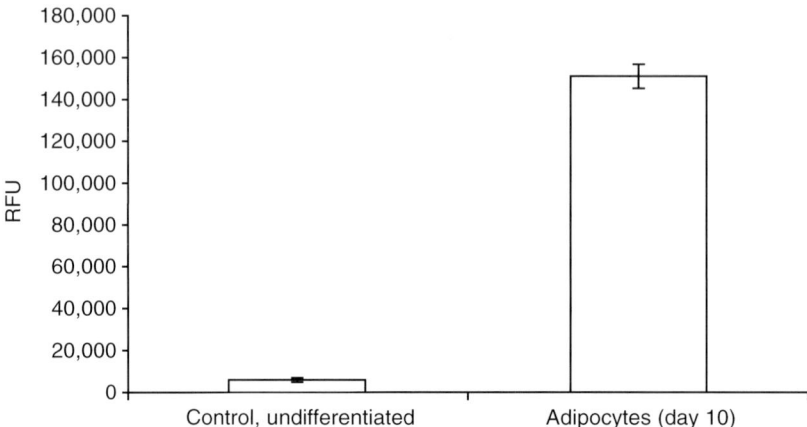

Fig. 1. An Adipored assay of intracellular triglyceride accumulation in differentiated primary human visceral adipocytes.

10. The differentiation of primary human adipocytes can also be assayed by measurement of the expression of a variety of "marker" genes such as those for leptin, AP2 or PPARγ (*see* **ref. *11***; **Note 13**).

3.2. Culture of Primary Human Osteoclast Precursors

1. Prepare Osteoclast Precursor Culture Medium by adding FBS, L-glutamine, penicillin, and streptomycin to the basal medium (The final concentrations of the supplements will be 10%, 2 m*M*, 100 U/mL, and 100 µg/mL, respectively). Warm 100 mL of Osteoclast Precursor growth medium in a 37°C water bath.
2. Quickly but completely thaw the vial of frozen cells in a 37°C water bath. Wipe the outside of the vial with 70% ethanol and aseptically transfer the cell suspension to a 50 mL conical tube. Rinse the cryovial with 1 mL of Osteoclast precursor culture medium. Add the rinse dropwise to the cells while gently swirling the tube (approx 1 min). Slowly add additional medium dropwise to the cells until the total volume is 5 mL, while gently swirling after each addition of several drops of medium (approx 3 min). Slowly bring the volume up to 40 mL by adding 1–2 mL volumes of medium drop wise, while gently swirling after each addition of medium (approx 10 min).
3. Centrifuge the cell suspension at 200*g* at room temperature for 15 min.
4. Carefully remove by pipet all but 1 mL of the wash. Gently resuspend the cell pellet in the remaining medium and count. When washing the cells, do not attempt to remove too much of the wash.
5. Dilute 20 µL of the cell suspension in 20 µL of 0.4% Trypan blue, do a cell count and determine percentage viability. Recovery should be greater than 90%.
6. Prepare the osteoclast precursor differentiation medium by the addition of 60 ng/mL M-CSF and 125 ng/mL soluble RANK ligand (*see* **Note 14**). This medium will be "2X" relative to the two cytokine concentrations; the cytokines will be diluted 1/1 on addition of the test samples (0.1 mL) to the seeded cells. Resuspend the precursors in the differentiation medium at 100,000 cells/mL and seed the cells at 0.1 mL/well of a 96-well OsteoAssay plate.
7. If the cells are to be treated with a series of test samples, set up several 24-well dilution plates with the appropriate volumes of osteoclast precursor Culture medium (without M-CSF or RANK ligand)/well and make the required serial dilutions of the test samples. Add 0.1 mL of each different concentration of test sample to wells of the precursors. "Control" wells should be set up, which contain **(A)** no added test sample (i.e., osteoclast precursor culture medium, without cytokines, only), **(B)** "solvent only" treatments if the test samples were dissolved in solvents such as DMSO, ethanol, and so on, and **(C)** culture medium with M-CSF but no soluble RANK ligand instead of complete differentiation medium; these wells will serve as "undifferentiated" controls. Each assay should be done in triplicate.
8. Incubate the cells at 37°C, in a humidified atmosphere of 5% CO_2 for 7 d (*see* **Note 15**).

10

Use of the CellCard™ System for Analyzing Multiple Cell Types in Parallel

Oren Beske, Daniel Bassoni, and Simon Goldbard

Summary

The CellCard™ system enables the analysis of multiple cell types within a single microtiter well. In doing so, the CellCard system not only determines the effect of an experimental condition on a cell type of interest, but also the relative selectivity of that response across nine other cell types. In addition, this approach of cellular multiplexing is a means of miniaturization without the necessity of microfluidic devices. The standard 96-well plate generates ten 96-well plates of data (or, the equivalent of a 960-well plate). Taken together, the CellCard technology enables multiple cell types to be assayed within a single microtiter well allowing for the simultaneous determination of cellular activity and compound selectivity. This chapter will describe a method by which multiple cell types can be simultaneously assayed for biological parameters of interest.

Key Words: Cell-based assay; CellCard carrier; selectivity.

1. Introduction

The introduction and adoption of "high content" assays has spurred a desire for technologies and assays that provide increasing amounts of biologically relevant data from a single assay reaction, typically from a microtiter well *(1–3)*. Automated high-resolution microscopy combined with sophisticated image analysis algorithms have gone a long way in providing the tools enabling this data to be accessible to the drug discovery industry. For example, the ability of cells to reduce a tetrazolium salt (e.g., MTT) has widely been used as a surrogate for cellular health *(4)*. However, there are many phenotypes and/or mechanisms of toxicity that can be missed or underestimated by this type of an assay. An ability to quantitatively assess the multiple cellular responses associated with toxicity enables the measurement of unique and/or subtle phenotypes. One high content approach in measuring cellular toxicity is to stain cells with a nuclear dye, a mitochondrial dye, and a lysosomal dye. When imaged and quantified these dyes can provide information regarding cell number, nuclear condensation and/or fragmentation, mitochondrial mass, mitochondrial function, lysosomal mass, and lysosomal pH and other relevant parameters. Clearly this approach provides a more comprehensive understanding of the nature of the cellular response to a potential toxin.

This assay strategy has enabled drug discovery scientists to better understand the effects of a potential therapeutic on cells *(5–8)*. Indeed, the main goal of these assay strategies and technologies is to make the drug discovery process more efficient (and informative) thereby reduce the large failure rate currently being experienced with investigational new drugs. It has been postulated that a reason for this high failure rate is the lack of compound selectivity. This selectivity can be viewed as cross reactivity with similar targets, unrelated targets, or other cell types. These off-target effects can lead to a lack of efficacy and/or undesired side effects. The CellCard™ System *(9–11)* was developed to provide a means by which the selectivity of an experimental condition (e.g., treatment with a small molecule) can be robustly and routinely determined by assaying multiple targets and/or cell types within a single microtiter well.

The CellCard System is a microparticle-based technology optimized for cell-based assays *(8)*. These particles, termed CellCard carriers, are designed to incorporate a colored barcode that allows different classes of carriers to be uniquely identified. Therefore, when different cell types are associated (e.g., grown on) with the CellCard carriers, the cell types become uniquely identifiable by virtue of the carrier code. Once the assay is performed and read with the CellCard reader, the cellular data is extracted, associated with the code, and returned to the researcher. It then follows that if multiple cell types are associated with uniquely coded carriers, then the data retrieved will provide insight into the cell type selectivity of the treatment. In addition, because the data is obtained from a single well the cellular data is unequivocally comparable. The selectivity observed is not subject to common sources of assay variation that is experienced with sequential approaches to selectivity determination. When multiple cell types are assayed sequentially, well-to-well or, more importantly, week-to-week variation in reagent stability and/or cellular physiology can lead to a false representation of the selectivity observed. However, obtaining these data points from a single reaction in a single microtiter well will remove these sources of variation.

In this chapter, we describe how the CellCard System can be used to profile the cell type selectivity of antiproliferative compounds. This is illustrated with an example in which 10 cell types were treated with antiproliferative compounds in an eight-point dose–response of in triplicate. A simple fluorescent assay designed to quantify the confluence of the cells (a marker of general cellular toxicity) was performed. In doing so, 40 triplicate dose–response curves were generated in a single 96-well microtiter plate.

2. Materials

1. *CellCard System*: the CellCard System (Vitra Bioscience Inc., Mountain View, CA) is a multi-component, commercially available platform. The system consists of tissue culture consumables, CellCard Carriers, CellCard Dispenser, CellCard Reader, and associated software.
2. *Cells*: all cells were provided by a development partner under confidentiality and are herein anonymized. The cells were grown adherently in 10-cm dishes using standard tissue culture techniques. Briefly, the cells were split using trypsin-EDTA (Invitrogen, Carlsbad, CA) and maintained in RPMI 1640 (Invitrogen) supplemented with 10% fetal bovine serum (Invitrogen), and an antibiotic/antimycotic (Invitrogen) in a humidified 5% CO_2 incubator at 37°C.
3. *Assay reagents*: the cells were stained with 2 µg/mL Hoechst (Invitrogen) in phenol red free, serum free RPMI. The potentially tested anticancer therapeutics was provided by a development partner under confidentiality and is herein anonymized.

3. Methods

3.1. Experiment Design and Plate Layout

A core software component of the CellCard system is the Experiment Manager application. This application provides an interface to design and manage all aspects of the multiplexed experiment to be performed. It stores information (i.e., fluorochromes, analysis parameters, number of plates, and so on) about the assay being run in order to appropriately configure the CellCard Dispenser and Reader. In addition, the software allows the user to log cellular and compound information (i.e., lot number, concentration), that are useful for downstream data analysis can also be managed with this software tool.

3.2. CellCard Preparation

Preparation of a CellCard CellPlex experiment requires standard tissue culture steps in which cells are seeded into wells previously loaded with CellCard carriers (*see* **Note 1**). The carriers are laid onto CellCard ladles, preinserted into wells of standard six-well tissue culture plates, which are designed to facilitate the subsequent mixing of CellCard carrier codes.

Users of the System must first determine an appropriate number of six well plates and CellCard ladles needed for an experiment during the initial experiment design. The Experiment Manager software contains a "ladle calculator" function to determine the total number of ladles, thus determining the number of CellCard carriers that will be required for the experiment. By inference, the number of ladles needed for the experiment will determine the amount of CellCard carriers that will be needed; carriers are packaged in vials containing a measured amount of one class (code) of carriers that will be transferred to a ladle. However, as a rule of thumb, 10 vials of CellCard carriers and 10 CellCard ladles are required for each 96-well assay plate. Therefore, if 10 cell types are to be assayed, a single vial of 10 separate carrier codes is sufficient; if five cell types are used in the CellPlex experiment, then two vials each of five carrier codes are required per 96-well plate.

Once the CellCard carrier and ladle requirements are established, the user is ready to begin the experiment. The disposable CellCard ladles are first placed in the wells of dry six-well tissue culture plates. Then 2.5 mL of complete media is added to each ladle-containing well. The ladles are now ready for the deposition of the carriers.

One vial of carriers will be deposited onto each ladle (*see* **Note 2**). It is impractical to transfer dry carriers to the ladles. Therefore, carrier transfer is facilitated by adding 1 mL of the complete media to the CellCard carrier vials. A sterile, disposable transfer pipet is used to aspirate the carriers out of the vial and then deposit them onto the ladles by simply touching the tip of the transfer pipette to the media above the ladles. Slight positive pressure and capillary action will result in the majority of the carriers falling out of the pipet and onto the ladle. Gentle "washing" of media through the pipet will remove carriers remaining in the pipette tip. The carriers are dispersed into a monolayer on the ladles by mounting the six-well plates on the CellCard Disperser and shaking for 45 s at 110 rpm. Dispersing the carriers is a critical step to ensure that a maximum number of carriers will be exposed to the cells that will be added to the wells.

3.3. Tissue Culture

The cell lines to be multiplexed are maintained using standard tissue culture protocols in the appropriate media. Cells are removed from plates; pellet in conical centrifuge tubes, resuspended and accurately counted before plating them on the CellCard carriers. Typically, $1–5 \times 10^5$ cells are seeded into the wells containing the carriers. The actual number is dependent on cell doubling time, total incubation time of cells on carriers, and desired final cell confluence. A range of seeding densities should be tested to ensure the proper confluence of cells for the particular assay that is to be performed. The cells should be allowed to adhere (under standard incubation conditions) for an appropriate amount of time to ensure robust adhesion to the particles. The rate of adhesion and cell spreading is cell-type specific and can range from 5 to 18 h. We recommend allowing the cells to adhere, spread, and initiate their cell cycle during an overnight incubation.

3.4. CellCard Carrier Mixing and Dispensing Into 96-Well Format

After the cells have adhered to the carriers, they are ready for mixing (*see* **Note 3**). This is a simple process that is started by attaching a purpose-built funnel to a standard 15-mL conical tube. The funnel-tube apparatus is filled with approx 55 mL media and the carriers are transferred to the funnel by simply gripping the peg-shaped handles of the CellCard ladles with forceps, lifting the ladles out of their wells, submerging the ladles in the funnel and allowing the CellCard carriers to fall off the ladle into the conical tube. Up to 40 ladles (or vials) worth of carriers, enough for four 96-well assay plates can be transferred to one conical tube.

After all carriers are transferred to the 15 mL tube, the excess media from the funnel is aspirated leaving the 15 mL tube full to its brim. The funnel is detached from the tube and capped. The carriers are then thoroughly mixed with a single 360° inversion of the conical tube. After mixing, the media is then aspirated from the tube down to the top of the stack of carriers to remove cells that might have been loosened from the carriers by the mixing process. The tube is refilled with fresh complete media to the 6 mL mark. The CellCard carriers are now ready for dispensing into assay plates.

CellCard carrier dispensing is accomplished by a liquid handling robot, called the CellCard Dispenser, designed to transfer CellCard carriers to 96-well assay plates. The Dispenser includes custom built pipet tips that are specially designed to aspirate a metered amount of carriers to be deposited into the 96-well assay plates. A dispense run is set up by mounting the 15-mL tube containing the CellCard carriers and the appropriate number of 96-well plates onto the Dispenser. A 50-mL tube containing a reservoir of media is also mounted on the Dispenser. The start cycle of the Dispenser requires the user to adjust the starting height of the dispensing tips through the software controller interface. From there, the Dispenser will automatically transfer carriers from the 15-mL conical tube to the assay plates. On completion of carrier dispensing, the carriers are then dispersed into a monolayer via the CellCard Disperser.

3.5. Compound Addition

Excess media is removed from each assay well with a wand aspirator or a plate washer to ensure that the same residual volume is present in all wells. We recommend leaving 50 µL of residual media in each well to minimize disturbance of the CellCard carriers while adding compound. The compound to be added is made up at 2X (twice the desired final) concentration, using the appropriate media, and an equal volume (50 µL in this case) of the 2X solution to each well to create a 1X final assay concentration. After the compounds have been added to each of the wells return the plates to the cell culture incubator for the appropriate incubation time. This incubation time is assay dependent and can range from 30 min for a signaling (i.e., nuclear translocation) assay to multiple days for a proliferation assay.

3.6. Assay Staining Protocols

In general, only minor alterations in standard assay protocols needs to be made when using the CellCard System. Generally, staining reagent concentrations and incubation times will be the same for assays run with or without CellCard carriers. For the Calcein AM/Hoechst assay presented here, the assay plate wells are first aspirated to 30 µL. The staining solution is made by mixing 4.5 µL of Calcein AM (4 mM stock) and 8 µL Hoechst (1 mg/mL stock) in 8 mL phenol red-free RPMI 1640. Add 75 µL of the staining solution to each assay well. Incubate the plate for 30 min at 37°C, 5% CO_2. At the completion of incubation, aspirate the wells to 50 µL. Top off the wells with approx 300 µL phenol red-free RPMI 1640 (*see* **Note 4**).

3.7. CellCard Reading

Before scanning the plates with the CellCard Reader, the appropriate plate description file (generated in the Experiment Manager application) must be loaded. This will configure the Reader to address the appropriate wells and acquire images with the appropriate assay specific fluorochromes.

Some preparation of the assay plates will also be necessary. The nature of the carrier recognition algorithm in the Image Analyzer software necessitates a noncolored solution in the assay wells. Any colored solutions must be washed out of the wells with solution such as 1X PBS (for fixed assays) or phenol red-free media (for live assays). In order to eliminate any shadowing within the bright field images, which can affect carrier code recognition, the wells must be completely topped off with the appropriate solution such that any meniscus is eliminated from each well.

After the microtiter plate has been placed in the Reader (either manually or through a robot) the image acquisition parameters must be set. The parameters for acquiring the bright field images are automatically set by the reader without user intervention. This will ensure that the colored bands of the CellCard codes will have the hues required by the Image Analysis software for robust code recognition. The parameters for fluorescent image acquisition are adjusted by the user. A detailed protocol for this is provided in the User's Guide. Briefly, exposure, gain, and offset parameters are set to produce high signal, low background images that can be robustly processed by the System's image analysis algorithms to provide the biological data. Once the acquisition parameters are set, the plate is scanned by the Reader and the images are saved to an appropriate file directory designated during the design of the experiment in Experiment Manager. Acquisition parameters are saved by the Reader software to alleviate any need to set exposure parameters for the remaining assay plates of the experiment.

3.8. Image Analysis and Data Visualization

To extract the cell type specific data, the images acquired by the CellCard Reader must be processed by a series of image analysis algorithms. The software will read in assay specific default image analysis parameters, including minimum pixel intensity thresholds, before proceeding with the analysis. The Image Analyzer software will allow the user to examine the effect of the threshold parameters on the analysis of the images. If necessary, the user can adjust these thresholds to ensure that the image analysis algorithms accurately extract biologically relevant data apparent in the images. We recommend that this threshold analysis be run on positive and/or negative control wells, with known and/or expected biological responses, before analyzing the entire experiment. The CellCard System User's Guide provides a detailed description of how to set proper image analysis parameters. Once appropriate analysis parameters are set, the image analysis of the entire experiment, typically a set of plates with the same assay, can be performed. The Image Analyzer will automatically analyze every well of every plate in the experiment and generate a fully annotated data table.

The data tables generated by the Image Analyzer software are saved as tab-delimited flat files. The files contain data generated from a variety of measures, including measures of stained area, spots, intensity of stains, and the like. These data files can then be opened and visualized in the Vitra Bioscience Data Analyzer software or other standard data visualization applications (i.e., Spotfire). The flat file structure also allows the data to be parsed and uploaded into an enterprise database for further analysis and storage.

3.9. Example Data

In this example 10 cell types were maintained using standard tissue culture protocols and seeded onto CellCard carriers as described earlier. The cells were then exposed to a dose–response of a small molecule compound, a known toxin and potential anti-cancer therapeutic, for 2 h. The compound was subsequently removed. The wells were then washed and fresh media (without the compound) was added to the wells for a 48 h chase period. At the end of this period the cells were stained with the nuclear dye and subsequently scanned and analyzed using the CellCard System (**Fig. 1**).

Compound A is known to be toxic and, therefore, was predicted to induce a dose-dependent decrease in the signal generated by the nuclear dye. Indeed that was the case. As shown in **Fig. 3B**, the relative number of cells decreased as the dose of the compound was increased. Interestingly, the 10 cell types partitioned into three classes of response profiles. The class-1 response is described as a gradual dose–response curve wherein cells begin to respond to the compound at a relatively low dose. This response gradually increases as a function of dose but does not reach a maximal response within the dose range tested (no right side plateau). The classes-2 and -3 responses, characterizing three and five cell types, respectively, result in sigmoid shaped curves

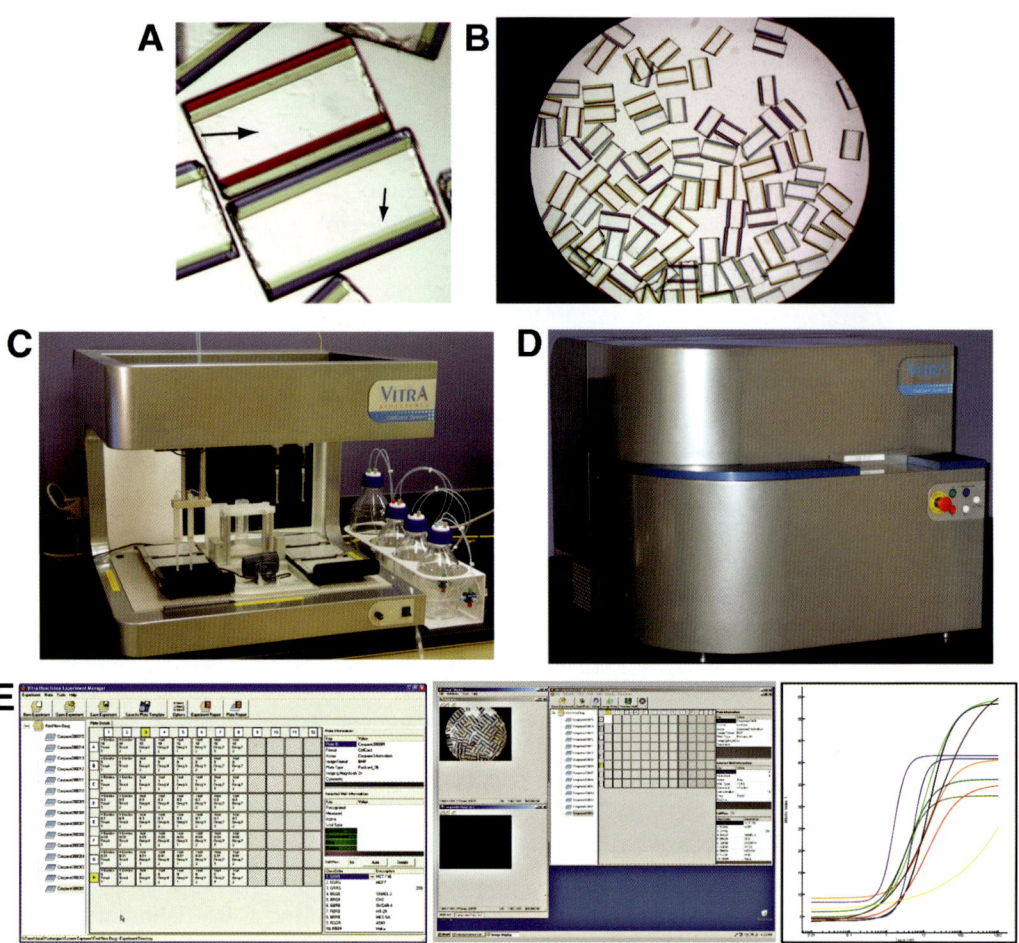

Fig. 1. System overview. (**A**) Image of CellCard carriers showing the coding bands (small arrow) and optically clear data read-out (big arrow) sections. (**B**) An image of a single well from a 96-well microtiter showing approx 100 CellCard carriers dispersed throughout the well. (**C**) The CellCard dispenser. It is designed to transfer approx 100 CellCards from a 15-mL conical tube into the wells of the microtiter assay plate. (**D**) The CellCard Reader. A CCD camera based reading device that acquires all the requisite bright field and fluorescent images to both decode the carrier codes as well as extract biological data. (**E**) A subset of the CellCard System suite of software. This figure shows screen shots from the Experiment Manager, CellCard Reader, and Data Analyzer software applications.

with plateaus on the low high dose ranges connected by a steep transition area. That is, at low doses there is no response of these cells to the compound, but, once a critical concentration is used, it does not require a large dose increase of the compound to elicit a maximal response (the high dose plateau). Although the mechanisms behind these different responses are not understood, by using the CellCard System multiple sources of assay artifact can be removed as potential explanations. For example, if the 10 cell types were assayed sequentially over a period of days to weeks, compound stability issues with multiple freeze-thaw cycles may result in the different cellular responses observed. By implementing the CellCard System, each of the cell

Fig. 2. *(Opposite page)* Schematic of the CellPlex assay workflow. CellCard carriers are first placed into 6-well format and cell types seeded onto the carriers. After mixing the carriers and attached cells they are dispensed into the 96-well microtiter plate, subjected to the experimental treatment of interest, and stained for the biological parameters of interest. Finally, the plates are imaged, analyzed, and the data plotted.

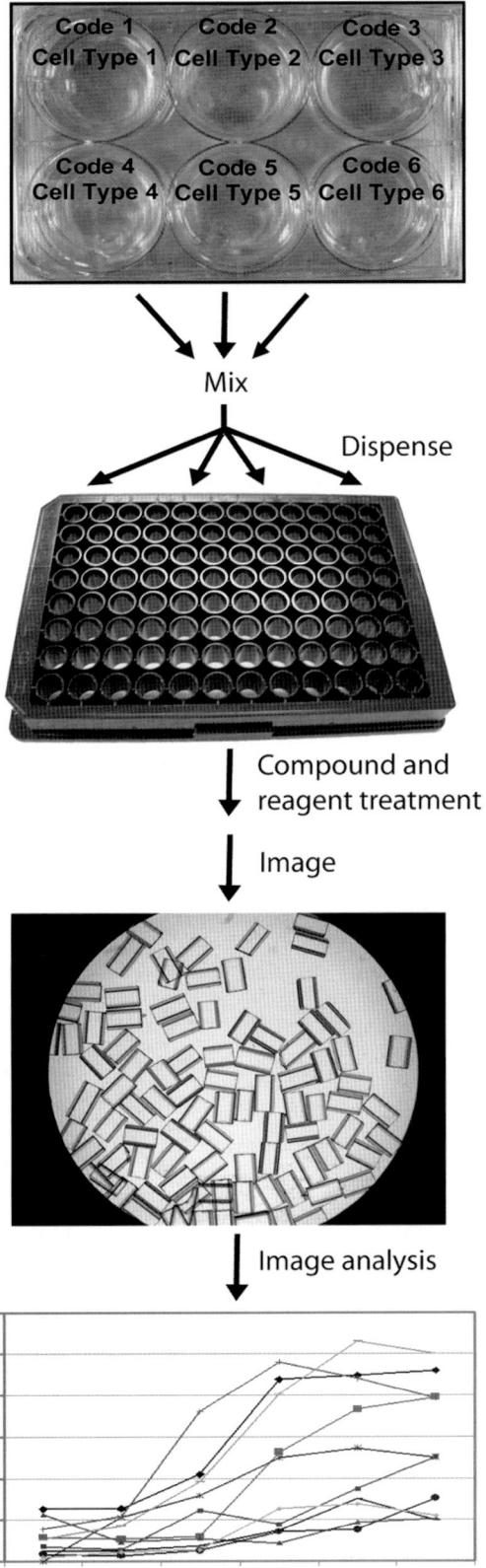

Fig. 2.

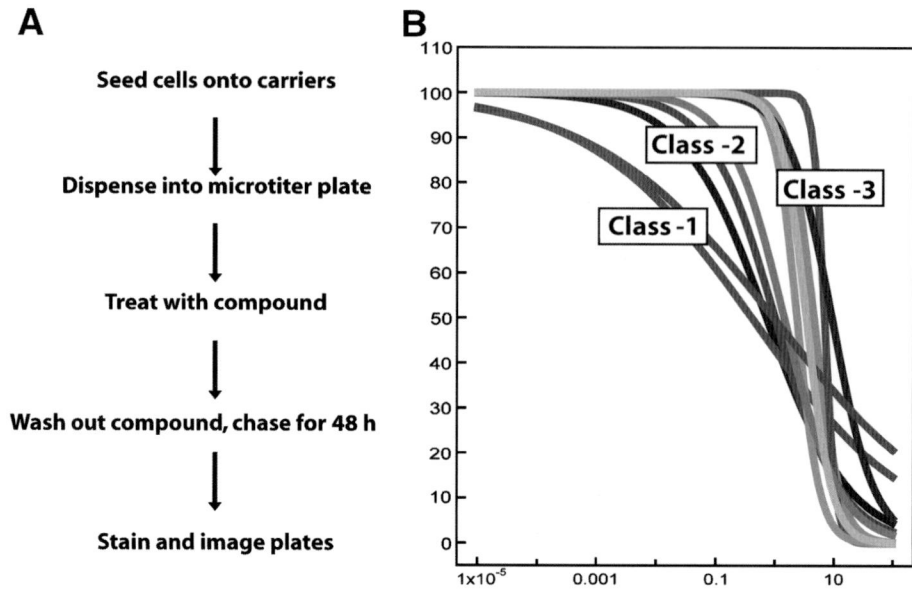

Fig. 3. Case study data example. (**A**) Outline of the protocol used. Ten cell types were introduced into the CellPlex experiment as described in the methods. They were treated with compound for 2 h, washed, then incubated in for another 48 h with standard growth media. (**B**) Dose–response of 10 cell types to a cytotoxic compound. These 10 cell types tended to separate into three distinct classes. The class-1 response in left shifted and shallow. The classes-2 and -3 responses are much more steep with the class-2 response being more potent.

Table 1
A Table Showing Examples of Cell Types Used and Assays Performed on the Cellcard System

Example cell types assayed on CellCard carriers	Example assays performed on CellCard carriers
A549, A431, MCF7, COLO205, SKMEL28, SW620, OVCAR3, OVCAR5, HCT116, HT29, T47D, LOVOS, HEK-293, CHO, HELA, COS, primary HUVEC, primary preadipocytes, primary renal tubule, and others	• β-lactamase reporter gene • Proliferation—cell count, BrdU incorporation, mitotic index • Toxicity—TUNEL, Caspase-3 activation, propidium iodide, C alcein-AM • Adipocyte differentiation—lipid accumulation and others

types were assayed simultaneously within a single well. Therefore, even if there were assay errors introduced by reagent stability issues, these cell types, in this assay, on this day, experienced identical assay conditions. The relative responses between the cell types within the well is unequivocal; at the low doses in which the class-1 group of cells show a slight but significant response and the classes-2 and -3 cells show no response, this result must be cell-type dependent.

The CellCard System provides a means by which ten unique cellular conditions can be assayed simultaneously. These conditions are typically represented by 10 different cell types (**Table 1**) but could also represent seeding densities, extra-cellular matrix coatings, and the like. When performing CellPlex assays, significant miniaturization is realized within the 96-well plate. That is, when assaying 10 cell types, the plate effectively becomes a 960-well plate. Finally, the number of cells analyzed per data point is significantly reduced. Moving forward, we have been exploring assays in which it is desirable to increase the number of data points derived from rare and/or valuable cells (i.e., human primary cells) *(12)*. We believe that this will result in a very powerful application of the CellCard System wherein patient samples could be assayed along side a clinical trial *(13,14)* leading to the development of in-vitro surrogate markers of compound efficacy and potential theranostic indicators.

4. Notes

1. The determination of which cell lines to be used in a CellPlex assay should follow a few guidelines. First of all, because the assay will be performed simultaneously across multiple cell types, the kinetics of the response to be measured should be similar. For example, if measuring activity across ten G protein-coupled receptors (GPCRs) simultaneously using a reporter gene assay, one should choose GPCRs with similar activation kinetics to be included in the same well *(10)*. An incubation time of 3 h could be chosen because the primary receptor for which the compound was identified showed a robust response within that time. However, if the maximal reporter gene expression for some of the cell lines in the CellPlexed experiment is 6 h, a lack of activity on that cell line would be interpreted as compound inactivity although the compound could be active albeit with slower kinetics. Similarly, when assessing experimental effects on the cell cycle the cells to be included in the same well should have similar doubling times. This will ensure that within the experimental incubation time, on average, the cell types will have experienced comparable cell cycle passage.
2. It is recommended only to dispense a single vial of carriers onto each ladle. However, with large experiments, it is advantageous to limit the total number of ladles and six-well plates that need to be handled. When placing more than a single vial of carriers onto a single ladle the probability of the carriers overlapping increases. If the carriers are overlapped when the cells are seeded, those on the underside will not be available for the cells to adhere to. When taken through to the assay output, for assay measures that depend on the overall cellular density, the data from CellCard carriers that were never seeded with cells often are identified as outliers. For those ratiometric assay measures that are not dependent on cellular density, it is worth noting that CellPlex assays derive data from very few cells (about 500/data point) and any process that decreases the size of the cellular population to be analyzed is more likely to increase the variance of the assay.
3. The generation of the array of cell types to be assayed requires that the cells (and CellCard carriers) be mixed. If this step is omitted, the carriers will be layered in the conical tube from which the dispensing into the 96-microtiter plate performed. Therefore, the last code put into the conical tube (the resulting top layer) will be the only carrier type dispensed in to the first few wells and the first code transferred to the conical tube (the bottom layer) will be the only code in the last few wells. Assuming that the experiment contains ten cell types, nine of these 10 cell types will be absent from a majority of the wells. The mixing procedure is simple requiring a single inversion of the conical tube (*see* **Subheading 3.4.**). If insufficient mixing is suspected, it is recommended that a second inversion of the tube be performed.
4. All cell washes must be done as gently as possible to limit CellCard carrier agitation or loss. In addition, wells should not be aspirated dry as carriers might be aspirated into the aspiration device resulting in a potential clog. The carriers also have a tendency to float when liquid is reintroduced to a well that has been aspirated dry. We recommend leaving 50 µL in each well between wash steps to minimize carrier disturbance or loss. For postwash aspiration steps, we recommend leaving more than 20 µL in each well; live assays (e.g., cells are not fixed) may require more than 30 µL. A 96-well plate washers can be used routinely, provided that they dispense gently enough to minimize carrier agitation; we recommend using a Tecan PW384 or a BioTek ELX405 Select CW for this purpose.

References

1. Abraham, V. C., Taylor, D. L., Haskins, J. R. (2004) High content screening applied to large-scale cell biology. *Trends Biotechnol.* **22(1),** 15–22.
2. Giuliano, K. A., Haskins, J. R., Taylor, D. L. (2003) Advances in high content screening for drug discovery. *Assay Drug Dev. Technol.* **1(4),** 565–577.
3. Liptrot, C. (2001) High content screening—from cells to data to knowledge. *Drug Discov. Today* **6(16),** 832–834.
4. Arnould, R., Dubois, J., Abikhalil, F., et al. (1990) Comparison of two cytotoxicity assays—tetrazolium derivative reduction (MTT) and tritiated thymidine uptake—on three malignant mouse cell lines using chemotherapeutic agents and investigational drugs. *Anticancer Res.* **10(1),** 145–154.

5. Rausch, O. (2005) Use of high content analysis for compound screening and target selection. *IDrugs* **8(7),** 573–577.
6. Perlman, Z. E., Mitchison, T. J., Mayer, T. U. (2005) High content screening and profiling of drug activity in an automated centrosome-duplication assay. *Chembiochem.* **6(2),** 218.
7. Borchert, K. M., Galvin, R. J., Frolik, C. A. (2005) High content screening assay for activators of the Wnt/Fzd pathway in primary human cells. *Assay Drug Dev. Technol.* **3(2),** 133–141.
8. Vogt, A., Cooley, K. A., Brisson, M., Tarpley, M. G., Wipf, P., and Lazo, J. S. (2003) Cell-active dual specificity phosphatase inhibitors identified by high content screening. *Chem. Biol.* **10(8),** 733–742.
9. Beske, O., Guo, J., Li, J., et al. (2004) A novel encoded particle technology that enables simultaneous interrogation of multiple cell types. *J. Biomol. Screen* **9(3),** 173–185.
10. Beske, O. A. G. (2002) High-thgoughput cell analysis using multiplexed array technologies. *Drug Discov. Today* **7(18),** S131–S135.
11. Beske, O., Goldbard, S., and Turpin, P. (2005) The CellCard System; a novel approach to assessing compound selectivity for lead prioritization. *Combinatorial Chemistry High-Throughput Screen* **8(4),** 293–299.
12. Borchert, K. M., Galvin, R. J., Frolik, C. A., et al. (2005) High content screening assay for activators of the Wnt/Fzd pathway in primary human cells. *Assay Drug Dev. Technol.* **3(2),** 133–141.
13. Di Nicolantonio, F., Knight, L. A., Whitehouse, P. A., et al. (2004). The ex vivo characterization of XR5944 (MLN944) against a panel of human clinical tumor samples. *Mol. Cancer Ther.* **3(12),** 1631–1637.
14. Terashima, M., Hayashi, K., Fukushima, M., et al. (1996). Drug sensitivity testing for clinical samples from oesophageal cancer using adhesive tumour cell culture system. *Br. J. Cancer* **74(1),** 73–77.

III

REAGENTS

11

Reagents to Measure and Manipulate Cell Functions

Kenneth A. Giuliano, D. Lansing Taylor, and Alan S. Waggoner

Summary

Reagents that are used as part of a discovery platform for the measurement and manipulation of cell functions are at the heart of single and multiplexed high content screening assays. Measurement reagents include physiological indicators, immunoreagents, fluorescent analogs of macromolecules, positional biosensors, and fluorescent protein biosensors. Recent developments in reagents that manipulate specific cell functions including small inhibitory RNAs, caged peptides, proteins, and RNAs, and gene switches complement measurement reagents, especially when both classes of reagents are used in the same living cells. The use measurement and manipulation reagents in multiplexed high content screening assays promises to enable a systems cell biology approach to drug discovery and biomedical research.

Key Words: Biosensors; cell-based assays; compound screening; fluorescent probes; photochemistry; RNAi.

1. Introduction

The cell is the first level of biological organization that exhibits life. A major challenge in the postgenomic era is to fully understand the functional dynamics of living cells. Detailed cellular knowledge is essential for understanding the normal development and function of organisms, the engineering of therapeutics for disease and the creation of novel diagnostics tools. With the human genome mostly in hand and strong activity in identifying and characterizing the proteome, there is a new wave of effort for mapping all the regulatory pathways in a cell, sorting out the mechanisms of regulation, and defining their roles in cell functions. This requires an understanding of the activities of organelles, metabolites, protein–protein interactions, protein modification, protein translocation, conformational changes, lipids, nucleic acids, and carbohydrates, as well as feedback, feedforward mechanisms. Activity is intense in the pharmaceutical industry because these regulatory pathways are obvious targets for drug discovery and a systems biology knowledge base is desired. There are many approaches in use for trying to figure out how cells work. This chapter will focus on the cellular reagent tools that fluorescence detection technologies offer for measurement, as well as some of the technologies that can be used for manipulations.

Fluorescent and luminescent labels have significantly replaced radioactive isotopes for screening of drug candidates and basic biomedical research. Many of these assays still require isolation of cell membranes or intracellular components that can be used to quantify binding of luminescent ligands. In recent years, intact cell-based assays have grown in importance. Although many cell-based assays require simple quantification of the amount or location of labeled macromolecules in cells, many other fluorescent cell-based assays are very sophisticated and require measurement of energy transfer changes, total internal reflectance excitation, confocal,

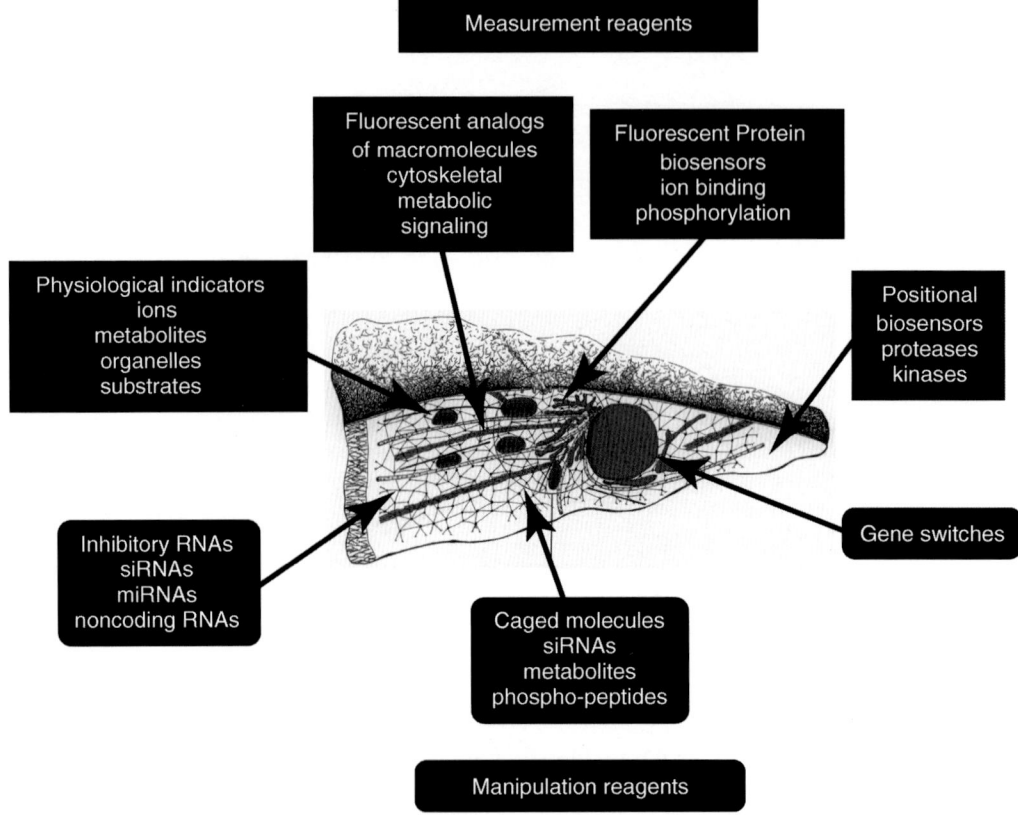

Fig. 1. Reagents to measure and manipulate cell functions. A large and growing repertoire of reagents is being used to measure temporal and spatial processes in living cells that involve ions, metabolites, macromolecules, and organelles. Moreover, new reagents continue to be developed that manipulate specific processes in living cells. A systems cell biology approach combines multiple measurement and manipulation reagents into multiplexed assays that can be used to build new cellular knowledge.

two-photon, or fluorescence correlation spectroscopy. The application of these methods has been widely reviewed *(1–8)*.

Our emphasis here is the evolution and latest developments for assays with fixed and living cells. Reagents that are used to measure and manipulate one or more specific cellular processes in the same living cells have become important tools in a systems cell biology approach to drug discovery and biomedical research (**Fig. 1**). Besides discussing the principles and applications of the fluorescent probes to measure and reagents to manipulate living cells, we will also point out technologies that remain weak and need improvement. This sets the stage for contemplation about what the future might hold.

2. The Evolution of Fluorescent Labels and Probes in Biomedical Research and HCS

More than a half century has passed as the intrinsic fluorescence of proteins and artificially attached fluorescent groups were initially used to measure the conformational dynamics associated with protein activity in vitro; early reviews have covered these accomplishments *(9,10)*. It is from these pioneering studies that a wide range of protein as well as small molecule-based physiological indicator fluorescent reagents evolved to measure cellular processes in living cells. The major developments in fluorescent labels and probes for cell-based assays are briefly

Table 1
Evolution of Reagents for the Measurement of Cell Functions

Year	Fluorescent reagent(s)	Cell function	References
1953	Intrinsic protein fluorescence and artificially attached fluorescent labels	Conformation dynamics associated with protein activity	*9,154,155*
1962	Fluorescently labeled immunoreagents	Antigen localization in tissue sections and tissue culture preparations	*10,19*
1970	Fluorescent membrane analogs	Conformational changes in biological membranes	*156*
1974	Small molecule probes of cellular physiology: fluorescent potential sensitive probes	Changes in cell and organelle membrane potentials	*69,70,157*
1978	FRET-based probes	Protein conformational changes associated with activity	*158–160*
1980	Fluorescent analogs of intracellular proteins	Structural and functional dynamics of the cytoskeleton and other constituents	*56,76,161,162*
1982	Multicolor immunoreagents for flow cytometry	Two-color analysis of lymphocytes with a single color laser	*20*
1982	Fluorescent probes of intracellular ion concentration	Intracelluar free calcium, magnesium, and pH	*163–165*
1987	Five-parameter labeling for live cell kinetic multimode microscopy	Nuclei, mitochondria, endosomes, actin-cytoskeleton, and cell volume	*166*
1989	Introduction of bright water-soluble fluorescent labeling reagents	Multicolor analysis of cellular constituents	*21–24*
1990	Fluorescent protein biosensors	Temporal and spatial measurements of calmodulin activity in living cells	*74,80*
1994	Applications of autofluorescent proteins as tags	Temporal and spatial measurements of endogenously synthesized fluorescent protein analogs in living cells	*102,167,168*
1997	Multiplexed HCS	High content measurements of multiple apoptosis events and kinetic analysis of glucocorticoid receptor translocation after drug treatments	*13*

summarized in **Table 1** and further discussed below. It is essential to understand that fluorescent reagents coevolved with advances in optical detection systems, including microscopy. Semiautomated microscopes with powerful software, multicolor filter sets, and environment controls have been essential for progress *(11,12)*. The development of high content screening (HCS) allowed the application of a high throughput approach to cell biology *(13)*. Reagents and imaging systems are interdependent and progress is optimized by a systems approach to design.

2.1. There are Essentially Three Groups of Fluorescent Reagents

First, covalent labels that allow localization and quantification of biomolecules. This class includes organic dyes with reactive groups. Fluorescein and rhodamine were developed early and became long-time standards, but the cyanine dye labels and then the Alexa dyes brought significant improvements in water solubility, spectral selection, photostability, and brightness. New labels seem to be reported monthly. The fluorescent proteins, such as green fluorescent protein (GFP), and other genetic labeling techniques like FlAsH (*see* **Table 2** and Chapters 14–16) would also have to be in this category because covalent bonds are formed between the fluorescent species and the labeled protein.

Second, noncovalent labeling reagents include membrane-associating dyes, like diI, and the DNA content probes like DAPI, Hoechst, and the intercalating dyes like propidium iodide. Their specificity depends on a noncovalent but high affinity for the target site. Nevertheless, these have found many applications and some are very widely used in cell biology, drug discovery assays, and diagnostics. Third, the fluorescent indicator dyes are designed to be sensitive to some local environment property such as pH, calcium level, membrane potential, protein conformation, or molecular proximity. These are often much more complicated in their use. They must be delivered to the cytoplasm, membrane, or specific compartment within the cell. Sometimes the indicator dye can also be intentionally localized to a target region of the cell by covalent labeling or by designing a specific noncovalent affinity for the site.

We will elaborate on the uses of these dyes in cell based assays below, but it is worth a brief discussion of the important properties a fluorescent dye must have to have utility in such assays. The chemists who create new fluorescent probes must be intimately aware of following principles and properties in order to synthesize useful probes *(11)*.

First, fluorescence brightness that results from high absorbancy and high quantum yield. Molar extinction coefficients above 80,000 are characteristic of the best fluorescent labels. There are many mechanisms by which a molecule might have low quantum yield and as a result most light absorbing dyes are not fluorescent. However, most useful labels have quantum yields greater than 0.1 and most are in the 0.3 to 0.8 range (one is the maximum possible). The quantum yields of many longer wavelength fluorophores are substantially lower than those of the best dyes in the visible and near UV region of the spectrum, but during detection this is compensated by reduced background interference at longer wavelengths.

Second, photodegradation (or photobleaching). Currently available fluorophores are all subject to photodegradation. This susceptibility increases at longer absorption wavelengths. Photobleaching is one of the most serious limitations of conventional fluorescent probes in quantitative imaging systems. High efficiency (numerical aperture) collection objectives and high sensitivity cameras in imaging systems are essential so that fluorescence signals can be acquired quickly before bleaching affects quantification. Photobleaching is more pronounced for dyes having longer wavelengths.

Third, chemical stability. On the whole, ultraviolet (UV) and visible dyes (coumarins, xanthenes, and cyanines) are relatively stable to acidic and basic conditions and to a range of redox environments. This simplifies the use of these fluorescent probes under many experimental conditions. These fluorophores can be incorporated by synthetic modification into useful reagents such as amidites, nucleotides, lipid analogs, protein analogs, drug analogs, and so on. However, the chemical stability of longer wavelength emitting fluorophores is lower. Instability of fluorophores in storage, stock solutions in room light or during sample handling can limit the utility and must be controlled.

Fourth, phototoxicity. Dye-sensitized phototoxicity to cells and tissues remains a significant problem in some assays, in which cells are under continuous illumination for more than a few seconds. For example, when forming image stacks either by stepping through a specimen followed by deconvolution or by confocal methods (single or multiphoton), or excessive illumination in live cell HCS studies, toxicity is frequently observed as indicated by the inability of cells to initiate or complete mitosis, membrane blebbing, and nuclear degradation. Production of singlet oxygen and its products is the main cause of phototoxicity. Care must be exercised in using the minimal dose of irradiation to generate the data.

Fifth, nonspecific binding. The more hydrophobic fluorophores are notorious for nonspecifically sticking to plastics and nonspecific cellular structures. Much progress has been made with the cyanine and Alexa dyes that contain sulfonic acid groups on the rings that reduce this complication.

In another example, mRNA molecules encoding a mutant glutamate receptor with an engineered reactive cysteine were microinjected in to oocytes *(124)*. After protein synthesis off of the microinjected mRNA was allowed to occur, the newly synthesized transmembrane glutamate receptors were labeled in a site-specific manner by reaction of the oocytes with a cysteine-reactive Alexa 546 fluorescent dye. Once labeling was complete, changes in the total fluorescence signal emanating from the biosensor was used to measure the kinetic conformational changes in the receptor on glutamate binding. Thus, these two recent examples of the many synthetic fluorescent protein biosensors produced over the last several years indicate how important it will be to rapidly solve issues of bulk loading of fluorescent protein biosensors into large populations of target cells for large scale HCS.

4.4.2. FRET-Based Fluorescent Protein Biosensors Constructed With Pairs of Autofluorescent Proteins

In the class of fluorescent protein biosensors of autofluorescent proteins, those that use FRET as the basis of their sensing and reporting capabilities are the most common *(125)*. These biosensors generally contain two complementary autofluorescent proteins whose FRET efficiency depends on the conformational state of the biosensor within living cells. In a study in which the quantification of FRET efficiency was carefully measured, it was reported that an optimal pair of autofluorescent proteins for FRET-based biosensors was the cyan fluorescent protein (CFP) coupled with a variant of the yellow fluorescent protein (YFP) call citrine *(126)*. The authors found that the CFP-citrine pair had a FRET efficiency twice that of a CFP-YFP pair at neutral pH. However, results from the same study point out at least two caveats to be considered when incorporating FRET-based fluorescent protein biosensors into demanding cell-based assays: (1) the optimal protein pair had a maximal FRET efficiency of 40%; and (2) fluorescent protein biosensors containing linked autofluorescent proteins exhibit some measure of FRET at all times owing to the relatively close proximity of the two fluorophores. Thus, one must consider the photophysical characteristics of FRET-based fluorescent protein biosensors that might limit the FRET assay response window. Nevertheless, FRET-based fluorescent protein biosensors have been designed to measure a wide range of biological activities and recent developments merit discussion.

A FRET-based biosensor was designed to measure changes in cAMP concentration in living cells *(127)*. The cAMP binding protein Epac was used as the basis of the endogenously expressed biosensor. The Epac domain, sandwiched between CFP and YFP, exhibited a conformational change on cAMP binding that induced a measurable change in the CFP/YFP fluorescence ratio. The biosensor was used to measure the rapid disappearance of cAMP in aldosterone-producing adrenal cells resulting from the activation of the phosphodiesterase PDE2. A similar approach was used to build a biosensor of histone H3 phosphorylation *(128)*. In this case, CFP and YFP coding sequences flanked the histone H3 phosphorylation site peptide and a phosphoserine-binding domain from the 14-3-3 protein. On phosphorylation of histone H3, which reversibly occurs during the early phases of mitosis, the biosensor exhibited a 15–25% change in the YFP/CFP emission ratio.

Several FRET-based biosensors have recently been used to study the dynamics of the cytoskeleton in living cells. In one example, a CFP-Citrine FRET pair was fused to the N- and C-terminal of the myosin II regulatory light chain *(129)*. When expressed in primary vascular smooth muscle cells, the biosensor was used to report the temporal and spatial kinetics of the calcium induced changes in myosin II phosphorylation that were associated with smooth muscle contraction. A 10% change in the CFP-Citrine fluorescence ratio was observed in cells stimulated with 80 mM potassium ion solution. In another example, the temporal and spatial regulation of neuronal Wiskott-Aldrich syndrome protein activity was measured in living cells using a CFP-YFP based FRET biosensor *(130)*. The authors showed that neuronal Wiskott-Aldrich syndrome protein was activated in filopodia and therefore played a role in regulating the cytoskeletal dynamics involved in membrane ruffling. Similarly, a CFP-YFP based FRET biosensor was used to determine that the Rho GTPases Rac1 and Cdc42 were differentially localized during NGF-induced neurite outgrowth *(131)*. A major challenge for FRET-based biosensor use in HCS is the relatively low Z' factor that is generally possible resulting from the small changes measured.

4.4.3. Fluorescent Protein Biosensors Based on a Single Species of Autofluorescent Protein

Autofluorescent proteins such as GFP have been engineered to detect and report biological activities through changes in their spectral properties *(111,118,132)*. In one recent application, EYFP was first engineered to have a fluorophore whose fluorescence signal was pH dependent *(133)*. Further modification of the protein through fusion with an amino acid sequence encoding a mitochondrial intermembrane targeting signal was used to direct the biosensor to a particular cellular compartment. The biosensor shows promise in providing temporal measurements of pH in the intermembrane space, which is key to the regulation of cellular energy metabolism, after activation of cell surface receptors or during the onset of apoptosis. Another fluorescent protein biosensor based on a single autofluorescent protein involved engineering a pair of redox sensitive cysteine residues into a GFP molecule at a site, in which their oxidation state would reversibly alter the protonation state, and therefore the fluorescence intensity of the protein-based fluorophore *(134)*. The signals from the biosensor, which were measured in living cells in a fluorescence ratio mode to correct for pathlength artifacts, were used to measure redox changes including during superoxide bursts in macrophage cells, hyper- and hypoxic-conditions, and in response to peroxide stimulating agents such as epidermal growth factor and lysophosphatidic acid.

4.4.4. Other Applications of Fluorescent Biosensors and Analogs

Recent applications of fluorescent biosensors and analogs involve the innovative use of cell-based reagents. In one case, insulin mediated translocation of an EGFP-Akt1 kinase fusion protein to the plasma membrane was measured by adding a fluorescence quencher to the cellular bathing solution *(135)*. When the EGFP-Akt1 analog translocated to the membrane, it was in close enough proximity to the extracellular quenching molecules to reduce the total fluorescence intensity of the entire population of cells expressing the biosensor, thus providing information on a cellular translocation, which could be measured in a high-throughput screening mode. In a second example, a biosensor of the phosphorylation dynamics of the EGF receptor was designed to report the translocation from the cytoplasm to the plasma membrane as a change in fluorescence intensity *(136)*. The biosensor contained the EGF receptor fused to both ECFP and EYFP with a phosphotyrosine-binding domain positioned between the ECFP and the EYFP. When the EGF receptor translocated to the membrane, in which the phosphotyrosine domain was phosphorylated, a conformational change in the biosensor occurred and induced fluorescence energy transfer between the ECFP and EYFP, providing a measurable fluorescence signal change.

In a final example, the EGF-induced nucleocytoplasmic shuttling of the stress kinase ERK was measured using a GFP derived from coral that had unusual photochromic properties *(137)*. The fluorescence intensity of the coral GFP-ERK fusion protein could be reversibly highlighted (photoactivated) or erased (photobleached) depending on the wavelength of light used to irradiate the cells. Irradiating only part of the cells allowed investigators to measure the relatively fast nucleocytoplasmic shuttling of the kinase on cell activation.

4.4.5. Future Potential of Protein-Based Biosensors

Technologies for the creation of new fluorescent protein biosensors plus the development of new applications for existing sensors, many of which will involve multiplexing, will ensure that this class of biosensors will continue to grow in both number and complexity. One of the most promising developments in the technology are new systems cell biology approaches that will be addressed with multiplexed live cell kinetic assays that incorporate two or more biosensors within the same cells.

5. Reagents to Combine Cellular Manipulation With HCS Assay Measurements

Reagents that manipulate the temporal and spatial regulation of specific cellular processes are essential tools for the use of HCS platforms as an approach to systems biology. Earlier reviews described several types of reagents for cellular manipulation *(57,58,138)*, but new technologies are also having a large impact. Recently, the use of photoremovable protecting groups have been used to "cage" phosphopeptides *(139)* and phosphoproteins *(140)*. It was shown that caged phosphopeptides, which were designed to target the 14-3-3 proteins involved in cell cycle regulation, could displace endogenous proteins from the complexes with 14-3-3 in living cells when irradiated with light. The uncaging of the phosphopeptides caused premature cell cycle entry, release of G1 cells from interphase arrest and loss of the S-phase checkpoint after DNA damage, accompanied by high levels of cell death *(139)*.

The use of RNA inhibition to modulate expression levels of key proteins in cells has escalated in recent years (*[141,142]; see also* Chapter 18). Furthermore, the activity of small inhibitory RNAs (siRNAs) can be controlled in time and space within living cells using photoremovable protecting groups (*see* Chapter 19).

Manipulation of gene expression in living cells also involves the upregulation of RNAs that encode proteins or other gene products including siRNAs and other noncoding RNAs. Systems based on tetracycline-based *(143)* activation of gene regulatory proteins have been used to modulate protein and siRNA levels in cells *(144,145)*. Thus, the coupling of RNA inhibition and gene switching reagents with HCS is an ideal combination for the manipulation and measurement the myriad integrated molecular processes that comprise living cells *(146–149)*. Recently, Cellumen, Inc. has created some switched molecules using an improved ecdysone receptor technology.

A discussion on reagents for the manipulation of cellular processes would not be complete without mention of the enormous collection of drugs that posses a multitude of biological activities. Strategies for drug discovery often include compound designs that modulate the activity of specific targets whether the compounds are synthetic *(150)* or derived from natural products *(151)*. It is specificity of action that makes many therapeutic drugs valuable as cellular manipulation reagents. For example, panels of drugs have been used to build phenotypic profiles of cellular responses using them as perturbagens *(152)* or in combination with other drugs *(153)* or in combination with other manipulation reagents such as siRNAs *(146)*.

6. Prospectus

Reagents to measure and manipulate cell functions are the foundation of cell-based high-throughput as well as high content assays. There are many areas in which technological advances in reagent development would significantly benefit drug discovery and biomedical research. New reagents with which to construct labels and probes are needed. Particularly useful would be low molecular weight, probably organic, dyes with narrow and limited excitation and emission spectra assembled into protective "sleeves." This would enable the development of multiplexed assays with much less crosstalk and would be extremely valuable for constructing FRET pairs to go along with multiplexed measurements. Photostable fluorophores in the near IR would extend the range of multiplexed reagents and would allow reduction of the interference from autofluorescence that is because of native cellular chromophores or to fluorescent candidate drugs. Probes with emission lifetimes longer than autofluorescent fluorophores but short enough for rapid emission photon acquisition would be valuable for reducing background fluorescence by time resolve imaging. The 10–100 ns time-scale for emission would be especially attractive. Dyes with larger two-photon absorption cross-sections, in addition to the properties above, would be useful for research in which optical sectioning is required.

The upcoming decades would greatly benefit from new ways to create fluorescent protein biosensors, and, therefore, the creation of systems biology knowledge, for all the regulatory events that take place in the networks and pathways of living cells. Development of these reagents and new technologies to simultaneously load multiple reagents into the same cells will require a lot of creativity.

We are in need of instrumentation advances to rapidly and cleanly readout the signals from fluorescent probes. More efficient photon capture, brighter excitation at many wavelengths, further improvements in optical filters to separate the excitation and crosstalk of the "new fluorophores" that will hopefully be developed (*see* Chapters 1 and 4).

References

1. Giuliano, K. A., Haskins, J. R., and Taylor, D. L. (2003) Advances in high content screening for drug discovery. *Assay Drug Dev. Technol.* **1,** 565–577.
2. Zemanova, L., Schenk, A., Valler, M. J., Nienhaus, G. U., and Heilker, R. (2003) Confocal optics microscopy for biochemical and cellular high-throughput screening. *Drug Discov. Today* **8,** 1085–1093.
3. Sekar, R. B. and Periasamy, A. (2003) Fluorescence resonance energy transfer (FRET) microscopy imaging of live cell protein localizations. *J. Cell Biol.* **160,** 629–633.
4. Hess, S. T., Huang, S., Heikal, A. A., and Webb, W. W. (2002) Biological and chemical applications of fluorescence correlation spectroscopy: a review. *Biochemistry* **41,** 697–705.
5. Thompson, N. L. and Lagerholm, B. C. (1997) Total internal reflection fluorescence: applications in cellular biophysics. *Curr. Opin. Biotechnol.* **8,** 58–64.
6. Rizzo, M. A. and Piston, D. W. (2005) High-contrast imaging of fluorescent protein FRET by fluorescence polarization microscopy. *Biophys. J.* **88,** L14–L16.
7. Errington, R. J., Ameer-Beg, S. M., Vojnovic, B., Patterson, L. H., Zloh, M., and Smith, P. J. (2005) Advanced microscopy solutions for monitoring the kinetics and dynamics of drug-DNA targeting in living cells. *Adv. Drug Deliv. Rev.* **57,** 153–167.
8. Taylor, D. L. and Wang, Y. L., eds. (1989) *Fluorescence Microscopy of Living Cells in Culture. Part A, vol. 29. Methods in Cell Biology,* Academic, New York.
9. Weber, G. (1953) Rotational Brownian motion and polarization of the fluorescence of solutions. *Adv. Protein Chem.* **8,** 415–459.
10. Steiner, R. F. and Edelhoch, H. (1962) Fluorescent protein conjugates. *Chem Rev.* **62,** 457–483.
11. Wang, Y. L. and Taylor, D. L., eds. (1989) *Fluorescence Microscopy of Living Cells in Culture. Part B, vol. 29. Methods in Cell Biology,* Academic, New York.
12. Farkas, D. L., Baxter, G., DeBiasio, R. L., et al. (1993) Multimode light microscopy and the dynamics of molecules, cells, and tissues. *Annu. Rev. Physiol.* **55,** 785–817.
13. Giuliano, K. A., DeBiasio, R. L., Dunlay, R. T., et al. (1997) High content screening: a new approach to easing key bottlenecks in the drug discovery process. *J. Biomol. Screen.* **2,** 249–259.
14. Wang, Y. L. and Taylor, D. L. (1980) Preparation and characterization of a new molecular cytochemical probe: 5-iodoacetamidofluorescein-labeled actin. *J. Histochem. Cytochem.* **28,** 1198–1206.
15. Taylor, D. L., Waggoner, A. S., Murphy, R. F., Lanni, F., and Birge, R. R., eds. (1986) *Applications of Fluorescence in the Biomedical Sciences,* Alan R. Liss, New York.
16. Simon, J. R. and Taylor, D. L. (1986) Preparation of a fluorescent analog: acetamidofluoresceinyl-labeled Dictyostelium discoideum α-actinin *J. Meth. Enz.* **134,** 487–507.
17. Taylor, D. L., Amato, P. A., Luby-Phelps, K., and McNeil, P. (1984) Fluorescent analog cytochemistry. *Trends Biochem. Sci.* **9,** 88–91.
18. Coons, A. H., Creech, H. J., and Jones, R. N. (1941) Immunological properties of an antibody containing a fluorescent group. *Proc. Soc. Exp. Biol. Med.* **47,** 200–202.
19. Dandliker, W. B. and Portmann, A. J. (1971) Fluorescent protein conjugates, in *Excited States of Proteins and Nucleic Acids,* (Steiner, R. F. and Weinryb, I., eds.), Plenum, New York, pp. 199–275.
20. Oi, V. T., Glazer, A. N., and Stryer, L. (1982) Fluorescent phycobiliprotein conjugates for analyses of cells and molecules. *J. Cell Biol.* **93,** 981–986.
21. Ernst, L. A., Gupta, R. K., Mujumdar, R. B., and Waggoner, A. S. (1989) Cyanine dye labeling reagents for sulfhydryl groups. *Cytometry* **10,** 3–10.
22. Mujumdar, R. B., Ernst, L. A., Mujumdar, S. R., and Waggoner, A. S. (1989) Cyanine dye labeling reagents containing isothiocyanate groups. *Cytometry* **10,** 11–19.

23. Southwick, P. L., Ernst, L. A., Tauriello, E. W., et al. (1990) Cyanine dye labeling reagents—carboxymethylindocyanine succinimidyl esters. *Cytometry* **11**, 418–430.
24. Panchuk-Voloshina, N., Haugland, R. P., Bishop-Stewart, J., et al. (1999) Alexa dyes, a series of new fluorescent dyes that yield exceptionally bright, photostable conjugates. *J. Histochem. Cytochem.* **47**, 1179–1188.
25. Kumar, R. K., Chapple, C. C., and Hunter, N. (1999) Improved double immunofluorescence for confocal laser scanning microscopy. *J. Histochem. Cytochem.* **47**, 1213–1218.
26. Berlier, J. E., Rothe, A., Buller, G., et al. (2003) Quantitative comparison of long-wavelength alexa fluor dyes to Cy dyes: fluorescence of the dyes and their bioconjugates. *J. Histochem. Cytochem.* **51**, 1699–1712.
27. Nagata, K., Izawa, I., and Inagaki, M. (2001) A decade of site- and phosphorylation state-specific antibodies: recent advances in studies of spatiotemporal protein phosphorylation. *Genes Cells* **6**, 653–664.
28. Irish, J. M., Hovland, R., Krutzik, P. O., et al. (2004) Single cell profiling of potentiated phosphoprotein networks in cancer cells. *Cell* **118**, 217–228.
29. Sachs, K., Perez, O., Pe'er, D., Lauffenburger, D. A., and Nolan, G. P. (2005) Causal protein-signaling networks derived from multiparameter single-cell data. *Science* **308**, 523–529.
30. De Rosa, S. C., Brenchley, J. M., and Roederer, M. (2003) Beyond six colors: a new era in flow cytometry. *Nat. Med.* **9**, 112–117.
31. Ramirez, S., Aiken, C. T., Andrzejewski, B., Sklar, L. A., and Edwards, B. S. (2003) High-throughput flow cytometry: validation in microvolume bioassays. *Cytometry A* **53**, 55–65.
32. Young, S. M., Bologa, C., Prossnitz, E. R., Oprea, T. I., Sklar, L. A., and Edwards, B. S. (2005) High-throughput screening with HyperCyt flow cytometry to detect small molecule formylpeptide receptor ligands. *J. Biomol. Screen.* **10**, 374–382.
33. Oertel, J. and Huhn, D. (2000) Immunocytochemical methods in haematology and oncology. *J. Cancer Res. Clin. Oncol.* **126**, 425–440.
34. Michalet, X., Pinaud, F. F., Bentolila, L. A., et al. (2005) Quantum dots for live cells, in vivo imaging, and diagnostics. *Science* **307**, 538–544.
35. Medintz, I. L., Uyeda, H. T., Goldman, E. R., and Mattoussi, H. (2005) Quantum dot bioconjugates for imaging, labeling, and sensing. *Nat. Mater.* **4**, 435–446.
36. Gao, X., Yang, L., Petros, J. A., Marshall, F. F., Simons, J. W., and Nie, S. (2005) In vivo molecular and cellular imaging with quantum dots. *Curr. Opin. Biotechnol.* **16**, 63–72.
37. Alivisatos, A. P., Gu, W., and Larabell, C. (2005) Quantum dots as cellular probes. *Annu. Rev. Biomed. Eng.* **7**, 55–76.
38. Mattheakis, L. C., Dias, J. M., Choi, Y. J., et al. (2004) Optical coding of mammalian cells using semiconductor quantum dots. *Anal. Biochem.* **327**, 200–208.
39. Ow, H., Larson, D. R., Srivastava, M., Baird, B. A., Webb, W. W., and Wiesner, U. (2005) Bright and stable core-shell fluorescent silica nanoparticles. *Nano Lett.* **5**, 113–117.
40. Lian, W., Litherland, S. A., Badrane, H., et al. (2004) Ultrasensitive detection of biomolecules with fluorescent dye-doped nanoparticles. *Anal. Biochem.* **334**, 135–144.
41. Zheng, J., Zhang, C., and Dickson, R. M. (2004) Highly fluorescent, water-soluble, size-tunable gold quantum dots. *Phys. Rev. Lett.* **93**, 077402.
42. Famulok, M. and Mayer, G. (2005) Intramers and aptamers: applications in protein-function analyses and potential for drug screening. *Chembiochemistry* **6**, 19–26.
43. Buerger, C. and Groner, B. (2003) Bifunctional recombinant proteins in cancer therapy: cell penetrating peptide aptamers as inhibitors of growth factor signaling. *J. Cancer Res. Clin. Oncol.* **129**, 669–675.
44. Lee, J. F., Hesselberth, J. R., Meyers, L. A., and Ellington, A. D. (2004) Aptamer database. *Nucl. Acids Res.* **32**, D95–D100, database issue.
45. Nimjee, S. M., Rusconi, C. P., and Sullenger, B. A. (2005) Aptamers: an emerging class of therapeutics. *Annu. Rev. Med.* **56**, 555–583.
46. Stojanovic, M. N. and Kolpashchikov, D. M. (2004) Modular aptameric sensors. *J. Am. Chem. Soc.* **126**, 9266–9270.
47. Jhaveri, S., Rajendran, M., and Ellington, A. D. (2000) In vitro selection of signaling aptamers. *Nat. Biotechnol.* **18**, 1293–1297.
48. Tan, W., Wang, K., and Drake, T. J. (2004) Molecular beacons. *Curr. Opin. Chem. Biol.* **8**, 547–553.

49. Knoll, E. and Heyduk, T. (2004) Unimolecular beacons for the detection of DNA-binding proteins. *Anal. Chem.* **76**, 1156–1164.
50. Heyduk, T. and Heyduk, E. (2002) Molecular beacons for detecting DNA binding proteins. *Nat. Biotechnol.* **20**, 171–176.
51. Toda, Y., Kono, K., Abiru, H., et al. (1999) Application of tyramide signal amplification system to immunohistochemistry: a potent method to localize antigens that are not detectable by ordinary method. *Pathol. Int.* **49**, 479–483.
52. Chao, J., DeBiasio, R., Zhu, Z., Giuliano, K. A., and Schmidt, B. F. (1996) Immunofluorescence signal amplification by the enzyme-catalyzed deposition of a fluorescent reporter substrate (CARD). *Cytometry* **23**, 48–53.
53. Hasui, K. and Murata, F. (2005) A new simplified catalyzed signal amplification system for minimizing non-specific staining in tissues with supersensitive immunohistochemistry. *Arch. Histol. Cytol.* **68**, 1–17.
54. Earnshaw, J. C. and Osbourn, J. K. (1999) Signal amplification in flow cytometry using biotin tyramine. *Cytometry* **35**, 176–179.
55. Ness, J. M., Akhtar, R. S., Latham, C. B., and Roth, K. A. (2003) Combined tyramide signal amplification and quantum dots for sensitive and photostable immunofluorescence detection. *J. Histochem. Cytochem.* **51**, 981–987.
56. Taylor, D. L., Amato, P. A., McNeil, P. L., Luby-Phelps, K., and Tanasugarn, L. (1986) Spatial and temporal dynamics of specific molecules and ions in living cells, in *Applications of Fluorescence in the Biomedical Sciences* (Taylor, D. L., Waggoner, A. S., Murphy, R. F., Lanni, F., and Birge, R. R., eds.), Alan R. Liss, New York, pp. 347–376.
57. Giuliano, K. A. and Taylor, D. L. (1995) Light-optical-based reagents for the measurement and manipulation of ions, metabolites, and macromolecules in living cells. *Methods Neurosci.* **27**, 1–16.
58. Giuliano, K. A. and Taylor, D. L. (1995) Measurement and manipulation of cytoskeletal dynamics in living cells. *Curr. Opin. Cell Biol.* **7**, 4–12.
59. Zhang, J., Campbell, R. E., Ting, A. Y., and Tsien, R. Y. (2002) Creating new fluorescent probes for cell biology. *Nat. Rev. Mol. Cell Biol.* **3**, 906–918.
60. Waggoner, A. S. (1985) Dye probes of cell, organelle, and vesicle membrane potentials, in *The Enzymes of Biological Membranes,* Second ed., vol. 4, (Martonosi, A., ed.), Plenum, New York, pp. 313–331.
61. Waggoner, A. S. (1986) Fluorescent probes for analysis of cell structure, function, and health by flow and imaging cytometry, in *Applications of Fluorescence in the Biomedical Sciences* (Taylor, D. L., Waggoner, A. S., Murphy, R. F., Lanni, F., and Birge, R. R., eds), Alan R. Liss, New York, pp. 3–28.
62. Waggoner, A., DeBiasio, R., Conrad, P., et al. (1989) Multiple spectral parameter imaging, in *Methods in Cell Biology, vol. 30* (Taylor, D. L. and Wang, Y. L., eds.), Alan R. Liss, New York, pp. 449–478.
63. Waggoner, A. S. (1990) Fluorescent probes for cytometry, in *Flow Cytometry and Sorting,* Second edition (Melamed, M. R., Lindmo, T., and Mendelsohn, M. L., eds.), Wiley-Liss, New York, pp. 209–225.
64. Chance, B., Schoener, B., and Elsaesser, S. (1964) Control of the waveform of oscillations of the reduced pyridine nucleotide level in a cell-free extract. *Proc. Natl Acad. Sci. USA* **52**, 337–341.
65. Tasaki, II., Carnay, L., and Watanabe, A. (1969) Transient changes in extrinsic fluorescence of nerve produced by electric stimulation. *Proc. Natl. Acad. Sci. USA* **64**, 1362–1368.
66. Davila, H. V., Salzberg, B. M., Cohen, L. B., and Waggoner, A. S. (1973) A large change in axon fluorescence that provides a promising method for measuring membrane potential. *Nat. New Biol.* **241**, 159–160.
67. Ross, W. N., Salzberg, B. M., Cohen, L. B., et al. (1977) Changes in absorption, fluorescence, dichroism, and Birefringence in stained giant axons: optical measurement of membrane potential. *J. Membr. Biol.* **33**, 141–183.
68. Cohen, L. B., Salzberg, B. M., Davila, H. V., et al. (1974) Changes in axon fluorescence during activity: molecular probes of membrane potential. *J. Membr. Biol.* **19**, 1–36.
69. Sims, P. J., Waggoner, A. S., Wang, C. H., and Hoffman, J. F. (1974) Studies on the mechanism by which cyanine dyes measure membrane potential in red blood cells and phosphatidylcholine vesicles. *Biochemistry* **13**, 3315–3330.
70. Waggoner, A. S. and Grinvald, A. (1977) Mechanisms of rapid optical changes of potential sensitive dyes. *Ann. NY Acad. Sci.* **303**, 217–241.

71. Loew, L. M. (1996) Potentiometric dyes: new modalities for optical imaging of membrane potential, in *Proceedings of Optical Diagnostics of Living Cells and Biofluids*, Proceedings of SPIE vol. 2678 (Farkos, D. C., Leif, R. C., Priezzhev, A. V., and Tromberg, B. J., eds.) SPIE, San Jose, CA, pp. 80–87.
72. Grinvald, A., Cohen, L. B., Lesher, S., and Boyle, M. B. (1981) Simultaneous optical monitoring of activity of many neurons in invertebrate ganglia using a 124-element photodiode array. *J. Neurophysiol.* **45,** 829–840.
73. Toutchkine, A., Kraynov, V., and Hahn, K. (2003) Solvent-sensitive dyes to report protein conformational changes in living cells. *J. Am. Chem. Soc.* **125,** 4132–4145.
74. Giuliano, K. A., Post, P. L., Hahn, K. M., and Taylor, D. L. (1995) Fluorescent protein biosensors: measurement of molecular dynamics in living cells. *Annu. Rev. Biophys. Biomol. Struct.* **24,** 405–434.
75. Tsien, R. Y., Ernst, L. A., and Waggoner, A. S. (2006) Fluorophores for confocal microscopy: photophysics and photochemistry, in *Handbook of Biological Confocal Microscopy*, third ed. (Pawley, J., ed.), Plenum, New York, in press.
76. Taylor, D. L. and Wang, Y. L. (1978) Molecular cytochemistry: incorporation of fluorescently labeled actin into cells. *Proc. Natl Acad. Sci. USA* **75,** 857–861.
77. Mason, W. T., ed. (1993) *Fluorescent and Luminescent Probes for Biological Activity*, Academic, San Diego, CA.
78. Taylor, D. L., Nederlof, M. A., Lanni, F., and Waggoner, A. S. (1992) The new vision of light microscopy. *Am. Sci.* **80,** 322–335.
79. Tanasugarn, L., McNeil, P., Reynolds, G. T., and Taylor, D. L. (1984) Microspectrofluorometry by digital image processing: measurement of cytoplasmic pH. *J. Cell Biol.* **98,** 717–724.
80. Hahn, K. M., Waggoner, A. S., and Taylor, D. L. (1990) A calcium-sensitive fluorescent analog of calmodulin based on a novel calmodulin-binding fluorophore. *J. Biol. Chem.* **265,** 20,335–20,345.
81. Hahn, K., Kolega, J., Montibeller, J., et al. (1993) Fluorescent analogues: optical biosensors of the chemical and molecular dynamics of macromolecules in living cells, in *Fluorescent and Luminescent Probes for Biological Activity* (Mason, W. T., ed.), Academic, San Diego, CA, pp. 349–359.
82. Post, P. L., DeBiasio, R. L., and Taylor, D. L. (1995) A fluorescent protein biosensor of myosin II regulatory light chain phosphorylation reports a gradient of phosphorylated myosin II in migrating cells. *Mol. Biol. Cell* **6,** 1755–1768.
83. DeBiasio, R. L., LaRocca, G. M., Post, P. L., and Taylor, D. L. (1996) Myosin II transport, organization, and phosphorylation: evidence for cortical flow/solation-contraction coupling during cytokinesis and cell locomotion. *Mol. Biol. Cell* **7,** 1259–1282.
84. Luby-Phelps, K. (1989) Preparation of fluorescently labeled dextrans and ficolls. *Methods Cell Biol.* **29,** 59–73.
85. Kolega, J. and Taylor, D. L. (1993) Gradients in the concentration and assembly of myosin II in living fibroblasts during locomotion and fiber transport. *Mol. Biol. Cell* **4,** 819–836.
86. Nelson, D. E., Ihekwaba, A. E., Elliott, M., et al. (2004) Oscillations in NF-kappaB signaling control the dynamics of gene expression. *Science* **306,** 704–708.
87. Almholt, D. L., Loechel, F., Nielsen, S. J., et al. (2004) Nuclear export inhibitors and kinase inhibitors identified using a MAPK-activated protein kinase 2 redistribution screen. *Assay Drug Dev. Technol.* **2,** 7–20.
88. McNeil, P. L. (1989) Incorporation of macromolecules into living cells, in *Methods in Cell Biology*, vol. 29 (Wang, Y. L. and Taylor, D. L., eds.), Alan R. Liss, New York, pp. 153–173.
89. Lee, K. D., Oh, Y. K., Portnoy, D. A., and Swanson, J. A. (1996) Delivery of macromolecules into cytosol using liposomes containing hemolysin from Listeria monocytogenes. *J. Biol. Chem.* **271,** 7249–7252.
90. McNeil, P. L. and Steinhardt, R. A. (2003) Plasma membrane disruption: repair, prevention, and adaptation. *Annu. Rev. Cell Dev. Biol.* **19,** 697–731.
91. Bright, G. R., Kuo, N. T., Chow, D., Burden, S., Dowe, C., and Przybylski, R. J. (1996) Delivery of macromolecules into adherent cells via electroporation for use in fluorescence spectroscopic imaging and metabolic studies. *Cytometry* **24,** 226–233.
92. McAllister, D. V., Wang, P. M., Davis, S. P., et al. (2003) Microfabricated needles for transdermal delivery of macromolecules and nanoparticles: fabrication methods and transport studies. *Proc. Natl Acad. Sci. USA* **100,** 13,755–13,760.

93. McAllister, D. V., Allen, M. G., and Prausnitz, M. R. (2000) Microfabricated microneedles for gene and drug delivery. *Annu. Rev. Biomed. Eng.* **2**, 289–313.
94. Hanania, E. G., Fieck, A., Stevens, J., Bodzin, L. J., Palsson, B. O., and Koller, M. R. (2005) Automated *in situ* measurement of cell-specific antibody secretion and laser-mediated purification for rapid cloning of highly-secreting producers. *Biotechnol. Bioeng.* **91**, 872–876.
95. Qian, Z. M., Li, H., Sun, H., and Ho, K. (2002) Targeted drug delivery via the transferrin receptor-mediated endocytosis pathway. *Pharmacol. Rev.* **54**, 561–587.
96. Ilk, N., Kupcu, S., Moncayo, G., et al. (2004) A functional chimaeric S-layer-enhanced green fluorescent protein to follow the uptake of S-layer-coated liposomes into eukaryotic cells. *Biochem. J.* **379**, 441–448.
97. Hariton-Gazal, E., Rosenbluh, J., Graessmann, A., Gilon, C., and Loyter, A. (2003) Direct translocation of histone molecules across cell membranes. *J. Cell. Sci.* **116**, 4577–4586.
98. Gariepy, J. and Kawamura, K. (2001) Vectorial delivery of macromolecules into cells using peptide-based vehicles. *Trends Biotechnol.* **19**, 21–28.
99. Joliot, A. and Prochiantz, A. (2004) Transduction peptides: from technology to physiology. *Nat. Cell Biol.* **6**, 189–196.
100. Kabouridis, P. S. (2003) Biological applications of protein transduction technology. *Trends Biotechnol.* **21**, 498–503.
101. Fischer, R., Kohler, K., Fotin-Mleczek, M., and Brock, R. (2004) A stepwise dissection of the intracellular fate of cationic cell-penetrating peptides. *J. Biol. Chem.* **279**, 12,625–12,635.
102. Chalfie, M., Tu, Y., Euskirchen, G., Ward, W. W., and Prasher, D. C. (1994) Green fluorescent protein as a marker for gene expression. *Science* **263**, 802–805.
103. Heim, R. and Tsien, R. Y. (1996) Engineering green fluorescent protein for improved brightness, longer wavelengths, and fluorescence resonance energy transfer. *Curr. Biol.* **6**, 178–192.
104. Tsien, R. Y. (2005) Building and breeding molecules to spy on cells and tumors. *FEBS Lett.* **579**, 927–932.
105. Verkhusha, V. V. and Lukyanov, K. A. (2004) The molecular properties and applications of Anthozoa fluorescent proteins and chromoproteins. *Nat. Biotechnol.* **22**, 289–296.
106. Miyawaki, A., Sawano, A., and Kogure, T. (2003) Lighting up cells: labelling proteins with fluorophores. *Nat. Cell Biol.* (**Suppl**), S1–S7.
107. Johnsson, N. and Johnsson, K. (2003) A fusion of disciplines: chemical approaches to exploit fusion proteins for functional genomics. *Chembiochemistry* **4**, 803–810.
108. Aumais, J. P., Lee, H. S., Lin, R., and White, J. H. (1997) Selective interaction of hsp90 with an estrogen receptor ligand-binding domain containing a point mutation. *J. Biol. Chem.* **272**, 12,229–12,235.
109. Chen, I. and Ting, A. Y. (2005) Site-specific labeling of proteins with small molecules in live cells. *Curr. Opin. Biotechnol.* **16**, 35–40.
110. Miller, L. W. and Cornish, V. W. (2005) Selective chemical labeling of proteins in living cells. *Curr. Opin. Chem. Biol.* **9**, 56–61.
111. Giuliano, K. A., Chen, Y. T., and Haskins, J. R. (2003) Fluorescent protein biosensors: a new tool for high content screening. *Mod. Drug Discov.* **August,** 33–37.
112. Ghosh, R. N., DeBiasio, R., Hudson, C. C., Ramer, E. R., Cowan, C. L., and Oakley, R. H. (2005) Quantitative cell-based high content screening for vasopressin receptor agonists using transfluor technology. *J. Biomol. Screen.* **10**, 476–484.
113. Almholt, K., Tullin, S., Skyggebjerg, O., Scudder, K., Thastrup, O., and Terry, R. (2004) Changes in intracellular cAMP reported by a redistribution assay using a cAMP-dependent protein kinase-green fluorescent protein chimera. *Cell Signal.* **16**, 907–920.
114. Barak, L. S., Ferguson, S. S. G., Zhang, J., and Caron, M. G. (1997) A β-arrestin/green fluorescent protein biosensor for detecting G-protein-coupled receptor activation. *J. Biol. Chem.* **272**, 27,497–27,500.
115. Knauer, S. K., Moodt, S., Berg, T., Liebel, U., Pepperkok, R., and Stauber, R. H. (2005) Translocation biosensors to study signal-specific nucleo-cytoplasmic transport, protease activity, and protein–protein interactions. *Traffic* **6**, 594–606.
116. Bartlett, P. J., Young, K. W., Nahorski, S. R., and Challiss, R. A. (2005) Single-cell analysis and temporal profiling of agonist-mediated Ins(1,4,5)P3, Ca^{2+}, diacylglycerol and protein kinase C signaling using fluorescent biosensors. *J. Biol. Chem.* **280**, 21,837–21,846.
117. Hahn, K. and Toutchkine, A. (2002) Live-cell fluorescent biosensors for activated signaling proteins. *Curr. Opin. Cell Biol.* **14**, 167–172.

118. Giuliano, K. A. and Taylor, D. L. (1998) Fluorescent-protein biosensors: new tools for drug discovery. *Trends Biotechnol.* **16,** 135–140.
119. Hahn, K., DeBiasio, R., and Taylor, D. L. (1992) Patterns of elevated free calcium and calmodulin activation in living cells. *Nature (London)* **359,** 736–738.
120. Post, P. L., Trybus, K. M., and Taylor, D. L. (1994) A genetically engineered, protein-based optical biosensor of myosin II regulatory light chain phosphorylation. *J. Biol Chem.* **269,** 12,880–12,887.
121. Adams, S. R., Harootunian, A. T., Buechler, Y. J., Taylor, S. S., and Tsien, R. Y. (1991) Fluorescence ratio imaging of cyclic AMP in single cells. *Nature (London)* **349,** 694–697.
122. Pertz, O. and Hahn, K. M. (2004) Designing biosensors for Rho family proteins—deciphering the dynamics of Rho family GTPase activation in living cells. *J. Cell Sci.* **117,** 1313–1318.
123. Nalbant, P., Hodgson, L., Kraynov, V., Toutchkine, A., and Hahn, K. M. (2004) Activation of endogenous Cdc42 visualized in living cells. *Science* **305,** 1615–1619.
124. Larsson, H. P., Tzingounis, A. V., Koch, H. P., and Kavanaugh, M. P. (2004) Fluorometric measurements of conformational changes in glutamate transporters. *Proc. Natl. Acad. Sci. USA* **101,** 3951–3956.
125. Pfleger, K. D. and Eidne, K. A. (2005) Monitoring the formation of dynamic G-protein-coupled receptor–protein complexes in living cells. *Biochem. J.* **385,** 625–637.
126. Hoppe, A., Christensen, K., and Swanson, J. A. (2002) Fluorescence resonance energy transfer-based stoichiometry in living cells. *Biophys. J.* **83,** 3652–3664.
127. Nikolaev, V. O., Bunemann, M., Hein, L., Hannawacker, A., and Lohse, M. J. (2004) Novel single chain cAMP sensors for receptor-induced signal propagation. *J. Biol. Chem.* **279,** 37,215–37,218.
128. Lin, C. W. and Ting, A. Y. (2004) A genetically encoded fluorescent reporter of histone phosphorylation in living cells. *Angew. Chem. Int. Ed. Engl.* **43,** 2940–2943.
129. Yamada, A., Hirose, K., Hashimoto, A., and Iino, M. (2005) Real-time imaging of myosin II regulatory light-chain phosphorylation using a new protein biosensor. *Biochem. J.* **385,** 589–594.
130. Ward, M. E., Wu, J. Y., and Rao, Y. (2004) Visualization of spatially and temporally regulated N-WASP activity during cytoskeletal reorganization in living cells. *Proc. Natl Acad. Sci. USA* **101,** 970–974.
131. Aoki, K., Nakamura, T., and Matsuda, M. (2004) Spatio-temporal regulation of Rac1 and Cdc42 activity during nerve growth factor-induced neurite outgrowth in PC12 cells. *J. Biol. Chem.* **279,** 713–719.
132. Taylor, D. L., Woo, E. S., and Giuliano, K. A. (2001) Real-time molecular and cellular analysis: the new frontier of drug discovery. *Curr. Opin. Biotechnol.* **12,** 75–81.
133. Porcelli, A. M., Ghelli, A., Zanna, C., Pinton, P., Rizzuto, R., and Rugolo, M. (2005) pH difference across the outer mitochondrial membrane measured with a green fluorescent protein mutant. *Biochem. Biophys. Res. Commun.* **326,** 799–804.
134. Dooley, C. M., Dore, T. M., Hanson, G. T., Jackson, W. C., Remington, S. J., and Tsien, R. Y. (2004) Imaging dynamic redox changes in mammalian cells with green fluorescent protein indicators. *J. Biol. Chem.* **279,** 22,284–22,293.
135. Lundholt, B. K., Linde, V., Loechel, F., et al. (2005) Identification of Akt pathway inhibitors using redistribution screening on the FLIPR and the IN Cell 3000 analyzer. *J. Biomol. Screen.* **10,** 20–29.
136. Offterdinger, M., Georget, V., Girod, A., and Bastiaens, P. I. (2004) Imaging phosphorylation dynamics of the EGF receptor. *J. Biol. Chem.* **279,** 36,972–36,981.
137. Ando, R., Mizuno, H., and Miyawaki, A. (2004) Regulated fast nucleocytoplasmic shuttling observed by reversible protein highlighting. *Science* **306,** 1370–1373.
138. Ottl, J., Gabriel, D., and Marriott, G. (1998) Preparation and photoactivation of caged fluorophores and caged proteins using a new class of heterobifunctional, photocleavable cross-linking reagents. *Bioconjugate Chem.* **9,** 143–151.
139. Nguyen, A., Rothman, D. M., Stehn, J., Imperiali, B., and Yaffe, M. B. (2004) Caged phosphopeptides reveal a temporal role for 14-3-3 in G1 arrest and S-phase checkpoint function. *Nat. Biotechnol.* **22,** 993–1000.
140. Rothman, D. M., Petersson, E. J., Vazquez, M. E., Brandt, G. S., Dougherty, D. A., and Imperiali, B. (2005) Caged phosphoproteins. *J. Am. Chem. Soc.* **127,** 846, 847.
141. Huppi, K., Martin, S. E., and Caplen, N. J. (2005) Defining and assaying RNAi in mammalian cells. *Mol. Cell* **17,** 1–10.
142. Juliano, R. L., Dixit, V. R., Kang, H., Kim, T. Y., Miyamoto, Y., and Xu, D. (2005) Epigenetic manipulation of gene expression: a toolkit for cell biologists. *J. Cell Biol.* **169,** 847–857.

143. Blau, H. M. and Rossi, F. M. (1999) Tet B or not tet B: advances in tetracycline-inducible gene expression. *Proc. Natl. Acad. Sci. USA* **96,** 797–799.
144. Wang, J., Tekle, E., Oubrahim, H., Mieyal, J. J., Stadtman, E. R., and Chock, P. B. (2003) Stable and controllable RNA interference: investigating the physiological function of glutathionylated actin. *Proc. Natl Acad. Sci. USA* **100,** 5103–5106.
145. Gupta, S., Schoer, R. A., Egan, J. E., Hannon, G. J., and Mittal, V. (2004) Inducible, reversible, and stable RNA interference in mammalian cells. *Proc. Natl. Acad. Sci. USA* **101,** 1927–1932.
146. Giuliano, K. A., Chen, Y. T., and Taylor, D. L. (2004) High content screening with siRNA optimizes a cell biological approach to drug discovery: defining the role of p53 activation in the cellular response to anticancer drugs. *J. Biomol. Screen.* **9,** 557–568.
147. Wheeler, D. B., Bailey, S. N., Guertin, D. A., Carpenter, A. E., Higgins, C. O., and Sabatini, D. M. (2004) RNAi living-cell microarrays for loss-of-function screens in Drosophila melanogaster cells. *Nat. Methods* **1,** 127–132.
148. Yarrow, J. C., Perlman, Z. E., Westwood, N. J., and Mitchison, T. J. (2004) A high-throughput cell migration assay using scratch wound healing, a comparison of image-based readout methods. *BMC Biotechnol.* **4,** 21.
149. Conrad, C., Erfle, H., Warnat, P., et al. (2004) Automatic identification of subcellular phenotypes on human cell arrays. *Genome Res.* **14,** 1130–1136.
150. Lipinski, C. and Hopkins, A. (2004) Navigating chemical space for biology and medicine. *Nature* **432,** 855–861.
151. Clardy, J. and Walsh, C. (2004) Lessons from natural molecules. *Nature* **432,** 829–837.
152. Ramanathan, A., Wang, C., and Schreiber, S. L. (2005) Perturbational profiling of a cell-line model of tumorigenesis by using metabolic measurements. *Proc. Natl Acad. Sci. USA* **102,** 5992–5997.
153. Giuliano, K. A. (2003) High-content profiling of drug-drug interactions: cellular targets involved in the modulation of microtubule drug action by the antifungal ketoconazole. *J. Biomol. Screen.* **8,** 125–135.
154. Weber, G. (1961) Excited states of proteins, in *Light and Life* (McElroy, W. D. and Glass, B., eds.), The Johns Hopkins Press, Baltimore, MD, pp. 82–107.
155. Konev, S. V. (1967) *Fluorescence and Phosphorescence of Proteins and Nucleic Acids*, Plenum, New York.
156. Waggoner, A. S. and Stryer, L. (1970) Fluorescent probes of biological membranes. *Proc. Natl Acad. Sci.* **67,** 579–589.
157. Waggoner, A. S. (1979) Dye indicators of membrane potential. *Annu. Rev. Biophys. Bioeng.* **8,** 47–68.
158. Stryer, L. (1978) Fluorescence energy transfer as a spectroscopic ruler. *Annu. Rev. Biochem.* **47,** 819–846.
159. Herman, B. (1989) Resonance energy transfer microscopy, in *Methods in Cell Biology*, vol. 30 (Taylor, D. L. and Wang, Y. L., eds), Academic, San Diego, CA, pp. 219–243.
160. Jovin, T. M. and Arndt-Jovin, D. J. (1989) FRET microscopy: digital imaging of fluorescence resonance energy transfer. Application in cell biology. In *Cell Structure and Function by Microspectrofluorometry*, vol. 30 (Kohen, E. and Hirschberg, J. G., eds.), Academic, San Diego, CA, pp. 99–117.
161. Taylor, D. L. and Wang, Y. L. (1980) Fluorescently labelled molecules as probes of the structure and function of living cells. *Nature (London)* **284,** 405–409.
162. Salmon, E. D. and Wadsworth, P. (1986) Fluorescence studies of tubulin and microtubule dynamics in living cells, in *Applications of Fluorescence in the Biomedical Sciences*, (Taylor, D. L., Waggoner, A. S., Murphy, R. F., Lanni, F., and Birge, R. R., eds.), Alan R. Liss, New York, pp. 377–403.
163. Tsien, R. Y., Pozzan, T., and Rink, T. J. (1982) Calcium homeostasis in intact lymphocytes: cytoplasmic free calcium monitored with a new, intracellularly trapped fluorescent indicator. *J. Cell Biol.* **94,** 325–334.
164. Rink, T. J., Tsien, R. Y., and Pozzan, T. (1982) Cytoplasmic pH and free Mg2+ in lymphocytes. *J. Cell Biol.* **95,** 189–196.
165. Grynkiewicz, G., Poenie, M., and Tsien, R. Y. (1985) A new generation of Ca^{2+} indicators with greatly improved fluorescence properties. *J. Biol. Chem.* **260,** 3440–3450.
166. DeBiasio, R., Bright, G. R., Ernst, L. A., Waggoner, A. S., and Taylor, D. L. (1987) Five-parameter fluorescence imaging: wound healing of living Swiss 3T3 cells. *J. Cell Biol.* **105,** 1613–1622.

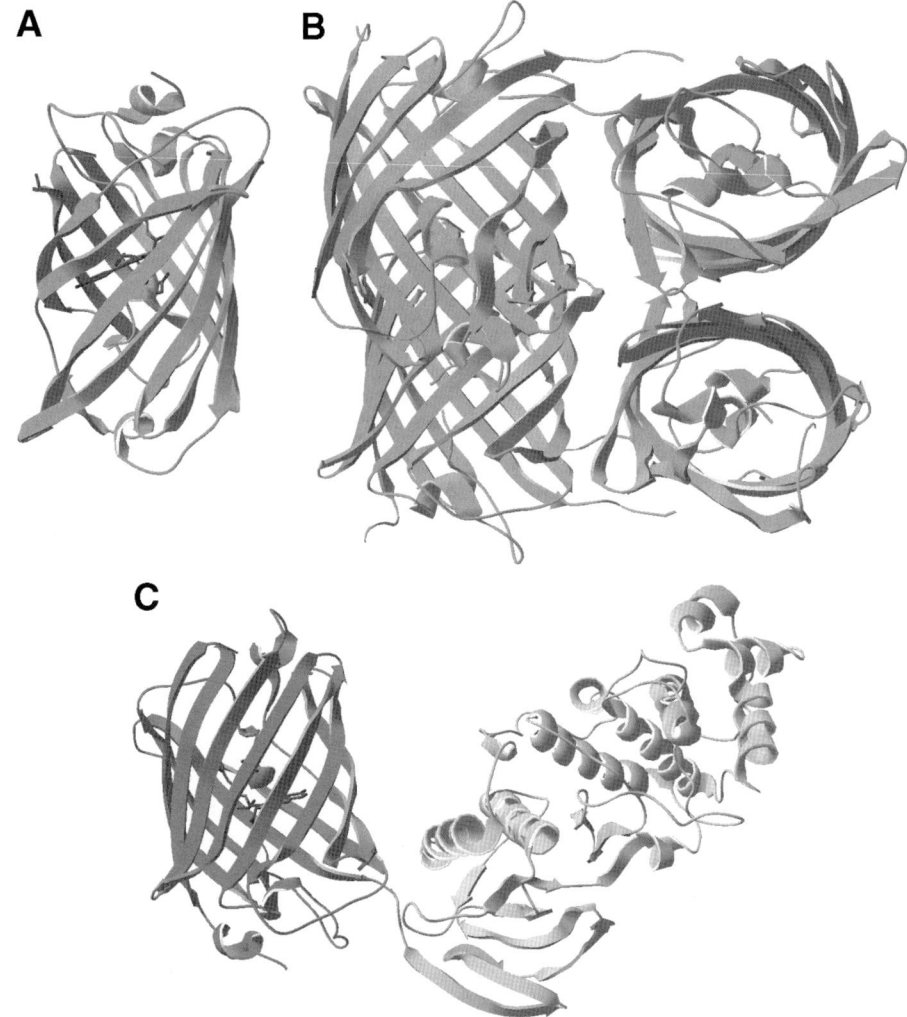

Fig. 1. Fluorescent protein structures. 3D models of (**A**) *A. victoria* GFP, (**B**) DsRed, and (**C**) GFP-p38 MAPK were visualized using Deepview/Swiss PdbViewer (www.expasy.org/spdbv) from coordinates obtained from the RSCB protein data bank *(122)* (www.rscb.org/pdb).

2. Evolution of High Content Screening Using Engineered Cells

The development of GFP and other fluorescent proteins suitable for expression in mammalian cells occurred in parallel with the emergence of the high throughput automated imaging and image analysis techniques *(37,38)*, which came to be known as high content screening (HCS). Although to a certain extent this was a chance coincidence in the emergence of two technologies, it was a very fortuitous alignment. The development of fluorescent proteins provided the means to tag proteins in living cells to directly and dynamically visualize biological processes that previously could only inferred from other techniques such as reporter gene assays. HCS instrumentation and analysis software provided the means to perform such experiments in sufficient number to make them applicable to the increasingly industrialized environment of drug discovery.

Initial HCS efforts focused on fixed cell assays using antibodies *(39,40)* but it was only a very short time between the first reports of cytoplasmic to nuclear protein translocation being visualized using glucocorticoid receptor-GFP fusion proteins *(41,42)* that HCS data were reported for HeLa cells transiently expressing glucocorticoid receptor-GFP fusion proteins *(43)*. Shortly thereafter use of a stable HEK293 cell line expressing a PTHR-GFP for HCS of G protein coupled receptor (GPCR) activation was reported *(44)*. From this point on engineered cell lines expressing GFP fusion proteins were rapidly adopted as a key tool for imaging gene expression and protein localization and redistribution *(45,46)* in drug development. This increase in the use of GFP fusion proteins continues to be accompanied by the mutually beneficial development of increasingly sophisticated high throughput fluorescence instrumentation capable of imaging and analyzing cellular events in live cells in real time *(47–49)*.

Since the development of optimum techniques for tagging proteins in mammalian cells with fluorescent proteins stable cells expressing GFP fusion proteins have been applied in studying a wide range of cellular molecules and processes including GPCR signaling *(50,51)*, cytoskeletal dynamics *(52)*, second messenger signaling *(53)*, protein kinase activity and localization *(54,55)*, chromatin structuring *(56)*, and protein trafficking *(57)*.

Although a number of publications present proof of principle data from HCS experiments for drug development *(58–63)*, for proprietary and other reasons publications describing the use of fluorescent proteins in full-scale screening are limited. One disclosed screen of 950,000 compounds *(64)* run on a GE Healthcare (Giles, UK) IN Cell Analyzer 3000 examined GPCR desensitization and internalization in a U2OS cell line stably expressing GFP-β-arrestin. Monitoring of GPCR activation through internalization of β-arrestin was initially reported *(65)* early in the development of fluorescent protein tagging, and was subsequently developed and industrialized for drug screening by Norak Biosciences (now Xsira Pharmaceuticals, Morrisville, NC) *(66)*. Recent developments in methods for gene knockdown with RNAi have combined with engineered cell lines expressing sophisticated fluorescent protein sensors and HCS to provide extremely powerful tools for elucidating gene function using RNAi screens (**Fig. 2**).

3. Designing Engineered Cells

Building a stable cell line that will withstand the rigors of HCS is a complex and often time-consuming business. In an ideal world all stable cell lines for HCS would be easy to engineer, select and maintain, brightly fluorescent and easy to image, free from any regulatory and patent constraints, an accurate model of a biological process, and provide robust and statistically valid data. In the real world this is rarely the case, and for most cell lines some compromises between opposing design factors inevitably have to be made to arrive at a cell line fit for the purpose for which it is intended. Some of the key design considerations involved in designing and engineering cells to express fluorescent fusion proteins are listed in **Table 2**. These elements are discussed in detail later with the aim of highlighting some of the key decision (or compromise) points involved in cell line design. Although many of the required engineering decisions can be made based on logic, preceding knowledge, or published data, it is inevitable, given the complex interrelationships between design options, that in many cases it is impossible to arrive at a single complete definitive design for a cell line. In these situations there is no alternative to empirical optimization; you have to get in the lab and figure out what works and what does not.

3.1. Target Protein Selection

The first choice to be made in designing a cell line to report on a cellular signaling pathway or process is to identify a suitable protein for fusion to a fluorescent protein. In some cases this will be a simple choice; if a signaling pathway of interest has been well characterized and reported in the

Fluorescent Proteins and Engineered Cell Lines

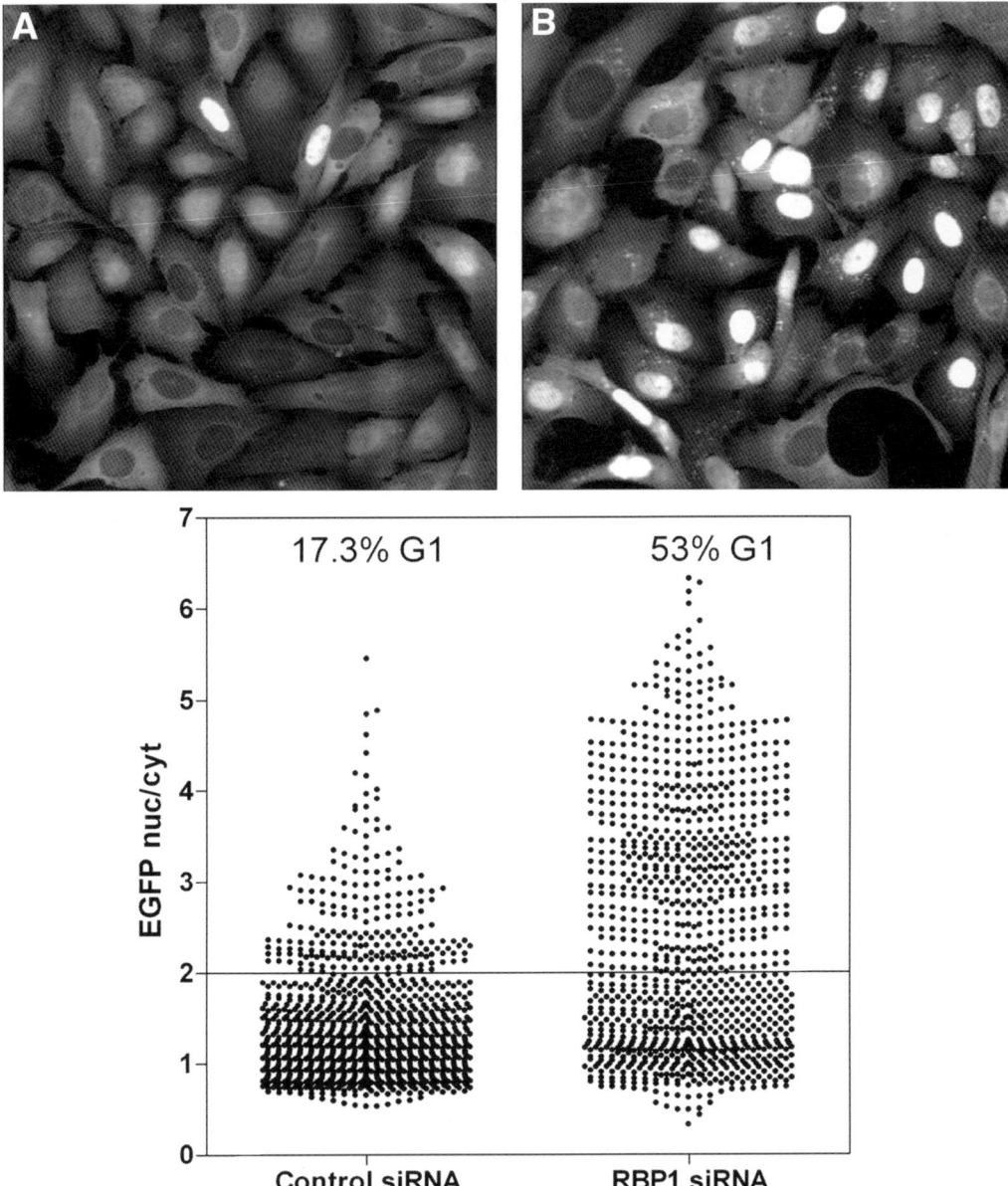

Fig. 2. HCS of siRNA gene knockdown screen in an engineered cell line. A stable U2OS cell line expressing an EGFP-helicase PSLD G1/S transition sensor was screened against a library of siRNAs for effects on G1/S transition. In G1 cells the EGFP-helicase PSLD is retained in the nucleus by a nuclear localization sequence. On transition to S-phase phosphorylation within the PSLD by Cyclin E/CDK2 unmasks a dominant nuclear export sequence leading to export of the sensor into the cytoplasm. Imaging and analysis of the nuclear/cytoplasmic distribution of the fusion protein allows quantitation of the G1/S block induced by siRNA knockdown of retinol binding protein 1 (RPB1) (**B**) relative to cells treated with a control siRNA (**A**). RBP1 controls cellular levels of retinol and retinol and its metabolites, including retinoic acid, are known to affect levels and activities of cell cycle control proteins. Retinoic acid reduces transcription of cyclins D and E and increases the activity of the CDK2 inhibitors p21 and p27. We hypothesize that in cells with reduced RBP1 these combined effects significantly reduce the ability of CyclinE-CDK2 and CyclinD-CDK4 to phosphorylate retinoblastoma protein and progress cells past the G1/S checkpoint.

Table 2
Design Elements for Engineered Cell Lines

Design element	Factors influencing choice
Target protein	Relevance to signaling pathway or process
	Reporting mechanism
	Timing of readout
	Specificity of readout
	Method of data abstraction
Host cell line	Ease of transfection
	Imaging quality
	Expression of pathway components
	Biological relevance
	Growth rate
	Biological containment category
	Tolerance and stability of fusion protein expression
Fluorescent protein	Suitable excitation and emission wavelengths for imaging instrumentation
	Compatibility with fusion partner folding and activity
	Sufficient brightness for imaging
	Color compatibility with cell staining
	Color compatibility with multiplexing
Construct order (FP-protein or protein-FP)	Protein terminus required for localization
	Exposed domain required for fusion protein function, interaction or processing
	Whole protein or domain
	Folding requirements of FP and fusion partner
Linker	Folding requirements of FP and fusion partner
Expression vector	Ease of cloning
	GMO containment category
	Selection marker
Promoter	Homologous or heterologous
	Level of expression compatible with biological relevance
	Level of expression required for imaging
Expression level	Interference or toxicity from over-expressed fusion protein
	Level of expression compatible with biological relevance
	Level of expression required for imaging

literature, perhaps with data from antibody staining showing changes in the localization of a key protein, engineering can proceed based on a fairly sound foundation. If on the other hand no imaging-based data is available for the pathway then it will be necessary to make some informed choices regarding one or more target proteins based on whatever background information is available.

At this stage in the design process a target can be any protein that undergoes a change in some characteristic, which is detectable by subcellular imaging, for example, the protein appears, or disappears or moves. The specificity of the readouts from candidate target proteins now has to be considered and weighed against the ease of obtaining quantitative data. In some cases a trade-off between specificity and analysis complexity might be required in order to arrive at a robust HCS assay. For example, choosing a target protein that undergoes a dramatic cytoplasmic to nuclear translocation on stimulation, such as nuclear factor (NF)-κB p65 *(67)*, will simplify the task of acquiring data by image analysis over the choice of a more subtle redistribution of fluorescence such as Rac movement to membrane ruffles *(68)*. However, this approach might not yield data of sufficient specificity if, like NF-κB p65, the target protein is involved in several cellular responses to external stimuli.

Similar considerations might have to be taken into account in choosing a target protein that undergoes a detectable change in intensity or location in a time-scale compatible with imaging. A protein that responds to a stimulus very quickly, for example, the rapid internalization of a membrane receptor *(69)*, might not allow sequential imaging of large numbers of tests and require more complex assay protocols requiring rapid imaging following stimulus.

In summary, the task at this point in the design process is to pick a target protein (or proteins) that will report a cellular response with the required specificity, with manageable kinetics for imaging, and with a change in fluorescence distribution and/or intensity that can be analyzed with available image analysis software.

3.2. Selecting a Host Cell Line

Picking the right cell type to engineer can be as important as picking the right protein; there is no point in making a detailed and informed selection of an assay target protein and then expressing it in a cell line that lacks a vital part of a signaling pathway, or that has morphology or growth characteristics that make it difficult or impossible to acquire images of sufficient quality for image analysis.

Historically cell lines for high-throughput screening using macro-imaging to measure luminescent reporter gene *(70)*, calcium flux *(71)*, and other low resolution cellular assays have been chosen principally on the basis of their ability to express large amounts of a drug target, typically a membrane receptor. In these assays the cell line is essentially a conveniently packaged collection of reagents *(72)* configured to allow drug activity at a massively overexpressed target to dominate over all other read-outs from the cell. This sledge–hammer approach to cell engineering has its place in primary drug screening; biological subtlety is not necessarily required or desirable when faced with weeding out hits from a million or so compounds.

However, moving from high-throughput screening to HCS (or indeed employing HCS in high-throughput screening) is generally motivated by the desire to acquire higher quality, more accurate, precise and informative, data on the effect of a drug, or other perturbation on a biological system. Obtaining higher quality information from a model system requires a proportionally higher level and quality of design.

Increasing development of stable cell lines for HCS has been accompanied by a move away from protein expression factories like HEK 293 and CHO to cells which are compatible with high throughput subcellular imaging such as U2OS *(73)*. Recent improvements in chemical transfection methods *(74)* and the use of retroviral *(75)* and baculovirus *(76)* vectors have largely removed transfection efficiency as a limitation on cell choice and allow many different cell lineages to be considered for engineering stable cell lines for HCS. If large numbers of cells are to be grown and screened consideration should be given to additional factors including the cell growth rate, the biological containment category for the parental cell and any restrictions on the use of genetically modified cells, all of which might impact on resources required to culture, manipulate and image the engineered cells.

The chosen host cell type should be tolerant of fusion protein expression both in the short term, for example, the target protein or domain should not be inherently toxic, and in the long term, for example, fusion expression should remain at a constant level over a useable number of passages. Insertion of transgenes into certain cell types is known to result in gradual loss of expression *(77)* either through transcriptional silencing or loss of the integrated transgene. If long-term stable expression is a key requirement *(78)* it might be necessary to test a number of cell types or clones for genetic stability and susceptibility to chromatin modification *(79,80)*. For discussion of cellular tolerance of fluorescent proteins and GFP in particular see the following section.

3.3. Selecting a Fluorescent Protein

Once the target protein and a suitable cell type have been identified for engineering the decision has to be made regarding which of the variety of available fluorescent proteins would be most suitable. The first and obvious choice is based on color, on the grounds that there's no point in making a fusion protein that's not compatible with one's HCS instrument setup. This explains the continued predominance of green fluorescent proteins; the widespread availability of instrumentation with laser or lamp illumination at fluorescein wavelengths makes these fluors an obvious first choice, and the availability of a red nuclear stain *(81)* allows rapid imaging in a single pass on multicamera systems.

Although instrument compatibility is obviously important, in some cases biological compatibility might have to be taken into account. If detection sensitivity is a key issue, either because the process being monitored requires it, or because it is desirable to minimize expression of the fusion protein to prevent overloading a cellular process, then it might be advantageous to select a fluorescent protein of a different color and increased brightness, for example, using enhanced yellow fluorescent protein (EYFP) instead of EGFP. Of course this selection only makes sense if any nominal gain in fluorescent protein brightness is not negated by physical factors, for example, use of suboptimal imaging filters, or biological factors, for example, less efficient protein folding.

All of the above assumes that choice of fluorescent protein color is not dictated by other assay design constraints. If the cell line is to be used in a multiplexed assay, or if particular fluorescent markers are required for image analysis, the colors of other probes or stains might influence the choice of fluorescent protein. In these cases all parameters should be evaluated; it might be that changing the color of another component might allow a preferred fluorescent protein to be used to maximize sensitivity and biological compatibility, for example, changing the fluor on a second antibody from fluorescein to Cy5 would allow EGFP to be used instead of a less sensitive red fluorescent protein.

3.4. Construct and Vector Design

Key factors here are whether a whole protein is required for fusion protein function and the orientation of the fusion protein construct, that is, whether the fluorescent protein is fused through the amino or carboxyl terminus. The optimum design for a fusion protein will depend on the properties of the fusion partner, for example, a particular terminus might need to be preserved to retain protein function and/or to ensure correct localization. However, retaining full protein function is not always possible, necessary, or desirable, for a functional assay. In some cases retaining full protein function, for example, enzymic activity and stimulus responsive translocation, might produce a poorer assay than a fusion that does not retain catalytic activity because of cellular perturbation resulting from overexpression of the enzyme. In such cases engineering the construct to mask the enzyme domain, or removal of the domain from the construct might produce a better fusion protein.

The decision of whether to use an intact protein or to remove unwanted activities from the fusion protein by using only part of the target protein will depend on the availability of published data. In some cases for well-characterized proteins it might be known that functional domains are portable, that is, they can be abstracted and appended to fusion proteins and retain partial or full functionality.

The two cell cycle phase marker constructs shown in **Fig. 3** use portable protein domains from well-characterized proteins. The key CRS and D-box domains *(82,83)* in the amino terminal region of Cyclin B1 used in engineering the G2/M CCPM provide all the necessary functionality to allow the EGFP fusion protein to shadow the localization and destruction of endogenous Cyclin B1 and, hence, report cell cycle position and progression. In this case use of domains from the target protein rather than the entire protein is imperative to allow engineering of a

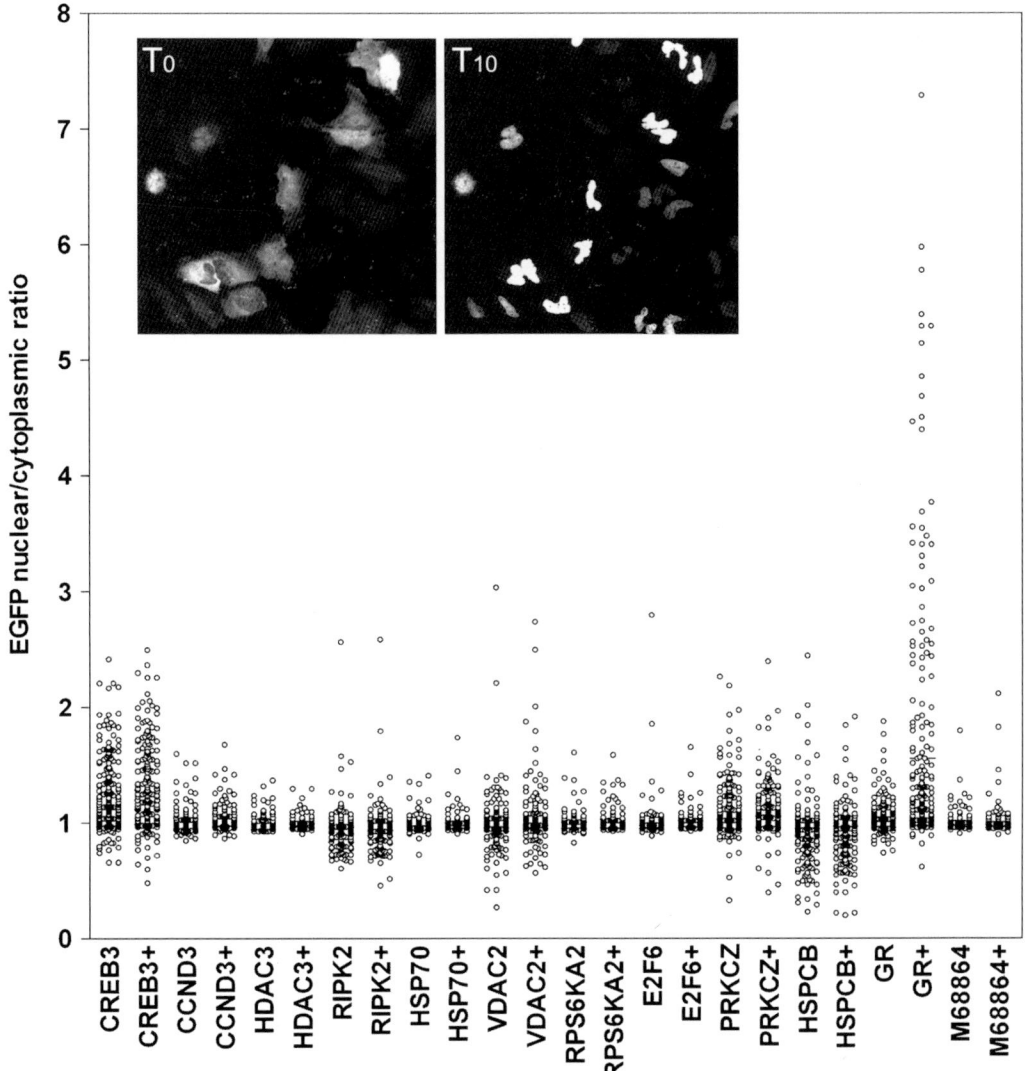

Fig. 6. Functional validation of EGFP fusions by screening transient expression. EGFP-cDNA fusion proteins previously screened for specific sub-cellular localization (**Fig. 4**) were evaluated for response to stimuli. HeLa cells expressing constructs were imaged before and after exposure to 10 μM dexamethasone and fusion protein translocation determined for individual cells by image analysis. Under these conditions an EGFP-glucocorticoid receptor fusion (GR) showed translocation from cytoplasm (T_0) to nucleus (T_{10}) in response to dexamethasone with significant increases in the ratio of nuclear/cytoplasmic EGFP fluorescence (GR+).

against a cellular signaling pathway for secondary screening or target validation will require extensive selection and validation to derive a cell line in which the biology under study is not perturbed by expression of the sensor.

4.1. Selection of Cells

Given that the primary task in cell selection is isolation of fluorescing cells from nonfluorescing cells it is not surprising that flow cytometry *(96)* is the method of choice for cell selection. Using fluorescence-activated cell sorting (FACS) to separate pools and individual cells with different levels of expression allows an efficient dual selection strategy to be used to combine some of the elements of assay validation with selection. In this approach cells are FACS sorted and pooled into populations with different levels of fluorescent protein expression, for example, high, medium, and low. Samples from these populations can then be used for cloning by limiting dilution into wells of a 96-well plate and the remaining cell populations expanded to test under assay conditions. If a FACS with single-cell separation capability is available cells can be directly isolated into wells of 96-well plates in a second run using the same gate settings.

Although cell clones are expanding the mixed populations can be tested under HCS conditions to determine the optimum level of fluorescent protein expression for the desired cellular response. This period also allows a certain amount of assay development work to be carried out to investigate different assay and analysis parameters, including devising a working image analysis strategy for abstracting data. Once cell clones are expanded, cells with suitable expression levels can be screened individually to select a cell line for HCS use. The final cell line should be reanalyzed by flow cytometry to check for uniform expression and if necessary FACS sorted before expansion and freezing down stock cells.

4.2. Validation of Stable Cells

Procedures used for validation of cells for HCS will vary widely depending on the nature of the fusion protein, the intended use of the cell line and the complexity of the biology involved, but generally involves two levels of testing; internal validation and external validation. First level internal validation tests the functionality of the engineered cell toward the assay it was designed to fulfill. This typically requires testing response to a series of known agonists/antagonists and establishing EC_{50}/IC_{50} values and order of potency, reflecting the use of the cell line in HCS. Data should be compared with published data available, but equivalence of EC_{50}/IC_{50} values should be interpreted with care, as these can vary considerably with assay procedure.

Second level external validation involves comparison of assay data generated using the fluorescent protein sensor and high content analysis against an independent assay measurement. This process can take a variety of forms, which may differ in their degree of independency depending on the analysis techniques available. The most stringent form of validation is to use a completely independent analysis process, for example, using propidium iodide staining of GFP expressing cells and flow cytometry to measure cell cycle distribution, which was the process used to validate the cell cycle sensors shown in **Fig. 3**. If a completely independent analysis procedure on a different instrument platform is not available, correlation of data from the fluorescent protein with data from an independent read-out in the same cells is a viable option, for example, correlation of translocation of the fusion protein with translocation of the endogenous target protein detected using antibody staining. For validation of the G1/S cell cycle sensor we used incorporation of bromodeoxyuridine as an independent marker for S-phase cells (**Fig. 7**). Further higher level biological validation will be needed in cell lines where it is required that introducing the fluorescent protein sensor does not perturb the very process it is designed to measure. Such studies will include establishing that the engineered cells display the same phenotype and growth characteristics as the parental cells (**Fig. 8**), and monitoring the effect of fluorescent protein expression on the biology of the engineered cells (*see* **Subheading 5.**).

5. Effect of Expressing Fluorescent Proteins on Cellular Biology

Expression of any recombinant protein in a cultured cell has the potential to perturb the biology of the cells, and might invalidate or obviate the purpose for which the engineered cell was created. Putting an additional load on cells by forcing them to express proteins fused to a large heterologous protein tag might disrupt or overload protein synthesis, trafficking, or proteolysis, leading to toxic effects. Despite the widespread use of fluorescent protein expressing cell lines, little work has been published on characterizing the effects of expressing tagged fusion proteins on the biology of the host cell.

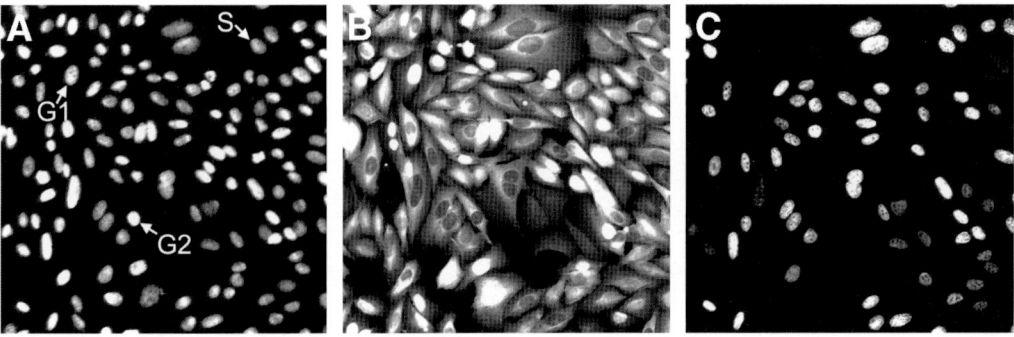

Fig. 7. Validation of EGFP fusion protein function against an independent sensor. U2OS cells expressing the G1/S cell cycle phase marker (**Fig. 3**) were pulse-labeled for 1 h with bromo-deoxyuridine (BrdU) and BrdU incorporation detected using a monoclonal anti-BrdU/nuclease and a Cy5 labeled second antibody. Three color imaging of (**A**) Hoechst stained DNA, (**B**) EGFP and (**C**) Cy5 allowed confirmation that the EGFP sensor was functioning as designed with intense nuclear EGFP fluorescence in nuclei of G1 cells (G1 arrowed), nuclear and cytoplasmic EGFP in BrdU positive S-phase cells (S arrowed), and predominantly cytoplasmic EGFP in G2 cells (G2 arrowed).

There is some historical evidence that in some circumstances very high levels of GFP expression might be toxic to mammalian cells *(97)*, and early work with novel fluorescent proteins such as DsRed revealed abnormal localization of the DsRed fusion proteins with aggregation and toxicity in some cases *(98,99)*.

However, it is now generally assumed that low to moderate expression of GFP or other monomeric fluorescent proteins is minimally perturbing to the host cell *(100)*. One study that examined the effect of stable expression of a GFP-α tubulin fusion in LLCPK-1 cells *(101)* showed that the mitotic index and doubling time were unchanged for cells expressing GFP-α tubulin as 17% of total tubulin, although endogenous tubulin expression was reduced compared to parental cells. A further study *(102)* examined the effects of expression of a GFP-estrogen receptor fusion in MCF-7 cells and demonstrated that GFP-estrogen receptor expression does not alter cell doubling time or cell cycle distribution and does not interfere with the induction of estrogen receptor responsive genes.

To investigate the cellular consequences of GFP expression we have examined the effect of GFP expression by analyzing cell cycle progression and gene expression in a GFP stable cell line and in parental U2OS cells *(103)*. The stable cell line *(104,105)* expresses a fusion of amino acids 1–170 from the amino terminus of Cyclin B1 coupled to EGFP, with expression under the control of the Cyclin B1 promoter. The EGFP fusion protein is consequently expressed and degraded in concert with endogenous Cyclin B1, but as the fusion protein lacks the C-terminal sequences necessary for Cyclin B1-CDK interaction, the EGFP fusion protein acts as a stealth sensor, shadowing endogenous Cyclin B1 expression and degradation without disturbing the cell cycle of cells in which it is expressed. To ensure minimal perturbation of the cell cycle the stable U2OS cell line was derived by screening a large number of clones and selecting a single clone with the minimal level of EGFP expression compatible with determination of cell cycle status by microscopy and image analysis. Measurement of EGFP fusion protein mRNA by quantitative RT-PCR showed the expression of the EGFP sensor to be equivalent to endogenous Cyclin B1 (7000 copies/cell in G2).

Analysis of cell cycle duration and cell cycle phase distribution by cell growth assays and flow cytometry revealed that the two cell lines had identical doubling times and cell cycle distributions. Microarray analysis showed a 0.9% (>twofold at $p < 0.001$ across 20,000 genes) difference in gene expression levels between parental and EGFP expressing U2OS cells, with no significant

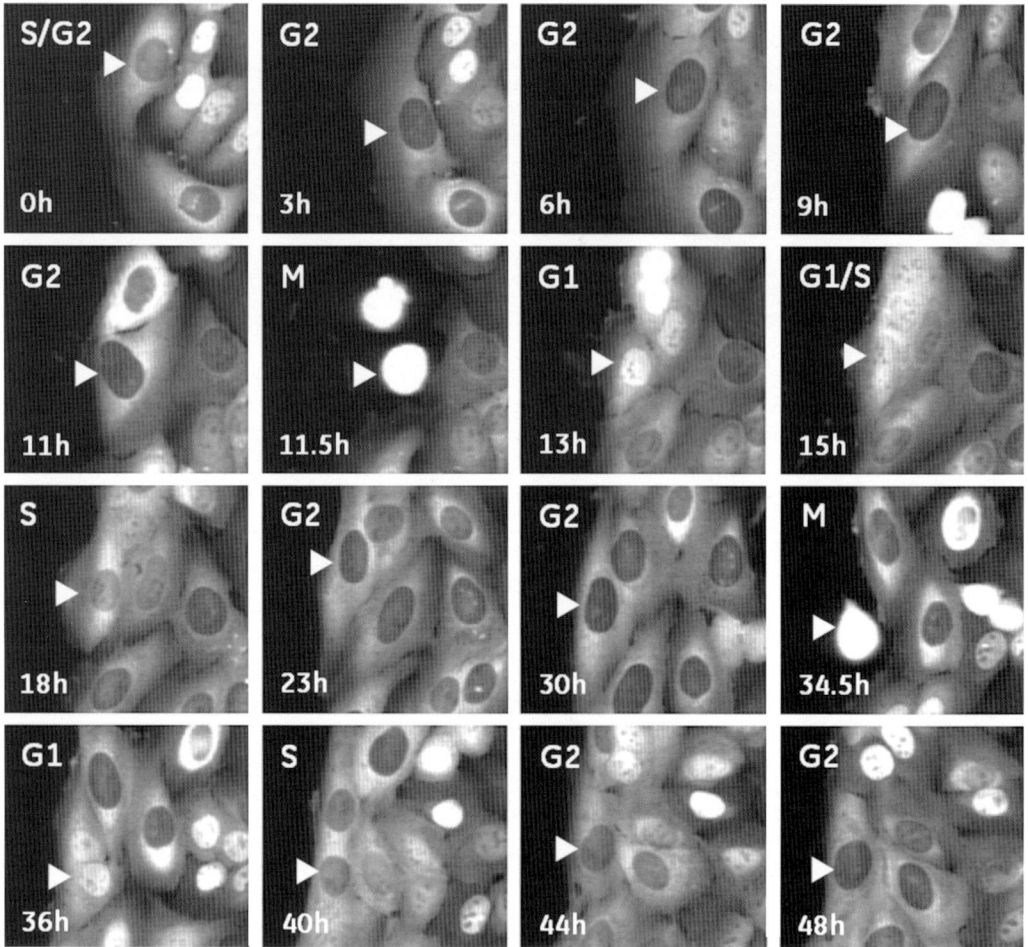

Fig. 8. Phenotype validation of EGFP expressing cells. U2OS cells expressing the G1/S cell cycle phase marker were imaged in culture over a 48 h period at intervals of 30 min. Examination of time lapse images showed a mitosis to mitosis time of 23 h (arrowed cell in frame 6 to frame 12 of those shown), an equivalent cell cycle time to parental U2OS cells, indicating that fusion protein expression does not prolong cell cycle transition.

differences in expression of Cyclins, CDKs, or CDK inhibitors between the two cell types. We conclude that engineering stable cell lines for low expression of EGFP fusion proteins is minimally perturbing to cellular biology.

6. Alternative Labels to Fluorescent Proteins for Engineering Cells for HCS

In recent years a number of genetically encoded labels have been developed for tagging proteins in living cells, which offer an alternative to fluorescent proteins. One method engineers recombinant proteins to insert four cysteines residues in an α-helix, which are subsequently labeled with arsenical fluorescein derivative *(106)*. The FlAsH label is cell permeable and nonfluorescent until it binds with high affinity and specificity to the tetracysteine domain. Although this protein tagging method adds less mass to the recombinant protein than the equivalent protein expressed as a GFP fusion, low affinity of binding requiring high expression of the target motif and high concentrations of the label with possible cellular toxicity limited adoption of the method. Recent developments of the technique *(107)* have increased the utility of the method by improving labeling affinity using a modified hairpin-helix binding site and provision of a red fluorescing resorufin label (ReAsH). Use of the improved method has been reported for examining gap junction formation *(108)*, AMPA receptor trafficking in neuronal cells *(109)* and HIV-1 Gag protein localization in a range of cell types *(110)*.

A second method, termed HaloTag *(111)*, uses a haloalkane dehalogenase from *Rhodococcus rhodochrous* (mDhaA), which has been mutated to form a stable covalent bond when it binds a chlorinated substrate analog. Fusion of the 33-kDa mDhaA to a cellular target protein allows tagging with cell-permeable rhodamine and fluorescein ligands. Although this method slightly increases the size of the tagged fusion protein relative to an equivalent GFP fusion, the ability to use different fluorescent labels with the same engineered cells might be advantageous in some circumstances. However, because in contrast to the FlAsH and ReAsH labels previously described mDhaA ligands are fluorescent in free solution cells must be washed to remove unbound ligand preceding to imaging.

A third recently developed method for labeling recombinant proteins in living cells SNAP-tag *(112)*, labels fusion proteins of a mutated form of human alkylguanine-DNA alkyltransferase (AGT) with fluorescent ligands. Labeling is based on the irreversible and specific reaction of AGT with O^6-benzylguanine derivatives, leading to the transfer of the label to a reactive cysteine residue within the active site of AGT. Fusion proteins produced for labeling with this method are slightly smaller than fusions with GFP or other fluorescent proteins; however, as with HaloTag labels, the fluorescein and rhodamine derivatives used to label AGT are fluorescent at all times and must be removed before imaging. Additionally, as the method uses a mutant form of a human enzyme, some labeling of endogenous AGT might occur in some cell types.

Although the labeling methods described above offer alternative approaches to visualizing proteins in living cells, and allow the flexibility of changing fluorescence wavelengths without the need to engineer new fusion proteins, they are as yet unproven in long-term culture and in stable cell lines. Moreover, this flexibility comes at the price of more complex experimental protocols with additional labeling and washing regimes. Finally, and perhaps most importantly, there is the issue of the compatibility of these synthetic labels with the dynamics of cellular processes. All cellular proteins go through a life cycle of expression, action and destruction, and all cells have the necessary machinery to accomplish this. The chromophore of fluorescent proteins is integral to the protein structure; therefore, provided that cells are not overloaded with recombinant protein, the fluorescence of the fusion protein will follow the natural life cycle of the protein. This is not the case for fusion proteins labeled with synthetic fluors; proteolysis of the fusion protein will not necessarily destroy the fluorescent label. This might be particularly problematic for proteins which are rapidly turned over. In such cases the population of specifically labeled proteins will be quickly depleted although the background fluorescence from liberated label increases, giving a limited period following labeling in which to carry out any analysis. This feature of synthetic labels might preclude their use in many areas of cell biology, for example, the cell cycle, where key proteins are constantly degraded and replenished. Synthetically, tagged proteins represent a finite and rapidly depleted cellular resource, fluorescent proteins are a constantly renewable resource.

7. Future Perspectives

Although to date the majority of HCS assays using fluorescent proteins have used stable cell lines, recent developments in viral vector systems are opening the way for use of transient expression as a flexible means of engineering cells for HCS.

Expression of fluorescent fusion proteins using adenoviral *(113–116)* and baculoviral *(117,118)* expression systems allows rapid development of HCS assays without the time delays inherent in producing stable cell lines, and permits assays to be run in a variety of different cell types. Our laboratory is currently generating a large panel of adenovirally encoded sensors using fluorescent fusion proteins and other sensors *(119)* for use in target validation and lead profiling. These constructs open the way to developing HCS assays in cell types, including primary cells, which offer a more physiologically relevant background for analysis of gene and drug function than standard transformed laboratory cell lines.

Ongoing research using stem cell lines engineered with GFP fusions *(120,121)* has the potential in the near future to generate a variety of differentiated cells expressing sensors for HCS in a cellular background, which closely matches the phenotype and physiology of diseased and normal tissue. Coupling of these new approaches to cellular engineering with ongoing advances in image analysis, which allow multiparameter phenotypic and morphological analysis will continue to advance the power of HCS to analyze complex cellular systems.

Acknowledgments

The author thanks June Davies, Helen Hay, Suzanne Hancock, and Simon Stubbs for the data in **Figs. 4–8**.

References

1. Prasher, D. C., Eckenrode, V. K., Ward, W. W., Prendergast, F. G., and Cormier, M. J. (1992) Primary structure of the *Aequorea victoria* green-fluorescent protein. *Gene* **111(2)**, 229–233.
2. Chalfie, M., Tu, Y., Euskirchen, G., Ward, W. W., and Prasher, D. C. (1994) Green fluorescent protein as a marker for gene expression. *Science* **263(5148)**, 802–805.
3. Inouye, S. and Tsuji F. I. (1994) *Aequorea* green fluorescent protein. Expression of the gene and fluorescence characteristics of the recombinant protein. *FEBS Lett.* **341(2–3)**, 277–280.
4. Chalfie, M. (1995) Green fluorescent protein. *Photochem. Photobiol.* **62(4)**, 651–656.
5. Stearns, T. (1995) Green fluorescent protein. The green revolution. *Curr. Biol.* **5(3)**, 262–264.
6. Prasher, D. C. (1995) Using GFP to see the light. *Trends Genet.* **11(8)**, 320–323.
7. Tsien, R. Y. (1998) The green fluorescent protein. *Annu. Rev. Biochem.* **67**, 509–544.
8. Heim, R., Cubitt, A. B., and Tsien, R. Y. (1995) Improved green fluorescence. *Nature* **373(6516)**, 663, 664.
9. Cormack, B. P., Valdivia, R. H., and Falkow, S. (1996) FACS-optimized mutants of the green fluorescent protein (GFP). *Gene* **173(1 Spec No)**, 33–38.
10. Zolotukhin, S., Potter, M., Hauswirth, W. W., Guy, J., and Muzyczka, N. (1996) A "humanized" green fluorescent protein cDNA adapted for high-level expression in mammalian cells. *J. Virol.* **70(7)**, 4646–4654.
11. Zhang, G., Gurtu, V., and Kain, S. R. (1996) An enhanced green fluorescent protein allows sensitive detection of gene transfer in mammalian cells. *Biochem. Biophys. Res. Commun.* **227(3)**, 707–711.
12. Miyawaki, A. (2002) Green fluorescent protein-like proteins in reef Anthozoa animals. *Cell Struct. Funct.* **27(5)**, 343–347.
13. Carter, R. W., Schmale, M. C., and Gibbs, P. D. (2004) Cloning of anthozoan fluorescent protein genes. *Comp. Biochem. Physiol. C. Toxicol. Pharmacol.* **138(3)**, 259–270.
14. Karasawa, S., Araki, T., Yamamoto-Hino, M., and Miyawaki, A. (2003) A green-emitting fluorescent protein from galaxeidae coral and its monomeric version for use in fluorescent labeling. *J. Biol. Chem.* **278(36)**, 34,167–34,171.
15. Campbell, R. E., Tour, O., Palmer, A. E., et al. (2002) A monomeric red fluorescent protein. *Proc. Natl. Acad. Sci. USA* **99(12)**, 7877–7882.
16. Bulina, M. E., Chudakov, D. M., Mudrik, N. N., and Lukyanov, K. A. (2002) Interconversion of Anthozoa GFP-like fluorescent and non-fluorescent proteins by mutagenesis. *BMC Biochem.* **3(1)**, 7.
17. Fradkov, A. F., Chen, Y., Ding, L., Barsova, E. V., Matz, M. V., and Lukyanov S. A. (2000) Novel fluorescent protein from discosoma coral and its mutants possesses a unique far-red fluorescence. *FEBS Lett.* **479(3)**, 127–130.
18. Tsien, R. Y. (2005) Building and breeding molecules to spy on cells and tumors. *FEBS Lett.* **579(4)**, 927–932.
19. Zhang, J., Campbell, R. E., Ting, A. Y., and Tsien, R. Y. (2002) Creating new fluorescent probes for cell biology. *Nat. Rev. Mol. Cell Biol.* **3(12)**, 906–918.

20. Shaner, N. C., Campbell, R. E., Steinbach, P. A., Giepmans, B. N., Palmer, A. E., and Tsien, R. Y. (2004) Improved monomeric red, orange and yellow fluorescent proteins derived from Discosoma sp. red fluorescent protein. *Nat. Biotechnol.* **22(12)**, 1567–1572.
21. Kendall, J. M. and Stubbs S. (2002) Fluorescent proteins in cellular assays. *J. Clin. Ligand Assay* **25**, 280–292.
22. Miyawaki, A., Sawano, A., and Kogure, T. (2003) Lighting up cells: labelling proteins with fluorophores. *Nat. Cell Biol.* **(Suppl)**, S1–S7.
23. Chudakov, D. M., Verkhusha, V. V., Staroverov, D. B., Souslova, E. A., Lukyanov, S., and Lukyanov, K. A. (2004) Photoswitchable cyan fluorescent protein for protein tracking. *Nat. Biotechnol.* **22(11)**, 1435–1439.
24. Lippincott-Schwartz, J. and Patterson G. H. (2003) Development and use of fluorescent protein markers in living cells. *Science.* **300(5616)**, 87–91.
25. Terskikh, A., Fradkov, A., Ermakova, G., et al. (2000) "Fluorescent timer": protein that changes color with time. *Science* **290(5496)**, 1585–1588.
26. Sampaio, K. L., Cavignac, Y., Stierhof, Y. D., and Sinzger, C. (2005) Human cytomegalovirus labeled with green fluorescent protein for live analysis of intracellular particle movements. *J. Virol.* **79(5)**, 2754–2767.
27. Southward, C. M. and Surette, M. G. (2002) The dynamic microbe: green fluorescent protein brings bacteria to light. *Mol. Microbiol.* **45(5)**, 1191–1196.
28. Kohlwein, S. D. (2000) The beauty of the yeast: live cell microscopy at the limits of optical resolution. *Microsc. Res. Tech.* **51(6)**, 511–529.
29. Bowerman, B. (2001) Cytokinesis in the *C. elegans* embryo: regulating contractile forces and a late role for the central spindle. *Cell Struct. Funct.* **26(6)**, 603–607.
30. Jacinto, A., Woolner, S., and Martin P. (2002) Dynamic analysis of dorsal closure in Drosophila: from genetics to cell biology. *Dev. Cell* **3(1)**, 9–19.
31. Zhu, H. and Zon L. I. (2004) Use of the DsRed fluorescent reporter in zebrafish. *Methods Cell Biol.* **76**, 3–12.
32. Jin, S., McKee, T. D., and Oprian, D. D. (2003) An improved rhodopsin/EGFP fusion protein for use in the generation of transgenic *Xenopus laevis*. *FEBS Lett.* **542(1–3)**, 142–146.
33. Hadjantonakis, A. K., Dickinson, M. E., Fraser, S. E., and Papaioannou, V. E. (2003) Technicolour transgenics: imaging tools for functional genomics in the mouse. *Nat. Rev. Genet.* **4(8)**, 613–625.
34. Cabot, R. A., Kuhholzer, B., Chan, A. W., et al. (2001) Transgenic pigs produced using in vitro matured oocytes infected with a retroviral vector. *Anim. Biotechnol.* **12(2)**, 205–214.
35. Chan, A. W., Chong, K. Y., and Schatten, G. (2002) Transgenic bovine embryo selection using green fluorescent protein. *Methods Mol. Biol.* **183**, 201–214.
36. Chan, A. W., Chong, K. Y., Martinovich, C., Simerly, C., and Schatten G. (2001) Transgenic monkeys produced by retroviral gene transfer into mature oocytes. *Science* **291(5502)**, 309–312.
37. Galbraith, W., Wagner, M. C., Chao, J., et al. (1991) Imaging cytometry by multiparameter fluorescence. *Cytometry* **12(7)**, 579–596.
38. Giuliano, K. A., Post, P. L., Hahn, K. M., and Taylor D. L. (1995) Fluorescent protein biosensors: measurement of molecular dynamics in living cells. *Annu. Rev. Biophys. Biomol. Struct.* **24**, 405–434.
39. Ding, G. J., Fischer, P. A., Boltz, R. C., et al. (1998) Characterization and quantitation of NF-kappaB nuclear translocation induced by interleukin-1 and tumor necrosis factor-alpha. Development and use of a high capacity fluorescence cytometric system. *J. Biol. Chem.* **273(44)**, 28,897–28,905.
40. Waggoner, A., Taylor, L., Seadler, A., and Dunlay, T. (1996) Multiparameter fluorescence imaging microscopy: reagents and instruments. *Hum. Pathol.* **27(5)**, 494–502.
41. Htun, H., Barsony, J., Renyi, I., Gould, D. L., and Hager, G. L. (1996) Visualization of glucocorticoid receptor translocation and intranuclear organization in living cells with a green fluorescent protein chimera. *Proc. Natl. Acad. Sci. USA.* **93(10)**, 4845–4850.
42. Carey, K. L., Richards, S. A., Lounsbury, K. M., and Macara, I. G. (1996) Evidence using a green fluorescent protein-glucocorticoid receptor chimera that the Ran/TC4 GTPase mediates an essential function independent of nuclear protein import. *Cell Biol.* **133(5)**, 985–996.

43. Giulano, K. A., DeBiasio, R. L., Dunlay, R. T., et al. (1997) High content screening: a new approach to easing key bottlenecks in the drug discovery process. *J. Biomol. Screen.* **2(4)**, 249–259.
44. Conway, B. R., Minor, L. K., Xu, J. Z., et al. (1999) Quantification of G-protein coupled receptor internalization using G-protein coupled receptor-green fluorescent protein conjugates with the arrayscan high content screening system. *J. Biomol. Screen.* **4**, 75–86.
45. Giuliano, K. A. and Taylor, D. L. (1998) Fluorescent-protein biosensors: new tools for drug discovery. *Trends Biotechnol.* **16(3)**, 135–140.
46. Kain, S. R. (1999) Green fluorescent protein (GFP): applications in cell-based assays for drug discovery. *Drug Discov. Today* **4(7)**, 304–312.
47. Thomas, N. (2001) Cell based assays–seeing the light. *Drug Discov. World* **3(1)**, 25–31.
48. Ramm, P. and Thomas, N. (2003) Image-based screening of signal transduction assays. *Sci. STKE* **(177)**, PE14.
49. Price, J. H., Goodacre, A., Hahn, K., et al. (2002) Advances in molecular labeling, high throughput imaging and machine intelligence portend powerful functional cellular biochemistry tools. *J. Cell. Biochem. Suppl.* **39**, 194–210.
50. Ferguson, S. S. and Caron, M. G. (2004) Green fluorescent protein-tagged beta-arrestin translocation as a measure of G protein-coupled receptor activation. *Methods Mol. Biol.* **237**, 121–126.
51. McLean, A. J. and Milligan, G. (2000) Ligand regulation of green fluorescent protein-tagged forms of the human beta(1)- and beta(2)-adrenoceptors: comparisons with the unmodified receptors. *Br. J. Pharmacol.* **130(8)**, 1825–1832.
52. Yoon, Y., Pitts, K., and McNiven, M. (2002) Studying cytoskeletal dynamics in living cells using green fluorescent protein. *Mol. Biotechnol.* **21(3)**, 241–250.
53. Zaccolo, M. and Pozzan, T. (2000) Imaging signal transduction in living cells with GFP-based probes. *IUBMB Life* **49(5)**, 375–379.
54. Filhol, O., Nueda, A., Martel, V., et al. (2003) Live-cell fluorescence imaging reveals the dynamics of protein kinase CK2 individual subunits. *Mol. Cell Biol.* **23(3)**, 975–987.
55. Mandell, J. W. and Gocan, N. C. (2001) A green fluorescent protein kinase substrate allowing detection and localization of intracellular ERK/MAP kinase activity. *Anal. Biochem.* **293(2)**, 264–268.
56. Kimura, H. and Cook, P. R. (2001) Kinetics of core histones in living human cells: little exchange of H3 and H4 and some rapid exchange of H2B. *J. Cell Biol.* 2001; **153(7)**, 1341–1353.
57. Waguri, S., Dewitte, F., Le Borgne, et al. (2003) Visualization of TGN to endosome trafficking through fluorescently labeled MPR and AP-1 in living cells. *Mol. Biol. Cell* **14(1)**, 142–155.
58. Torrance, C. J., Agrawal, V., Vogelstein, B., and Kinzler, K. W. (2001) Use of isogenic human cancer cells for high-throughput screening and drug discovery. *Nat. Biotechnol.* **19(10)**, 940–945.
59. Lundholt, B. K., Linde, V., Loechel, F., et al. (2005) Identification of Akt pathway inhibitors using redistribution screening on the FLIPR and the in cell 3000 analyzer. *J. Biomol. Screen.* **10(1)**, 20–29.
60. Almholt, K., Tullin, S., Skyggebjerg, O., Scudder, K., Thastrup, O., and Terry, R. (2004) Changes in intracellular cAMP reported by a Redistribution assay using a cAMP-dependent protein kinase-green fluorescent protein chimera. *Cell Signal.* **16(8)**, 907–920.
61. Almholt, D. L., Loechel, F., Nielsen, S. J., et al. (2004) Nuclear export inhibitors and kinase inhibitors identified using a MAPK-activated protein kinase 2 redistribution screen. *Assay Drug Dev. Technol.* **2(1)**, 7–20.
62. Schlag, B. D., Lou, Z., Fennell, M., and Dunlop, J. (2004) Ligand dependency of 5-hydroxytryptamine 2C receptor internalization. *J. Pharmacol. Exp. Ther.* **310(3)**, 865–870.
63. Ghosh, R. N., Chen, Y. T., DeBiasio, R., et al. (2000) Cell-based, high content screen for receptor internalization, recycling and intracellular trafficking. *Biotechniques* **29(1)**, 170–175.
64. Cooke, E-L., Ainscow, E., Hargreaves, A., et al. (2003) G-protein-coupled receptor high-throughput screen using norak Transfluor® technology and the IN cell analyser 3000, *The Society for Biomolecular Screening 9th Annual Conference,* Portland, Oregon, 22, September, 2003.
65. Barak, L. S., Ferguson, S. S., Zhang, J., and Caron, M. G. (1997) A beta-arrestin/green fluorescent protein biosensor for detecting G protein-coupled receptor activation. *J. Biol. Chem.* **272(44)**, 27,497–27,500.
66. www.xsira.com. Accessed Feb, 27, 2006.
67. Ariga, A., Namekawa, J., Matsumoto, N., Inoue, J., and Umezawa, K. (2002) Inhibition of tumor necrosis factor-alpha-induced nuclear translocation and activation of NF-kappa B by dehydroxymethylepoxyquinomicin. *J. Biol. Chem.* 2002, **277(27)**, 24,625–24,630.

13

Optimizing the Integration of Immunoreagents and Fluorescent Probes for Multiplexed High Content Screening Assays

Kenneth A. Giuliano

Summary

Immunoreagents formed the basis of early fixed end point high content screening (HCS) assays and their use in HCS applications in drug discovery will continue to increase. One important application of immunoreagents is their incorporation into multiplexed HCS assays in which multiple physiological features are simultaneously measured and related in the same cells. However, creating multiplexed HCS assays that incorporate multiple immunoreagents presents issues such as reagent compatibility, spectral signal overlap, and reproducibility that must be addressed. Here, an example multiplexed fixed end point HCS assay is used to guide potential assay developers on how to optimize complex, yet cellular information rich, multiplexed HCS assays although avoiding some common pitfalls.

Key Words: Cell cycle; fluorescence-based immunoassays; high-throughput screening; microtubule cytoskeleton; systems cell biology; tumor suppressors.

1. Introduction

Fluorescent immunoreagents have become important tools for fixed end point high content screening (HCS) because of their exquisite specificity and sensitivity. Furthermore, fluorescent immunoreagents can also be combined with additional immunoreagents as well as other fluorescent physiological indicators and biosensors in the same cells to produce multiplexed HCS assays.

With the commercial availability of thousands of immunoreagents and fluorescent probes, large numbers of multiplexed fixed end point HCS assays are possible. However, the compatibility of reagents when integrated into a single assay can be problematic. Furthermore, multiplexed HCS assays involving the use of several immunoreagents have an inherent complexity that demands strict attention to detail to achieve a validated assay for HCS. Nevertheless, the systems cell biology knowledge gained from multiplexed fixed end point HCS assays more than compensates for the extra assay development effort required.

In the drug discovery process, simultaneous measurement of the effects of potential lead compounds on multiple cellular pathways has become a valuable tool, especially when applied early on in the discovery process to create systems cell biology knowledge *(1)*. Thus, the following multiplexed HCS assay, which involves the use of six immunoreagents plus a fluorescent DNA-binding probe provides information on how libraries of compounds affect cell cycle regulation, the degradation of DNA and changes in nuclear morphology associated with apoptosis, the stability of the microtubule cytoskeleton, the activation of the tumor suppressor protein p53, and the activity of a signaling kinase associated with chromatin structure. A brief description of how to design and perform a four-color multiplexed HCS assay is presented. In addition, several reagent issues unique to multiplexed HCS assay development are also considered and addressed.

2. Materials

1. The cell permeabilization reagent Triton X-100, the DNA-binding fluorescent probe Hoechst 33342, mouse anti-α-tubulin antibody (no. 6), dimethyl sulfoxide (DMSO), 5-fluorouracil, and vinblastine are from Sigma Chemical Company (St. Louis, MO).
2. Cell culture medium, antibiotics, fetal bovine serum, trypsin, 37% formaldehyde, and Hank's balanced salt solution (HBSS), and 384-well microplates (Falcon, cat. no. 3962) are obtained from Fisher Scientific (Pittsburgh, PA).
3. Sheep antihuman p53 pantropic polyclonal antibody (cat. no. PC35) is purchased from Calbiochem (La Jolla, CA) (see **Note 1**).
4. Rabbit antiphospho histone H3 antibody (cat. no. 70) is obtained from Upstate, Inc. (Charlottesville, VA).
5. The fluorescently labeled secondary antibodies, which include fluorescein isothiocyanate (FITC)-labeled donkey antimouse IgG (cat. no. 715-095-150), Cy3-labeled donkey antirabbit IgG (cat. no. 65-152), and donkey antisheep IgG (cat. no. 713-175-147) are from Jackson ImmunoResearch Laboratories, Inc. (West Grove, PA) (see **Note 2**).

3. Methods

1. For multiplexed HCS assays, many primary cell types and established cell lines have been used. In this example, A549 lung carcinoma cells (CCL-195; ATCC, Manassas, VA) were used. The cells were cultured in Ham's F12 medium plus 10% fetal bovine serum and penicillin/streptomycin. For this type of multiplexed HCS assay, it was important to use actively growing and dividing cells. The most reproducible way to obtain these cells was to allow them to grow in a culture vessel for not more than 36 h. We often use a "shuffling" procedure in which cells are trypsinized and plated into large culture vessels (e.g., T-150 or T-175 flasks) on the day before transferring them to microplates.
2. For the multiplexed HCS assay, cells were trypsinized from flasks and plated at a density of 7500–8000 cells per well (40 µL) in 384-well microplates that were coated with collagen I using an automated liquid handling system (Biomek® 2000; Beckman-Coulter, Inc., Fullerton, CA). This alternative to precoated microplates was preferred for two main reasons: (1) economic and (2) manually coating microplates provided cell morphologies much more amenable to high-content imaging than did the precoated microplates. Furthermore, it was useful and relatively easy to prepare rat tail collagen I solutions using established procedures *(2)*.
3. Cells were exposed to drugs within 2–8 h of plating. Concentrated stocks of all drugs were diluted into solutions of HBSS plus 10% fetal bovine serum and added to the microplates (10 µL per well). For this assay, which is optimized to measure the effects of compounds on the regulation of the cell cycle, cells were incubated in the presence of compounds for 20–24 h (see **Note 3**).
4. After incubation with compounds, the solution was removed from the microplates by shaking them out and immediately replaced with a solution of HBSS containing 4% formaldehyde to fix the cells.
5. After incubation at room temperature for 30 min, the solution was removed from each well and replaced with HBSS (100 µL/well). At this point, microplates could be sealed and stored at 4°C overnight.
6. After removing the HBSS from each well, 0.5% (w/w) Triton X-100 in HBSS was added (10 µL/well) and the plate incubated for 5 min at room temperature to detergent extract a fraction of the soluble cellular components including destabilized tubulin.
7. The wells were washed with HBSS (100 µL/well) followed by the addition of a primary antibody solution containing mouse anti-α-tubulin (1/3000), rabbit antiphospho histone H3 (1/500), and sheep antihuman p53 (1/400) in HBSS (10 µL/well) (see **Notes 4** and **5**).
8. After a 1 h incubation at room temperature, the microplate wells were washed with HBSS as in **step 7** followed by the addition of a secondary antibody solution containing FITC-labeled donkey antimouse (1/300), Cy3-labeled donkey antirabbit (1/300), and Cy5-labeled donkey antisheep (1/300) antibodies diluted in HBSS containing 10 µg/mL Hoechst 33342 (10 µL/well).
9. After a 1 h incubation at room temperature, the microplate wells were washed as above and HBSS was added (100 µL/well) before sealing the microplates. Labeled microplates could be stored at 4°C for up to 2 wk before high content analysis.

10. HCS of microplates prepared using the above method is platform independent provided that the HCS reader employed has the capability to image all four of the fluorescent labels. HCS was routinely performed with a V3.1 ArrayScan® HCS Reader (Cellomics, Inc., Pittsburgh, PA). In this example, the instrument was used to scan multiple optical fields, each with multiplexed fluorescence, within a subset of the wells of a 384-well microplate. Typically, 1000 cells per well were measured and was usually accomplished by scanning three to four fields per well. However, significant effects induced by compounds could also be measured by scanning only one field per well, which typically provided 250–300 cell measurements. Thus, a trade-off exists between the sample scanning rate and the confidence with, which cellular responses can be measured.

11. Methods for the detailed interpretation of multiplexed HCS assay data have been previously described *(3–7)*. For this example, which is focused on immunoreagent optimization, some representative multiplexed HCS assay images of cells that were either untreated or were treated with two drugs, vinblastine and 5-fluorouracil are provided (**Fig. 1**). These images show that the signals from each of the fluorescent labels were well separated and balanced such that crosstalk between the fluorescence channels was minimal. Visual inspection of the images from this assay revealed the profiling capabilities of the multiplexing approach, but the resulting conclusions only a fraction of the information that was extracted using imaging algorithms. For example, 100 nM vinblastine induced nuclear condensation, microtubule destabilization (note the loss of microtubule structure), and increased histone H3 phosphorylation, whereas 10 μM 5-fluorouracil had little visually detectable effect on nuclear morphology and microtubule stability. However, it is easy to discern that 5-fluorouracil produced a considerable activation of p53 as well as inhibition of histone H3 phosphorylation to levels below that of cells treated with DMSO. Therefore, this brief example shows how multiplexed immunoreagents have the potential to generate large amounts of systems cell biology knowledge when coupled with the appropriate bioinformatics tools *(7,8)*.

4. Notes

1. Storage of primary antibody solutions in a liquid form at –20°C provided the most flexibility in developing and executing HCS assays. Thus, primary antibody stock solutions as received from the manufacturer were routinely diluted 1:2 with glycerol, if not done so by the manufacturer, and stored at –20°C in the liquid form until use.

2. Secondary antibodies optimized for multicolor labeling is used for multiplexed HCS assays are highly recommended. Fluorescently labeled secondary antibody solutions were stored in the dark at 4°C as recommended by the manufacturer.

3. In designing multiplexed fixed end point HCS assays, careful attention must be paid to building an acceptable suite of immunoreagents. In the example we showed here, a primary–secondary antibody labeling approach was used that relied on three primary antibody reagents raised in three types of animals. In general, choosing compatible primary immunoreagents is the most difficult part of multiplexed fixed end point HCS assay development. However, the increasing commercial availability of fluorescently labeled primary antibodies and other fluorescent physiological indicators and biosensors is reducing the complexity of multiplexed HCS assay design.

4. Once the primary antibodies have been chosen, the obvious controls in which cells are labeled with individual primary–secondary antibody pairs must be performed. It is important during this step to balance the fluorescence signals from each primary–secondary antibody pair. Balancing the signals can be accomplished by varying the dilution strengths of each of the primary antibodies as well as the secondary antibodies, if necessary. The fluorescence signal from a single primary–secondary antibody pair is deemed too strong if it overflows into one or more of the other fluorescence channels. In the example presented here, a combination of the quality of the antitubulin antibody and the high-intracellular concentration of tubulin dictated that a high dilution of the primary antitubulin antibody relative to the dilutions of the other primary antibodies was necessary.

5. After optimization of the primary–secondary antibody pairs in isolation, mixtures of these reagents can be tested for compatibility. In the procedure presented here, it was possible to incubate the reagents with cells using a mixture of all three primary antibodies followed by incubation with a mixture of all three secondary antibodies. Occasionally, immunoreagent incompatibilities arise and are often solved by doing serial labeling where necessary. Finally, nonimmunoreagents (e.g., Hoechst 33342 in this example, labeled phalloidins, fixable small molecule physiological indicators, and so on) are added to the

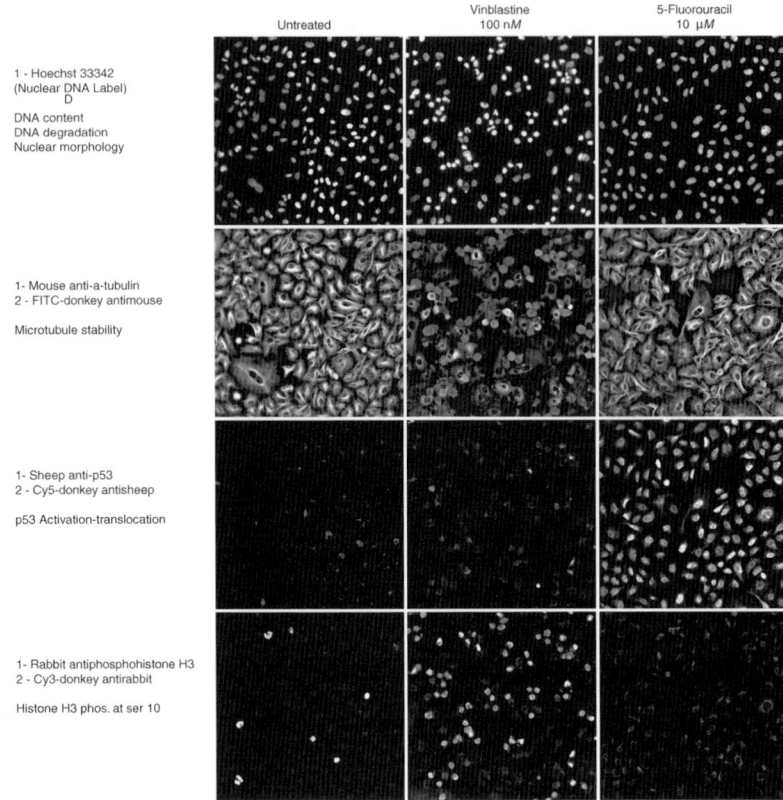

Fig. 1. An example multiplexed HCS assay that incorporates multiple immunoreagents and a physiological indicator probe. A549 human lung tumor cells were treated with the anticancer agents vinblastine and 5-fluorouracil for 24 h and then labeled for multiplexed HCS using the reagents indicated. Example cellular feature measurements are shown in bold next to the images in which the information was extracted using the HCS bioapplication. Thus, visual inspection alone shows the compatibility of the multiplexed reagents and that each drug produced an individual response profile.

labeling protocol where necessary. For example, Hoechst 33342 could be incubated with cells along with the labeled secondary antibodies.

References

1. Taylor, D. L. and Giuliano, K. A. (2005) Multiplexed high content screening assays create a systems cell biology approach to drug discovery. *Drug Discov. Today: Technol.* **2,** 149–154.
2. Strom, S. C. and Michalopoulos, G. (1982) Collagen as a substrate for cell growth and differentiation. *Methods Enzymol.* **82 A,** 544–555.
3. Vogt, A., Tamewitz, A., Skoko, J., Sikorski, R. P., Giuliano, K. A., and Lazo, J. S. (2005) The benzo (C) phenanthridine alkaloid, sanguinarine, is a selective, cell-active inhibitor of mitogen-activated protein kinase phosphatase-1. *J. Biol. Chem.* **280,** 19,078–19,086.
4. Giuliano, K. A., Chen, Y. T., and Taylor, D. L. (2004) High content screening with siRNA optimizes a cell biological approach to drug discovery: defining the role of p53 activation in the cellular response to anticancer drugs. *J. Biomol. Screen.* **9,** 557–568.
5. Giuliano, K. A. (2003) High-content profiling of drug-drug interactions: cellular targets involved in the modulation of microtubule drug action by the antifungal ketoconazole. *J. Biomol. Screen.* **8,** 125–135.

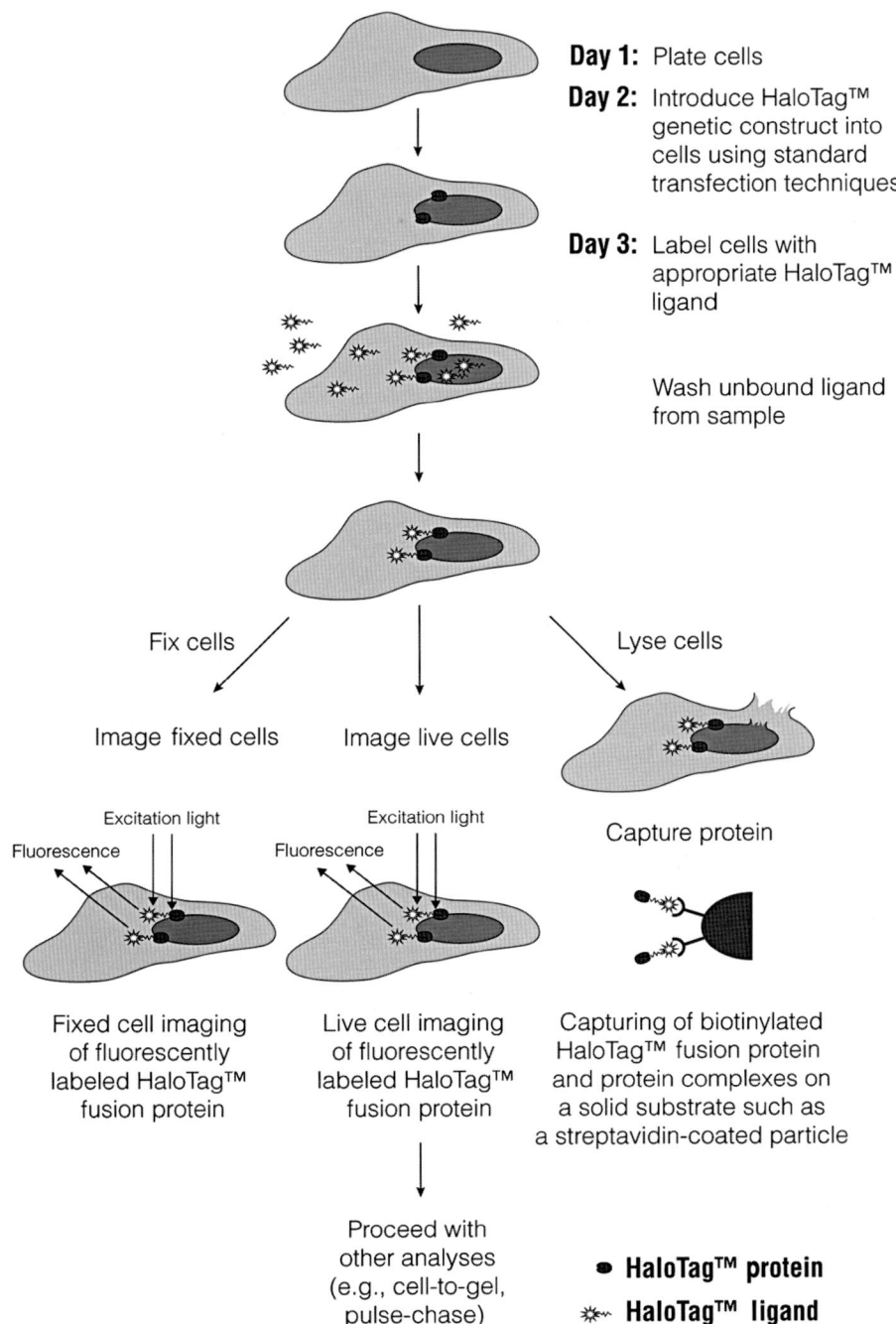

Fig. 3. Overview of cell-based applications for the HaloTag™ technology.

stability of the bond permits imaging of live cells during long time periods, imaging of fixed cells, and multiplexing with different cell/protein analytical techniques *(1)*. The general process for cell-based application includes: (1) making a vector encoding a fusion of the HaloTag protein to a protein, protein-domain, or peptide sequence of interest; (2) expressing the fusion chimera in cells; (3) labeling the cells with the HaloTag ligand; and (4) imaging the sample, either as live or fixed cells (**Fig. 3**).

2. Materials

1. HaloTag pHT2 Vector (Promega, Madison, WI).
2. HaloTag Ligands (Promega).
3. HeLa cells (ATCC, cat. no. CCL-2).
4. CHO-K1 cells (ATCC, cat. no. CCL-61).
5. Fetal bovine serum.
6. Serum-free cell culture medium.
7. 8-well Lab-Teck^{R}II chamber cover glass (Nalgen, Nunc).
8. Transfection reagents: Lipofectamine 2000 (Invitrogen) and LT1 transfection reagent (Mirus, Madison, WI).
9. Endotoxin-free (transfection grade) plasmid DNA.
10. PBS (37°C).
11. Wide-field or confocal fluorescent microscope.
12. 37°C cell culture incubator.
13. 3.7% paraformaldehyde containing 0.5 mM sucrose.
14. Triton® X-100.
15. 0.1% sodium azide/PBS solution.
16. SDS-polyacrylamide gels and associated buffers and stains, as well as electroblot transfer buffers (BioWhittaker Molecular Applications; Rockland, ME).
17. Protein molecular weight standards (Invitrogen).
18. Anti-p65 antibody (BD Biosciences).
19. Anti-IκB antibody (BD Biosciences).
20. Antiβ-III Tubulin Antibody (Promega).
21. Antimouse IgG, conjugated with AlexaFluor™-488 (Invitrogen).
22. HRP-conjugated goat anti-mouse IgG (Promega).
23. Enhanced chemiluminescence system (Pharmacia-Amersham).
24. Sample buffer: 1% SDS, 10% glycerol, and 1 mM β-mercaptoethanol, pH 6.8.
25. Transfer buffer: 25 mM Tris base/188 mM glycine pH 8.3, 20% (v/v) methanol.
26. TBST buffer: 10 mM Tris-HCl, 150 mM NaCl, pH 7.6, 0.05% Tween-20.
27. Blocking solution: 3% dry milk or 1% BSA in TBST buffer.

3. Methods

The methods described next outline: (1) the description and design of expression plasmids; (2) culture, transfection, labeling with the HaloTag Ligands, fixing, and immunocytochemical analysis of mammalian cells; (3) imaging of live and fixed cell; (4) capture and characterization of protein.

3.1. Expression Plasmids

This section describes the HaloTag pHT2 Expression Vector, and a design of the expression plasmids for a HaloTag·(NLS)$_3$ (the HaloTag Protein targeted to nucleus) and p65-HaloTag fusion protein, which has cytosolic localization (i.e., excluded from nucleus) in nonstimulated cells.

3.1.1. HaloTag pHT2 Expression Vector

The HaloTag pHT2 Expression Vector (**Fig. 4A,B**) contains the open reading frame of the modified hydrolase gene cloned into a mammalian expression vector with the following features:

- Kozak sequence for translation initiation was added to the beginning of the gene.
- The following restriction sites were added for convenience:

    o *Bam*HI and *Nae*I restriction sites were added within the HaloTag gene coding region to allow convenient creation of protein fusions. The *Bam*HI site is located immediately after the ATG start codon and the *Nae*I site is located just before the stop codon.
    o The other restriction sites were added just outside of the coding region. Near the N-terminus are *Nhe*I, *Pvu*II, *Eco*RV, and a nonunique *Nco*I; near the C-terminus are *Pac*I and *Not*I. The *Eco*RV site cuts the pHT2 Vector in frame with the HaloTag coding region.

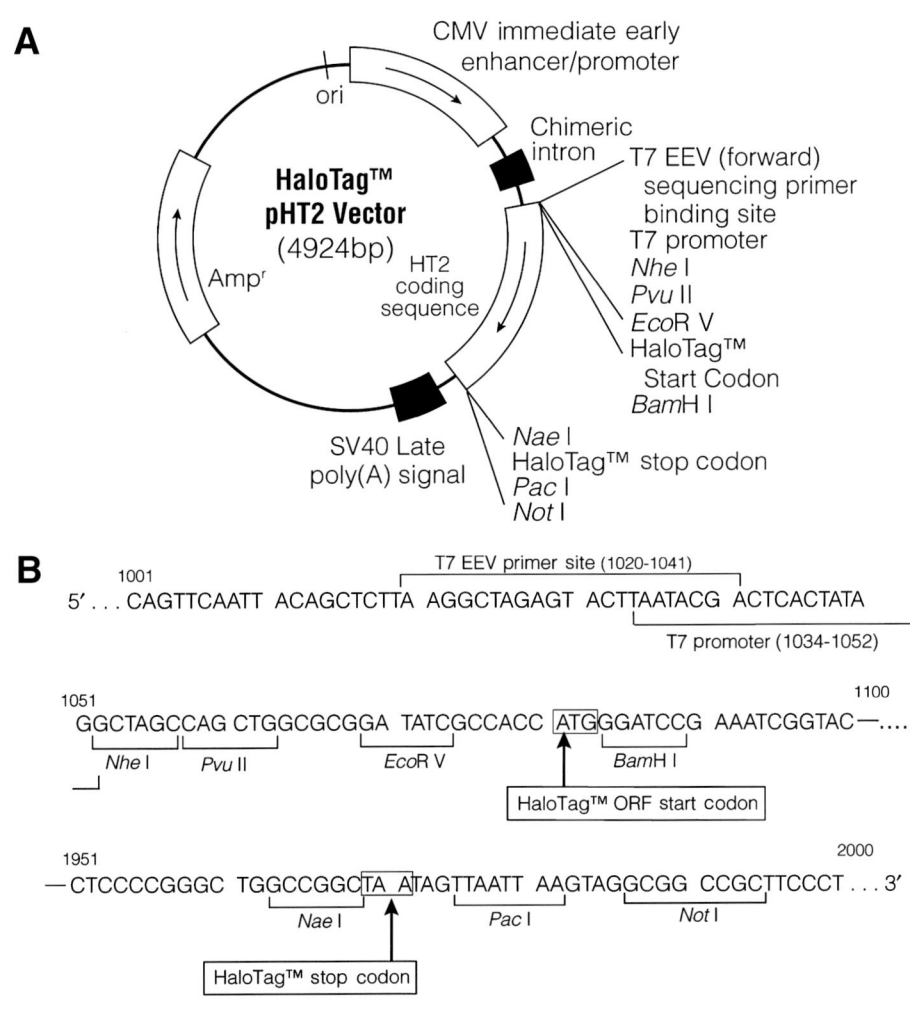

Fig. 4. (**A**) HaloTag™ pHT2 vector circle map and sequence reference points, and (**B**) potential cloning sites and primer hybridization near the start and stop codons of the coding region.

When designing fusions to the HaloTag Protein, we recommend inserting a polypeptide linker between the fusion partners to reduce the potential for structural hindrance. The size and the sequence of the polypeptide linker should be determined empirically, although a 17-amino acid linker comprising a repeated series of glycines and serines has served as a good starting point in our experiments.

3.1.2. Design of the HaloTag·(NLS)$_3$ and p65- HaloTag Expression Vectors

The HaloTag Protein can be fused to a protein of interest, such as one that will direct the fusion protein to a specific subcellular location. To demonstrate this capability we generated expression vectors encoding the HaloTag·(NLS)$_3$ and the p65-HaloTag·FLAG fusion proteins (**Fig. 5**). An (NLS)$_3$ is three repeats of a nuclear localization sequence from simian virus large T-antigen (2). The p65 protein (also known as RelA and NF-κB3) is a member of the eukaryotic nuclear factor-κB (NF-κB)/Rel transcription factor protein family. The NF-κB proteins contain a nuclear-localization sequence (NLS) that is rendered inactive in nonstimulated cells through

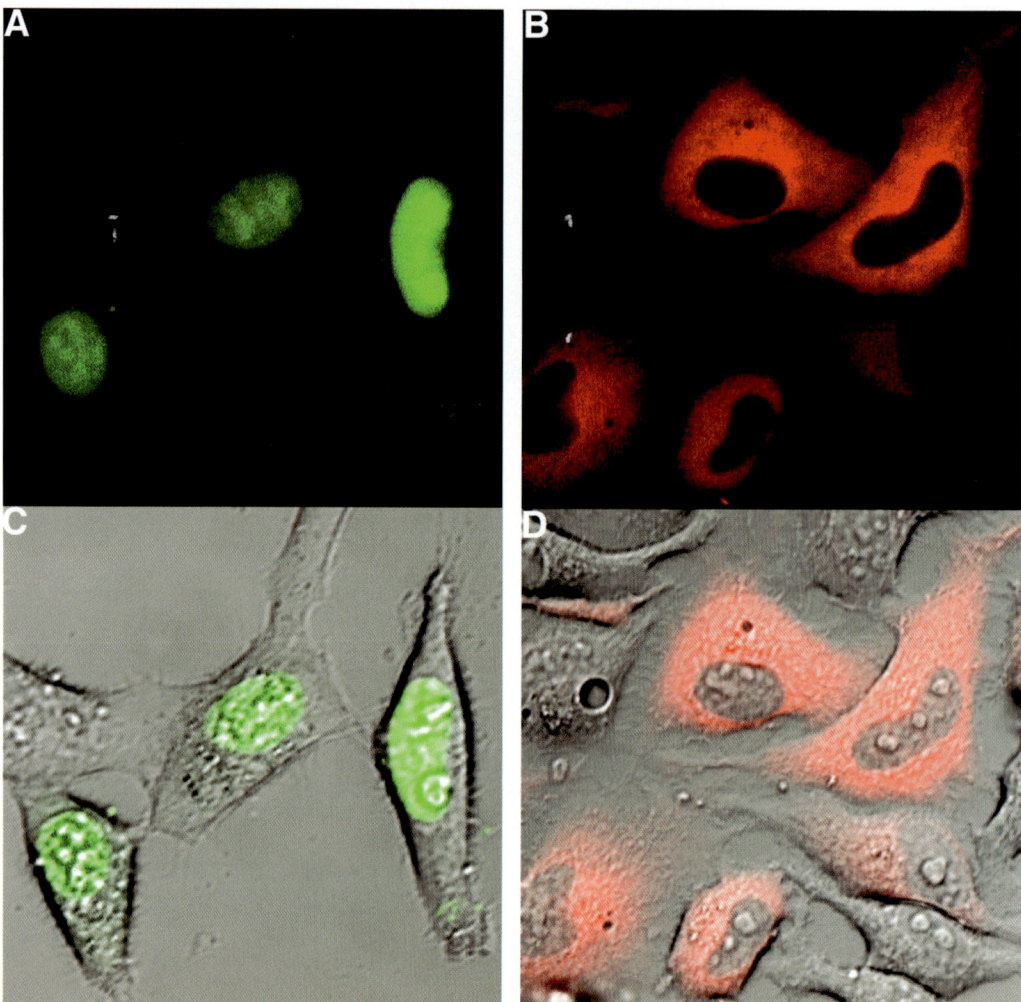

Fig. 5. Nuclear localization of the HaloTag™ (NLS)$_3$ protein and cytosolic localization of the p65-HaloTag·FLAG fusion protein labeled with the HaloTag Ligands. HeLa cells were transiently transfected with the vector encoding the HaloTag·(NLS)$_3$ (**A,C**) or the p65-HaloTag·FLAG (**B,D**) fusion protein were labeled with 5 μ*M* HaloTag diAcFAM Ligand (**A,C**) or HaloTag TMR Ligand (**B,D**) for 15 min at 37° C. Images were generated on an Olympus FV500 confocal microscope using appropriate filter sets for TMR and FAM or transmitted light. (**C**) Overlaid FAM fluorescence and transmitted light. (**D**) Overlaid TMR fluorescence and transmitted light.

the binding of specific NF-κB inhibitors, known as the IκB proteins. Binding of IκB masks the NLS, which leads to retention of NF-κB proteins (including the p65-p50 heterodimer) in the cytoplasm of the cells *(3–5)*.

The expression vector encoding the HaloTag·(NLS)$_3$ fusion protein was generated by subcloning the coding sequence of (NLS)$_3$ (5´-GA TCCAAAAAAG AAGAGAAAGG TAGATC- CAAA AAAGAAGAGA AAGGTAGATC CAAAAAAGAA GAGAAAGGTA-3´) in frame at the C-terminus of the HaloTag Protein.

Plasmid encoding p65 was generously provided by Dr. Johannes A. Schmid *(6)*. The expression vector encoding the p65-HaloTag·FLAG fusion protein was generated by subcloning the coding sequence for the sequence of p65 in frame at the N-terminus of the HaloTag Protein. To reduce potential three-dimensional structural hindrance effects of the fusion partners, a 15-amino acid flexible polypeptide linker was inserted between p65 and the HaloTag Protein. The sequence of the linker is: LDPLVTRGTSRVDAA (5´-TTG GAT CCA CTA GCT ACG CGT GGT ACC TCT AGA GTC GAC GCC GCC-3´). The resulting clones were confirmed by DNA sequencing.

3.2. Culture, Transfection, and Labeling With the HaloTag Ligands

1. Maintaining healthy cell culture is essential for all mammalian cell-based applications. Culturing conditions for many cell lines (e.g., CHO-K1 or HeLa) are available from the cell supplier (e.g., ATCC). For additional information, consult **refs. 7** and **8**.
2. A variety of methods might be used for transient transfection for example lipofection *(9,10)*, calcium phosphate *(11,12)*, electroporation *(13,14)*, or viral *(15,16)*. Stable cell lines expressing HaloTag Protein or HaloTag Protein-based fusions also can be generated (*see* **Notes 1** and **2**).
3. All procedures described in this chapter were performed at the standard cell growing conditions, i.e., at 37°C, in atmosphere of 5% CO_2. HeLa cells or CHO-K1 cells were maintained in DMEM/F12 media supplemented with 10% fetal bovine serum (Invitrogen) according to ATCC recommendations.
4. To study transient expression of different proteins, cells were plated on 8-well Lab-Teck^{R}II chamber cover glass (Nalgen, Nunc) at seeded density of $7.5–10 \times 10^3$ cells/cm^2 ($9–12 \times 10^3$ cells/cm^2) in 400 µL growth medium, and allowed to grow to approx 85% confluency (approx 24–48 h).
5. Cells were transfected using Lipofectamine 2000 or LT1 transfection reagents according to manufacturer's protocols.
6. The HaloTag TMR, HaloTag diAcFAM, and HaloTag Biotin Ligands readily cross the cell membrane, allowing labeling and detection of the HaloTag Protein in living mammalian cells (*see* **Note 3**).
7. The HaloTag Ligands are commercially available as 5 or 10 mM stock solutions in DMSO. Ligands are dispensed into aliquots and stored at –20°C, desiccated and protected from light.
8. The growth medium on cultured cells was replaced with growth medium containing the appropriate ligand (200 µL/well). The recommended working concentration in medium are: 1–5 µM for HaloTag TMR Ligand; 1–10 µM for HaloTag diAcFAM Ligand, and 5–10 µM for HaloTag Biotin Ligand and HaloTag Coumarin Ligand.
9. After 15 min, cells were rinsed twice with PBS (pH 7.4), incubated in fresh media for 30 min, then media was replaced with fresh media or PBS, and cells were used for imaging or protein capture.

3.3. Fixing Mammalian Cells and Immunocytochemistry

The stability of the covalent bond between the HaloTag Protein and the HaloTag Ligands allows imaging the HaloTag fusion proteins in fixed cells (**Fig. 6**). In addition, the resistance of the fluorescent signal to cell fixatives also allows multiplexing the HaloTag Technology with different immunocytochemical and, potentially, immunohistochemical techniques (**Fig. 7**). In this section we describe fixing and immunocytochemical analysis of cells expressing HaloTag Proteins.

1. HeLa cells transiently transfected with the plasmid encoding the HaloTag fusion protein were labeled with the HaloTag TMR, the HaloTag diAcFAM, or HaloTag Coumarin Ligand as described earlier.
2. Cells were rinsed with PBS (37°C, 400 µL/well of 8-well Lab-Teck^{R}II chamber), PBS was replaced with 400 µL freshly prepared 3.7% paraformaldehyde containing 0.5 mM sucrose, and cells were incubated for 10 min at RT in the dark.
3. The fixative was replaced with PBS containing 0.1% Triton X-100, and cells were incubated for 30 min at RT in the dark.
4. The Triton X-100 solution was replaced with 400 µL PBS, 15 min, RT.
5. The cells were incubated in a blocking solution (1% BSA in TBST buffer: 10 mM Tris-HCl, 150 mM NaCl, pH 7.6, containing 0.05% Tween-20) for 30 min at room temperature.
6. Cells were washed with TBST buffer (1.0 mL/cm^2, 15 min, RT) and incubated with mouse anti-βIII tubulin antibody at 1 µg/mL in TBST buffer for 45 min at RT.
7. Cells were washed with TBST buffer (1.0 mL/cm^2, three times, 15 min each, RT).
8. Primary antibody was visualized by incubation of the cells with Alexa Fluor™-488 conjugated goat-antimouse IgG (dilution 1:1000, 30 min, RT) and then the washing procedure was repeated. The cells were imaged immediately or were stored in a 0.1% sodium azide/PBS solution protected from light.

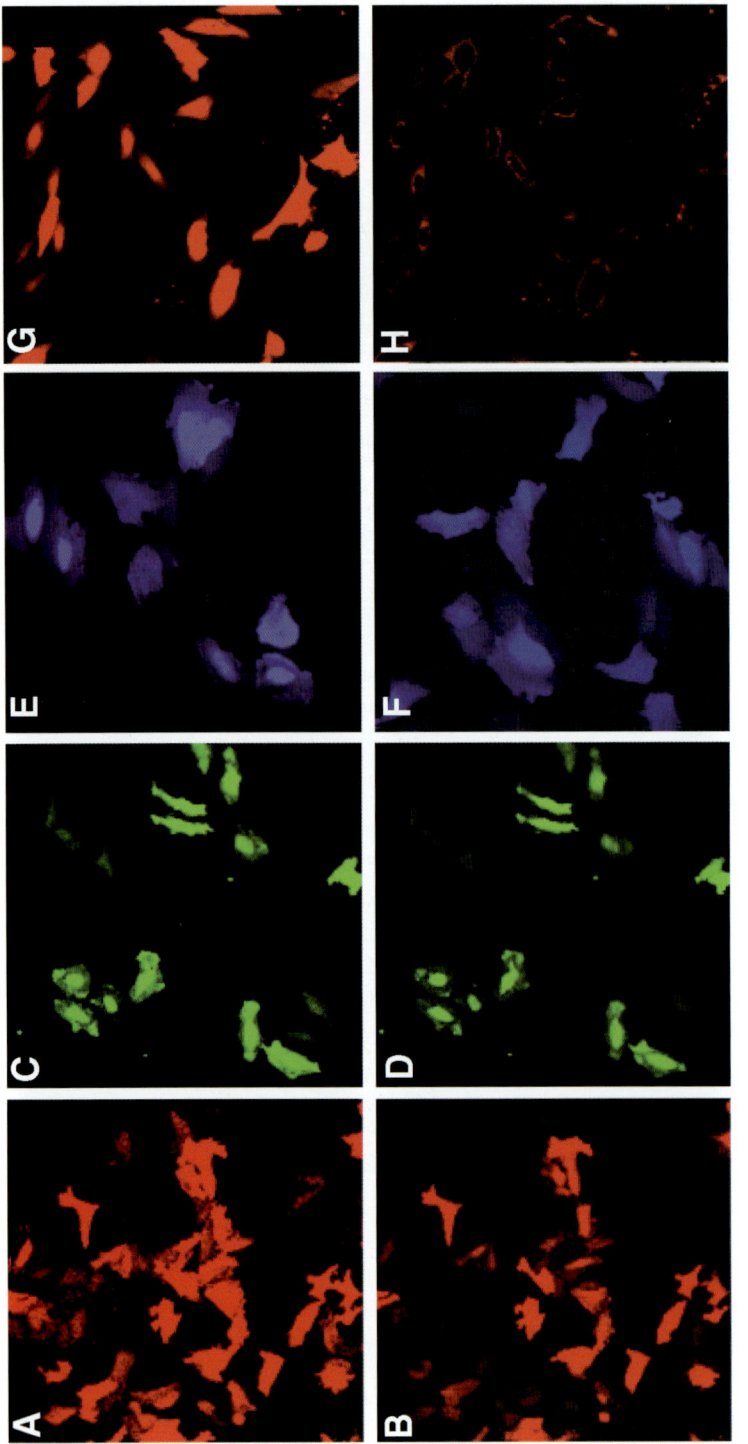

Fig. 6. The HaloTag™ ligand withstands fixation. HeLa cells were transiently transfected with the HaloTag pHT2 Vector (**A–F**) or a DsRed2 (**G,H**). Twenty-four hours later, cells expressing the HaloTag Protein were labeled with 5 µM HaloTag TMR ligand (**A,B**); 10 µM HaloTag diAcFAM ligand (**C,D**); or 25 µM HaloTag Coumarin ligand (**E,F**), for 15 min at 37°C/5% CO_2. Cells were imaged, fixed with 3.7% paraformaldehyde, and imaged again using identical microscope settings. Images in **A–D**, **G,H** were generated on an Olympus FV500 confocal microscope using appropriate filter sets for TMR or FAM, and show identical fields before and after fixation. Images in **E,F** were captured with Hamamatsu CCD camera with appropriate filter set for coumarin, and represent different fields of view within the same culture well.

The HaloTag™

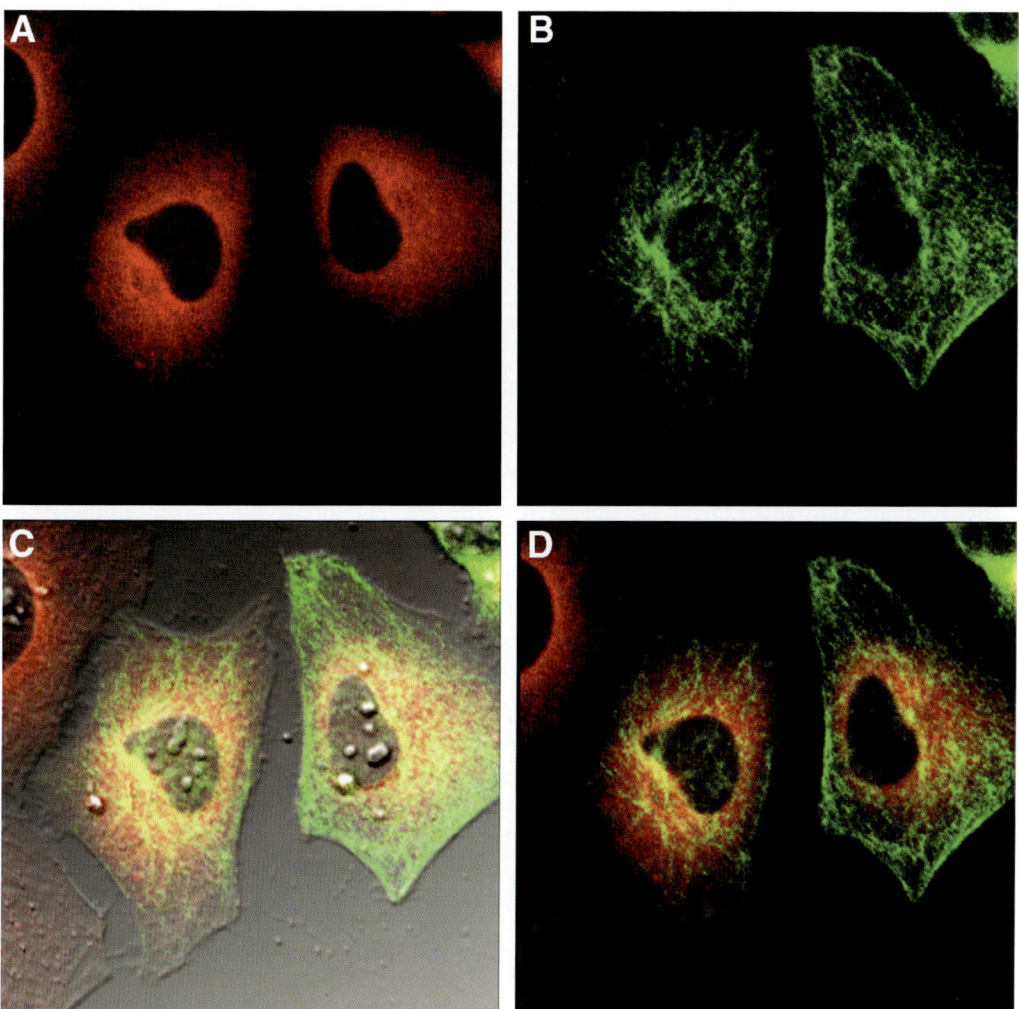

Fig. 7. Immunocytochemistry of the cells expressing the p65-HaloTag™ FLAG fusion protein and labeled with the HaloTag TMR ligand. HeLa cells transiently transfected with the vector encoding the p65-HaloTag FLAG fusion protein were labeled with 5 μM HaloTag TMR Ligand. Cells were fixed with 3.7% paraformaldehyde and processed for immunocytochemistry using primary mouse Antiβ-III Tubulin Antibody (1μg/mL) and secondary goat antimouse IgG, conjugated with AlexaFluor™-488. Images were generated on an Olympus FV500 confocal microscope in sequential mode using appropriate filter sets for TMR, AlexaFluor™-488, or transmitted light. (**A**) TMR fluorescence. (**B**) AlexaFluor™-488 fluorescence. (**C**) Overlaid AlexaFluor™-488 and TMR fluorescence and transmitted light. (**D**) Overlaid AlexaFluor™-488 and TMR fluorescence.

3.4. Fluorescent Detection of HaloTag Protein Fusions in Living and Fixed Mammalian Cells, and by Protein Gel Electrophoresis

Although subcellular location and translocation of HaloTag protein fusions might be determined by fluorescence microscopy, the ability to resolve subcellular structures or fluorescently labeled proteins within specific cell compartments varies with instrument capabilities. We recommend using a confocal microscope with high numerical aperture objectives for achieving good results *(17,18)*.

Fluorescently labeled proteins can also be detected and quantified using fluorescence scanners, flow cytometry or fluorescence plate readers. Accurate quantification of the fluorescence could be complicated by a number of factors including low protein expression, instrument sensitivity, and quality of cell growth surfaces.

3.4.1. Fluorescent Detection of HaloTag Protein Fusions in Living Cells

1. Cells were imaged on a confocal microscope FV500 (Olympus, Japan) using a 488 nm Ar/Kr laser line or a 543- or 633-HeNe laser line. Scanning speed and laser intensity were adjusted to avoid photobleaching of the fluorophores and damage of the cells.
2. The cells labeled with the HaloTag-Coumarin Ligand were imaged on Olympus IX81epifluorescent microscope (Olympus, Japan) equipped with Chroma filter set no. 31,000 DAPI using Hg-lamp and an Orca CCD camera (Hamamatsu, Japan). The microscope was equipped with a microenvironmental chamber to maintain physiological conditions during long-term experiments.

3.4.2. Cell to Gel Analysis: Detection of HaloTag Fusion Proteins Using SDS-PAGE and a Fluorescence Imager

Because the HaloTag Ligand is held by a stable covalent bond, the fluorescently labeled HaloTag Protein can be boiled with sample buffer and resolved by SDS-PAGE without loss of the fluorescence signal. Analyzing labeled HaloTag Protein can be combined with other protein analysis techniques such as Western blotting. Our preliminary data indicate that this approach also can be used successfully to study posttranslational modification of the HaloTag Protein-based fusions (e.g., proteolytic cleavage, data not shown).

1. Cells transiently expressing the HaloTag-based fusion proteins and labeled with the HaloTag TMR Ligand were solubilized in a sample buffer, boiled for 5 min, and resolved on SDS-PAGE (4–20% gradient gels; BioWhittaker Molecular Applications).
2. Gels were analyzed on a fluorescence imager Typhoon 9400 (Hitachi, Japan) at an E_{ex}/E_{em} appropriate for TMR, were processed for Western blot analysis, or were stained with Coomassie blue (Promega).

3.4.3. Western Blot Analysis

1. Electrophoretic transfer of proteins to a nitrocellulose membrane (0.2 µm, Scheicher and Schuell, Germany) was carried out in transfer buffer for 2.0 h with a constant current of 80 mA (at 4°C) in Xcell II Blot module (Invitrogen).
2. The membranes were rinsed with TBST buffer and incubated in blocking solution for 30 min at room temperature or overnight at 4°C.
3. Membranes were washed with 50 mL of TBST buffer and incubated with anti-p65 antibody (dilution 1:5000), anti-IkB antibody (dilution 1:10,000), or anti-HaloTag antibody (dilution 1:50,000) for 45 min at room temperature.
4. The membranes were washed with TBST buffer (50 mL, 5 min, three times).
5. The membranes were then incubated with HRP-conjugated donkey antimouse IgG (30 min, room temperature) and then the washing procedure was repeated.
6. The proteins were visualized by the enhanced chemiluminescence system (Pharmacia-Amersham) according to the manufacturer's instructions. Levels of proteins were quantified using computer-assisted densitometry.

3.5. Conclusions

The HaloTag Interchangeable Labeling Technology provides a means for attaching small synthetic molecules onto a specific fusion protein within living mammalian cells. The synthetic molecules are attached covalently to the fusion protein through a short PEG-like tether capable of penetrating cellular membranes. The kinetics of binding are rapid, and the covalent bond securely and specifically locks the synthetic molecules in place. The molecules are retained on the fusion protein even under denaturing conditions, such as in fixed cells or in SDS electrophoresis.

The technology enables fusion proteins within cells to be labeled with a range of standard fluorophores for image analysis. This provides the benefits of allowing different colors to be used interchangeably on the same genetic construct or transgenic cells; of allowing conditional switching of the colors, thus enabling "pulse-chase" type experiments; and of providing a range of colors in fixed cells, in which other methods tend to be inefficient or unstable. Because the fluorophore can also be exchanged with an affinity handle, such as biotin, the fusion proteins might also be captured from cells for in vitro analysis. Fusion proteins might also be specifically and stably attached directly onto surfaces containing the reactive tether without requiring preceding sample purification.

The unrestricted architecture of the binding interaction opens the possibility for incorporating other types of synthetic molecules. Examples might include fluorophores that can respond to subcellular microenvironments or indicate physiological changes occurring within cells. The synthetic molecules might also mediate interactions with other proteins within the cells, or trigger specific subcellular events. The efficiency and specificity of the interaction even within the extreme complexity of living cells might offer new technical opportunities, in which current alternatives have not been sufficient.

4. Notes

1. The level of the HaloTag Protein expression depends on many factors including cell type, efficiency of transfection, type of promoter, and protein coding sequence. Transfection can be toxic to cells, which frequently correlates with the transfection efficiency. Balance between sufficient protein expression, transfection efficiency, and low toxicity is essential for generating reliable data. Cells should be actively proliferating. The recommended cell density for most cell lines at transfection time is approx 80–90% confluency. Preliminary experiments should be done to optimize cell density, amount of DNA, and transfection reagent for transient transfection. Efficiency of transfection might be affected by the specific cell line, cell culture and transfection conditions, and specific DNA constructs.
2. The concentration of the HaloTag Ligands, and cell labeling and washing protocols, should be optimized for different cell lines and different applications. HaloTag TMR, HaloTag diAcFAM, and HaloTag Biotin Ligands have exhibited no detectable toxicity or morphological side effects at recommended labeling conditions in the cell lines tested (HeLa, CHO-K1). HaloTag Ligands can be added to serum-containing medium directly. The HaloTag TMR and HaloTag Biotin Ligands can be premixed with medium. In contrast, the HaloTag diAcFAM Ligand must be mixed with media immediately before adding it to the cells to avoid premature hydrolysis of the diacetyl groups by serum esterases. The HaloTag diAcFAM Ligand can cross-cellular membranes, but the deacetylated FAM derivative cannot. Because fluorescent dyes are light sensitive, light exposure of the cells should be avoided during ligand loading and washing procedures.

Appendix—HaloTag Mechanism

The native hydrolase is a monomer with MW approx 33 kDa. The activity of enzyme cleaves carbon–halogen bonds in aliphatic halogenated compounds involving a hydrolytic triad at the active site.

$$R - Cl + Enz + HOH \rightarrow R - Enz + Cl- + HOH \rightarrow R- OH + Enz + H^+ + Cl-$$

In the reaction catalyzed by the native enzyme, an enzyme–substrate complex is formed by a nucleophilic attack involving Asp106 and the formation of an ester intermediate; His272 activates H_2O that hydrolyzes this intermediate, releasing product from the catalytic center (**Fig. 8**; **refs. *19–21***). A point mutation in the gene resulting in a His272Phe substitution impairs the hydrolysis step, leading to formation of a covalent bond between protein and ligand containing the functional reporters (**Fig. 9**).

The amino acid sequence of the protein was further optimized to provide better access to the active site by the ligand. These changes result in a dramatic increase in the kinetics of ligand

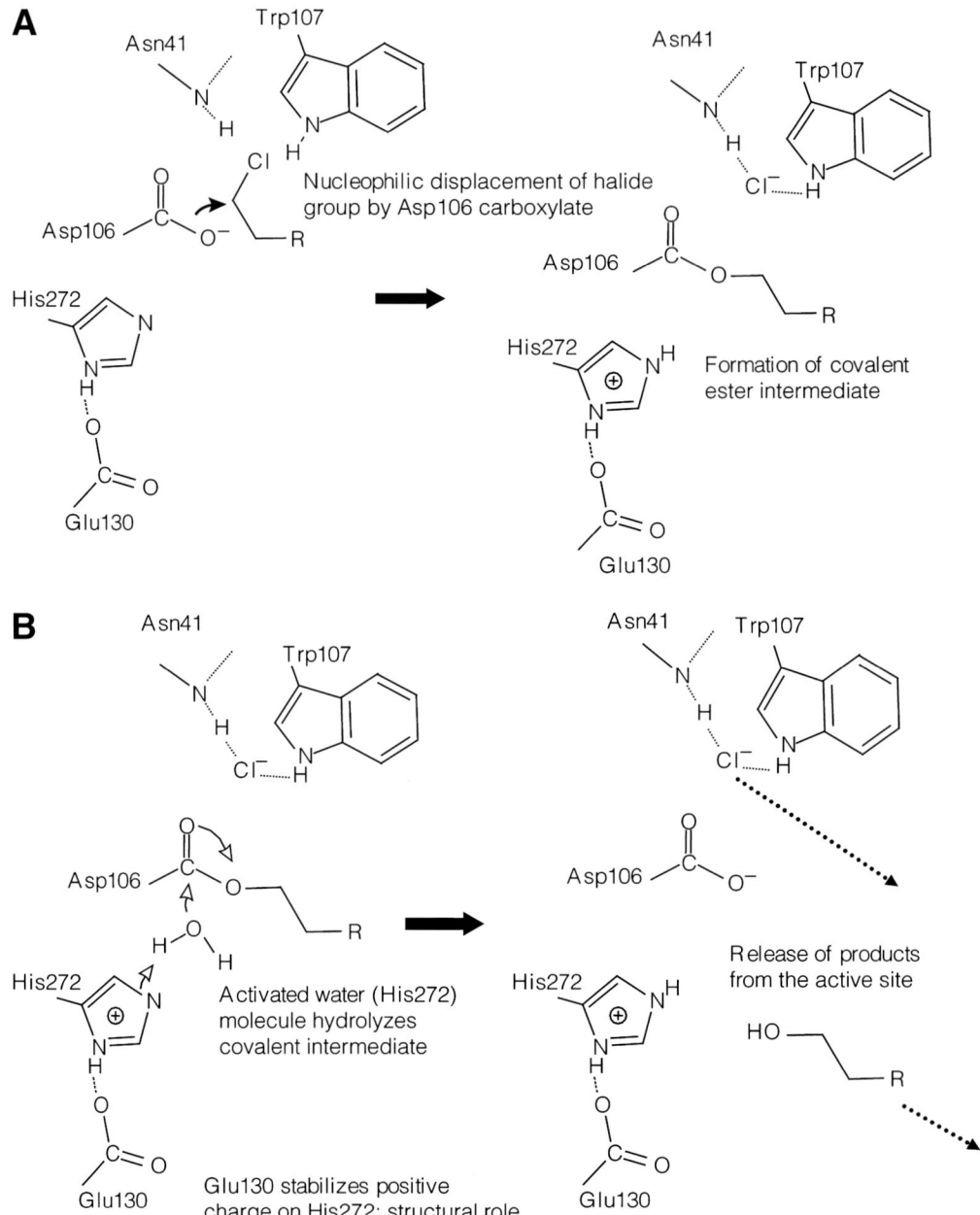

Fig. 8. The catalytic mechanism of wild-type hydrolase.

binding rate by several 1000-fold, leading to almost immediate binding for HaloTag TMR Ligand binding to GST-HaloTag Protein fusion.

Acknowledgments

The HaloTag technology presented in this chapter represents the creative insights and efforts from many scientists. The authors wish to acknowledge a long list of contributors from Promega Corporation and Promega Biosciences for this accomplishment.

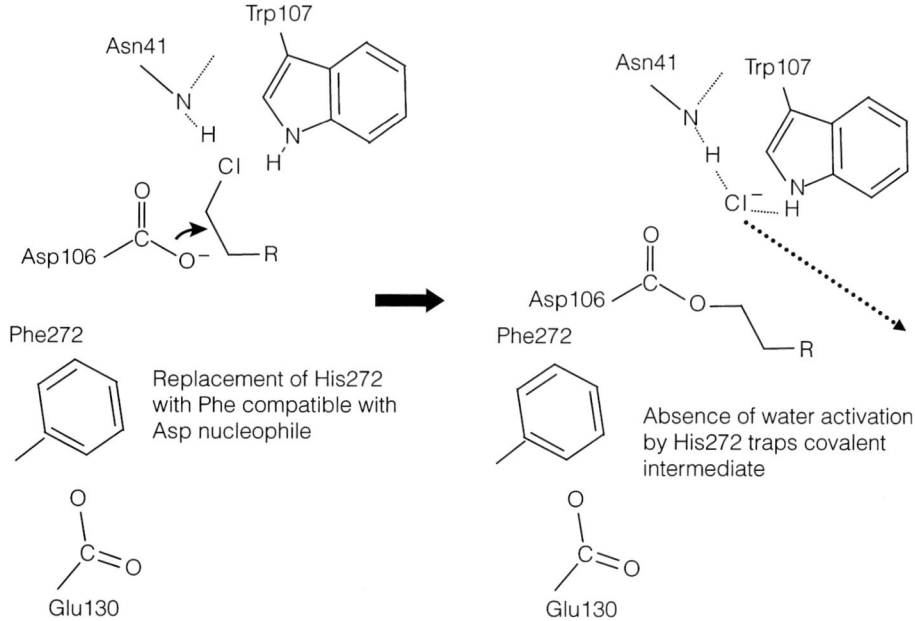

Fig. 9. The engineered protein encoded by the HaloTag™ pHT2 vector includes the His272Phe substitution, which is critical for covalent bond formation.

References

1. Los, G. et al. (2005) HaloTag™ interchangeable labeling technology for cell imaging and protein capture. *Cell Notes* **11**, 2–6.
2. Kalderon, D., Roberts, B. L., Richardson, W. D., and Smith, A. E. (1984) A short amino acid sequence able to specify nuclear location. *Cell* **39(3 Part 2)**, 499–509.
3. Ghosh, S., May, M. J., and Kopp, E. B. (1998) NF-κB and Rel proteins: evolutionarily conserved mediators of immune responses. *Annu. Rev. Immunol.* **16**, 225–260.
4. Karin, M., Yamamoto, Y., and Wang, Q. M. (2004) The IKK NF-kappa B system: a treasure trove for drug development. *Nature Rev.* **3**, 17–26.
5. Burnstein, E. and Duckett, C. S. (2003) Dying for NF-kappaB? Control of cell death by transcriptional regulation of the apoptotic machinery. *Curr. Opin. Cell Biol.* **15**, 732–737.
6. Schmid, J. A., Birbach, A., Hofer-Warbinek, R., et al. (2000) Dynamics of NF kappa B and Ikappa Balpha studied with green fluorescent protein (GFP) fusion proteins. Investigation of GFP-p65 binding to DNa by fluorescence resonance energy transfer. *J Biol. Chem.* **275 (22)**, 17,035–17,042.
7. Freshney, R. I. (1986) *Animal Cell Culture: A Practical Approach*, IRL Press, Oxford, UK.
8. Ausubel, F. M. et al. (1994) *Current Protocols in Molecular Biology*, Green Publishing Associates and Wiley-Interscience, New York.
9. Felgner, P. L., Holm, M., and Chan, H. (1989) Cationic liposome mediated transfection. *Proc. West. Pharmacol. Soc.* **32**, 115–216.
10. Felgner, P. L. and Ringold, G. M. (1989) Cationic liposome-mediated transfection. *Nature* **337**, 387–388.
11. Chen, C. and Okayama, H. (1987) High-efficiency transformation of mammalian cells by plasmid DNA. *Mol. Cell Biol.* **7**, 2745–2752.
12. Wigler, M. et al. (1977) Transfer of purified herpes virus thymidine kinase gene to cultured mouse cells. *Cell* **11**, 223–232.
13. Chu, G., Hayakawa H., and Berg, P. (1987) Electroporation for the efficient transfection of mammalian cells with DNA. *Nucl. Acids Res.* **15**, 131–226.
14. Shikegawa, K. and Dower, W. J. (1988) Electroporation of eukaryotes and prokaryotes: a general approach to the introduction of macromolecules into cells. *Biotechniques* **6**, 742–751.

15. Ramezani, A. and Hawley, R. G. (2002) Overview of the HIV-1 lentiviral vector system, in *Current Protocols in Molecular Biology,* (Ausubel, F.M. et al., ed.), John Wiley and Sons, Inc., Hoboken, NJ, pp. 16.21.1–16.21.15.
16. Moss, B. and Earl, P. (1998) Overview of the vaccinia virus expression system, in *Current Protocols in Molecular Biology,* (Ausubel, F.M. et al., ed.) John Wiley and Sons, Inc., Hoboken, NJ, pp. 16.15.1–16.15.5.
17. Herman, B. (2001) *Fluorescence Microscopy*, 2nd ed., Springer-Verlag, New York.
18. Perassamy, A., ed. (2001) *Methods in Cellular Imaging,* Oxford University Press, New York.
19. Kulakova, A. N., Larkin, M. J., and Kulakov, L. A. (1997) The plasmid-located haloalkane dehalogenase gene from Rhodococcus rhodochrous NCIMB 13064. *Microbiology* **143,** 109–115.
20. Schindler, J. F. et al. (1999) Haloalkane dehalogenase: steady-state kinetics and halide inhibition. *Biochemistry* **38,** 5772–5778.
21. Newman, J. et al. (1999) Haloalkane dehalogenase: structure of a *Rhodococcus* enzyme. *Biochemistry* **38,** 16,105–16,114.

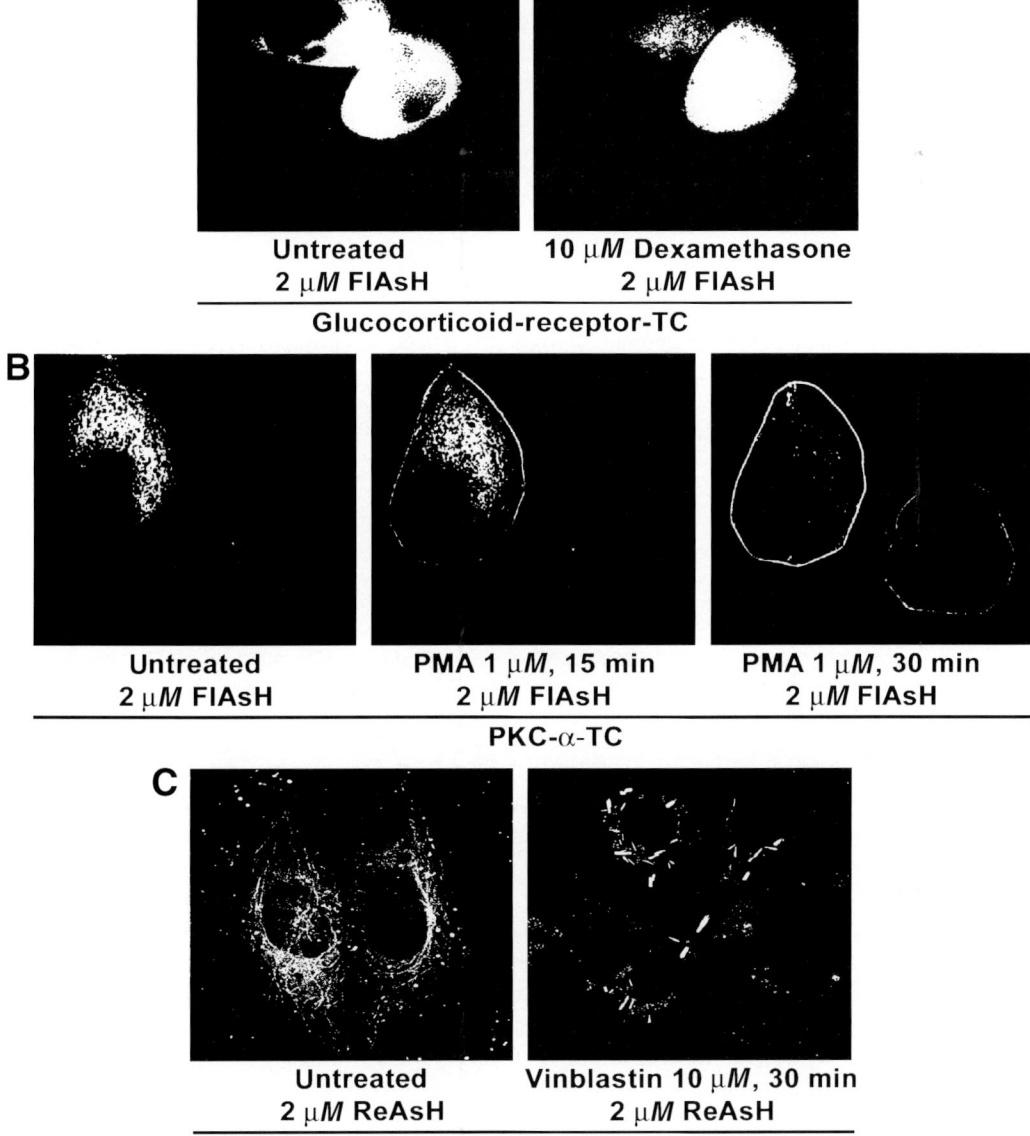

Fig. 2. Examples for imaging protein dynamics in living cells using FlAsH/ReAsH staining. CHO cells were transiently transfected with (**A**) Glucocorticoid receptor-TC (GR-TC), (**B**) PKC-α-TC or (**C**) TC-tubulin-α. After transferring the cells onto chambered cover slips the cells were labeled with 2 μ*M* FlAsH or ReAsH (as indicated) for 60 min at 37°C and subsequently washed three times with Opti-MEM containing 200 μ*M* EDT2. (**A**) The cells transfected with Glucocorticoid receptor were left untreated or treated for 20 min with either 10 μ*M* Dexamethasone. Dexamethasone binds and activates GR resulting in the translocation of GR-TC from cytoplasm to the nucleus. (**B**) The cells transfected with PKC-a were left untreated or treated for 20 min with 1 μ*M*. PMA treatment leads to activation of PKC-a followed by translocation from the cytoplasm to the plasma membrane. (**C**) The cells transfected with TC-β-tubulin were left untreated or treated for 60 min with 10 m*M* vinblastine which induces microtubule depolymerization and the appearance of short tubulin aggregates. The two images show different fields within the sample.

Images were aquired on a Deltavision Image Restoration System (Olympus IX-71 inverted microscope equipped with a Plan-Apo ×60, NA 1.4 objective). All images were subjected to deconvolution processing.

3.2. Generation of TC-Tagged PKC-α Expression Construct

We are using PKC-α as an example to illustrate the practical steps required for generating, expressing, and labeling TC-tagged proteins. For this example, the DNA of PKC-α was cloned through PCR directly from cDNA generated from HeLa cell mRNA using standard molecular biology techniques.

3.2.1. Cloning

1. For the generation of C-terminal TC-fusion constructs we modified Invitrogen's mammalian expression vector pcDNA6™/V5-His by replacing the His tag with the TC tag.
2. The full-length PKC-α fragment was cloned into the *Hind*III and *Xho*I sites of the modified pcDNA6™/V5(-TC) vector. The inclusion of the V5 tag did not interfere with the labeling or function of the target protein in cells and was additionally used to independently verify expression and correct localization of the PKC-α-fusion protein through indirect immunofluorescence.
3. The plasmid DNA was transformed into the TOP10™ *E. coli* strain and plated on LB plates containing ampicillin (50 mg/mL).
4. Individual colonies were selected and grown in LB (+ ampicillin).
5. The plasmid DNA was purified and checked for proper insert size and verified by sequencing before further use.
6. The modification of the expression vector and subsequent insertion of the PKC-α DNA was carried out following standard molecular biology protocols. A detailed description would exceed the scope of this article.

We also had success using the previously mentioned Gateway cloning system to generate N-terminal TC-tagged expression constructs for numerous different proteins (e.g., TC-tubulin-α, **Fig. 2C**). For these constructs we utilized Invitrogen's Ultimate™ ORF collection, which offers more than 4000 sequence validated open reading frames in Gateway entry vectors. The desired entry clones were transferred by recombination into the Gateway N-terminal Lumio destination vector (pCDNA™6.2/Lumio-DEST).

3.2.2. Expression of TC-Fusion Proteins in CHO-K1 Cells

We have expressed many distinct TC-tagged fusion proteins (*see* **Subheading 3.1.**) in different cellular backgrounds without encountering any problems with insufficient expression or toxicity that could be attributed to the addition of the TC tag. The transfection protocol provided serves as a working example. It is recommended to optimize transfection protocols for each novel experimental setup (e.g., different expression systems, cell lines, and transfection methods), before performing large-scale experiments (*see* **Note 3**).

We used Lipofectamine 2000 as transfection reagent for our transient and stable transfections. Other transfection reagents were used with similar overall results regarding transfection efficiency and cell viability. For detailed information regarding principles of transfection and application protocols we refer the reader to the manufacturer's protocol for Lipofectamine 2000. Later we describe a standard protocol routinely employed in our laboratory for the transfection of CHO-K1 cells.

3.2.2.1. Day 1

1. Plate 0.5×10^6 CHO-K1 cells in 3 mL Dulbecco's modified Eagle's medium (DMEM) + 10% FCS in a 6-well plate and incubate for 16 h under cell culture conditions. The CHO-K1 cells used in this experiment were cultivated in DMEM supplemented with nonessential amino acids, penicillin, streptomycin, 10 mM HEPES and 10% FCS. The cells were kept at 37°C in a humidified incubator and an atmosphere containing 5% CO_2.

3.2.2.2. Day 2

1. Remove medium from cells and replace with 2.5 mL of fresh complete DMEM medium + 10% FCS.
2. Prepare two microtubes with 250 µL OPTI-MEM I (tubes A and B). Add 4 µg of the plasmid DNA to tube A and 10 µL of Lipofectamine 2000 to tube B and mix gently. Mix content of both tubes by pipeting, and incubate the transfection mixture for at least 15 min at room temperature.

3. Add 500 µL transfection mixture to the cells and incubate for 24 h under cell culture conditions. We generally achieve between 30 and 60% transfection efficiency using this particular cell line and transfection protocol.
4. Plate the cells according to imaging application. For applications requiring the use of oil immersion objectives (and limited working distance) we used either chambered-cover slips (Nunc Labtek II, cover slips made of German glass 1.5 thickness [0.16–0.19 mm]), which can be purchased with 1, 2, 4, or 8 individual chambers. For applications that require better accessibility (e.g., for subsequent microinjection, and so on), 35 mm glass bottom dishes are more suitable. The glass surface normally provides an excellent substrate for most adherent cells. If necessary, the cover slips/dishes can be coated with Matrigel (or any other adherence promoting factors, such as collagen, and poly-L-lysine) to improve adherence. A coating of poly-L-lysine will allow the attachment of suspension cells to glass or plastic surfaces. We also used 96-well plates for experiments, which allow the handling and observation of large number of samples and the use HCS/HTS instrumentation. However, the use of plastic materials for imaging purposes requires objective with long working distance and inherently limits resolution and image quality and is therefore not suitable for applications that require the acquisition of high-quality images of subcellular structures. For general imaging, the cell density should not exceed 70–80% confluency in order to achieve optimal cell morphology. After plating, the cells should be incubated for additional 24 h under cell culture conditions to allow for cell adherence.

3.3. Staining With FlAsH Reagent

The staining protocol provided here should be used as general guidance for the staining of living cells and should be modified as required to achieve optimal and consistent results. In general, staining cells with FlAsH is a fast and uncomplicated process and will provide in most cases highly specific and sensitive staining of TC-tagged proteins. However, a couple of issues need to be addressed in order to obtain the best results with this labeling technology.

The single most important problem using TC/FlAsH is unspecific background staining *(27,28)*, which might lead to poor signal-to-noise ratio especially for low-expressing targets (*see* **Note 4**). Although the FlAsH reagent has considerable specificity for the TC motif over individual cysteine residues *(7,8)*, the sheer number of available cysteines within a cell will result in some background labeling. It is therefore important to adjust incubation and wash conditions to optimize signal-to-noise ratio. This might require some additional experimentation, especially for low-expressing proteins. The most commonly used method for background reduction is the inclusion of small dithiols, such as EDT_2 and BAL, during labeling and washing procedures. These dithiols effectively compete with single cysteines for FlAsH and, therefore, prevent most nonspecific intracellular labeling. However, it should be kept in mind that at high concentrations (low millimolar) EDT_2 and BAL will eventually displace most of the TC-bound FlAsH/ReAsH. It is therefore recommended to carefully assess wash conditions. An additional source of background staining is the nonspecific association of FlAsH with proteins independent of dithiols *(28)*, presumably owing to interaction with exposed hydrophobic sites. This type of background is particularly pronounced in damaged and dead cells, which often appear as extremely bright rounded up cells in the sample. The inclusion of the uncharged dye Disperse Blue 3 (provided with Invitrogen TC-FlAsH/ReAsH Kit) will suppress most of this type of dithiol-independent background. Further background suppression can be achieved by the addition of high concentrations of a quenching dye such as Patent blue *(28)*.

The second important issue to be considered when using TC/FlAsH for labeling experiments is the requirement for a strongly reducing environment. FlAsH will not bind to oxidized cysteines. The use of TC/FlAsH should be therefore limited to proteins/protein domains that reside in the cytoplasm and nucleus (*see* **Note 5**). Membrane spanning proteins should be fused to the TC tag at their cytoplasmic portion. The use of TC/FlAsH for extracellular proteins (or domains) or for proteins that reside in the lumen of intracellular compartments, such as ER, Golgi, or endosomes should be avoided. The use of FlAsH for in vitro application requires complete reduction of the

FlAsH-binding cysteines. A helpful and informative discussion of in vitro FlAsH applications can be found in Griffin et al. *(28)*. The reader is also referred to the Invitrogen website (www.invitrogen.com) for further information about FlAsH as in vitro protein labeling reagent (Lumio green).

3.3.1. Procedure for Loading Cells With FlAsH

1. Prepare 2 µM FlAsH staining solution by diluting the FlAsH stock solution 1/1000 (Invitrogen TC-FlAsH In-Cell labeling reagent is provided as 2 mM stock solution in DMSO) in Opti-MEM I. This staining solution yielded satisfactory results in all our experiments. The use of Opti-MEM I or any other low/no serum medium or salt solution (e.g., Hank's balanced salt solution) is highly recommended because FlAsH reagent is known to bind serum proteins *(28)*. We did not observe significantly improved staining results (for imaging applications) after using higher concentrations of FlAsH reagent, which might also lead to increased background. As previously discussed, the inclusion of dithiols greatly reduced background staining. We found it generally beneficial for the signal-to-noise ratio to include some EDT_2 in the staining solution. As shown in **Fig. 2**, the inclusion of increasing concentrations of EDT_2 led to considerable reduction in nonspecific staining. Nevertheless, higher concentrations resulted in a reduction of specific signal. We recommend the addition of a 5- to 50-fold excess of EDT_2 in the staining solution to suppress nonspecific staining. The exact concentration of EDT_2 should be determined empirically in a pilot experiment for each construct and cell line. The required amount of staining (and subsequently washing) solution is dependent on the used cultivation vessel (35 mm dish: 2 mL/8-well LabtekII: 0.4 mL/96-well plate: 100 µL/well).
2. Remove DMEM medium from cells and wash once with Opti-MEM I to remove detached (=dead/damaged) cells and cell debris, which are typically present at slightly elevated levels after transfection. Dead cells and debris can be a considerable source for background because of the high level of nonspecific FlAsH uptake.
3. Add FlAsH staining solution and incubate for 30 min at 37°C under cell culture conditions. The incubation time can be varied according to need and labeling result. We incubated the samples between 30 and 60 min with the staining solution and found little difference in staining intensity and quality. Longer periods of incubations might result in increased background.
4. Remove staining solution and wash once with Opti-MEM I to remove residual dye.

3.4. Reduction and Suppression of Background Staining in Cells

The inclusion of EDT_2 (or BAL) in the staining solution will substantially but not completely prevent background staining (the Invitrogen TC-FlAsH In-Cell labeling kit includes BAL wash buffer). For further reduction of background, it is recommended that the samples be washed once or more times with Opti-MEM I containing EDT_2 (or BAL) to remove most nonspecifically bound FlAsH. The concentration of dithiols and the number of washes required for optimal background reduction needs to be established empirically in pilot experiments for each construct, cell line, and application. **Figure 3** provides an example for the application of multiple rounds of washes with different concentrations of EDT_2. We found in our experiments that two or three wash steps with EDT_2 concentrations between 100 and 500 µM yield the best results (with respect to signal-to-noise ratio). The application of higher concentrations of EDT_2 will result in a gradual loss of specific signal. It should also be mentioned that BAL is a more effective competitor for FlAsH than EDT_2. The use of BAL needs therefore to be adjusted accordingly in order to prevent the loss of specific signal. However, the intense odor of EDT_2 makes the use of BAL more convenient for the end-user.

3.4.1. Standard Procedure for Background Reduction/Suppression

The following protocol yielded satisfactory results for most of our experiments and is routinely applied to new experimental settings (e.g., new target protein or cell line).

1. Preparation of wash medium. We routinely use Opti-MEM to prepare wash medium, but most other cell culture media and salt solutions (including supplements) will be also suitable for this purpose. Our standard wash medium is prepared by adding EDT_2 to a final concentration of 200 µM. The wash medium

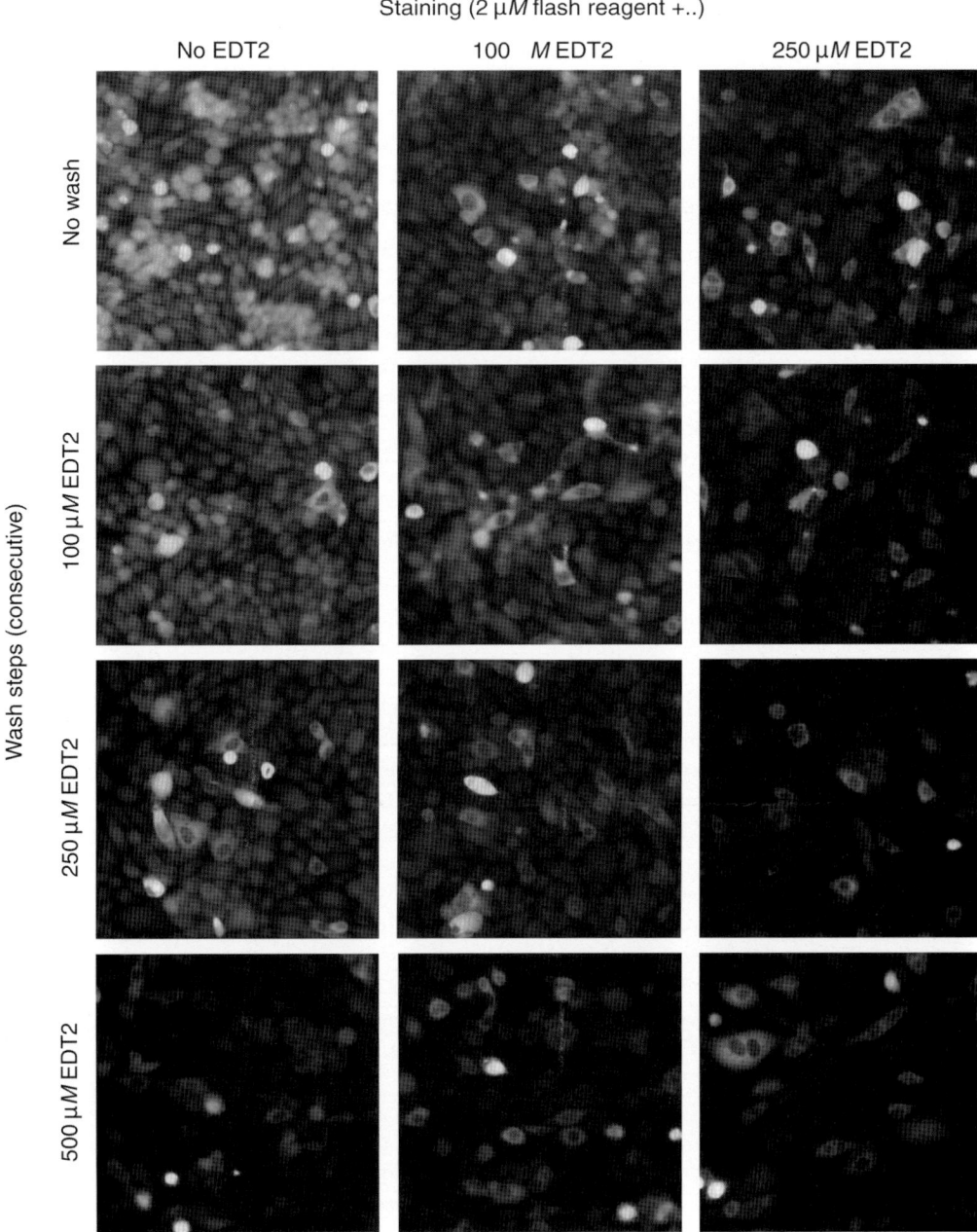

Fig. 3. Use of EDT_2 for background reduction during FlAsH loading and washing. CHO cells were transiently transfected with PKC-α-TC. The cells were labeled with 2 μM FlAsH in Opti-MEM in absence or presence of EDT_2 (100 or 250 μM, as indicated in the top row) for 60 min at 37 min. The samples were then subjected to consecutive wash steps with increasing concentrations of EDT_2 (100, 250, and 500 μM in Opti-MEM) to remove background staining (as indicated on the left side). Images were taken with a Zeiss Axiovert 25 using a ×10 objective, using identical exposure times.

is always prepared fresh (because of the instability of EDT_2/BAL). As previously mentioned EDT_2 has a strong odor and should be handled in the fume hood. We usually prepare a small amount (100 μL) of a 100 mM EDT_2 stock solution in DMSO to carry out an experiment. We discard all plastic ware that comes in contact with EDT_2 in separate plastic bags to avoid unnecessary odor development.

2. Add wash solution to the sample and incubate for 5–10 min at 37°C. The required amount of wash and assay medium is dependent on the used cultivation vessel (35 mm dish: 2 mL/8-well LabtekII: 0.4 mL/ 96-well plate: 100 µL/well).
3. Repeat wash three times.
4. Wash once with Opti-MEM to remove residual traces of EDT_2.
5. Add assay medium to the sample. The assay medium can be any phenol red-free cell culture medium (including supplements and FCS). To reduce dithiol-independent background we include in the final assay medium 20 µM disperse blue 3 and 0.5 mM patent blue V.

3.5. Microscopy

The images shown in this report were taken either with a Zeiss Axiovert 25 microscope (Fluar ×10, NA 0.5 objective) equipped with a Pixera Penguin 600CL CCD camera or with a Deltavision Image Restauration System (Olympus IX 71 inverted microscope, ×60 Planapo, NA 1.4 objective) equipped with a Photometrix Coolsnap HQ camera. The excitation and emission filters were chosen according to the spectral properties of FlAsH (ex_{max} 508 nm/em_{max} 528 nm) and ReAsH (em_{max} 593 nm/em_{max} 608 nm) (*see* **Note 6**). High-resolution images (**Figs. 1** and **2**) were subjected to deconvolution processing.

3.6. Application Examples for FlAsH Staining in Living Cells

As mentioned before, there are currently no published examples for the use of FlAsH/Lumio labeling in HCS. We tested a number of well-known biological models used in our laboratory as practical examples for the use of FlAsH to investigate protein dynamics in living cells and its potential use for the development of high content assays. These examples also demonstrate the functionally inert nature of the TC tag (*see* **Note 7**). The proteins were tagged either on the N- (TC-β-tubulin) or C-terminus (PKC-α-TC, glucocorticoid receptor-TC) with the TC motif. Both PKC-α and glucocorticoid receptor (**Fig. 2A,B**), show a stimulus (PMA and dexamethasone, respectively) dependent translocation of the target proteins between different compartments. The stimulation of PKC-α with phorbol ester PMA induces conformational changes followed by the translocation of PKC-α from the cytoplasm to the plasma membrane, whereas treatment with dexamethasone results in the activation of glucocorticoid receptor and its movement from the cytoplasm to the nucleus. The third example (**Fig. 2C**) shows vinblastine (a well-known antimicrotubule compound) dependent dissolution of microtubules to small crystal like tubulin aggregates as an example for drug-induced phenotypic changes. The staining of β-tubulin was conducted using ReAsH instead of FlAsH, demonstrating that both dyes are suitable for monitoring protein dynamics in living cells (*see also* **refs. *13*** and ***14***).

4. Notes

1. In a recent publication from Roger Tsien's laboratory two new TC sequences *(29)* with higher affinity for biarsenical dyes and better quantum yield have been reported (sequence 1: FLNCCPGCCMEP, sequence 2: HWRCCPGCCKTF). The enhanced properties of these TC tags allow for more stringent washing conditions and therefore lower background staining, resulting in improved signal-to-noise ratio. The authors claim an at least 20-fold increase in contrast compared with the standard CCPGCC motif. These claims have not yet been evaluated in our laboratory.
2. Toxicity: In general we did not observe any obvious signs of compromised-cell health during the labeling procedures with FlAsH and ReAsH. However, it has been reported that the use of ReAsH might cause temporary changes (24–48 h post-labeling) in cell morphology. These phenotypical changes are presumably caused by the generation of singlet oxygen when ReAsH is illuminated by a high-intensity light source *(11)*. The generation of reactive oxygen species makes ReAsH an excellent choice for fluorophore-assisted light inactivation experiments *(11)*, but could potentially lead to altered behavior of the target protein in imaging applications that require high-intensity illumination. We did not observe any signs of toxicity caused by the application of dithiols (EDT_2 and BAL) for background reduction up to concentrations of 500 µM.

3. Sensitivity: According to R. Tsien, FlAsH labeling is at least one magnitude less sensitive than GFP *(1,8)*. FlAsH might not the best method to label weakly expressing or unstable proteins. It is a good idea to validate expression levels by an independent method (e.g., Western blot or immunofluorescence using a protein-specific antibody).
4. Specificity: Background is a frequent issue with FlAsH labeling. The need for background reduction was stressed before. In addition, the choice of cell type/line needs to be considered. It has been frequently reported that some cell lines (e.g., 293T and CHO) show relatively high residual background, which might interfere with the detection of weakly expressing proteins. It is therefore recommended to test expression and labeling in multiple cell lines and to select the best suited cell line.
5. Reducing environment: Staining with FlAsH requires a reducing environment *(28)*. It is therefore not recommended to use FlAsH for labeling of extracellular proteins (or extracellular domains of proteins) or intracellular proteins that reside or translocate into the lumen of the ER, Golgi, and vesicular compartments.
6. Selection of labeling reagent: For imaging applications it is in general preferable to use FlAsH as labeling reagent. One reason is that the earlier-mentioned potential for phenotypic changes induced by ReAsH under certain conditions, the other reason is the substantially higher photostability of FlAsH as compared to ReAsH.
7. The use of genetically encoded tags always includes the possibility of interference of the tag with the function and localization of the tagged protein. This is rarely reported, but we have occasionally encountered this problem, especially while working with proteins that are part of larger assemblies (e.g., FP-tagged tubulin). Although the TC tag provides a unique size-advantage over every other live-cell labeling technology, commercially available or described in literature, there is still the possibility of interference, especially if used as integrated tag or in combination with linkers. A recent report by L. Rudner et al. shows that HIV-1 Gag would localize differently depending on the use of either TC alone or a TC-linker combination *(18)*. It is therefore extremely important to validate proper function and localization of the tagged-protein against wild-type protein (e.g., by immunohistochemistry/fluorescence).

References

1. Tsien, R. Y. (2005) Building and breeding molecules to spy on cells and tumors. *FEBS Lett.* **579,** 927–932.
2. Zhang, J., Campbell, R. E., Ting, A. Y., and Tsien, R. Y. (2002) Creating new fluorescent probes for cell biology. *Nat. Rev. Mol. Cell Biol.* **3,** 906–918.
3. Andresen, M., Schmitz-Salue, R., and Jakobs, S. (2004) Short tetracysteine tags to beta-tubulin demonstrate the significance of small labels for live cell imaging. *Mol. Biol. Cell* **15,** 5616–5622.
4. Shaner, N. C., Campbell, R. E., Steinbach, P. A., Giepmans, B. N., Palmer, A. E., and Tsien, R. Y. (2004) Improved monomeric red, orange and yellow fluorescent proteins derived from Discosoma sp. red fluorescent protein. *Nat. Biotechnol.* **22,** 1567–1572.
5. Keppler, A., Gendreizig, S., Gronemeyer, T., Pick, H., Vogel, H., and Johnsson, K. (2003) A general method for the covalent labeling of fusion proteins with small molecules in vivo. *Nat. Biotechnol.* **21,** 86–89.
6. Miller, L. W., Cai, Y., Sheetz, M. P., and Cornish, V. W. (2005) In vivo protein labeling with trimethoprim conjugates: a flexible chemical tag. *Nat. Methods* **2,** 255–257.
7. Griffin, B. A., Adams, S. R., and Tsien, R. Y. (1998) Specific covalent labeling of recombinant protein molecules inside live cells. *Science* **281,** 269–272.
8. Adams, S. R., Campbell, R. E., Gross, L. A., et al. (2002) New biarsenical ligands and tetracysteine motifs for protein labeling in vitro and in vivo: synthesis and biological applications. *J. Am. Chem. Soc.* **124,** 6063–6076.
9. Marek, K. W. and Davis, G. W. (2002) Transgenically encoded protein photoinactivation (FlAsH-FALI): acute inactivation of synaptotagmin I. *Neuron* **36,** 805–813.
10. Hoffmann, C., Gaietta, G., Bunemann, M., et al. (2005) A FlAsH-based FRET approach to determine G protein-coupled receptor activation in living cells. *Nat. Methods* **2,** 171–176.
11. Tour, O., Meijer, R. M., Zacharias, D. A., Adams, S. R., and Tsien, R. Y. (2003) Genetically targeted chromophore-assisted light inactivation. *Nat. Biotechnol.* **21,** 1505–1508.
12. Nakanishi, J., Nakajima, T., Sato, M., Ozawa, T., Tohd, K. and Umezawa, Y. (2001) Imaging of conformational changes of proteins with a new environment-sensitive fluorescent probe designed for site specific labeling of recombinant proteins in live cells. *Anal. Chem.* **73,** 2920–2928.

13. Nakanishi, J., Maeda, M., and Umezawa, Y. (2004) A new protein conformation indicator based on biarsenical fluorescein with an extended benzoic acid moiety. *Anal. Sci.* **20**, 273–278.
14. Gaietta, G., Deerinck, T. J., Adams, S. R., et al. (2002) Multicolor and electron microscopic imaging of connexin trafficking. *Science* **296**, 503–507.
15. Sosinsky, G. E., Gaietta, G. M., Hand, G., et al. (2003) Tetracysteine genetic tags complexed with biarsenical ligands as a tool for investigating gap junction structure and dynamics. *Cell Commun. Adhes.* **10**, 181–186.
16. Park, H., Hanson, G. T., Duff, S. R., and Selvin, P. R. (2004) Nanometre localization of single ReAsH molecules. *J. Microsc.* **216**, 199–205.
17. Thorn, K. S., Naber, N., Matuska, M., Vale, R. D., and Cooke, R. (2000) A novel method of affinity-purifying proteins using a bis-arsenical fluorescein. *Protein Sci.* **9**, 213–217.
18. Rudner, L., Nydegger, S., Coren, L. V., Nagashima, K., Thali, M., and Ott, D. E. (2005) Dynamic fluorescent imaging of human immunodeficiency virus type 1 gag in live cells by biarsenical labeling. *J. Virol.* **79**, 4055–4065.
19. Goldstein, J. C., Munoz-Pinedo, C., Ricci, J. E., et al. (2005) Cytochrome c is released in a single step during apoptosis. *Cell Death Differ.* **12**, 453–462.
20. Panchal, R. G., Ruthel, G., Kenny, T. A., et al. (2003) In vivo oligomerization and raft localization of Ebola virus protein VP40 during vesicular budding. *Proc. Natl. Acad. Sci. USA* **100**, 15,936–15,941.
21. Ignatova, Z. and Gierasch, L. M. (2004) Monitoring protein stability and aggregation in vivo by real-time fluorescent labeling. *Proc. Natl. Acad. Sci. USA* **101**, 523–528.
22. Ignatova, Z. and Gierasch, L. M. (2005) Aggregation of a slow-folding mutant of a beta-clam protein proceeds through a monomeric nucleus. *Biochemistry* **44**, 7266–7274.
23. Cavagnero, S. and Jungbauer, L. M. (2005) Painting protein misfolding in the cell in real time with an atomic scale brush. *TIBS* **23**, 157–162.
24. Robia, S. L., Flohr, N. C., and Thomas, D. D. (2005) Phospholamban pentamer quaternary conformation determined by in-gel fluorescence anisotropy. *Biochemistry* **44**, 4302–4311.
25. Feldman, G., Bogoev, R., Shevirov, J., Sartiel, A., and Margalit, I. (2004) Detection of tetracysteine-tagged proteins using a biarsenical fluorescein derivative through dry microplate array gel electrophoresis. *Electrophoresis* **25**, 2447–2451.
26. Chen, B., Mayer, M. U., and Squier, T. C. (2005) Structural uncoupling between opposing domains of oxidized calmodulin underlies the enhanced binding affinity and inhibition of the plasma membrane Ca-ATPase. *Biochemistry* **44**, 4737–4747.
27. Stroffekova, K., Proenza, C., and Beam, K. G. (2001) The protein-labeling reagent FLASH-EDT2 binds not only to CCXXCC motifs but also non-specifically to endogenous cysteine-rich proteins. *Pflugers Arch.* **442**, 859–866.
28. Griffin, B. A., Adams, S. R., Jones, J., and Tsien R. (2000) Fluorescent labeling of recombinant proteins in living cells with FlAsH. *Meth. Enzym.* **327**, 565–578.
29. Martin, B. R., Giepmans, B. N. G., Adams, S. R., and Tsien, R. Y. (2005) Mammalina cell-based optimization of the biarsenical-binding tetracysteine motif fo improved fluorescence and affinity. *Nat. Biotechnol.* **23**, 1309–1314.

16

Exploiting Network Biology to Improve Drug Discovery

Marnie L. MacDonald and John K. Westwick

Summary

Mammalian signal transduction occurs in the context of multiprotein complexes, yet currently available drug discovery strategies do not reflect this fact. We present a strategy for screening drugs and targets in living human cells by utilizing high content protein-fragment complementation assays. Synthetic fragments of a mutant fluorescent protein ("Venus" and/or enhanced yellow fluorescent protein) are used for protein-fragment complementation assay construction, allowing us to measure spatial and temporal changes in protein complexes in response to drugs that activate or inhibit particular pathways. Here we describe the utility of this novel strategy for high-throughput screening of known targets, and for screening previously undrugable targets and profiling drug leads for improved selectivity and safety.

Key Words: Cell-based assay; chemical biology; drug discovery; fluorescence microscopy; G protein-coupled receptor; high content screening; kinase; network; nuclear receptor; pathway; protein complexes; protein-fragment complementation assay; signal transduction; transcription factor.

1. Introduction

The modern drug discovery paradigm involves selecting a target, establishing an in vitro assay, and then identifying a molecule that binds to or regulates the target. This is accomplished by high-throughput screening of a chemical library and/or by rational design of a compound based on crystal structure of the target-binding site. Current approaches treat the cell as a black box, and cell-based methods are usually limited to secondary assays, which are designed primarily to verify the bioavailability and potency of the lead compound. However, an isolated protein in a test tube is far removed from its native cellular and subcellular context, and from the many other molecules that influence its structure, activity, and ultimate function. Cells are complex systems in which a multitude of biochemical reactions take place at any one time. Protein targets of drugs often exist as components of dynamic multiprotein complexes, connected to many other proteins and macromolecules. Further, the same protein often exists in different forms, as components of different complexes and in different cellular compartments (**Fig. 1**). An example of this phenomenon is found in the Akt signaling pathway. Akt is found primarily at the plasma membrane, but also exists in the nucleus. We have observed that PI3K-phosphoinositide-dependent kinase (PDK)/Akt complexes at the plasma membrane are regulated by drugs (such as PI3 kinase inhibitors), whereas nuclear complexes containing Akt and p27 are regulated by other drugs (schematized in **Fig. 1**). Also, even a highly potent, targeted compound might regulate the activity of dozens of proteins apart from its intended target or even its target class. Thus, identifying the full spectrum of activity of any drug or drug candidate is not possible with in vitro assays, regardless of the size of the panel. We

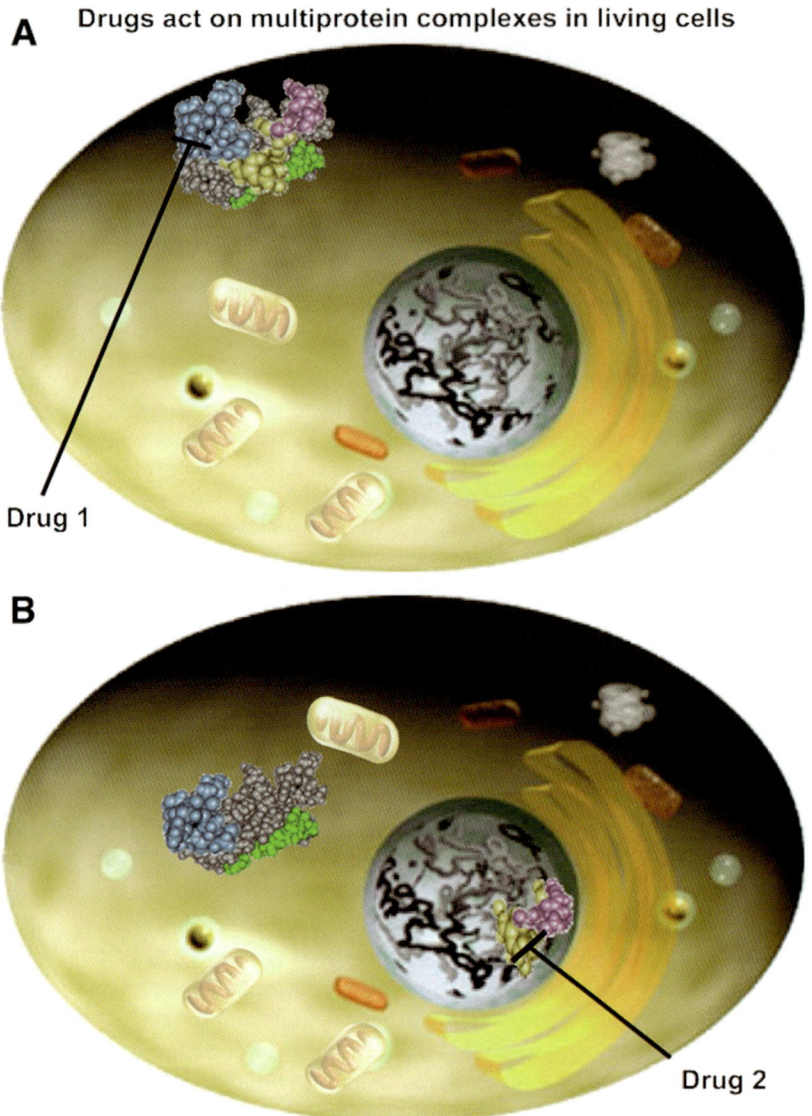

Fig. 1. Proteins in cells exist as large, multimolecular structures, and drugs act in the context of these complexes. A representation of a protein complex is shown (**A**). Drug 1 acts on a protein complex located in the plasma membrane. In (**B**), a portion of the complex is localized in the nucleus. In this case, drug two regulates the nuclear protein, but drug one is not active in this context.

believe that the living human cell is the central, missing component in preclinical drug discovery, and that both potential drug liabilities as well as beneficial applications might be lost because of the unforeseen actions of drugs within the cellular milieu.

The need to examine global effects of drugs on human cells has inspired the applications of DNA microarrays, including the field of toxicogenomics, which involves the use of complex populations of mRNA to understand toxicity. Cells, or whole animals, are treated with drugs; messenger RNA is isolated from the cell or tissue; and the gene expression patterns of the mRNA in the absence and presence of a drug are compared. Such transcriptional profiling can identify subtle differences between compounds, in which the compounds affect the ultimate transcriptional activity of one or more pathways *(1)*. Identifying specific genes that are stimulated or repressed in response to specific conditions or treatments is a useful way to begin to unravel the cellular mechanisms of disease and of drug response. However, changes in the level of individual mRNA molecules correlate poorly with the level of the corresponding protein, and even less with protein activity *(2)*. Furthermore, many proteins undergo numerous posttranslational modifications and protein interactions, which might affect the functions and activities of proteins within a tissue or cell. Thus, simply measuring the mRNA species present at a particular time does not yield an informative picture of a drug. Finally, a targeted drug might affect the transcription of dozens or hundreds of genes, making interpretation of drug mechanism of action difficult to deconvolute *(3)*.

Direct temporal and spatial measurements of specific chemical transitions within biochemical pathways could eliminate the problems associated with interpretation of drug perturbations of transcriptional profiles. Among the most obvious and general transitions that could be measured are the induction and dissociations of protein complexes. Unlike transcriptional reporter assays, the information obtained by monitoring protein complexes reflects the effect of a drug on a particular branch or node of a biochemical pathway, not its end point *(4,5)*. For example, stimulation of a pathway by an agonist could lead to an increase in the association of an intracellular protein (such as a kinase) with a cognate binding partner (such as a substrate). The drug effect could therefore be assessed by quantifying the amount or location of the kinase/substrate complex in the absence and presence of the drug. In this example the kinase/substrate complex serves as a "sentinel" of pathway activity. A drug acting either at the beginning of the pathway (such as a receptor antagonist) or acting on another target downstream of the receptor but upstream of the sentinel could alter either the amount or location of the kinase/substrate complex within the cell. Thus, assessing complexes in the absence or presence of a chemical compound could reveal whether or not the drug acts on that pathway.

In addition to being composed of large multiprotein complexes, biochemical pathways are highly organized in subcellular space. For instance, in signal transduction pathways, membrane receptors that receive signals at the plasma membrane transmit the signals to the cell nucleus, resulting in the activation or repression of gene transcription. Thus, the subcellular compartment of a particular complex reflects the function of that complex at a particular point in time (*see* **Figs. 1–3D**). High content methods that allow for the visualization and quantification of protein complexes in subcellular compartments can provide far more information than simple bulk measurements of fluorescence.

In this chapter we review high content assays for studying the spatial and temporal dynamics of protein complexes based on protein-fragment complementation assays (PCA) and provide examples of their applications to drug discovery.

2. Localization and Quantification of Protein Complexes in Cells

A protein–protein interaction assay is capable of measuring either the interaction between two proteins (in network biology terms, the "transition") or the complex that is formed as a result of an interaction (the "state" of the proteins that results from the interaction). PCA represents a particularly useful strategy for the measurement of protein complexes in living cells. At its basic level, PCA is a universal and flexible strategy that allows for the detection, quantification, and localization of protein–protein complexes in intact, living cells.

Michnick and coworkers (*see* **ref. 6** herein) first showed that a reporter protein can be intentionally fragmented in such a way that the two polypeptide fragments, when fused separately to two other molecules that interact, are capable of refolding and reconstituting the activity of the original reporter. Because all of the information for protein folding is encoded in the primary amino acid sequence of the protein, close apposition of the fragments—as a result of the interaction of the molecules to which the fragments are fused—is sufficient to cause the folding of fragments into an active protein, thereby recapitulating the unimolecular folding reaction of the

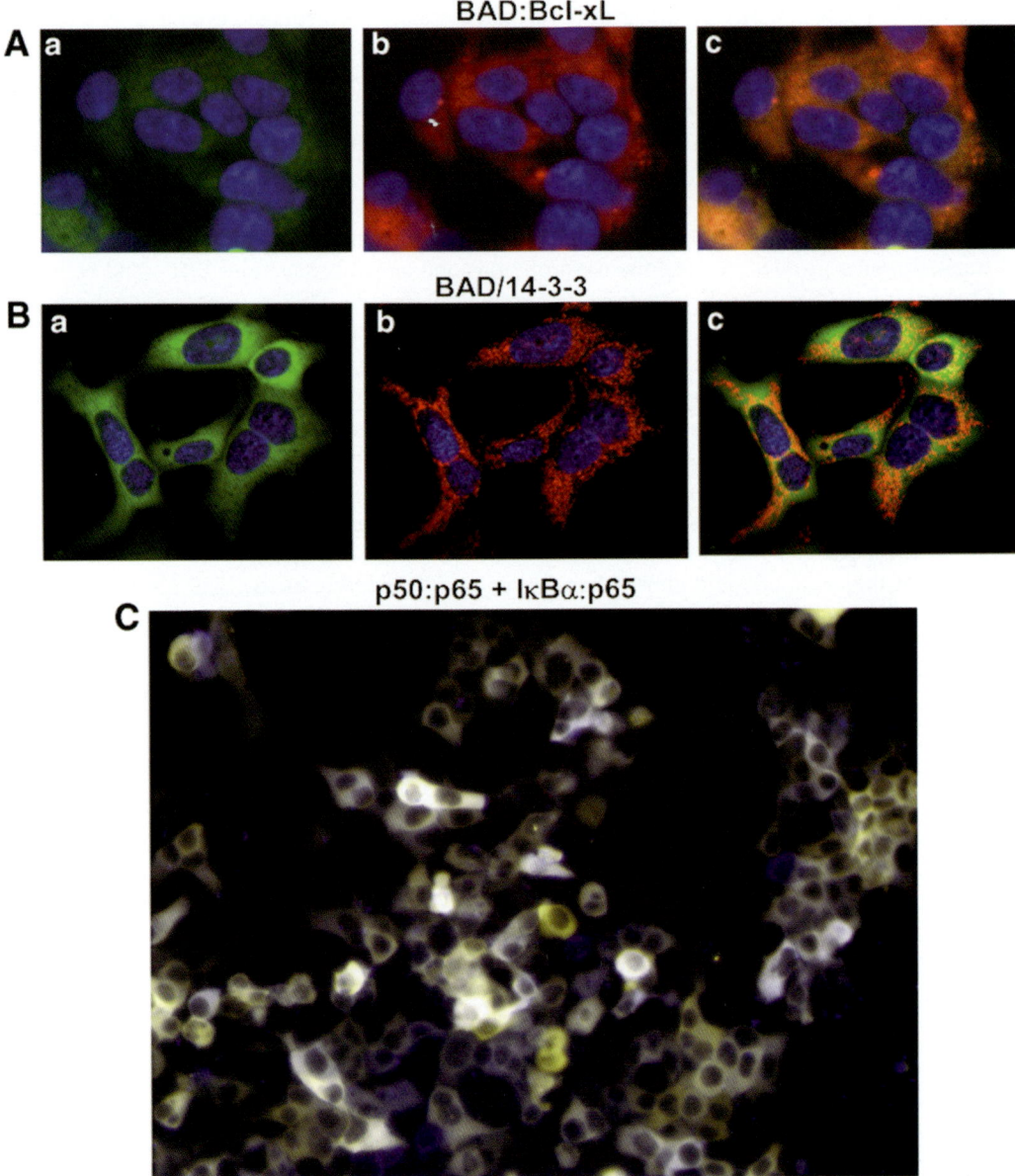

Fig. 2. (**A,B**) BAD complexes in cells as assessed with a fluorescence (enhanced yellow fluorescent protein) PCA. (**A**) BAD/Bcl-XL: (**B**), BAD/14-3-3-sigma. Shown are: (a) the PCA signal (green; nuclei are stained with Hoechst), (b) a mitochondrial dye (Mitotracker red, Molecular Probes), and (c) overlay of (a) and (b). (**C**) Dual-color PCAs for the NFκB pathway, showing IκBα/p65 (yellow fluorescence) and p50/p65 (blue fluorescence). Cells were transiently transfected with the three fragment constructs. Different colors are generated according to the interactions of the proteins to which the respective fragments are fused. Fluorescence images were acquired with an SP Nikon fluorescence microscope using a Chroma CFP filter (excitation: 426–446 nm; emission: 460–500 nm; dichroic mirror: 455LP), and a FITC filter (excitation: 460–500 nm; emission: 505–560 nm; dichroic mirror: 505LP). 16-bit monochrome images were acquired with a CoolSnap HQ CCD camera. CFP and FITC images for each PCA were subsequently pseudo-colored and overlaid using Metamorph software (Molecular Devices, Sunnyvale, CA).

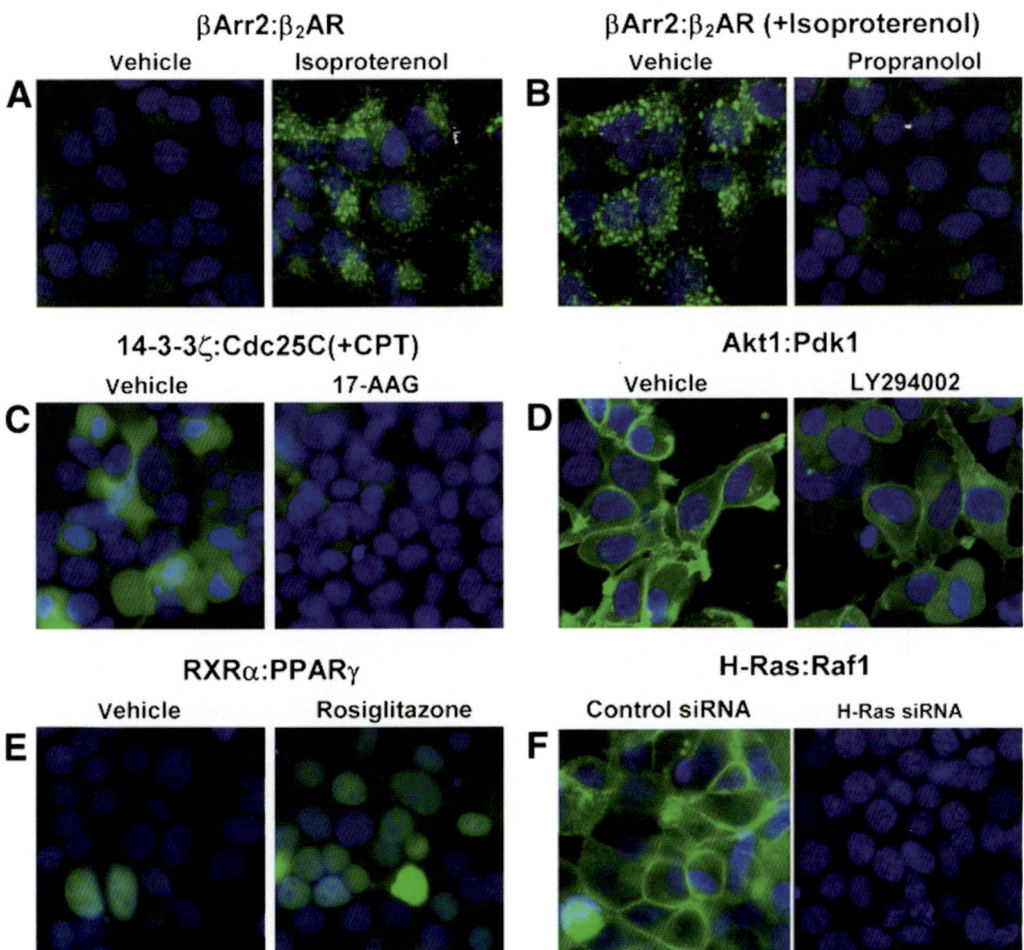

Fig. 3. Examples of dynamic changes that occur in response to drugs or siRNAs include increases, decreases, or change in localization of the fluorescence signal. PCAs and pretreatment conditions, in which used, are as follows: (**A**): β-arrestin2/β$_2$-adrenergic receptor; (**B**) β-arrestin2/β$_2$-adrenergic receptor (+isoproterenol); (**C**) 14-3-3-ζ/Cdc25C (+CPT); (**D**) AKT1/PDK1; (**E**) RXRα/PPARγ; (**F**) H-Ras/Raf1.

original reporter and the reconstitution of enzymatic activity. Subsequent studies suggested that, if a flexible linker of 10 amino acids is included between the interacting protein(s) of interest and the complementary fragments, a distance of 80 Å or less allowed refolding of the reporter fragments and reconstitution of reporter activity. Thus, protein-fragment complementation serves as a direct assay of complex formation between molecules to which the fragments are fused. PCAs based on nearly 100 different enzymes and fluorescent proteins have been conceived of, and many of these have been constructed, including GFP and a number of mutant fluorescent proteins (cyan, yellow, citrine, red fluorescent proteins, and variants thereof), luciferases from various species, dihydrofolate reductase, β-lactamase, hygromycin phosphotransferase, GAR transformylase, and others *(6–15)*. Recently, we have constructed PCAs based on intense and rapidly maturing variants of yellow fluorescent protein (YFP) to probe hundreds of different known and novel protein complexes, which include all major classes of drug targets and biochemical pathways, different subcellular compartments, and pathways that are both general and specific to differentiated cell types. In this review, we discuss examples taken largely from signal transduction pathways.

Since the original invention in 1997, the PCA technology has sometimes been dubbed with other names, including interaction-dependent enzyme activation, bimolecular fluorescence complementation, trimolecular fluorescence complementation, "half-zyme," "split-GFP," "split-luciferase," and "luciferase complementation imaging," all of, which refer to the same basic technology *(16–20)*. Owing to the diversity of reporter proteins that can be used to construct PCAs, assays can be tailored to the particular demands of the cell type, target, signaling process, affinity of the binding partners, and screening instrumentation. Finally, the ability to choose among a wide range of reporter fragments generating fluorescence signals provides a choice of high-content or high-throughput assay formats.

PCA has unique features that make it an important tool in drug discovery:

- The PCA strategy is applicable to any cell into which DNA constructs or their encoded protein products can be introduced, including prokaryotic and eukaryotic cells and primary cells.
- Molecular interactions are detected directly, not through secondary events such as transcriptional activation or second messenger production.
- Tagging of proteins with large molecules, such as intact fluorescent proteins, is not required. In contrast to assays utilizing intact reporters, many PCAs reveal dramatic drug- or agonist-induced changes in total fluorescence, not just changes in subcellular localization (*see,* e.g., **Figs. 3** and **4**).
- Proteins are expressed in the relevant cellular context, reflecting the native state of the proteins with their correct post-translational modifications and in the presence of other enzymatic and structural components that regulate the complexes.
- PCA fragments can be synthesized and/or genetically engineered to create assays with specific properties (signal intensity, stability, spectral properties, wavelength, and so on).
- Flexibility in expression vector design enables the choice of various gene orientations, linker lengths, constitutive or inducible promoters, plasmid or viral vectors, and various selectable marker strategies depending on the assay demands.
- PCA can be used to map proteins and drugs into signaling pathways and validate novel targets by determining whether the protein–protein complex(es) can be modulated in response to an agonist, antagonist, inhibitor, or siRNA.

3. Measuring Distinct Complexes Formed in Different Specific Subcellular Locations

Considering that all proteins exist in complexes, and interact with multiple proteins, it is not surprising that any particular protein might function simultaneously or sequentially in different subcellular compartments in conjunction with different interacting partners. These relationships are usually inferred from colocalization studies, but such studies only show that two proteins are localized in the same compartment at the same time—not that they exist in a physical complex. Direct analysis of these interactions can be achieved with high content PCAs. **Figure 2(A,B)** shows two different protein complexes, both of which involve the protein Bad. Bad is a member of the BH3-only subfamily of apoptosis-regulating proteins. When Bad is phosphorylated it is found in the cytosol, where it is bound to 14-3-3 proteins and where it is incapable of inducing cell death. Protein 14-3-3 is a family of multifunctional phosphoserine peptide-binding molecules that serve as effectors of survival signaling. BAD is known to form complexes with the protein 14-3-3 as well as with other members of the BAD family such as Bcl-xL. As shown in **Fig. 2A**, Bad/Bcl-xL complexes are located in the mitochondria, as the PCA signal (green) is colocalized with the mitochondrial dye (red). In contrast, Bad/14-3-3 complexes are clearly located throughout the cytosol of the cells (**Fig. 2B**). Importantly, the different BAD complexes carry out different, highly specialized, functions in the cell. Differentiation of these functions requires analysis of specific BAD complexes, and can only be studied at the subcellular level using high content methods.

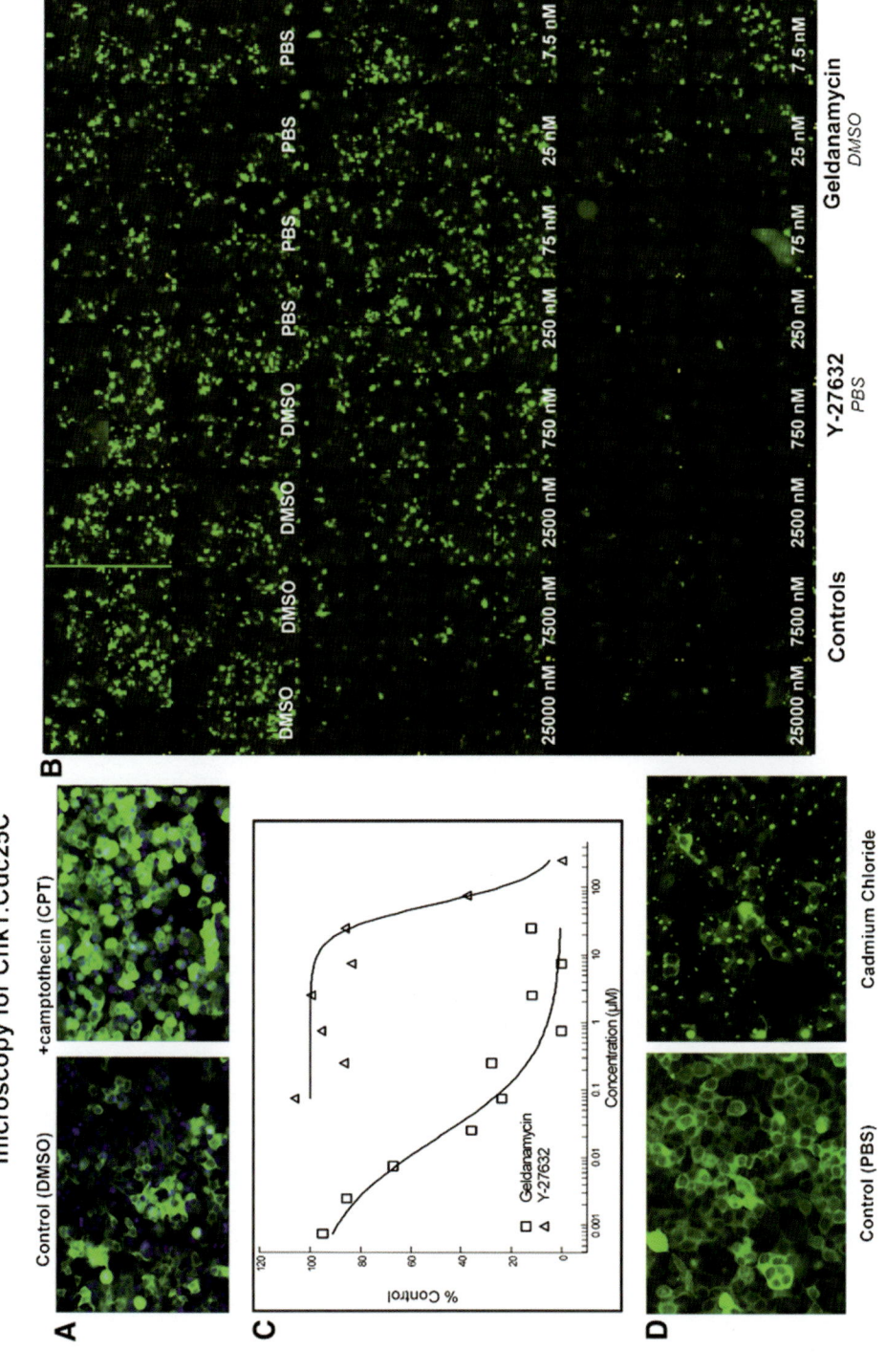

Fig. 4. Dose-dependent effects of drugs on protein interactions, as assessed by automated microscopy for Chk1/Cdc25C in HEK293 cells stably expressing the two constructs. (**A**) CPT induces the interaction of Chk1 with Cdc25C in the cell nucleus. (**B**) Y27632 and geldanamycin suppress Chk1/Cdc25C complexes in a dose-dependent manner, as seen in this 96-well plate view which shows eight different scans for each drug dose. (**C**) Dose-response curves for the drugs based on total fluorescence. (**D**) High content changes in signal pattern showing the formation of bright subnuclear foci in response to cadmium chloride.

We see similar compartmentalization with dozens of other well-characterized protein complexes. This means that simple tagging of a protein with a fluorophore such as GFP—which is done to monitor translocation of individual proteins or colocalization of two different proteins—only illuminates the bulk of the tagged protein in the cell, and completely ignores the specialized functions of proteins in the context of their subcellular compartments and interacting partners. Drug effects might be missed if the bulk changes in individual proteins are studied, but can be readily detected if dynamic protein complexes are monitored. Moreover, monitoring of different complexes in the same pathway allows for the determination of the mechanism of action of a drug at different steps, allowing for the site of action of the drug to be narrowed down to a range of, or to a specific, protein. **Figure 2C** provides an example of multiplexing this type of analysis with multicolor PCA in living cells. The NFκB p65 subunit forms complexes with the p50 subunit, and also with the protein IκBα. In resting cells, IκBα binds to NFκB and retains the complex in the cytoplasm. Thus, p65 forms cytoplasmic protein–protein complexes with p50 and also with IκBα. We cotransfected HEK293 cells simultaneously with three PCA constructs: CFP[1]-p50; CFP[2]-p65; and IκBα-YFP[1]. If a protein–protein complex forms between p50 and p65, the CFP[1] fragment should complement the CFP[2] fragment, producing blue fluorescence. Alternatively, if a protein–protein complex forms between IκBα and p65, the YFP[1] fragment should complement the CFP[2] fragment, producing a yellow fluorescence. As shown in **Fig. 2C**, both p65/p50 (blue) and IκBα/p65 (yellow) complexes could be detected in the cytoplasm as expected. Cells displaying a lighter yellow (almost white) cytoplasmic signal are expressing both p65/p50 and IκBα/p65 complexes. The ability to construct multicolor PCAs allows for the detection and quantification of multiple distinct protein–protein complexes within the same cells.

4. Measuring the Dynamics of Protein Complexes With High-Content PCAs

A key feature of PCA, as compared with in vitro, proteomics-based approaches, is the ability to capture the *dynamics* of complexes in their native context. In conjunction with high content cell imaging, detailed information on the levels and localization of signaling complexes can be obtained. Thus, PCA can be used not only to identify the components of signaling networks, but also to probe the activity or flux of pathways within these networks. The examples below demonstrate that it is possible to assay discrete signaling nodes for agents that act directly on the signaling proteins of the interaction, as well as for targets and agents that act "upstream" of the interaction. The strategy works equally well for assessing effects of small molecules or genetic probes such as dominant negative/active clones and RNAi on signal transduction pathways.

Six different examples of dynamic interactions are shown in **Fig. 3**. The examples include a GPCR/GPCR modulator interaction (**A,B**); a cell cycle phosphatase/regulatory protein interaction (**C**); a kinase/kinase interaction (**D**); a nuclear hormone receptor/coactivator interaction (**E**); and a small GTPase/kinase complex (**F**). β-Arrestins are regulatory proteins that form complexes with most GPCRs and play a central role in receptor desensitization, sequestration, and down-regulation (for a review *see* Luttrell and Lefkowitz *[21]*). β-arrestin binding to GPCRs both uncouples receptors from their cognate G proteins and targets the receptors to clathrin-coated pits for endocytosis. β-arrestins might also function as GPCR signal transducers. β-Arrestin movement from the plasma membrane to intracellular vesicles has been visualized by tagging β-arrestin with GFP and monitoring the subcellular distribution of the fluorescence signal in living cells *(22)*. We sought to create a quantitative fluorescence assay for which changes in GPCR activation would be detected by an increase or decrease in the reconstituted fluorescent signal generated by binding of the receptor to β-arrestin. A stable HEK293 cell line expressing the constructs β-arrestin2-IFP(1) and β2AR-YFP(2) was generated. Cells were treated at 37°C with 10 μM isoproterenol for 60 min. Cells were fixed and stained and images of the reconstituted fluorescent signal were acquired on a Discovery-1 instrument (Molecular Devices, Sunnyvale, CA) using a ×20 objective. **Figure 3A** demonstrates the induction of fluorescence in response to agonist (isoproterenol) and the appearance of signal in intracellular granules. The appearance of the protein–protein complex in intracellular granules (corresponding to endosomes) is consistent with the process of receptor internalization in response to agonist, and the effect can be blocked by pretreatment of cells with the reverse agonist, propranolol (**Fig. 3B**). It is important to note the key distinctions between this PCA and previous cell-based assays involving a GFP-tagged β-arrestin, which measures a different phenomenon. Direct tagging of a protein, such as β-arrestin, with a fluorescent protein or other optically detectable molecule only enables imaging of its subcellular localization *(22)*. In contrast, PCA quantifies the association of two proteins and the subcellular location of the protein–protein complex. Other examples of dynamic changes of protein complexes are shown in **Fig. 3**. **Figure 3C** shows a complex between a 14-3-3 protein and Cdc25C. These complexes are induced by camptothecin (CPT). Treatment of the cells with an HSP90 inhibitor such as 17-AAG (17-allylaminogeldanamycin) causes destruction of the complex, because of ubiquitin-mediated proteolysis of PCA-components that are client proteins of HSP90. We have seen similar effects of HSP90 inhibitors on other complexes including Chk1/Cdc25C (discussed later; **Fig. 4**) and H-Ras1/Raf1 (data not shown). Thus, PCA provides a useful method to screen small molecule libraries for HSP90 inhibitors, to identify the panoply of cellular proteins that are regulated by HSP90, and to study their regulation in living cells. **Figure 3D** illustrates dynamics of complexes containing Akt1 and PDK1. These complexes reside at the cell membrane in cells grown in serum, and the localization of fluorescence at the plasma membrane can easily be seen in the vehicle-treated cells. Inhibition of PI3 kinase, by drugs such as LY294002, rapidly causes the redistribution of the complex to the cytosol (**Fig. 3D**). This drug effect can be observed within seconds following treatment with LY294002 and is maximal within 7 min. This result also shows that drug effects, which occur "upstream" from a particular interaction can be detected by monitoring interactions "downstream" from the site of action of the drug, an important principle which allows mapping of drugs into pathways by using various interactions as "sentinels" of pathways. **Figure 3E** shows the complexes between the nuclear hormone receptor, PPARγ, and its coactivator, RXRα. Antidiabetic drugs such as rosiglitazone are PPARγ ligands that induce the formation of transcriptional complexes containing PPARγ and coactivators including RXRα (retinoid X-receptor-α). Formation of these complexes precedes transcriptional regulation of genes involved in glucose and lipid homeostasis. These PCAs are useful in screening small molecule libraries to identify other small molecule agonists that regulate various members of the PPAR family. In addition, the selectivity of various lead compounds can be determined by studying the effects of agonists on assays that report the activity of PPAR in association with other coactivators such as SRC-1 and SRC-3, because the preference of PPAR for a particular coactivator is determined by the conformation adopted on agonist binding. Also, similar assays can be constructed for known and orphan nuclear hormone receptors, making this drug target class eminently tractable for novel drug discovery. **Figure 3F** shows complexes of the small GTPase, H-Ras, with the kinase, Raf1. Small GTPases of the Ras and Rho families are among the most studied signaling proteins, and represent promising therapeutic targets for human neoplastic disease. Despite the high level of interest in these proteins, direct analysis of most aspects of Ras protein biology in living cells has not been possible. Much of our knowledge, therefore, has been derived from in vitro analyses, or from functional assays reporting on a downstream effect of Ras activity (such as cellular transformation or gene expression). Drug discovery efforts to date have focused on upstream enzymatic regulators of Ras pathway activation (e.g., screens for receptor tyrosine kinase inhibitors) or on Ras posttranslational modification (e.g., farnesyl transferase inhibitor screens) or on downstream kinase-regulated signaling events (e.g., screens for Raf kinase inhibitors). Identification of probes directly regulating Ras family protein activity would enhance our understanding of this area of biology, and possibly lead to identification of novel therapeutic agents. It is notable that the Ras/Raf complexes are localized at the plasma membrane; Raf proteins are known to associate with the effector domain of active Ras proteins at the plasma membrane *(23)*. By probing the Ras/Raf complex (as compared with, e.g., fluorescently labeled Raf protein alone) this assay focuses exclusively on active signaling complexes. Cotransfection of siRNA targeting H-Ras clearly obliterates the complex. Thus PCA can be used to detect gene silencing in living cells. We recently have applied this approach to detect targets linked to validated drug targets and pathways, where the validated target is used to construct the PCA and the effects of inhibiting individual proteins on the PCA are assessed with high content assays.

5. Quantitative Analysis of Drug Effects on Protein Complexes

When combined with the appropriate image analysis algorithms, PCAs can be used to screen small molecule libraries and to establish dose–response and structure–function relationships. **Figure 4** shows an example of a PCA that reports on DNA damage response pathways. Chk1/Cdc25C complexes are induced by cell treatment with the topoisomerase inhibitor, CPT, for 18 h (**Fig. 4A**). Test compounds that block the pathway from topoisomerase to this signaling node can be identified by preincubating cells with the test compounds preceding treatment with CPT. Two such agents, with different mechanisms of action, are the natural product geldanamycin, and the small molecule kinase inhibitor, Y27632. Chk1 is a client protein for HSP90 *(24)*, and geldanamycin suppresses Chk1/Cdc25C complexes as a result of destabilization of Chk1. These results are easily visualized and quantified, as shown in the plate view and the corresponding dose–response curve (**Fig. 4C**). Y27632, previously described as an inhibitor of Rho kinases, also suppresses these complexes through an unknown mechanism and with a different dose dependency (**Fig. 4C**). Thus, total fluorescence can be used to assess the effects of test compounds acting through these mechanisms. In other cases, however, drugs and toxicants cause changes in fluorescence signal pattern that are not obvious based on total fluorescence. For example, cadmium chloride induces the formation of bright punctuate signals at discrete foci within the nucleus of the cell (**Fig. 4D**). This effect would be masked by measurements of bulk fluorescence but can readily be quantified using algorithms designed to quantify granularity of signal. These examples illustrate the fact that high content methods are capable of detecting not only total changes in signal intensity but also more subtle changes in signal pattern that can occur.

6. Pharmacological Profiling in Living Cells

Even a selective chemical compound that binds to a therapeutic target might have completely unexpected or "off-target" effects when it contacts a living cell, resulting in expensive preclinical and clinical failures. An improved understanding of drug action in living cells would accelerate attrition of compounds (a "fail-fast" strategy) and improve productivity in pharmaceutical research. This goal is best accomplished by profiling lead compounds broadly across cellular targets and pathways in living cells. We, therefore, constructed diverse panels of high content PCAs covering major drug target classes and mechanisms of action and used these assay panels to analyze known drugs and known toxicants, and to perform lead optimization in an iterative fashion. In addition, we have used these assay panels to identify new therapeutic indications for known drugs. The process of evaluating numerous known, marketed drugs, clean and potent lead compounds, and compounds that have failed in preclinical and clinical development has given us a view concerning what constitutes a selective drug vs a nonselective drug. **Figure 5** shows an example of a set of 20 compounds that includes five highly nonselective compounds (1–5) vs other, more selective, lead compounds (6–20). All of these compounds were potent in cells, but the nonselective compounds subsequently failed in animal models resulting from toxicity in various organs. Our data suggests that cellular off-target activity is a common, but underappreciated component of drug failure. In addition to identifying the extent of off-target activity, the underlying pathways contributing to the off-target activity can be pinpointed with these assay panels.

Exploiting Network Biology

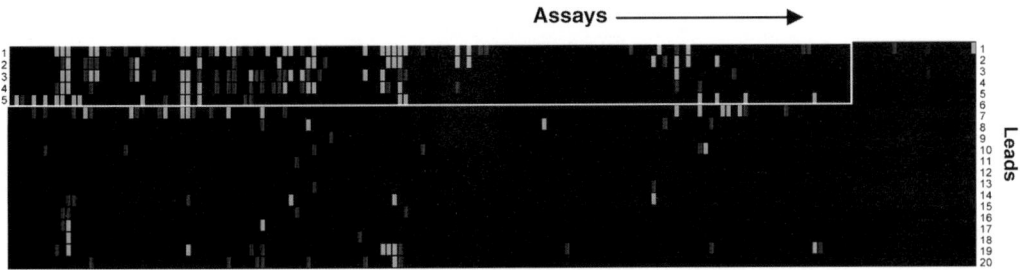

Fig. 5. Cellular selectivity of drugs in human cells. Each of 20 different small molecule lead compounds were analyzed with 191 unique assays in HEK293 cells in 384-well plates, using automated confocal imaging systems (Evotec Opera, Hamburg, Germany). Each assay was a particular protein interaction at a particular time and pretreatment condition. The results are represented as a matrix enabling visualization of significant hits, in which the assays are on the *x*-axis and the compounds are on the *y*-axis. Leads 1–5 were highly nonselective, as is apparent from the number of hits across the assay panel, and these compounds proved to be toxic in preclinical studies.

For example we have seen unintended, off-pathway effects of supposedly selective kinase inhibitors on GPCR pathways. In addition, we have identified new therapeutic uses of known drugs by pinpointing off-target activity on pathways linked to therapeutic efficacy, and by combining PCA with siRNA, we have validated novel drug targets. In sum, these high content assays provide a discovery platform with the potential to impact many areas of pharmaceutical discovery—from target validation to lead optimization—in addition to basic biomedical research.

7. Conclusions

Classical drug discovery paradigms treat protein targets as isolated entities, and pathways as serial and parallel "circuit diagrams." These concepts do not reflect reality and perpetuate faulty decisions in drug discovery and development. A better understanding of the physical nature of cell signaling and cellular networks, and of drug effects in the context of those networks, is essential to improving drug selectivity and specificity. This growing recognition has led to the call for a pathway-based approach to drug discovery *(25)*.

High-content, cell-based methods enable drug discovery in the context of the biochemical pathways of living cells. We have employed protein complexes as dynamic sensors of cellular pathways. The ability to localize and quantify these events directly in cells allows for the translation of a more realistic view of cell biology into improved productivity of biomedical research and development. As we have shown, these approaches can be adapted to high throughput discovery on a large scale. Moreover, by profiling drugs and lead compounds against a wide array of cellular pathways, we can identify the on-target and off-target activities of drugs in human cells. Strategies such as those described here, and in other chapters in this volume, open the door to a more effective paradigm for drug discovery.

References

1. Stoughton, R. B. and Friend, S. H. (2005) How molecular profiling could revolutionize drug discovery. *Nat. Rev. Drug Discov.* **4,** 345–350.
2. Tian, Q., Stepaniants, S. B., Mao, M., et al. (2004) Integrated genomic and proteomic analyses of gene expression in Mammalian cells. *Mol. Cell Proteomics* **3,** 960–969.
3. Miklos, G. L. and Maleszka, R. (2004) Microarray reality checks in the context of a complex disease. *Nat. Biotechnol.* **22,** 615–621.

4. Remy, I. and Michnick, S. W. (2001) Visualization of biochemical networks in living cells. *Proc. Natl. Acad. Sci. USA* **98,** 7678–7683.
5. Yu, H., West, M., Keon, B. H., et al. (2003) Measuring drug action in the cellular context using protein-fragment complementation assays. *Assay Drug Dev. Technol.* **1,** 811–822.
6. Michnick, S. W., Remy, I., Valois, F. X., Vallee-Belisle, A., Galarneau, A., and Pelletier, J. N. (2000) Detection of protein–protein interactions by protein fragment complementation strategies, Parts A and B, in *Methods in Enzymology,* (Abelson, J. N., Emr, S. D., and Thorner, J., eds.), Academic, San Diego, CA, pp. 208–230.
7. Pelletier, J. N., Remy, I., and Michnick, S. W. (1998) Protein-fragment complementation assays: a general strategy for the in vivo detection of protein–protein interactions. *J. Biomol. Tech.* **10,** 32–39.
8. Galarneau, A., Primeau, M., Trudeau, L. -E., and Michnick, S. W. (2002) A protein fragment complementation assay based on TEM1 β-lactamase for the detection of protein-protein interactions. *Nat. Biotech.* **20,** 619–622.
9. Remy, I., Wilson, I. A., and Michnick, S. W. (1999) Erythropoietin receptor activation by a ligand-induced conformation change. *Science* **283,** 990–993.
10. Remy, I. and Michnick, S. W. (1999) Clonal selection and in vivo quantitation of protein interactions with protein fragment complementation assays. *Proc. Natl. Acad. Sci. USA* **96,** 5394–5399.
11. Remy, I. and Michnick, S. W. (2004) Regulation of apoptosis by the Ft1 protein, a new modulator of protein kinase B/Akt. *Mol. Cell Biol.* **24,** 1493–1504.
12. Remy, I., Montmarquette, A., and Michnick, S. W. (2004) PKB/Akt modulates TGF-beta signalling through a direct interaction with Smad3. *Nat. Cell Biol.* **6,** 358–365.
13. Remy, I. and Michnick, S. W. (2004) Mapping biochemical networks with protein-fragment complementation assays. *Methods Mol. Biol.* **261,** 411–426.
14. Remy, I. and Michnick, S. W. (2004) A cDNA library functional screening strategy based on fluorescent protein complementation assays to identify novel components of signaling pathways. *Methods* **32,** 381–388.
15. Remy, I., Pelletier, J. N., Galarneau, A., and Michnick, S. W. (2002) Protein interactions and library screening with protein fragment complementation strategies, in *Protein–Protein Interactions: A Molecular Cloning Manual,* (Golemis, E. A., ed.), Cold Spring Harbor Laboratory Press, Cold Spring Harbor, NY, pp. 449–475.
16. Spotts, J. M., Dolmetsch, R. E., and Greenberg, M. E. (2002) Time-lapse imaging of a dynamic phosphorylation-dependent protein–protein interaction in mammalian cells. *Proc. Natl. Acad. Sci. USA* **99,** 15,142–15,147.
17. Fang, D. and Kerppola, T. K. (2004) Ubiquitin-mediated fluorescence complementation reveals that Jun ubiquitinated by Itch/AIP4 is localized to lysosomes. *Proc. Natl. Acad. Sci. USA* **101,** 14,782–14,787.
18. Hu, C. D., Chinenov, Y., and Kerppola, T. K. (2002) Visualization of interactions among bZIP and Rel family proteins in living cells using bimolecular fluorescence complementation. *Mol. Cell* **9,** 789–798.
19. Paulmurugan, R. and Gambhir, S. S. (2003) Monitoring protein–protein interactions using split synthetic renilla luciferase protein-fragment-assisted complementation. *Anal. Chem.* **75(7),** 1584–1589.
20. Luker, K. E., Smith, M. C., Luker, G. D., Gammon, S. T., Piwnica-Worms, H., and Piwnica-Worms, D. (2004) Kinetics of regulated protein–protein interactions revealed with firefly luciferase complementation imaging in cells and living animals. *Proc. Natl. Acad. Sci. USA* **101(33),** 12,288–12,293.
21. Luttrell, L. M. and Lefkowitz, R. J. (2002) The role of β-arrestins in the termination and transduction of G-protein-coupled receptor signals. *J. Cell Sci.* **115(3),** 455–465.
22. Barak, L. S., Ferguson, S. S., Zhang, J., and Caron, M. G. (2001) A beta-arrestin/green fluorescent protein biosensor for detecting G-protein-coupled receptor activation. *J. Biol. Chem.* **272,** 27,497–27,500.
23. Stokoe, D., Macdonald, S. G., Cadwallader, K., Symons, M., and Hancock, J. F. (1994) Activation of Raf as a result of recruitment to the plasma membrane. *Science* **264,** 1463–1467.
24. Arlander, S. J., Eapen, A. K., Vroman, B. T., McDonald, R. J., Toft, D. O., and Karnitz, L. M. (2003) HSP90 inhibition depletes Chk1 and sensitizes tumor cells to replication stress. *J. Biol. Chem.* **278(52),** 52,572–52,577.
25. Fishman, M. A. and Porter, J. A. (2005) A new grammar for drug discovery. *Nature* **437,** 491–493.

17

Physiological Indicators of Cell Function

Michael J. Ignatius and Jeffrey T. Hung

Summary

Successful high content screening (HCS) assays place large demands on the cell-based reagents used in their development and deployment. Fortunately, there is a wide range of fluorescent physiological indicators from which to choose that are continually increasing in size and variety. Ideal fluorescent reagents for cell-based assays exhibit optimal selectivity, signal intensity, and cell solubility, yet will be easily incorporated into assays across multiple detection platforms. The repertoire of existing fluorogenic and color changing dyes that indicate physiological changes in cells for live cell kinetic and fixed end-point assays are surveyed as well as newly developed reagents for the next generation of HCS assays.

Key Words: Apoptosis; assays; calcium; cell counts; cell biology; drug discovery; expression reporters; fluorescence; high content screening; high content analysis; HTS; high-throughput screening; imaging; membrane voltage; microplate; organelle; pathway analysis.

1. Introduction

The growing potential of fluorescent reagents for cell-based discovery is most apparent in the scaled up imaging approach offered by high content screening *(1–3)*. The sacrifice in throughput compared with nonimage-based high-throughput screening (HTS) is offset by both the quality and content of the data and ability to query multiple parameters per well. Coupled with RNAi or more routine gene transfections, it is now feasible to study the cellular function of hundreds of genes in one well-conceived assay *(4)*—a staggering leap in technology from just 5 yr ago. Gone are the legions of radioactivity based applications, which were limited by costs, single analytes, and cumbersome disposal and health risks. Bright field and chromogenic dyes with their poor dynamic range, low sensitivity, nonquantitative nature, and limited targets are being superceded by fluorescent biosensing reagents. In a recent survey, novel reagents and probes were identified as having the biggest potential impact on the field of high content screening (HCS) over the next few years *(1)*. Clearly, work remains.

2. The Automated Imager's Innovator Tool Box

Automated imaging, in which the information and conclusions are extracted through software analysis, demands much of the protocol and reagent, as well as the instrument. Thresholding requirements, contrast settings, cell segmentation, background binding all confound the process, as discussed in the section on Informatics and Bioinformatics. Add in the diversity of instruments out there, which differ in magnification, light source, confocality, capture device, analysis software, and then cell type, and it is nearly impossible to insure that any dye or protocol will work in all systems. In addition to the range of instruments and cells types, reagent choice is further constrained by spectral limits—for example, green is often crowded with fluorescent protein (FP)-based sensors.

However, the motivated assay designer will find solace in the range of options in color and probes from which to solve their individual problem. To aid assay developers, this chapter provides an overview of the dyes that have been successfully qualified for whole-cell, microplate assays, while anticipating others that might prove useful in the future. Of practicing HCS users, assay development was listed as the most restrictive limitation *(1)*. Although this creates an opportunity for reagent companies, it is impossible to make kits for all potential targets. Fortunately, the ease of exploring the available options through the web and the many outstanding tech services departments make the search easier.

Molecular Probes, a part of the growing family of Invitrogen-owned discovery companies since 2003, has been principally a chemistry-driven company. The applications in this space have been developed to a large degree by others. Companies like Cellomics (now a part of Fisher Scientific International, Hampton, NH) and Molecular Devices (Sunnyvale, CA) have pioneered the use of many of the Molecular Probes dyes in the HCS/HTS space. Although others might be using Molecular Probes technology in these applications, Probes has traditionally qualified reagents in simple single-cell applications. It has always been up to others to establish their suitability in any HCS/HTS type protocol. Scalability is addressed internally primarily by testing the uniformity of the response, the intensity of the signal, wash steps required, media compatibility, fixability, and general ease-of-use. Clearly a response that is crisp and bright with zero background will work on any device—including those with ×10 dry optics. Those applications with the need for automation, loosely adherent cells, and general cost reduction, minimizing wash steps are critical—benefits also enjoyed in any protocol. In the end, dedicated reagent kits for one device are not sought, rather, dyes that might enjoy widespread utility on all instruments are the focus.

This chapter will highlight the dyes that indicate physiological changes in cells. Although many are traditionally used in HTS mode, through plate readers that integrate the entire well response, these reagents can also be used in any imaging based protocols. For the purpose of this chapter, reagents that are traditionally used in HTS on live intact cells will be covered, provided they could also be used in imaging based HCS modes as well. In designing any physiological probes, three parameters are considered: selectivity, signal intensity, and cell permeability. The perfect probe can be used across platforms—from cells, to wells, to multiwells, to gels.

3. Collection Modality Determines Reagent Options

Fluorescent dyes and probes can be viewed as working in one of three modalities. First and the most challenging to the instrument are assays on live cells that are repeatedly imaged live (kinetic). Second and the most challenging to the reagents are assays that can be used in live-cell protocols but whose response can be preserved or retained with standard fixation. The third and most common are assays in which the cells are fixed then labeled, primarily with antibodies, or expression tags. These latter two are typically described as end-point analysis. With the range of instruments available, it is sensible for any reagent provider, unconstrained by their own proprietary box design, to qualify their reagents in these differing modalities. **Tables 1–3** shows lists of dyes and the general conditions for their use.

The live-cell experiments can be further broken down into (1) single time-point, (2) continuous time-series, and, finally, (3) staggered time-series in which each well of multiplate is imaged sequentially and time-lapse series extracted later. When considering the diversity of physiological events that can be induced in phenotypic screens of RNAi screens, or any cellular event that is represented in only a fraction of a cells' daily life, the need for multiple time domains is apparent. These live cell refinements are the cutting edge of instrument features, requiring environmental,

Table 1
Small Molecule-Based Ion Indicators

Calcium	fluo-4 AM, fluo-4 NW, fura-2,AM, Calcium Green™-1, rhod-2 AM, Indo1 AM[a]
Magnesium	mag-fluo-4 AM, mag-fura-2 AM
Sodium	SBFI (crown ether), Sodium Green, CoroNa Red
Potassium	PBFI (crown ether)
Chloride	SPQ (6-methoxy-*N*-(3-sulfopropyl) quinolinium, MQAE (*N*-(ethoxycarbonylmethyl)-6-methoxyquinolinium bromide), MEQ (6-methoxy-*N*-ethylquinolinium iodide)
pH	SNARF®-1, BCECF-AM (2′,7′-bis(2-carboxyethyl) -5-(and-6)-carboxyfluorescein), fluorescein, CypHer™ 5 (GE/Amersham)
Heavy metals (lead and zinc)	FluoZin™-1, Phen Green™, Leadmium™ Green AM dye
Membrane voltage sensors	
Fast	Di-4- and Di-8-ANEPPS, (pyridinium, 4-(2-(6-(dibutylamino)-2-naphthalenyl) ethenyl)-1-(3-sulfopropyl)-, hydroxide)
	VSP [voltage sensing probe (oxonol and coumarin based)]
Slow	DiSBAC$_2$(3) bis-(1,3-diethylthiobarbituric acid)trimethine oxonol
	DiBAC$_4$(3) bis-(1,3-dibutylbarbituric acid)trimethine oxonol

[a]For complete list please *see* The Handbook of Fluorescent Probes and Research Chemicals (probes.invitrogen.com).

control liquid handling, and extensive memory. Such factors have limited their acceptance and driven the popularity of many end-point assays, such as translocation assays in which movement of a reporter molecule is detected either green fluorescent protein (GFP) *(5)*, or an antibody *(2)* often to the nucleus. These include the TransFluor assay by NORAK *(6)* and nuclear factor of activated T-cells (NFAT), NF-κβ, PKC-α activation, signal transducers and activators of transcription STAT *(2,7–10)* in which events that required several hours to days to appear, are captured by the permanent movement of a reagent. Yet, the number of physiological events that are covered by this approach remain rare. Live cell reagents represent the greatest hope of parsing out these more complex systems biology types of interactions, and are the primary focus of this chapter on physiological indicators.

4. Physiological Reagents for Automated Imaging Platforms Including HCS
4.1. Calcium Sensing

A rise in intracellular Ca^{2+} can occur directly or indirectly from a variety of critically important drug targets. Ca^{2+} signaling is vital to processes as diverse as memory, cell proliferation, apoptosis, and muscle action *(11–16)*. Indirect release is produced with stimulation of any of the G protein coupled receptors linked to adenylate cyclase, phospholipase C, or some ion channels. Direct release of Ca^{2+} occurs through the many ligand gated channels *(17)*. Nature and chemist have shown that this divalent cation is the most tractable of ions for designing affinity scaffolds—with the added benefit of eliciting large structural changes on binding. The high affinity chelation complexes formed create long lasting structural changes useful in transmitting signaling cascades in nature and fluorogenic changes in dyes for discovery scientist.

The first set of calcium indicators monitor changes in their peak excitation or emission maxima with calcium binding *(18)*. For fura-2, by scanning with excitation wavelengths in the UV, a shift in absorption toward the blue can be seen when Ca^{2+} rises, whereas indo-1 shifts its emission toward the blue with Ca^{2+} loading *(19)*. Although these ratiometric dyes produce the most accurate absolute values of Ca^{2+} concentrations, their ratiometric and UV excitation make them difficult to use in high-throughput imaging screens. Dedicated beam splitters, fast filter wheels, UV sources, and other technologies are needed.

The nonratiometric, visibly light excited fluo-3, fluo-4, and most recently fluo-4NW (NW for no-wash) are ideal HCS/HTS/microplate substitutes. These dyes exhibit up to a 10-fold signal increase upon binding Ca^{2+}. Although non-ratiometric and, therefore, not able to determine absolute levels, some HTS and HCS companies have aggressively adopted these dyes (Molecular Devices, Perkin Elmer [Shelton, CT], Hamamatsu [Okayama City, Japan] for HTS, Cellomics, Evotec, Atto/Becton Dickson [San Jose, CA], and many others for HCS). Currently a popular reagent for calcium sensing is the Calcium-3 kit from Molecular Devices (Sunnyvale, CA), which includes a quencher dye in solution to reduce background emission and to obviate the need to wash the dye-loading medium prior to analysis.

The fluo-4 NW improves on Calcium 3 and standard fluo-4 reagents by better delivery formulation and by achieving robust S/N changes without the need for quencher dyes that can often confound pharmacology. Although fluo-4 NW backgrounds are slightly higher in whole media one initial media removal step with no subsequent washing steps will produce even better results. Efficient delivery of the dye solution into a broad range of cell types with the new delivery vehicle has been confirmed as well. The reduction in work flow and the absence of quenchers that might confound pharmacological interpretations make fluo-4 NW an ideal reagent for automated protocols. fluo-4 NW is typical of an ideal add and read reagent, in which fluorogenic, cell-permeant, robust signals are produced with minimal wash steps and few associated components.

4.2. Protein Expression Reporter Based Calcium Sensors

A number of expression reporter based calcium sensors have been described *(20–22)*. Requiring no reagent addition these expression tags are ideal for creating stable reporter lines. Like fluo-4, and other nonratiometric dyes, absolute values for calcium concentration remain a challenge. To address this, Miyawaki et al. *(23,20)* joined two GFP colored variants together with a composite linker sequence containing calmodulin and calmodulin-binding peptide. On Ca^{2+} binding a structural change is induced that bends the complex, increasing the fluorescence resonance energy transfer (FRET) signal. Based on rises in green emission with Ca^{2+} binding, this radiometric fluorescent protein-based cameleon reporter can be used directly or possibly targeted to a particular subcellular domain with targeting vectors to measure absolute Ca^{2+} flux values.

β-lactamase (BLA)-based Ca^{2+} sensors are also available (Invitrogen LiveBLAzer™ FRET B/G assay kit). This cell permeable substrate for BLA is a β-lactamase cleavable dye pair complex that before cleavage transfers the blue emission from a donor dye to a green emitting acceptor dye *(24,25)*. Cells with uncleaved substrate retain the FRET pairing and continue to glow green. Cells expressing LiveBLAzer™ BLA and therefore β-lactamase activity, cleave the substrate, uncoupling FRET and glow blue. The BLA construct has been made with two promoter elements that are downstream of GPCR activation and Ca^{2+} entry: the CRE-BLA (cAMP responsive element) *(26)* and NFAT *(9,10)*. Both of these transcription factors are expressed when a stable increase in Ca^{2+} has occurred, either through Ca^{2+} influx or efflux from intracellular stores.

Although distal to the calcium sensing and voltage sensor response reagents, the CRE-based reporter provides more information about the pathway that triggered the Ca^{2+} release. Moreover, the bright signals produced are long lived. Neither construct would be considered the top choice for kinetic or dose–response studies, because the response tends to have limited dynamic range. In their favor, they can be done in live or live-fixed samples, are very robust, changing a brief calcium transient into a permanent signal. These features make it a perfect choice in studies where fluidics or live cell imaging are not available but an understanding that a pathway was activated during a certain time period is sought.

MTT (3-[4,5-dimethylthiazol-2-yl]-2, 5-diphenyltertrazolium bromide) or the orange and more water-soluble version XTT (2,3-bis-[2-methoxy-4-nitro-5 sulfophenyl]-2H-tetrazolium-5-carboxyanilide) and CyQUANT NF would be the best safe guard.

Knowing cell number and shape, especially in dense cultures, is a challenge for any software system. Reagents that hasten cell segmentation or boundary determination are critical in most every assay, physiology-based or otherwise. One common trick is to use nucleic acid stained nuclei to infer cell number, whereas cell segmentation or cell masks are created by cytosolic markers like cytoskeleton or plasma membrane stains. However, a single dye for both nuclei and generalized cell masks could achieve both ends. The red emitting Draq-5 (BioStatus, Leicestershire, UK) is a popular reagent for staining DNA in live cells, and can be detected in multiple emission channels from green to red *(45)*. But at most concentrations, faint cytoplasmic staining is observed and can serve this dual role as cell mask and nuclear stain.

In one of those ironic cases of not knowing when a flaw is indeed a feature, we have traditionally failed nucleic acid stains that showed this type of faint cytoplasmic staining and broad excitation maxima. But indeed compounds exist that have exactly these features and improve on Draq-5 by being fixable, even working in fixation solutions and are stable, working well in cells and tissue preparations. Our new dye CellMask Red will help in assays in which, nuclear and cytoplasmic dimensions are sought in one step on fixed cells and tissues as well as live cells. By comparison Draq 5 is poorly retained in fixed cells or tissue. In our studies it has been quite useful in cell counts and cell spreading, segmentation and similar cell masking protocols.

There are additional physiological indicators that space here does not permit covering. For our part, we continue to qualify reagents based on the uniformity of their response, fluorogenic always preferred, cell permeability and fidelity to a given response a requirement. The mitochondrial superoxide sensor, MitoSOX^{TM} is a prime example. Overburden a cell with oxidative stress and this dye will glow—likely indicative of cell stress or toxicity. This has not been shown yet to be HCS friendly—nor is it fixable but it reads out a key new therapeutic target in cancer, aging, fertility, ADME-Tox, and more *(46)*. Additional reagents for lipid metabolism accumulation, glucose, nitric oxide, and others await testing under the demanding auspices of HCS paradigms.

4.6. Standards and Optimizing Photon Output

Translocation based assays built on expression tags are emerging as critical tool in this field, despite requiring cell transfections and often heavy licensing fees *(8,6,47)*. The biggest contribution to this set of assays are the collection of FPs and expression tags, discussed in great detail in Chapters 14 and 15.

Due in large part to the abundance of GFP reagents, dyes that fall outside of this excitation/emission range can facilitate multiplexing. Expression reporter compatible dyes are blue- and red-shift and when partnered with a FP indicator can make valued tools for unraveling complex physiological pathways.

For calibrating your imaging instrumentation Molecular Probes offers excellent bead-based dye standards, for intensity, size, spectral accuracy, and even spectral unmixing. Flow Cytometrists have long appreciated the critical need of a calibrated instrument, understanding the value of both accuracy and precision of their device, especially in any longitudinal studies. For that reason microspheres are routinely used in many flow labs every day to calibrate their instruments. The case for equivalent vigilance in imaging has been made as well *(48,49)*. At present these microspheres are available on slides or in solution and provide useful means to establish signal-intensity level evaluate spectral accuracy, and alignment, and assist in size determination, Z-resolution, and more. It would be nice to know that an ever dimming signal is in fact a machine drifting out of specs, a bulb or lamp source failing, a thicker dish impairing light delivery or a score of other maladies, and not the assay itself.

The majority of existing HCS applications are fixed end-point reads, that incorporate one or two cell markers and an analyte selective reagent, usually an antibody or expression probe. Picking the right secondary antibody detection reagent is critical, with cost and brightness being the most important criteria to consider when scaling up for HCS, especially considering the comparatively light-starved dry objectives used in most instruments. To expand the dynamic range of antisera visualization, bright, photostable dyes are essential. Brightness is achieved through many mechanisms, initial brightness, maintained brightness, and amount of fluor per antibody. Although often expressed in terms of quantum yield and extinction coefficients, real-life sample-based determinations are more reasonable expressed in the simple values of how bright and for how long. The Alexa Fluor dyes (Molecular Probes/ Invitrogen) start out bright and are more photostable than equivalent organic dyes. Of additional virtue is the ability to load more dyes per IgG molecule than can be achieved with other dyes. Dye overloading leading to intramolecular quenching is a common failing of other dyes, whereas multiple molecules of Alexa Fluor dyes can be incorporated. Labeled secondaries are offered in nearly every variety. For the researcher interested in avoiding the extensive wash steps incumbent on secondary detection protocols, directly labeled primaries with high degree of labeling with these proprietary Alexa Fluor dyes might prove convenient. New microscale antibody kits optimized to label antibodies from 20 μg to 1 μg are available to expedite such labeling (Molecular Probes/Invitrogen).

In October of 2005, driven by activities on the Eugene Campus, Invitrogen purchased two companies making semiconductor nanocrystals, BioPixel and Quantum Dot Corp Inc. (Hayward, CA). Many features of these nanoparticles have intriguing possibilities in automated imaging. First, their initial brightness is equivalent to if not better than standard organic dyes. Second, they can all be excited at a single wavelength, removing much of the instrument costs and variability in excitation light. Third, their narrower spectral widths allow more analytes in the visible and near infrared region. Finally, their nearly concrete photostability allows for more constant signal strength for accurate intensity determinations, longer exposure times for dim signals, and storage convenience for later testing and retesting if needed.

In the near term we will continue to add parameters in characterizing our dyes that more accurately predict their utility in this promising area of automated, scaled up imaging. Wash steps, a trivial inconvenience in small samples can present an enormous barrier to automation and cost containment. Moreover, wash steps confound any discovery assays, in which perturbants of cell adhesion are sought. Serum in media confounds analysis by binding up dye and blocking their uptake. We are seeking ways to reduce this effect—and can recommend the Advanced D-MEM offered by our affiliate, GIBCO (Invitrogen). This DMEM substitute is designed to reduce serum formulations down to 1% thereby avoiding some of the confounding effects and costs of serum. In addition we are considering the strain on budgets and workflow that our current packaging and pricing directed at small throughput users is creating. Molecular Probes and Invitrogen as a whole is hoping to better enable all aspects of image-based discovery—from lead discovery to optimization to ADME-Tox to animals.

For more complete information on many of the Molecular Probes/Invitrogen Detection Technology products mentioned, please view the website (www.probes.invitrogen.com), or request a free copy of the most recent Molecular Probes/Invitrogen Handbook, volume 10.

References

1. Comley, J. (2005) High content screening emerging importance of novel reagents/probes and pathway analysis. *Drug Discov. World* **Summer,** 31–53.
2. Taylor, D. L. and Giuliano, K. A. (2005) Multiplexed high content screening assays create a systems cell biology approach to drug discovery. *Drug Discov. Today: Technol.* **2,** 149–154.
3. Hertzberg, R. P. and Pope, A. J. (2000) High-throughput screening: new technology for the 21st century. *Chem. Biol.* **4,** 445–451.

4. Conrad, C., Erfle, H., Warnat, P., et al. (2004) Automatic identification of subcellular phenotypes on human cell arrays. *Genome Res.* **14,** 1130–1136.
5. Kain, S. R. (1999) Green fluorescent protein (GFP): applications in cell-based assays for drug discovery. *DDT* **4,** 304–312.
6. Oakley, R. H., Hudson, C. C., Cruickshank, R. D., et al. (2002) The cellular distribution of fluorescently labeled arrestins provides a robust, sensitive, and universal assay for screening G protein-coupled receptors. *Assay Drug Dev. Technol.* **1,** 21–30.
7. Giuliano, K. A., Haskins, J. R., and Taylor, D. L. (2003) Advances in high content screening for drug discovery. *Assay Drug Dev. Technol.* **1,** 565–577.
8. Vakkila, J., DeMarco, R. A., and Lotze, M. T. (2004) Imaging analysis of STAT1 and NF-κB translocation in dendritic cells at the single cell level. *J. Immunol. Methods* **294,** 123–134.
9. Rao, A., Luo, C., and Hogan, P. G. (1997) Transcription factors of the NFAT family: regulation and function. *Annu. Rev. Immunol.* **15,** 707–747.
10. Crabtree, G. R. and Olson, E. N. (2002) NFAT signaling: choreographing the social lives of cells. *Cell* **109,** S67–S69.
11. Ashcroft, F. M. (ed) (2000) *Ion Channels and Disease.* Academic, San Diego, CA.
12. Doyle, J. L. and Stubbs, L. (1998) Ataxia, arrhythmia and ion-channel gene defects. *Trends Genet.* **14,** 92–98.
13. Marks, A. R. (1997) Intracellular calcium-release channels: regulators of cell life and death. *Am. J. Physiol.* **272,** H597–H605.
14. Kuriyama, H., Kitamura, K., Itoh, T., and Inoue, R. (1998) Physiological features of visceral smooth muscle cells, with special reference to receptors and ion channels. *Physiol. Rev.* **78,** 811–920.
15. Berridge, M. J., Bootman, M. D., and Lipp, P. (1998) Calcium—a life and death signal. *Nature* **395,** 645–648.
16. Plank, D. M. and Sussman, M. A. (2005) Impaired intracellular Ca2+ dynamics in live cardiomyocytes revealed by rapid line scan confocal microscopy. *Microsc. Microanal.* **11,** 235–243.
17. Moreno, D. H. (1999) Molecular and functional diversity of voltage-gated calcium channels. *Ann. NY Acad. Sci.* **868,** 102–117.
18. Minta, A., Kao, J. P., and Tsien, R. Y. (1989) Fluorescent indicators for cytosolic calcium based on rhodamine and fluorescein chromophores. *J. Biol. Chem.* **264,** 8171–8178.
19. Haugland, R. P., Spence, M. T. Z., and Johnson, I., eds. (2005) *The Handbook: A Guide to Fluorescent Probes and Labeling Technologies.* Printed by Invitrogen, CA.
20. Miyawaki, A. (2004) Fluorescent proteins in a new light. *Nat. Biotechnol.* **11,** 1374–1376.
21. Miyawaki, A. (2005) Innovations in the imaging of brain functions using fluorescent proteins. *Neuron* **48,** 189–199.
22. Romoser, V. A., Hinkle, P. M., and Persechini, A. (1997) Detection in living cells of Ca^{2+}-dependent changes in the fluorescence emission of an indicator composed of two green fluorescent protein variants linked by a calmodulin-binding sequence. A new class of fluorescent indicators. *J. Biol. Chem.* **272,** 13,270–13,274.
23. Miyawaki, A., Llopis, J., Heim, R., et al. (1997) Fluorescent indicators for Ca^{2+} based on green fluorescent proteins and calmodulin. *Nature* **388,** 882–887.
24. Gao, W., Xing, B., Tsien, R. Y., and Rao, J. (2003) Novel fluorogenic substrates for imaging beta-lactamase gene expression. *J. Am. Chem. Soc.* **125,** 11,146, 11,147.
25. Zlokarnik, G., Negulescu, P. A., Knapp, T. E., et al. (1998) Quantitation of transcription and clonal selection of single living cells with β-lactamase as reporter. *Science* **279,** 84–88.
26. Johannessen, M., Delghandi, M. P., and Moens, U. (2004) What turns CREB on? *Cell Signal* **16,** 1211–1127.
27. Drews, J. (2000) Drug discovery: a historical perspective. *Science* **287,** 1960–1964.
28. England, P. J. (1999) Discovering ion-channel modulators—making the electro-physiologist's life more interesting. *DDT* **4,** 391–392.
29. Velicelebi, G., Stauderman, K. A., Varney, M. A., Akong, M., Hess, S. D., and Johnson, E. C. (1999) Fluorescence techniques for measuring ion channel activity. *Methods Enzmol.* **294,** 20–47.
30. Gonzalez, J. E., Oades, K., Leychkis, Y., Harootunian, A., and Negulescu, P. A. (1999) Cell-based assays and instrumentation for screening ion-channel targets. *Drug Disc. Today* **4,** 431–439.
31. Denyer, J., Worley, J., Cox, B., Allenby, G., and Banks, M. (1998) HTS approaches to voltage-gated ion channel drug discovery. *DDT.* **7,** 323–332.

32. Wuskell, J. P., Boudreau, D., Wei, M. D., et al. (2006) Synthesis, spectra, delivery and potentiometric responses of new styryl dyes with extended spectral ranges. *J. Neurosci. Methods* **151,** 200–215.
33. Gonzalez, J. E. and Tsien, R. Y. (1995) Voltage sensing by fluorescence resonance energy transfer in single cells. *Biophys. J.* **69,** 1272–1280.
34. Gonzalez, J. E. and Tsien, R. Y. (1997) Improved indicators of cell membrane potential that use fluorescence resonance energy transfer. *Chem. Biol.* **4,** 269–277.
35. González Jesús, E. and Maher, M. P. (2002) Cellular fluorescent indicators and voltage/ion probe reader (VIPR™): tools for ion channel and receptor drug discovery. *Receptors Channels* **8,** 283–295.
36. Epps, D. E., Wolfe, M. L., and Groppi, V. (1994) Characterization of the steady-state and dynamic fluorescence properties of the potential-sensitive dye bis-[(1,3-dibutylbarbituric acid) trimethine oxonol Dibac$_4$(3)] in model systems and cells. *Chem. Phys. Lipids* **69,** 137–150.
37. Schroeder, K. S. and Neagle, B. D. (1996) FLIPR: a new instrument for accurate, high-throughput optical screening. *J. Biomol. Screen.* **1,** 75–80.
38. Sullivan, E., Tucker, E. M., and Dale, I. L. (1999) Measurement of [Ca2+] using the fluorometric imaging plate reader (FLIPR). *Methods Mol. Biol.* **114,** 125–133.
39. Bell, T. W. and Hext, N. M. (2004) Supramolecular optical chemosensors for organic analytes. *Chem. Soc. Rev.* **33,** 589–598.
40. Green, D. R. (2005) Apoptotic pathways: ten minutes to dead. *Cell.* **121,** 671–674.
41. Spierings, D., McStay, G., Saleh, M., et al. (2005) Connected to death: the (unexpurgated) mitochondrial pathway of apoptosis. *Science* **310,** 66, 67.
42. Watanabe, M., Hitomi, M., van der Wee, K., et al. (2002) The pros and cons of apoptosis assays for use in the study of cells, tissues, and organs. *Microsc. Microanal.* **8,** 375–391.
43. Gasparri, F., Mariani, M., Sola, F., and Galvani, A. (2004) Quantification of the proliferation indes of human dermal fibroblast cultures with the ArrayScan™ high content screening reader. *J. Biomol. Screen.* **9,** 232–243.
44. Petronilli, V., Miotto, G., Canton, M., et al. (1999) Transient and long-lasting openings of the mitochondrial permeability transition pore can be monitored directly in intact cells by changes in mitochondrial calcein fluorescence. *Biophys. J.* **76,** 725–734.
45. Smith, P. J., Blunt, N., Wiltshire, M., et al. (2000) Characteristics of a novel deep red/infrared fluorescent cell-permeant DNA probe, DRAQ5, in intact human cells analyzed by flow cytometry, confocal and multiphoton microscopy. *Cytometry.* **40,** 280–291.
46. Hansen, J. M., Go, Y. M., and Jones, D. P. (2006) Nuclear and mitochondrial compartmentation of oxidative stress and redox signaling. *Annu. Rev. Pharmacol. Toxicol.* **46,** 215–234.
47. Granas, C., Lundholt, B. K., Heydorn, A., et al. (2005) High content screening for G protein-coupled receptors using cell-based protein translocation assays. *Comb. Chem. High-Throughput Screen.* **8,** 301–309.
48. Zucker, R. M. and Price, O. T. (1999) Practical confocal microscopy and the evaluation of system performance. *Methods* **18,** 447–458.
49. Zucker, R. M. and Price, O. (2001) Evaluation of confocal microscopy system performance. *Cytometry.* **44,** 273–294.

18

The Use of siRNA to Validate Immunofluorescence Studies

K. Gregory Moore, Wayne Speckmann, and Ronald P. Herzig

Summary

Cellular immunofluorescence studies can be validated by using either specific small interfering RNA (siRNA) duplexes or expression plasmids that induce the expression of specific siRNAs. The usage of either siRNA tool reduces the expression of the specific protein being studied, thus reducing substantially or abolishing the immunofluorescence detected when using a fluorescent antibody that recognizes the protein.

Key Words: Antibody validation; immunofluoresence; RNA duplexes; RNAi; short hairpin RNA; siRNA.

1. Introduction

RNA interference (RNAi) is a cellular process common to most eukaryotes. The role this process plays in the life of a cell has been extensively investigated *(1–3)*. Functions such as protection against viral infection, regulation of chromatin remodeling, and regulation of gene expression have been determined. RNAi contributions to the etiology of cancer are aggressively being researched. Although the mechanism of RNAi-mediated gene silencing remains to be fully elucidated, the use of RNAi has become a valuable tool for analysis of gene function and target validation. RNAi leads to the inhibition of protein expression by utilizing sequence-specific, dsRNA-mediated degradation of the target messenger RNA (mRNA) *(4)*. In 2001, Tuschl and his colleagues showed that when short RNA duplexes (19–23 bases in length) were introduced into mammalian cells in culture, sequence-specific inhibition of target mRNA was effected without inducing an interferon response *(5)*. These short dsRNA, referred to as small interfering RNA (siRNA), act catalytically at submolar concentrations and can cleave up to 95% of the target mRNA in the cell, substantially reducing expression of the encoded protein. The siRNA-mediated effect has been shown to be relatively stable over time and silencing might be observed through several cell generations *(6)*.

One of the most powerful applications of RNAi is its use in functional genomic screens for gene target identification. Through such programs, it is possible to correlate specific gene targets with specific cellular phenotypes more accurately and quickly than ever experienced in the history of life science. The most critical factor in these screens is the assay that reports the results. Bringing together RNAi and fluorescent antibodies makes possible high content screening (HCS) assays that enable rapid and accurate assessment of cellular phenotypes. With these HCS assays, the specific genes responsible for a cellular phenotype can be rapidly identified in a cell-based format in which proteins and protein modifications are being monitored. Research programs using this technology have been responsible for identifying gene targets that drug discovery programs the world over are utilizing as targets in inhibitory compound screens. RNAi can be used to validate the specificity of the antibodies being used in immunofluorescent HCS assays.

It is important when evaluating cellular immunofluorescence studies to confirm the specificity of an antibody's reaction with its respective antigen and assess the presence of any cross-reactivity. In the past, specificity has been evaluated in cellular immunofluorescence studies by using immunizing peptides as inhibitory controls. These peptide reagents block the action of the antibody, but they do not validate the antibody as being specific for its respective protein antigen. siRNA duplexes and expression plasmids that induce the expression of siRNAs specific for a single gene target can be effectively used to validate immunofluorescence studies. The use of siRNA reagents to knock down the expression of a specific protein antigen can then be observed by a lack of fluorescence when using a specific antibody in a cellular immunofluorescence study. This chapter describes the use of siRNA duplexes and expression plasmids that encode specific siRNAs in cellular immunofluorescence studies to validate an antibody's immunoreactivity to its specific protein antigen.

2. Materials
2.1. Plating of Cells for Transfection
1. Human HeLa cells or another adherent cell line model.
2. Sterile six-or 96-well Costar® tissue culture plates (Corning, Corning NY).
3. Round glass cover slips for six-well plate studies (Fisher Scientific Int., Pittsburgh PA).
4. Dulbecco's minimal essential medium (DMEM) culture media (Mediatech Inc., Herndon VA).
5. Tissue culture sterile fetal bovine serum (FBS) and penicillin-streptomycin (HyClone, Logan UT).
6. Cell culture incubator with 5% CO_2.

2.2. siRNA Duplexes Method
1. Specific siRNA duplexes (5 nmoles SMARTpool® from Dharmacon, Boulder, CO or Upstate Group LCC; Charlottesville, VA).
2. Nonspecific control siRNA duplexes (1 nmole from Dharmacon or Upstate).
3. 5X siRNA buffer: 100 mM KCl, 1 mM $MgCl_2$, 30 mM HEPES, pH 7.5; or available from Upstate.
4. Sterile RNAase-free water.
5. Transfection reagent of choice, such as siIMPORTER™ (do not freeze) (Upstate).
6. 1.6-mL microcentrifuge tubes.

2.3. siRNA Expression Plasmid Method
1. siRNA plasmid DNA preferably in pKD™ vector (Upstate).
2. Transfection reagent of choice, such as FuGene6 transfection reagent (Roche Diagnostics, Alameda CA).

2.4. Cellular Immunofluorescence Method
1. Phosphate-buffered saline (PBS): 150 mM NaCl, 100 mM phosphate buffer, pH 7.2.
2. Appropriate cell fixative for retaining protein antigencity (either 95% ethanol/5% acetic acid, or 50% methanol 50% ethanol, or 1% paraformaldeyde in PBS (Sigma, St. Louis, MO).
3. Triton X-100 (Sigma).
4. Primary antibody for protein of interest, preferably directly conjugated to a red fluorophore (e.g., AlexaFluor 555®; Molecular Probes/Invitrogen, Eugene OR).
5. An appropriate secondary antibody conjugated to a fluorophore if the primary antibody is not directly conjugated to a fluorophore (Molecular Probes).
6. ProLong Gold® slide mounting medium for immunofluorescence (Molecular Probes). This reagent contains DAPI to stain the nuclei blue.
7. Fluorescent microscope or HCS instrument with appropriate filters.

3. Methods
3.1. Plating of Cells for Transfection
1. Plate HeLa cells or a desired cell line model in DMEM supplemented with 10% FBS and 1% penicillin–streptomycin in either a 96-well tissue culture plate (2K cells/well) or onto glass cover slips in a 6-well tissue culture plate (80K cells/well) (*see* **Note 1**).

3.2. The Usage of Specific siRNAs Duplexes

The selection of functional siRNAs is one of the major issues confronting the RNAi application. This selection requires sophisticated selection criteria to identify highly active siRNAs. It is recommended that scientists wishing to use siRNA as a validation tool strongly consider using commercial providers for these reagents. We have been exclusively using Dharmacon's pooled siRNA reagents. Dharmacon has developed two programs (SMARTselection and SMARTpooling) that have been successful at designing effective siRNAs. SMARTselection uses an algorithm consisting of 33 criteria and parameters that effectively eliminate nonfunctional siRNAs. SMARTpooling uses a sophisticated algorithm to combine four or more SMARTselected siRNA duplexes in a single pool. Each Dharmacon siRNA pool reduces mRNA levels by at least 70% and many will reduce mRNA levels by 95%. We routinely use the siIMPORTER™ reagent to transfect siRNA duplexes into mammalian cells (*see* **Note 3**). The siIMPORTER transfection reagent is a cationic lipid formulation that has been developed for efficient transfection of siRNA duplexes into mammalian cells.

It is important to determine the optimal time-point for knockdown of the target protein's expression after transfection of cells with siRNA reagents. The half-life of cellular proteins can vary considerably. In general, the mRNA levels of target proteins are usually significantly reduced by 24–48 h post-transfection. The best time period for showing the knockdown of most protein targets; however, generally occurs between 72 and 96 h post-transfection.

3.2.1. Transfection of Cells Using siRNA Duplexes

1. Dilute the 5X siRNA Buffer to 1X by mixing four volumes of sterile RNase-free water with one volume of 5X siRNA Buffer.
2. Each Dharmacon SMARTpool® contains 5 nmoles of material and each nonspecific control contains 1 nmole of material. The siRNA should be resuspended using 250 µL of 1X siRNA Buffer for a recommended concentration of 20 µM (20 pmol/µL). Final concentration ranges from 1 to 200 nM should be used in initial experiments so that the optimal concentration for the knock down of the protein target can be determined for the assay.
3. The siRNA nonspecific control pool should be resuspended using 50 µL of 1X siRNA Buffer for a recommended concentration of 20 µM (20 pmol/µL). We recommend using a negative control siRNA in every set of transfection studies at the same concentration as the experimental siRNA (*see* **Note 4**). We also recommend including untransfected or mock transfected cells as an additional negative control in siRNA studies.
4. For lipid complex formation and subsequent transfection, we recommend following the instructions provided by the transfection reagent manufacturer and taking measures to test and optimize the conditions best suited for the cell line of choice. We recommend using cell densities at approx 70–90% confluent (approx 1×10^5 cells/mL density for 96-well plates) at the time of transfection. The optimal cell number necessary to achieve this amount of confluence will vary with the growth characteristics of the cells.
5. For transfection of cells with siIMPORTER, use a microcentrifuge tube to first mix the siIMPORTER reagent with serum-free medium (Tube 1). For 96-well experiments, mix 0.5 µL of siIMPORTER with 2.5 µL of serum-free medium. For six-well plate experiments, mix 5.0 µL of siIMPORTER with 25 µL of serum-free medium.
6. In a second microcentrifuge tube (Tube 2), mix siRNA diluent and serum-free medium together and then add either the specific siRNA pool or the nonspecific control siRNA. The siRNA diluent promotes complex formation between siRNA and siIMPORTER. For 96-well experiments, mix 2.0 µL of siRNA diluent with 1 µL of serum-free medium followed by the addition of the siRNA. Adding 0.5 µL of a 20 µM siRNA preparation will give a final concentration in the reaction mixture of 100 nM. For six-well plate experiments, mix 25.0 µL of siRNA diluent with 10 µL of serum-free medium. To this mixture, add 5.0 µL of the 20 µM siRNA preparation to achieve a 100 nM final concentration. Mix gently by pipetting. Do not vortex.

7. Add the siRNA solution prepared in **step 6** (Tube 2) to the diluted siIMPORTER solution prepared in **step 5** (Tube 1). Incubate this mixture for 5 min to allow siRNA/lipid complexes to form. Do not incubate this mixture for longer than 30 min before use or transfection efficiency might be diminished.
8. For 96-well experiments, add 7.0 µL of the siRNA/siIMPORTER mixture with 93 µL of media. For 6-well plate experiments, add 70 µL of the siRNA/siIMPORTER mixture with 930 µL of media. With some cell types, higher transfection efficiencies are seen if serum is not present during the first 4 h of incubation. In these experiments, a small aliquot of media containing 20% FBS can be added after 4 h of incubation.
9. For immunofluorescence studies, incubate the cells for 72–96 h at 37°C and 5% CO_2 to detect siRNA-induced knockdown of protein targets.

3.2. The Usage of Specific siRNA Expression Plasmids
3.2.1. siRNA Oligonucleotide Design and Cloning Into pKD

We routinely use an Upstate-developed expression vector, pKD. This vector was designed to receive double stranded DNA oligonucleotides so that when the resulting plasmid is transfected into mammalian tissue culture cells, the cloned sequence gets transcribed and processed into a functional siRNA. A schematic of the pKD plasmid is shown in **Fig. 1**. A brief description of the design and usage of this expression vector is as follows: a double-stranded, annealed DNA oligonucleotide is generated that corresponds to the target gene mRNA sequence such that the target gene sense sequence is represented 5′ of its antisense and is separated by a 8 bp "loop" region. The DNA oligonucleotides are designed with the first 22 nucleotides being the sequence used in the siRNA for the target gene using a highly advanced search algorithm to identify nonredundant sequences in the genome. This oligo is then cloned into the pKD expression vector, which uses the human HI, RNA polymerase III-based promoter to express the cloned sequence. The RNA transcript produced by the pKD vector's HI promoter is terminated by the dT5 sequence immediately 3′ of the cloned oligos. The transcript is then able to fold onto itself as the sense and antisense regions are able to basepair. The eight nucleotide "loop" region allows for the short hairpin RNA to form. Cellular ribonucleases process the short hairpin RNA into a siRNA, which is fully functional for RNAi-mediated degradation of a particular mRNA target.

3.2.2. Transfection Complex
1. Plate cells for 24 h before transfection as described in **Subheading 3.1.**
2. In a sterile microcentrifuge tube, combine the following in the prescribed order, as the order of addition of components to the complex mixture is important (*see* **Note 5**):

	96-well	6-well
Serum-free DMEM	9.425 µL	94.3 µL
FuGene6 transfection reagent	0.075 µL	0.75 µL
siRNA plasmid DNA (0.1 mg/mL)	0.5 µL	5 µL
Total	10 µL	100 µL

3. Mix tube contents by gently tapping (three to four times).
4. Incubate at room temperature for a minimum of 15 min but not for more than 45 min as this might affect transfection efficiency.

3.2.3. Plasmid Transfection
1. Do not remove the old media from the cell culture.
2. Slowly add the transfection complex to each tissue culture well while gently swirling the plate. Use 10 µL of the transfection complex for each well of a 96-well plate and 100 µL for each well of a 6-well plate containing a glass cover slip.
3. Incubate the cells at 37°C and 5% CO_2 for 72–96 h before using cells for immunofluorescence studies.

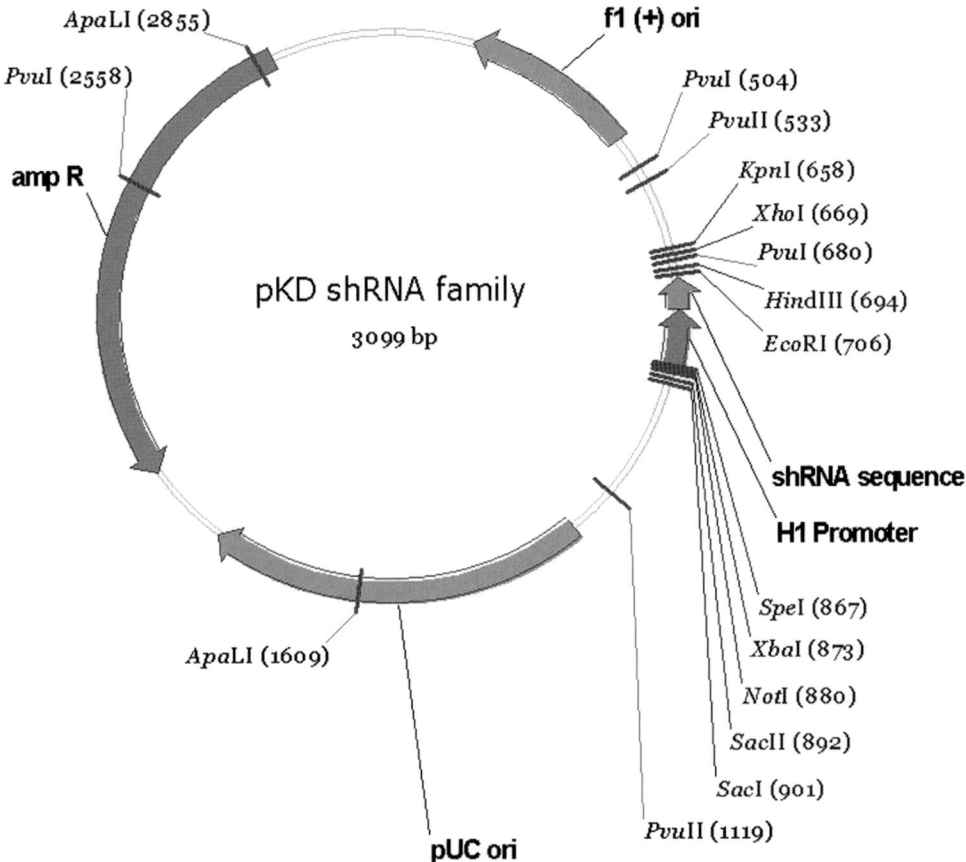

Fig. 1. A schematic drawing of the pKD expression plasmid used to induce the expression of specific short hairpin RNAs in mammalian cells. Cellular ribonucleases process the short hairpin RNAs into functional, siRNAs. (Please *see* the companion CD for the color version of this figure.)

3.3. Immunofluorescence of siRNA-Treated Cells

1. Carefully aspirate the culture media and rinse cells carefully with PBS.
2. Aspirate the PBS and then carefully add fixative. The suitable fixative must be determined empirically for each cell system studied. Three of the most commonly used fixatives are: (1) 95% ethanol/5% acetic acid; (2) 50% methanol/50% ethanol, and (3) 3.7% formaldehyde in PBS. We routinely fix cells at room temperature for 20 min using 3.7% formaldehyde.
3. Immediately wash the cells twice for 5 min with PBS. Do not shake.
4. Aspirate the PBS and add a cellular permeabilization agent. We routinely use 0.5% Triton X-100 (diluted in PBS) for 2 min. Permeabilization conditions and the reagents used, however, it might need to be modified for each cell type.
5. Optional blocking step: with some antibodies an optional blocking step might be needed to reduce background cellular fluorescence. In those instances, cover cells with 8% BSA in PBS and incubate for 1 h at room temperature. Perform the incubation in a sealed humidity chamber to prevent air drying of the cells. Wash cells twice for 5 min with PBS afterwards if this step is necessary.
6. Gently remove excess PBS and cover cells with the primary antibody of choice diluted in PBS. The antibody can be diluted in 1% BSA if background fluorescence is a concern. Incubate for 1 h at room temperature in a humidity chamber in the dark. Ideally, use a primary antibody directly conjugated to a red fluorophore (e.g., AlexaFluor 555). The optimal concentration of the primary antibody will need to be empirically determined (*see* **Note 6**).

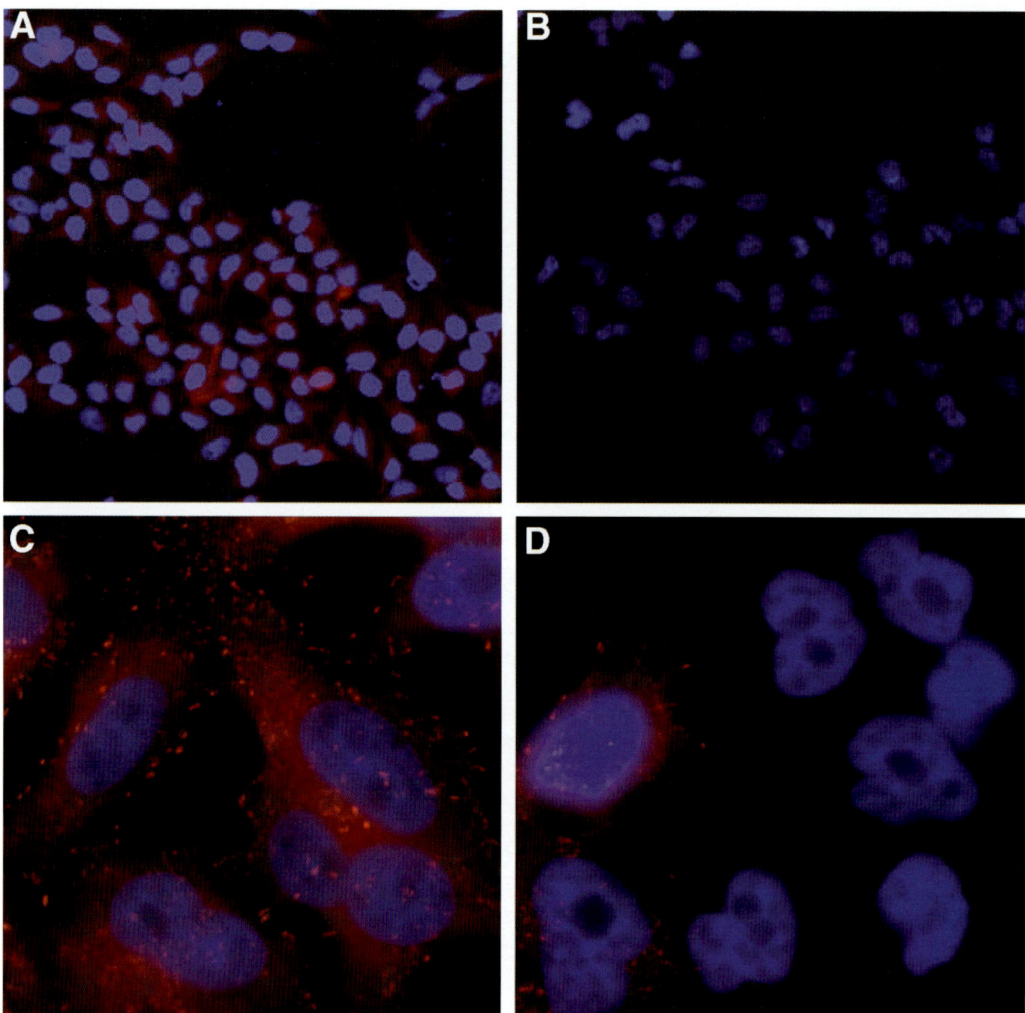

Fig. 2. Use of a pKD expression plasmid to block the expression of focal adhesion kinase (FAK) in Human HeLa cells. (**A,C**) HeLa cells were transfected with pKD-NegCon-v1, a nonspecific pKD expression plasmid (Upstate) and incubated for 96 h at 37°C. The cellular expression of FAK was then evaluated by immunofluorescence using an AlexaFluor 555®-conjugated monoclonal FAK antibody (2 µg/mL; Upstate). Note the fine particulate membrane staining of FAK. (**B,D**) HeLa cells were transfected with pKD-Fak-v1, a short hairpin RNA expression plasmid specific for Fak (Upstate) and incubated for 96 h at 37°C. The cellular expression of FAK was then evaluated by immunofluorescence as in **A** and **C**. The exposure time for each image was equivalent. The cellular level of FAK protein is dramatically reduced and only occasional cells (likely untransfected) show FAK staining similar to that seen in control reactions. Cell nuclei were stained blue using DAP. Images **A** and **B** are at ×20 magnification; images **C** and **D** are at ×100 magnification.

7. Wash the cells three times for 5 min with PBS. If a fluorophore-conjugated primary antibody was used, go directly to **step 10**; if a secondary antibody conjugated to a fluorophore is needed to visualize the target protein, go to **step 8**.
8. If a fluorophore-conjugated primary antibody is not available, gently remove excess PBS and incubate cells with a fluorescent-conjugated secondary antibody of choice in PBS for 1 h at room temperature in the dark. Add 1% BSA as a blocking reagent to the antibody preparation if necessary. The optimal concentration of the fluorescent-conjugated antibody will need to be empirically determined. Perform the incubation in a darkened, humidity chamber.

Table 1
Caged Substrates, Caging Chromophores, and Efficiencies Applied in Biological Studies

Classes of released substrates	Representative examples of substrate	Caging chromophores[a]	Approximately λ_{excit} (nm) ranges[b]	Quantum yields[c] (Φ)	References
Phosphates, nucleotides, and so on	H_3PO_4	pHP	300–340	0.3–0.38	28
		DMCM, DMACM	385	0.08	29
		BNZ (pH dependent)	300–365	0.01–0.15	30
	ATP	pHP	300–340	0.3	31
		oNB	300–370	0.19	32,33
		oNP	320	0.63	34
		DMACM	385	0.07–0.09	35
	GTP	pHP	300–330	na	36
		oNP	300–350	na	37
	Thymidine	oNB, oNP	365	0.2	38
		oNBP	350	na	39
	cAMP	BNZ	360	0.33	8,40
		ACM; MCM	340	0.07	41
	NADP	NV	>300	na	42
	DNA	oNP	360	na	43
	RNA	BHC	350–365	na	44
		oNB	308	na	45
	siRNA	oNB			46
	Phosphopeptides	oNP	300–365	0.26–0.33	47,48
C-terminus carboxylic acids	Glu	pHP	>300	0.14	49
Amino acids, oligopeptides, and proteins		HCM-carbamate	740 (2 hv)	1 GM	50
		MNI	350	0.085	51
		N-sub-6-oNP-7-coumaryl-3-carboxyl	300–400 740 (2 hv)	0.33 1 GM	52
	GABA	pHP, m-substituted pHP analogs	300–390	0.03–0.38	53
	Serine	oNP-carbamate	350	0.65	54,55
	Aspartate	MNI	334–364	0.09	56
		oNB	315	na	57
		β-DCNB	308	0.14	58
	Alanine				59
	NMDA	DNBH	345	na	60
	Capsaicin	oNB, NV	300–375	na	61
	Ala-Ala	pHP	313	0.26	62
	Leu-leu-Me	4-gluco-oNB	375	N/A	63
	Acetate	DMBNZ	>300	0.64	64
	Bradykinin	pHP	>300	0.22	38
	Fluorescein	oNB	350	na	65
N-terminus amines, and so on	Phenylephrine	oNB, NV	300–400	0.1–0.4	66
Amino acids, oligo-peptides, and proteins	Epinephrine	oNB, NV	300–400	0.1–0.4	42
	Isoproterenol	oNB, NV	300–400	0.1–0.4	42
	NADP	CNB	>320	0.09–0.19	67

(Continued)

Table 1 (*Continued*)

Classes of released substrates	Representative examples of substrate	Caging chromophores[a]	Approximately λ_{excit} (nm) ranges[b]	Quantum yields[c] (Φ)	References
	Nitrous oxide	5,8-dimethoxy-1-allylnaphthyl, others	350	0.66	**68,69**
Amino acid side chains	C-Kemptide (cysteine)	oNB	300–365	0.62	**70**
	Cysteine, tyrosine	oNB	300–350	na	**71,72**
	Aspartate	DNBH	300–400	0.6	**73**
	Arginine	DMoNB	300–400	0.1–0.4	**74**
	Tamoxifen	NV	365		**75**
Ca^{2+}	EGTA	oNB	347		**11**

[a]Abbreviations for the chromophores are: pHP = *p*-hydroxyphenacyl; oNB = *o*-nitrobenzyl; oNP = *O*-nitrophenethyl; NV = 4,5-dimethoxy-*O*-nitrobenzyl; CNB = α-carboxy-*O*-nitrobenzyl; BNZ = benzoin; DMBNZ = 3′,5′-dimethoxybenzoyl; HCM = 7-hydroxycoumarylmethyl; ACM = 7-acetoxyCM; MCM = 7-methoxyCM; DMACM = 7-dimethylaminoCM; DMCM = 6,7-dimethoxycoumarylmethyl; MNI = 4-methoxy-7-nitroindoline; DNBH = *o*,*o*′-dinitrobenzhydryl; BHC = 6-bromo-7-hydroxycoumarin-4-ylmethyl.

[b]The wavelength or wavelength range is based on data provided from known UV-vis spectra reported in the references or is estimated based on available data from other sources.

[c]Efficiencies vary with substituents on the chromophore and with changes in the reaction media and conditions. GM = Goppert-Meyer units for two photon (2 hv) excitation.

Oligofecamine) and peptide based systems (e.g., Pep1 *[13]* and MPG *[14]*) are also available to deliver substrates such as oligonucleotides, peptides, and proteins through the cell membrane. Among these, the most effective delivery agents are those that transport the cargo into the cell and avoid endosomal pathways. The Express™ reagent (Panomics) is such a delivery reagent system, which is MPG-based and thus successfully evades endosomal pathways *(15)*.

Despite the numerous advantages these delivery reagents offer over microinjection or standard pipet techniques, spatial and temporal control of cell activation frequently remains elusive to those implementing in commercially available HCS system. In this aspect, HCS would benefit greatly from a photoactivated caged initiation process.

Batch transfection of caged molecules offers the advantage that equal amounts of silent or inactive antagonists, agonists or substrates can be delivered to all cells and thereby makes the transfection independent of the assay outcome. Thereafter, the uniform illumination to multiple cell arrays with transfected caged reagents under prescribed conditions enables initial null ($t = 0$) measurements followed by precise regulation of the substrate release for HCS. In this way, caged compounds will yield far greater information than simple batch experiments with a group of cells that produce repeatable responses after a period of recovery. This is illustrated with the recent development of controllable siRNA (csiRNA *[16]*).

Caged siRNA is a timely example because RNAi has quickly become one of the most exciting arenas *(17)*, owing in large part to its potential in drug discovery and therapeutics *(18)*. In addition, recent studies have incorporated RNAi into HCS assays *(19)*. csiRNA is therefore at the forefront of application of caged substrates to large-scale biology.

csiRNA is a caged siRNA that is incapable of catalyzing the normal gene expression knockdown process. The biologically benign caged substrate remains dormant and inactive until absorption of 365 nm light in which siRNA is released. The uncaged siRNA is capable of participating in the normal RNAi process. The following sections will highlight two of the controllable features of caged reagents, temporal, and dosage control, to illustrate the potential use of caged reagents in the high content arena.

1.1. Equipment and Materials

This section lists the materials and equipment needed to conduct gene expression knockdown experiments using csiRNA. Although portions of the experimental design are not described as high throughput, the technology is quite amendable to this technique as well.

2. Materials

1. HEK 293 and HeLa cells (ATCC, Manassas, VA).
2. Growth medium: 10% FBS, DMEM, nonessential amino acids, sodium pyruvate, prepared fresh.
3. PC Phosphoramidite (Glenn Research, Sterling, VA).
4. csiGAPDH™ (Panomics, Fremont, CA), light sensitive, store at –20ºC.
5. 5′-phosphate-GAPDH antisense oligonucleotide (TriLink Biotechnologies, San Diego, CA).
6. GAPDH siRNA negative control (Ambion, Austin, TX).
7. Lipofectamine 2000 (Invitrogen, Carlsbad, CA).
8. Standard annealing solution (Panomics).
9. Clear-bottom, black-wall, 96-well microtiter plates.
10. UCOM Microplate Photoactivator (Panomics).
11. QuantiGene Reagent System (Panomics).
12. QuantiGene Probesets (Panomics).

3. Methods

The application of controllable siRNA (csiRNA) to inhibit gene expression will be described herein under five separate headings:

1. The design and synthesis of csiRNA strands (**Subheading 3.1.**).
2. The quantum efficiency to establish a working curve for variable gene knockdown and establish the maximum energy required for 100% release of the siRNA's activity (**Subheading 3.2.**).
3. Delivery of csiRNA into cells cultured in a 96-well format (**Subheading 3.3.**).

This chapter will describe two different experiments to illustrate two features of csiRNA: temporal control and dosable activation.

1. Light-activation of csiRNA at t = 4 or 24 h post-transfection, followed by gene expression analysis at t = 4, 24, and 48 h (**Subheading 3.4.**).
2. Increased activation of csiRNA through increasing energy of light, followed by gene expression analysis at t = 24 h post-transfection (**Subheading 3.5.**).

Finally, we will close with a few concluding remarks (**Subheading 3.6.**). Throughout the discussion, items that require specific care or particular attention will be described in **Subheading 4**.

3.1. Reagent Preparation

SiRNA oligonucleotides were designed in accordance with guidelines set forth by Tuschl *(20)*. For this discussion, GAPDH was used as the gene of interest. The general structure of siRNA molecules is a double-stranded 21-mer ribooligonucleotide with TT-overhangs on each 3′-terminus. The sense and complementary antisense strand syntheses were carried out using standard phosphoramidite chemistry. The GAPDH negative control siRNAs were obtained from Ambion. The sequence of the GAPDH siRNA sense strand is 5′-caucaucccugccucuacuTT-3′.

The mode of action of siRNA has been well studied *(21)*, and several reports have noted the importance of the phosphorylation of the 5′-antisense strand during gene expression knockdown *(22)*. As such, the 5′-end was targeted for protection with a photoactivatable protecting group (*see* **Note 1**). The photolabile phosphoramidite, [1-*N*-(4,4′-dimethoxytrityl)-5-(6-biotinamido-capro-amidomethyl)-1-(2-nitrophenyl)ethyl]-2-cyanoethyl-(*N′,N′*-diisopropyl)-phosphoramidite [obtained from Glenn Research, Sterling, Virginia] was coupled to the 5′ terminus of the antisense strand of a 21-mer siRNA using standard phosphoramidite chemistry during the normal oligonucleotide synthesis. The modified, 21-mer antisense strand was purified using RNase-free

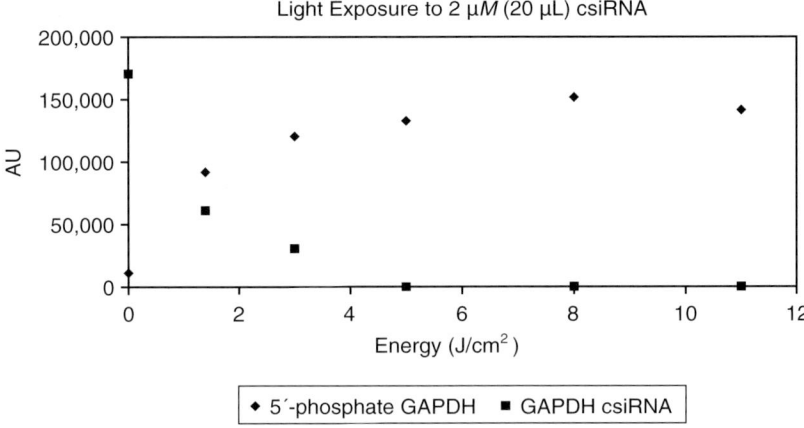

Fig. 1. Light dose–response curve for photorelease of β-actin csiRNA.

HPLC and the purity verified by gel electrophoresis and mass spectrometry (*see* **Note 2**). The sense and antisense strands were annealed:

1. Dissolve the oligonucleotide pellet in standard annealing buffer to a concentration of 300 μM.
2. Confirm the concentration through UV absorption and dilute the sample to 100 μM stock solution (*see* **Note 3**).
3. Combine equal volumes of each oligonucleotide in a 500 μL amber vial.
4. Vortex and centrifuge the sample for several seconds, heat the solution at 85°C for 5 min, and allow the sample to cool to room temperature over 4 h.
5. Vortex and centrifuge the sample for several seconds. The final concentration for the stock solution of annealed csiRNA or siRNA is 50 μM, which is confirmed using UV absorption.
6. Samples might be aliquoted and diluted for working stock solutions.
7. Annealed samples can be stored at –20°C for up to 6 mo and thawed for desired use.

3.2. Analysis

HPLC analysis of the antisense csiRNA was used to establish a light–dosage working curve for csiRNA. Concentration curves of pure starting material (GAPDH csiRNA) and photoproduct (5′-phosphate GAPDH siRNA) were established. Samples of csiRNA were exposed to 365 nm light using the UCOM while monitoring the amount of caged and released csiRNA through HPLC analysis (**Fig. 1**). The energy light flux to uncage 100% of csiRNA at 2 μM was found to be 5 J/cm^2. The initial energy light flux of 1.4 J/cm^2 released approx 26 pmol of csiRNA, which is nine times greater than the amount of csiRNA exposed to cells. For in vivo release of csiRNA cells will be exposed to 1.4 J/cm^2 of 365 nm light (*see* **Note 5**).

3.3. Cell Preparation: Transfection of csiRNAs

HEK 293 or HeLa cells were transfected using Lipofecamine 2000 in a 96-well clear-bottom, black-wall microtiter plate in accordance with the csiRNA manual (*see* **Note 4**). Approximately 5000 cells were plated. The final concentration of csiRNA delivered to the cells was 3 nM. At $t = 4$ h post-transfection, the complexes were removed and replaced with 120 μL of fresh complete growth medium (*see* **Note 6**).

3.4. Preliminary Studies: GAPDH Expression Knockdown Through csiRNA Activation at Different Time-Points

A key advantage of caged materials is the ability to keep the substrates silent until it is required, experimentally, to activate the substrate. Here, this feature is demonstrated with

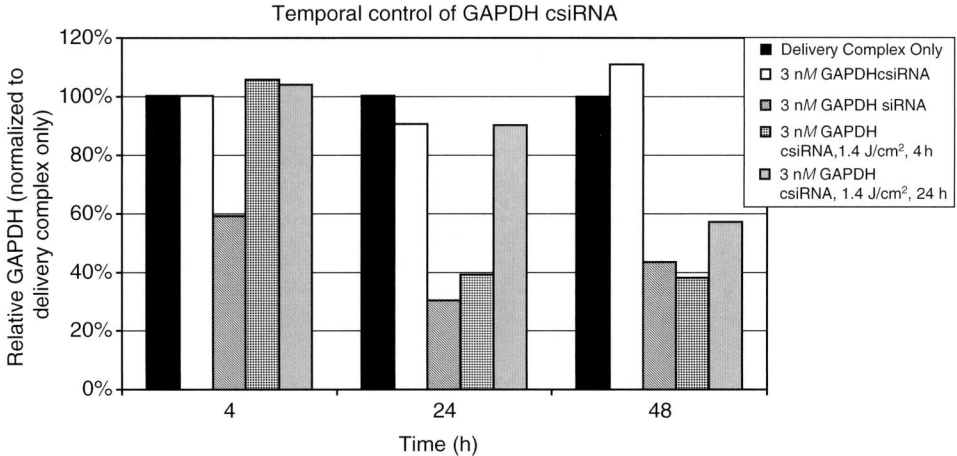

Fig. 2. GAPDH expression knockdown at various time-points. GAPDH csiRNA was transfected into HeLa cells and exposed to 365 nm light at $t = 4$ or 24 h posttransfection. Control conditions include cells exposed to transfection reagent only, GAPDH csiRNA without light activation, and GAPDH siRNA. Expression levels were measured at $t = 4$, 24, and 48 h for all conditions. Control experiments (delivery complex only and csiRNA without 365 nm light exposure) shows GAPDH expression continues unimpeded. GAPDH expression levels are knocked down for GAPDH siRNA and for csiRNA *only* after exposure to 365 nm light.

csiRNA by transfecting cells as described in **Subheading 3.3.** and incubating cells for 4, 24, or 48 h posttransfection. Cells were exposed to 365 nm light using UCOM, according to the UCOM user manual, at $t = 4$ or 24 h time-points. GAPDH expression levels were analyzed using Quantigene, according to the user manual (**Fig. 2**).

Cells that were not transfected with GAPDH siRNA or csiRNA maintained their normal GAPDH expression levels. Cells that were transfected with csiRNA maintained their normal GAPDH expression levels until csiRNA was activated with the UCOM. The most important advantage is allowing substrates to remain dormant in cells until the desired time to activate them, illustrated with cells irradiated at 24 h. Prior to irradiation, GAPDH expression levels were normal. However, at $t = 24$ h, cells were exposed to 365 nm light and GAPDH expression was knocked down ($t = 48$ h) to less than 60% below normal GAPDH levels.

3.5. Light Dosable Photo-Activation of csiRNA

The importance of controlling dose release in biological studies cannot be overstated (*see* **Note 6**). Kinetic studies, as well as phenotypic assays, are greatly enhanced when modulators introduced to cells can be activated with a high degree of accuracy and precision. For most phenotypic assays, especially, it is highly desirable to accurately titrate the amount of material required to elicit a phenotypic response. In this context, we demonstrate the activity of csiRNA can be tuned by controlling the energy exposed to transfected cells. By exposing the cells to increasing light energy, an increase in siRNA activity is achieved. The beauty of this system (as with all caged systems) is that a known number of photons (i.e., energy) will trigger a known quantity of siRNA precisely because there is a single caging group positioned at the 5′-end per siRNA molecule.

Cells were prepared as previously described in **Subheading 3.3.**

1. The cells were exposed to 0.0–1.4 J/cm^2 of 365 nm light using a UCOM Microplate Photo-Activator (Panomics) according to the UCOM manual.
2. The cells were incubated at 37°C for $t = 24$ h post-transfection, and lysed using QuantiGene lysis buffer according to the QuantiGene user manual. Replicates of three wells were run for all conditions tested. Gene expression levels were measured using QuantiGene according to the user manual.

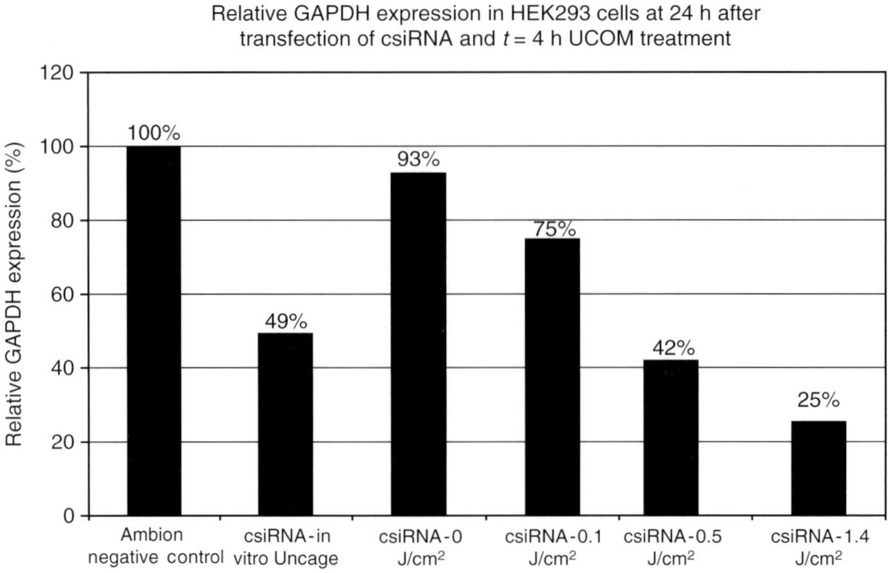

Relative GAPDH expression in HEK293 cells at 24 h after transfection of csiRNA and $t = 4$ h UCOM treatment

Sample	Relative GAPDH expression (%)
Ambion negative control	100%
csiRNA-in vitro Uncage	49%
csiRNA-0 J/cm²	93%
csiRNA-0.1 J/cm²	75%
csiRNA-0.5 J/cm²	42%
csiRNA-1.4 J/cm²	25%

Fig. 3. In vivo light–dosage exposure to HEK 293 cells transfected with GAPDH csiRNA, negative control siRNA at $t = 4$ h post-transfection. GAPDH csiRNA that was previously exposed to 1.4 J/cm² 365 nm light was also transfected into HEK 293 cells as a positive control. Cells were incubated at 37°C for $t = 24$ h and the GAPDH expression levels were measured.

As **Fig. 3** illustrates, increasing light dosage results in more csiRNA uncaged to release active siRNA effectively reducing residual mRNA levels through the normal RNAi pathway *(23–27)*. The activity of caged reagents is not simply "on" or "off." By exposing the appropriate energy dosage on the UCOM it is possible to tune in the amount of active reagent available in cells.

3.6. Concluding Remarks

There are an infinite number of applications for including cell survival, cell cycle regulation and cell development. Caging technology offers experimentalists a wide array of control in temporal, spatial, and concentration parameters. And with the advent of tools designed to bring light control to multiplexed assay systems, caged compounds may now be implemented in high content and high-throughput screens. We are, in fact, witness to several technologies that have been available for quite some time, be integrated in complementing fashion. These integrated technologies will surely help to better understand cellular pathways, off-target and downstream effects, and substrate effects on these pathways.

4. Notes

1. Attachment of photolabile groups to RNA and DNA using postsynthetic methods has been reported in literature; however this method does not take siRNA active sites into account. The design described here requires only a single caging group per siRNA molecule to take full advantage of the caging agent (*vide supra*). The postulated methodic placement of the caging group on siRNA molecules is limited to the following locations: 5′-, 3′-, or 2′-hydroxy groups; on the phosphate backbone; or on an individual nucleotide base. It was hypothesized that the 5′-hydroxy group was the most accessible synthetically and would cause the greatest disruption to the RNAi process. To test this, a GAPDH siRNA was synthesized with derivatives that permanently modified 5′-terminus. 5′-*O*-methyl siRNA analogs were shown to have zero activity compared to normal siRNA analogs. In addition, an *O*-alkyl phosphate modified siRNA (5′-C6-amine-GAPDH) also failed to catalyze gene expression knockdown for GAPDH. These experiments indicated a caging group on the 5′-phosphate would also block siRNA action. There are reports of other photoactivatable siRNA systems, which do not cage the 5′-end

Table 2
Purity of Antisense Csirna™ Compared to Background Gene Expression Knockdown: The Lower the Purity of the Antisense Strand, the Higher the siRNA Activity

Lot no.	Purity of antisense strand (%)	GAPDH expression level (normalized to cyclophilin expression levels [%])
1A	99.4	91
2B	97.1	80
3C	96.9	73
4D	96.3	70
5E	93.8	74
6F	78	44

exclusively (28). These systems are not as potent for a number of reasons. Either there are more caging groups per siRNA molecule leading to decreased sensitivity to light, and hence less dosable, or siRNA is caged at random locations. The end result is a system that is not completely silent and might posses some or all normal activity. In addition, to remove all of the caging groups requires substantially more light energy, which can lead to cell death (75).

2. Highly purified csiRNA antisense strands are extremely important for successful controllable knockdown experiments. We investigated the correlation between the purity of csiRNA and the activity of csiRNA in cells, measured by HPLC chromatograms of the antisense strand. Six different lots of various purities were transfected into HeLa cells according to the csiRNA manual and incubated for 24 h at 37°C. The cells were not exposed to 365 nm light in order to keep the csiRNA caged and unreactive. The cells were lysed and GAPDH mRNA expression levels were measured using Quantigene detection system. Any drop in GAPDH expression levels prior to light activation is viewed as a less efficient csiRNA. According to **Table 2**, there is a drop in caging efficiency below 97% purity, and an even more dramatic drop below 94% purity. This is most likely because of $(n-1)$ residues that make up the majority of impurities from oligonucleotide syntheses. For csiRNA $(n-1)$ residues are fully active, complete siRNA molecules.

3. Spectroscopic determination of concentration was carried out for two purposes. First, a 1:1 ratio of sense to antisense should be used to achieve the highest activity. Second, an accurate measure of csiRNA concentration is needed to yield optimal delivery to cells. This will result in the highest potential knockdown activity with the lowest background.

4. It is essential to use clear-bottom microtiter plates, as the UCOM Microplate Photoactivator delivers light from the bottom. The UCOM has been tested to be compatible with the following microtiter plates:
   a. Corning, Costar® (cat. no. 3904).
   b. BD, Falcon (cat. no. 353948).
   c. Greiner (cat. no. 655090).
   d. Nunc, Nalgene (cat. no. 237105).

5. Cytotoxicity experiments for UCOM 365 nm light exposure on HeLa cells showed ED50 values of 21 J/cm^2. It is vital to maintain energy doses lower than the ED50 level. Detrimental effects to cells, including cell death, are evident above the ED50 level. Cells show a very good tolerance to 365 nm light at energy levels below the ED50 value.

6. It is vital that the media be replaced following the transfection protocol. Although transfection methods might be highly efficient, it is impossible to have delivered all csiRNA into the cells. As light is completely unselective toward csiRNA inside or outside of cells, it is necessary to remove the undelivered extra cellular caged reagents.

Acknowledgments

We would like to thank Frank Witney, Gary K. McMaster, and Quan Nguyen for helpful advice and discussions. Support from the National Institutes of Health (Grant no. GM72910 [RSG]) is gratefully acknowledged.

References

1. Porter, G. (1972) Flash photolysis and some of its applications-Nobel lecture December 11, 1967, in *Nobel Lectures, Chemistry*, Elsevier Publishing Company, Amsterdam, pp. 1963–1970.
2. Givens, R. S., Conrad, P. G. II, Yousef, A. L., and Lee, J. -I. (2004) Photoremovable protecting groups in *The CRC Handbook of Organic Photochemistry and Photobiology, 2nd Ed.* (Horspool, W., and Lenci, F., ed.), CRC Press, Boca Raton, FL, pp. 69.1–69.46.
3. Kozlowski, D. J. and Weinberg, E. S. (2000) Photoactivatable (caged) fluorescein as a cell tracer for fate mapping in the zebrafish embryo. *Methods Mol. Biol.* **135**, 349–355.
4. Dirks, R. W., Molenaar, C., and Tanke, H. J. (2001) Methods for visualizing RNA processing and transport pathways in living cells. **115**, 3–11.
5. Wang, Q., Scheigetz, J., Roy, B., Ramachandran, C., and Gresser, M. J. (2002) Novel caged fluoresce in diphosphates as photoactivatable substrates for protein tyrosine phosphatases. *Biochim. Biophys. Acta* **1601**, 19–28.
6. Kozlowski, D. J., Murakami, T., Ho, R. K., and Weinberg, E. S. (1997) Regional cell movement and tissue patterning in the zebrafish embryo revealed by fate mapping with caged fluorescein. *Biochem. Cell Biol.* **75**, 551–562.
7. Pelliccioli, A. P. and Wirz, J. (2002) Photoremovable protecting groups: reaction mechanism and applications. *Photochem. Photobiol. Sci.* **1**, 441–458.
8. Pirrung, M. C. (1997) Spatially addressable combinatorial libraries. *Chem. Rev.* **97**, 473–488.
9. Walker, J. W. and Marriott, G. (1999) Caged peptides and proteins: new probes to study polyeptide function in complex biological systems. *Trends Plant Sci.* **4**, 330–334.
10. Shigeri, Y., Tatsu, Y., and Yumoto, N. (2001) Synthesis and application of caged peptides and proteins. *Pharmacol. Ther.* **91**, 85–92.
11. Mitchison, T. J., Sawin, K. E., Theriot, J. A., Gee, K., and Mallavarapu, A. (1998) Caged fluorescent probes. *Methods Enzymol.* **291**, 63–78.
12. Zhao, Y., Zheng, Q., Dakin, K., Xu, K., Martinez, M. L., and Li, W. -H. (2004) New caged coumarin fluorophores with extraordinary uncaging cross sections suitable for biological imaging applications. *J. Am. Chem. Soc.* **126**, 4653–4663.
13. Kasai, H., Matsuzaki, M., and Ellis-Davies, G. C. R. (2005) Two-photon uncaging microscopy, in *Imaging in Neuroscience and Development: A Laboratory Manual, Second ed.*, (Yuste, R. and Konnerth, A., eds.), Cold Spring Harbor Laboratory Press, Cold Spring Harbor, NY.
14. Fischer, P. M., Krausz, E., and Lane, D. P. (2001) Cellular delivery of impermeable effector molecules in the form of conjugates with peptides capable of mediating membrane translocation. *Bioconjug. Chem.* **12**, 825–841.
15. Borman, A., Howell, M. T., Patton, J. G., and Jackson, R. J. (1993) The involvement of a spliceosome component in internal initiation of human rhinovirus RNA translation. *J. Gen. Virol.* **74**, 1775–1788.
16. May, C., Morris, J. D., Jean, M., Frederic, H., and Divita, G. (2001) A peptide carrier for the delivery of biologically active proteins into mammalian cells. *Nature* **19**, 1173–1176.
17. Morris, M. C., Vidal, P., Chaloin, L., Heitz, F., Divita, G. (1997) A new peptide vector for efficient delivery of oligonucleotides into mammalian cells. *Nucl. Acids Res.* **25**, 2730–2736.
18. Federica, S., May, C., Morris, F. H., and Divita, G. (2003) Insight into the mechanism of the peptide-based gene delivery system MPG: implications for delivery of siRNA into mammalian cells. *Nucl. Acids Res.* **31**, 2717–2724.
19. Nguyen, Q. N., Chavli, R. V., Marques, J. T., et al. (2006) Light controllable siRNAs regulate gene suppression and phenotypes in cells. *Biochim. Biophys. Acta*, epub ahead of print.
20. Hannon, G. J. (2002) RNA interference. *Nature* **418**, 244–251.
21. Lu, P. Y., Xie, F. Y., and Woodle, M. C. (2003) siRNA-mediated antiturorigenesis for drug target validation and therapeutics. *Curr. Opin. Mol. Ther.* **5**, 225–234.
22. Giuliano, K. A., Chen, Y. -T., Lansing, D. L. (2004) High content screening with siRNA optimizes a cell biological approach to drug discovery: defining the role of p53 activation in the cellular response to anticancer drugs. *J. Biomol. Screen.* **9**, 557–568.
23. Elbashir, S. M., Harborth, J., Lendeckel, W., Yalcin, A., Weber, K., and Tuschl, T. (2001) Duplexes of 21-nucleotide RNAs mediate RNA interference in mammalian cell culture. *Nature* **411**, 494–498.
24. McManus, M. T. and Sharp, P. A. (2002) Gene silencing in mammals by small interfering RNAs. *Nat. Rev.* **3**, 737–747.

25. Nykänen, A., Haley, B., and Zamore, P. D. (2001) ATP requirements and small interfering RNA structure in the RNA interference Pathway. *Cell* **107**, 309–321.
26. Hidekei, A., Furata, T., Tsien, R., and Okamoto, H. (2001) Photo-mediated gene activation using caged RNA/DNA in zebra fish embryos. *Nat. Genet.* **28**, 317–325.
27. Shah, S., Rangarajan, S., and Friedman, S. H. (2005) Light-activated RNA interference. *Angew. Chem. Int. Ed.* **44**, 1328–1332.
28. Park, C. -H. and Givens, R. S. (1997) New Photoactivated Protecting Groups. 6. *p*-hydroxyphenacyl: a phototrigger for chemical and biochemical probes. *J. Am. Chem. Soc.* **119**, 2453–2463.
29. Schmidt, R., Geissler, D., Hagen, V., and Bendig, J. (2005) Kinetic study of the photocleavage of (coumarin-4-yl)methyl esters. *J. Phys. Chem. A* **109**, 5000–5004.
30. Givens, R. S., Athey, P. S., Matuszewski, B., Kueper, L. W., Xue, J., and Fisher, T. (1993) Photochemistry of phosphate esters: α-keto phosphates as a photoprotecting group for caged phosphate. *J. Am. Chem. Soc.* **115**, 6001–6012.
31. Givens, R. S. and Park, C. -H. (1996) Hydroxyphenacyl ATP: a new phototrigger. V. *Tetrahedron Letts.* 6259–6263.
32. Kaplan, J. H., Forbush, G. III, and Hoffman, J. F. (1978) Rapid photolytic release of adenosine 5′-triphosphate from a protected analogue: utilization by the Na:K pump of human red blood cell ghosts. *Biochemistry* **17**, 1929–1935.
33. Zimmermann, B., Somlyo, A. V., Ellis-Davies, G. C. R., Kaplan, J. H., and Somlyo, A. P. (1995) Kinetics of prephosphorylation reactions and myosin light chain phosphorylation in smooth muscle. Flash photolysis studies with caged calcium and caged ATP. *J. Biol. Chem.* **270**, 23,966–23,974.
34. Walker, J. W., Reid, G. P., McCray, J. A., and Trentham, D. R. (1988) Photolabile 1-(2-nitrophenyl)ethyl phosphate esters of adenine nucleotide analogs. Synthesis and mechanism of photolysis. *J. Am. Chem. Soc.* **110**, 7170–7177.
35. Geissler, D., Kresse, W., Wiesner, B., Bendig, J., Kettenmann, H., and Hagen, V. (2003) DMACM-caged adenosine nucleotides: ultrafast phototriggers for ATP, ADP and AMP activated by long-wavelength irradiation. *Chembiochemical* **4**, 162–170.
36. Du, X., Frei, H., and Kim, S. H. (2001) Comparison of nitrophenylethyl and hydroxyphenacyl caging groups. *Biopolymers* **62**, 147–149.
37. Dolphin, A. C., Wootton, J. F., Scott, R. H., and Trentham, D. R. (1988) Photoactivation of intracellular guanosine triphosphate analogues reduces the amplitude and slows the kinetics of voltage-activated calcium channel currents in sensory neurones. *Pflugers Arch: Eur. J. Physiol.* **411**, 628–636.
38. Walbert, S., Pfleiderer, W., and Steiner, U. E. (2001) Photolabile protecting groups for nucleosides: mechanistic studies of the 2-(2-nitrophenyl)ethyl group. *Helv. Chim. Acta* **84**, 1601–1611.
39. Schlichting, I., Rapp, G., John, J., Wittinghofer, A., Pai, E. fl, and Goody, R. S. (1989) Biochemical and crystallographic characterization of a complex of c-Ha-ras p21 and caged GTP with flash photolysis. *Proc. Nat. Acad. Sci. USA* **86**, 7687–7690.
40. Givens, R. S., Athey, P. S., Kueper, L. W. III, Matuszewski, B., and Xue, J. -Y. (1992) Photochemistry of α-keto phosphate esters: photorelease of a caged cAMP. *J. Am. Chem. Soc.* **114**, 8708–8710.
41. Furuta, T. and Iwamura, M. (1998) New caged groups: 7-substituted coumarinylmethyl phosphate esters. *Methods Enzymol.* **291**, 50–63.
42. Salerno, C. P., Magde, D., and Patron, A. P. (2000) Enzymatic synthesis of caged NADP cofactors: aqueous NADP photorelease and optical properties. *J. Org. Chem.* **65**, 3971–3981.
43. Dussy, A., Meyer, C., Quennet, E., Bickle, T. A., Giese, B., and Marx, A. (2002) New light-sensitive nucleosides for caged DNA strand breaks. *Chembiochemical* **3**, 54–60.
44. Ando, H., Furuta, T., Tsien, R. Y., and Okamoto, H. (2001) Photo-mediated gene activation using caged RNA/DNA in zebrafish embryos. *Nat. Genet.* **28**, 317–325.
45. Chaulk, S. G. and MacMillan, A. M. (1998) Caged RNA: photo-control of a ribozyme reaction. *Nucl. Acids Res.* **26**, 3173–3178.
46. Shah, S., Rangarajan, S., and Friedman, S. H. (2005) light activated RNA interference. *Angew. Chem. Int. Ed. Engl.* **44**, 1328–1332.
47. Nguyen, A., Rothman, D. M., Stehn, J., Imperiali, B., and Yaffe, M. B. (2002) Caged phosphopeptides reveal a temporal role for 14-3-3 in G1 arrest and S-phase checkpoint function. *Nat. Biotechnol.* **22**, 993–1000.
48. Rothman, D. M., Vazquez, M. E., Vogel, E. M., and Imperiali, B. (2002) General method for the synthesis of cyclic phosphopeptides: tools for signal transduction pathways. *Org. Lett.* **4**, 2865–2868.

49. Givens, R. S., Weber, J. F. W., Conrad, P. G. II, Oroz, G., Donahue, S. L., and Thayer, S. A. (2000) New phototriggers 9: p-hydroxyphenacyl as a C-terminal photoremovable protecting group for oligopeptides. *J. Am. Chem. Soc.* **122**, 2687–2697.
50. Furuta, T., Wang, S. S. -H., Dantzker, J. L., et al. (1999) Brominated 7-hydroxycoumarin-4-ylmethyls: photolabile protecting groups with biologically useful cross-sections for two photon photolysis. *PNAS* **96**, 1193–1200.
51. Canepari, M., Nelson, L., Papageorgiou, G., and Corrie, J. E. T. (2001) Photochemical and pharmacological evaluation of 7-nitroindolinyl-and 4-methoxy-7-nitroindolinyl-amino acids as novel, fast caged neurotransmitters. *J. Neurosci. Methods* **112**, 29–42.
52. Zhao, Y., Zheng, Q., Dakin, K., Xu, K., Martinez, M. L., and Li, W. -H. (2004) New caged coumarin fluorophores with extraordinary uncaging cross sections suitable for biological imaging applications. *J. Am. Chem. Soc.* **126**, 4653–4663.
53. Conrad, P. G. II, Givens, R. S., Weber, J. F. W., and Kandler, K. (2000) New phototriggers: extending the p-hydroxyphenacyl π-π* Absorption Range. *Org. Lett.* **2**, 1545–1547.
54. Khan, S., Amoyaw, K., Spudich, J. L., Reid, G. P., and Trentham, D. R. (1992) Bacterial chemoreceptor signaling probed by flash photorelease of a caged serine. *Biophys. J.* **62**, 67–68.
55. Veldhuyzen, W. F., Nguyen, Q., McMaster, G., and Lawrence, D. S. (2003) A light-activated probe of intracellular protein kinase activity. *J. Am. Chem. Soc.* **125**, 13,358–13,359.
56. Huang, Y. H., Sinha, S. R., Fedoryak, O. D., Ellis-Davies, G. C. R., and Bergles, D. E. (2005) Synthesis and characterization of 4-methoxy-7-nitroindolinyl-D-aspartate, a caged compound for selective activation of glutamate transporters and N-methyl-D-aspartate receptors in brain tissue. *Biochemistry* **44**, 3316–3326.
57. Mendel, D., Ellman, J. A., and Schultz, P. G. (1991) Construction of a light-activated protein by unnatural amino acid mutagenesis. *J. Am. Chem. Soc.* **113**, 2758–2760.
58. Schaper, K., Mobarekeh, S. A. M., and Grewer, C. (2002) Synthesis and photophysical characterization of a new, highly hydrophobic caging group. *Eur. J. Org. Chem.* 1037–1046.
59. Niu, L., Wieboldt, R., Ramesh, D., Carpenter, B. K., and Hess, G. P. (1996) Synthesis and characterization of a caged receptor ligand suitable for chemical kinetic investigations of the glycine receptor in the 3-μs time domain. *Biochemistry* **35**, 8136–8142.
60. Maier, W., Corrie, J. E. T., Papageorgiou, G., Laube, B., and Grewer, C. (2005) Comparative analysis of inhibitory effects of caged ligands for the NMDA receptor. *J. Neurosci. Methods.* **142**, 1–9.
61. Katritzky, A. R., Xu, Y. -J., Vakulenko, A. V., Wilcox, A. L., and Bley, K. R. (2003) Model compounds of caged capsaicin: design, synthesis, and photoreactivity. *J. Org. Chem.* **68**, 9100–9104.
62. Givens, R. S., Jung, A., Park, C. -H., Weber, J., and Bartlett, W. (1997) New photoactivated protecting groups. 7. p-hydroxyphenacyl: a phototrigger for excitatory amino acids and peptides. *J. Am. Chem. Soc.* **119**, 8369–8370.
63. Mizuta, H., Watanabe, S., Sakurai, Y., et al. (2002) Design, synthesis, photochemical properties and cytotoxic activities of water-soluble caged L-Leucyl-L-leucine methyl esters that control apoptosis of immune cells. *Bioorg. Med. Chem.* **10**, 675–683.
64. Sheehan, J. C., Wilson, R. M., and Oxford, A. W. (1971) Photolysis of methoxy-substituted benzoin esters. Photosensitive protecting group for carboxylic acids. *J. Am. Chem. Soc.* **93**, 7222–7228.
65. Girdham, C. H. and O'Farrell, P. H. (1994) The use of photoactivatable reagents for the study of cell lineage in drosophila embryogenesis. *Methods Cell Biol.* **44**, 533–543.
66. Muralidharan, S. and Nerbonne, J. M. (1995) Photolabile "caged" adrenergic receptor agonists and related model compounds. *Photochem. Photobiol. B: Biol.* **27**, 123–137.
67. Cohen, B. E., Stoddard, B. L., and Koshland, D. E., Jr. (1997) Caged NADP and NAD. Synthesis and characterization of functionally distinct caged compounds. *Biochemistry* **36**, 309–315.
68. Bushan, K. M., Xu, H., Ruane, P. H., et al. (2002) Controlled photochemical release of nitric oxide from O2-naphthylmethyl- and O2-naphthylallyl-substituted diazeniumdiolates. *J. Am. Chem. Soc.* **124**, 12,640–12,641.
69. Pavlos, C. M., Xu, H., and Toscano, J. P. (2004) Controlled photochemical release of nitric oxide from O2-substituted diazeniumdiolates. *Free Radic. Biol. Med.* **37**, 745–752.
70. Pan, P. and Bayley, H. (1997) Caged cystiene and thiophosphoryl peptides. *FEBS Lett.* **405**, 81–85.
71. Philipson, K. D., Gallivan, J. P., Brandt, G. S., Dougherty, D. A., and Lester, H. A. (2005) Incorporation of caged cysteine and caged tyrosine into a transmembrane segment of the nicotinic ACh receptor. *Am. J. Physiol. Cell Physiol.* **281**, C195–C206.

72. Wu, N., Deiters, A., Cropp, T. A., King, D., and Schultz, P. G. (2004) A genetically encoded photo-caged amino acid. *J. Am. Chem. Soc.* **126,** 14,306–14,307.
73. Jasuja, R., Keyoung, J., Reid. G. P., Trentham, D. R., and Khan, S. (1999) Chemotactic responses of *Escherichia coli* to small jumps of photoreleased L-aspartate. *Biophys. J.* **76,** 1706–1719.
74. Wood, J. S., Koszelak, M., Liu, J., and Lawrence, D. S. (1998) A caged protein kinase inhibitor. *J. Am. Chem. Soc.* **120,** 7145, 7146.
75. Link, K. H., Shi, Y., and Koh, J. T. (2005) Light activated recombination. *J. Am. Chem. Soc.* **127,** 13,088–13,089.

IV

INFORMATICS AND BIOINFORMATICS

20

Overview of Informatics for High Content Screening

R. Terry Dunlay, Wallace J. Czekalski, and Mark A. Collins

Summary

With the growing use of high content screening (HCS) and analysis in drug discovery and systems biology, informatics has come to the forefront as a critical technology to effectively utilize the massive volumes of high content data and images being generated. Informatics technologies are required to transform HCS data and images into useful information and then into knowledge to drive decision making in an efficient and cost effective manner. In this chapter, we provide an overview of informatics tools and technologies for HCS, discuss some of the challenges of harnessing the huge and growing volumes of HCS data, and provide insight to help toward implementing or selecting, and utilizing a high content informatics solution to meet your organization's needs.

Key Words: Data integration; data management; data mining; databases; high content screening; image management; informatics; N-tier architecture; visualization.

1. Introduction

High content screening (HCS) systems generate enormous amounts of data and images that are pushing the limits of conventional information technologies. The massive volumes of feature-rich data and images being generated by these systems and the effective management and use of information from the data have created a number of challenges. These challenges lie not only in the capabilities of the software and hardware technologies, but also in educating users in the optimal use of informatics tools. In addition, partnerships between researchers and their counterparts in information technology (IT) are critical to effectively manage HCS data, share it, and integrate it, so that it can be used in meaningful ways. To fully exploit the potential of data and images from modern high content systems, it is therefore crucial to understand the key factors in determining a suitable high content informatics solution to fit your organization's needs.

HCS systems typically scan a multiwell plate with cells or cellular components in each well, acquire multiple images of cells, and extract multiple features (or measurements) relevant to the biological application, resulting in a large quantity of data and images. The amount of data and images generated from a single microtiter plate can range from hundreds of megabytes (MB) to multiple gigabytes (GB). Large numbers of plates are typically analyzed in screening operations and large-scale system biology experiments, often resulting in billions of features and millions of images with a need for multiple terabytes (TB) of storage in a short period of time.

High content informatics tools are needed to manage the large volume of HCS data and images generated for collection, storage, retrieval, analysis, and display to enable understanding of the samples under investigation. The importance of informatics for HCS is briefly discussed in **refs.** *1* and *2*.

Our goal in this chapter is to provide an overview of the key aspects of informatics tools and technologies needed for HCS, including characteristics of HCS data; data models/structures for storing HCS data; HCS informatics system architectures, data management approaches, hardware and network considerations, visualization, data mining technologies, and integrating HCS data with other data and systems.

2. Characteristics of HCS Data

HCS data is characterized as having large numbers of parameters, massive data sets, and large numbers of high resolution images that require significant amounts of storage, especially in drug discovery, and systems biology applications. In order to better understand these characteristics, we will provide some background. We should note that when we refer to typical values here and throughout the rest of the chapter, we are basing these on our experience and they by no means cover the full range of possibilities.

HCS systems typically scan and analyze multiwell microtiter plates. These "plates" typically have 96, 384, or 1536 wells. Each "well" is a container in the plate that typically contains an individual sample of cells. Each well is divided into multiple fields. Each "field" is a region of a well that represents an area to image (this is also sometimes referred to as a "field-of-view," "frame," or "scene"). Each field typically consists of multiple images, one for each individual wavelength of light (referred to as a "channel" or "color"), corresponding to the fluorescent markers/probes used for the biology/dye of interest (e.g., Hoechst). There are typically between one and four channels per field (e.g., one channel may show the nuclei, another the cytoplasm, another the cell membrane, and so on). In each field, a certain number of cells are selected to be analyzed by the HCS system. The number of cells per field varies depending on the experiment, but typically ranges between 10 and 500 cells. For each cell, multiple cell features (or measurements) are calculated by the HCS system's image analysis algorithms. The cell features include measurements such as size, shape, intensity, and so on. The number of cell features calculated varies depending on the assay, but typically ranges between 5 and 50. In addition, cell features are often aggregated to the well level to provide well level statistics familiar to discovery scientists. The well features include measurements such as average size, standard deviation of size, average shape, total intensity, and so on. The number of well features varies depending on the assay, but typically ranges between 5 and 50. In kinetics assays, the above measurements are taken at multiple points in time, from a few seconds to minutes or hours, and additional features (e.g., rate changes, min, max, and so on) are also calculated. The number of time-points again varies, but typically ranges between 1 and 10. Thus, a large amount of data is collected for just one well of a single plate. In addition, other associated information about the assay or experiment, such as protocol information, is also typically recorded.

We define three categories of HCS data:

1. *Image data*—these are the images acquired at each channel for each field within a well.
2. *Derived data*—these are the measurements that result from performing an analysis on an image with image analysis algorithms (e.g., well features, cell features, and so on).
3. *Metadata*—these are the associated data that provide context for the other two categories of data (i.e., metadata is data that describes other data). For example, assay type, plate information, protocols, operators, calculated data such as dose–response values, as well as annotations imported from other systems (e.g., sample identifiers and properties).

From a data volume perspective, the data to be saved per plate is primarily based on the image data and the derived data. The size of the Metadata in comparison is negligible. For each well of a plate, the data is estimated by the number of feature records needed to store the derived data and the number of images acquired. The number of images acquired can be estimated by: [number of wells × number of fields × images per field (i.e., the number of channels × number of time-points)]. The typical size of an image ranges between 262 kb (for a 512 × 512 × 1 byte image)

Table 1
Example Data Volumes for Different HCS Application Scenarios.

HCS application scenarios (assuming 100 cells per field and an image size of 0.5 MB ([= 0.05 GB] for all examples)	Image data, number of images (storage GB)	Derived data (GB), number of records (storage GB)
One hundred 96-well plates, one field/well, two channel/field, 20 feature/well, 50 features/cell, one time-point	192 images (9.6 GB)	48.2 million records (1.5 GB)
One hundred 96-well plates, two field/well, three channel/field, 50 feature/well, 25 features/cell, one time-point	576 images (28.8 GB)	48.5 million records (1.6 GB)
One hundred 96-well plates, four field/well, four channel/field, 20 feature/well, 50 features/cell, one time-point	1536 images (76.8 GB)	192.2 million records (6.2 GB)
One hundred 96-well plates, 10 fields/well, two channels/field, 50 feature/well, 50 features/cell, one time-point	1920 images (96 GB)	480.5 million records (15.4 GB)
One hundred 384-well plate, two fields/well, two channels/field, 20 feature/well, 100 features/cell, one time-point	1536 images (76.8 GB)	769.9 million records (24.6 GB)
One hundred 384-well plate, four fields/ well, two channels/ field, 50 feature/well, 50 features/cell, one time-point	6144 images (307.2 GB)	769.9 million records (24.6 GB)
One hundred 96-well plates, one field/well, two channel/field, 20 feature/well, 50 features/cell, 10 time-point	1920 images (96 GB)	480.2 million records (15.4 GB)
One hundred 96-well plates, four field/well, three channel/field, 50 feature/well, 50 features/cell, 20 time-point	11,520 images (576 GB)	960.5 million records (30.7 GB)

Shown are data volumes for Image Data and Derived Data together with the "number of images" and associated storage requirements and the "number of records" for cell and well features stored in the database and estimated storage requirements.

and 2 MB (for a 1024 × 1024 × 2 byte image). Images are often compressed using some form of lossless compression, which usually results in a 25–50% storage reduction. For derived data, the number of cell feature records can be estimated by (number of wells × number of fields × number of cells × number of features per cell) and the number of well features can be estimated by (number of wells × number of features per well).

The amount of data generated in a period varies depending on a number of factors including the biological assay, the types of experiments or tests to be run, the throughput of the instrument or analysis application, the number of instruments, and so on. **Table 1** shows ranges of possibilities for different types of example assays (*see* **Note 1** for detailed example calculations). This data could be generated in days or weeks leading to tens of TB of storage requirements in a few months.

Although similar to other informatics modalities in some aspects, high content informatics has some unique characteristics. The requirements for management of high content data and images are different than the requirements for purely managing images with simple annotated data. In high content informatics, the data is supported by the images as opposed to the images supporting annotated data. As we can see from the **Table 1**, high content data is far more complex and voluminous than simple image annotations. Any high content informatics solution therefore needs to be able to efficiently handle the relationships between the various levels of feature data and the associated images.

3. Data Model/Structure for HCS Data

To enable effective decision making in HCS, data and images and associated information must be stored with high integrity in a retrievable form. HCS data should be stored in a manner that takes advantage of the characteristics of this type of data to enable full access and exploitation of the data. The underlying data model (or database structure or database schema) should be flexible to handle the various HCS data types (i.e., image data, derived data, and metadata) and a wide range of changes in the data (e.g., different numbers of wells, cells, features, images, different image sizes and formats, different number of time-points [in kinetic assays], and so on).

The structure of the metadata is also important. The metadata provides a means of describing data and the relationships within the data, enabling data to be better organized, cataloged, and searched effectively. Metadata enables joining of related data to allow meaningful visualization, analysis, and data mining. The metadata is also important for integration with other systems and data sources and defined vocabularies should be used for metadata whenever possible. For example, using defined lists and consistent words for describing assays, samples cell lines, and so on, rather than free comments. This is an area where standards across the HCS field would be helpful, but should at least be consistent within an organization.

4. System Architecture

Managing the collection, storage, retrieval, analysis, and display of huge volumes of HCS data demands a system architecture that utilizes best practices from the world of IT. The system architecture defines the fundamental organization of the system, the underlying structure of the various components and their interrelationships, and the principles governing the overall design.

A key component of any high content informatics solution is the data management component and this is best handled by some form of database technology, because managing HCS data via file based systems does not provide a scalable solution. In contrast, databases (e.g., relational, object oriented, or object-relational) are designed to provide efficient access to large amounts of data. Relational databases are the most commonly used databases for HCS data. Relational databases are available from many vendors (e.g., Oracle, www.oracle.com; Microsoft SQL Server, www.microsoft.com; and so on) and high quality open source databases also exist (e.g., MySQL, www.mysql.com and PostgreSQL, www.postgresql.org). Relational databases address two of the three categories of HCS data, derived data, and metadata. The remaining HCS data category (i.e., image data), are the images usually stored outside the relational database with only pointers to the images being stored in the database, because the database simply grows too large to be efficiently managed with traditional tools if the images are stored directly in the database. In contrast, there are a wide variety of options available when storing the images outside the database. All that is really necessary is a large amount of disk space. However, as the needs of the system grow, more complex technologies may be employed such as Network Attached Storage, Storage Area Networks, Content Addressed Storage, and Hierarchical Storage Management, which are available from various vendors including IBM (www.ibm.com), EMC (www.emc.com), and Network Appliance (www.netapp.com). The key to scaling image storage is that a pointer in the database to the externally stored image must exist in the relational database in order to retrieve the image at a later time.

For the relational database to be useful, the HCS data must be entered into the database via some automated collection, transfer and integration processes. This is an essential and complex task for all but the smallest usage scenarios. This requirement is best handled by a Utility Service *(3)*. Utility Services provide features like processing results in an unattended manner, running even when the computer is not logged on, and sending notifications of important events or status.

Retrieving data from the relational database is equally important as getting it in. Once again, a Utility Service offers the best approach to accomplish this requirement. Such a service will allow the configuration of permissions to create, view, update, and delete data to be consolidated within the service. This is a key feature that allows the architecture to scale within an enterprise.

Collectively, moving the data in and out of the system comprises the underlying business rules/logic or middleware. Often this logic is exposed to both the given system and external systems by means of a dedicated application server. An application server may expose some or all of the underlying business rules via a Web Service *(4)*. The Web Service is the key integration point that allows the HCS data to be integrated within a customer's own enterprise data repository.

Client applications are the software applications that are used to visualize, analyze and mine information from HCS data. For users, the client applications are usually the most important component, as these tools are what they interact with on a daily basis. In follow-on sections of this chapter, we review two client applications, visualization tools and data mining tools in more detail.

Combining the relational database, application server, HCS instruments, and client application components together form the basis of a traditional "N-Tier" architecture *(5,6)*. This can be seen in **Fig. 1** with the associated HCS system components.

An N-Tier architecture refers to a system that has at least three tiers (or "layers") that are separate and each tier interacts only with the tier below (or above) and has a specific function that it is responsible for:

- *Presentation tier*—The presentation tier is for displaying the user interface and driving that interface. Essentially these interfaces are the user facing parts of HCS instrument software and the client applications. This is also sometimes referred to as the "user tier."
- *Middle tier*—The middle tier provides the automated transfer of data from the instruments to the data tier and moves data back from the data tier back to the presentation tier. This tier is also responsible for processing the data retrieved and sent. The middle tier is also sometimes referred to as the "application tier" or the "business tier."
- *Data tier*—The database for the HCS data and the repository for the images reside in the data tier. This is where the three categories of HCS data (image data, derived data, and metadata) are stored.

These tiers can be physically together but conceptually separate. They can also be located on physically different servers even if the servers are in different geographical locations. Separating the logic and processing contributes to the major benefits of N-tier, which are robustness, maintainability, and scalability. The scalability part is especially important, allowing improvements to be applied where needed (e.g., additional or more powerful database servers can be used as data volumes grow).

Because each tier can be located on one computer or physically different computers, each can be scaled to the needs of an organization (i.e., number of users, number of instruments, amount of data, and so on). This approach is provided by Cellomics, Inc. (Pittsburgh, PA). in their HCi™ informatics platform (www.cellomics.com). In limited usage scenarios, all three of these tiers may be physically installed onto a single computer. Usually though, at least the application server and relational database are installed on different computers than the client applications. At the upper end of the scale, multiple computers may be used at each tier. This will generally be the case at sites with multiple HCS instruments and/or client applications.

5. Hardware and Network Considerations

There are a wide variety of ever evolving options for server hardware, storage hardware, and networking capabilities for an organization's informatics solution. The number of HCS instruments, number of users, the number of sites, and the network bandwidth within a site (i.e., Local Area Network) and between sites (i.e., Wide Area Network), are a few of the key factors impacting the hardware requirements for an informatics solution.

Sizing and scoping the optimal hardware for an informatics solution is an area where having professional IT support is critical. Each organization is unique in their HCS usage scenarios, which

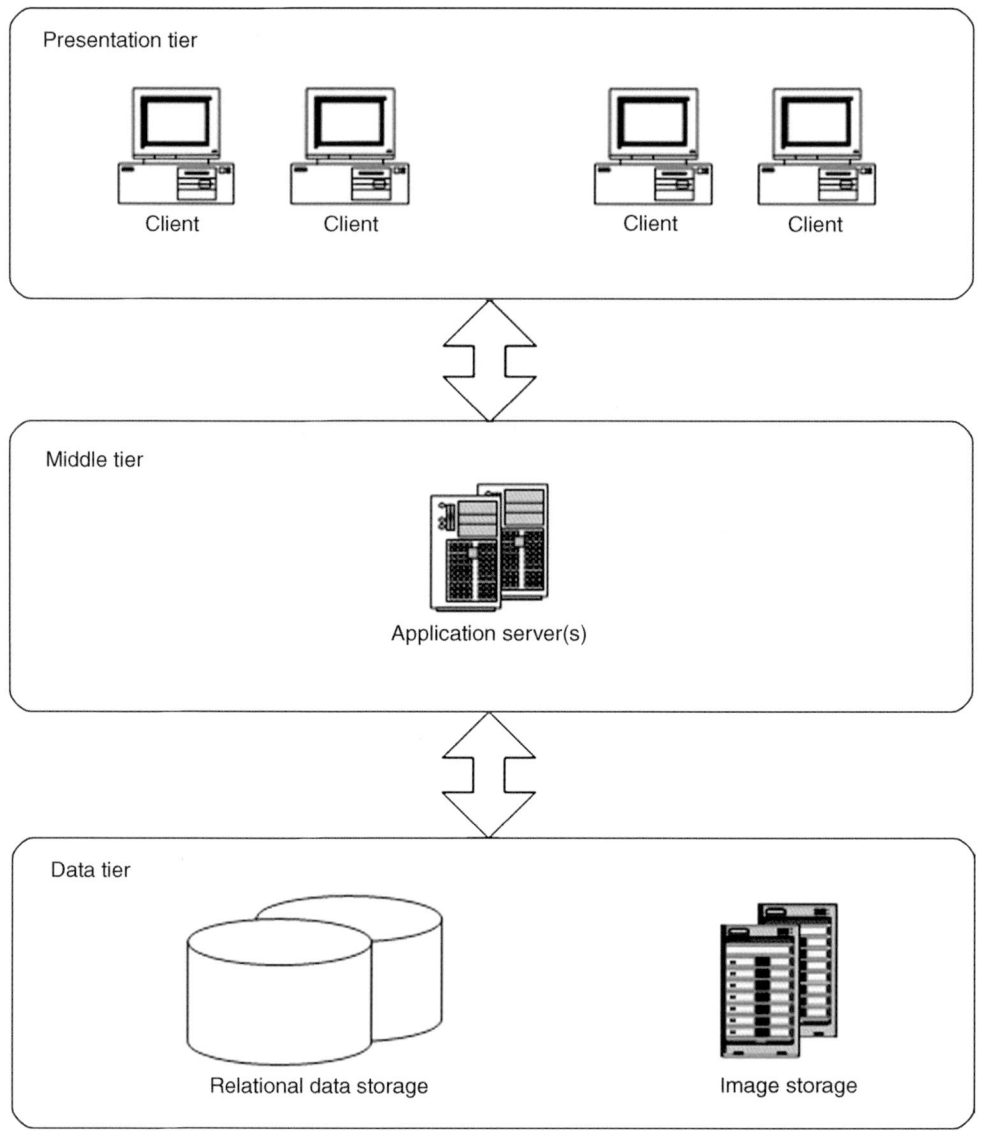

Fig. 1. "N-Tier" Informatics system architecture.

directly impacts the requirements put on an informatics solution. In general, it is best to identify an informatics solution with a system architecture that can scale as the organization's HCS needs evolve over time. For example, a user might start with one instrument, one combined database and image storage server, one application server, and a couple of client applications. Then it may grow to multiple instruments at multiple sites with multiple database servers, multiple image storage systems, multiple applications servers, and multiple client applications. In general, a vendor's informatics solution should be flexible and scalable to fit a variety of hardware configurations and usage scenarios.

A key factor impacting the configuration of the system architecture across multiple sites is the bandwidth of the network that connects the various computers together. HCS instruments typically generate data and images at rate of 1–5 GB (or more) per hour and there are limits to current network and server technology that can support instruments writing this amount of data across networks at multiple sites. Tradeoffs between network bandwidth, server, and storage system configuration, and each organization's unique use cases for how information will be accessed and shared, all need to be taken into account in order to optimize overall system performance.

6. Data Management

Over a period of time, a tremendous amount of HCS data will be collected. An effective workflow must be developed to manage the data. This workflow will include such things as who is allowed to view and manage the data, how the data will be backed up, and when to archive or delete data.

Policies or procedures for storing HCS data need to be determined by each organization. Specifically what HCS data (Image Data, Derived Data, and Metadata) will be stored for how long (e.g., 5, 10, or 15 yr or more) directly impacts the workflow as well as the overall data volume, scope or data management, and cost. This is currently an area where policies are still being formulated by organizations, and in many cases they just decide to play it safe and store everything, further fueling the need to store, and manage even more data.

Regulatory compliance (e.g., FDA 21 CFR part 11 [www.fda.gov/ora/compliance_ref/part11]) is another issue to consider, in particular, in pharmaceutical, and other highly regulated industries. A high content informatics product that can help an organization establish and maintain regulatory compliance could be critical. Although no product itself can make an organization compliant without the proper policies, a properly designed, developed, and supported product can make complying with regulations significantly easier.

Regarding who can view or manage the data, some forethought must occur just to get the system up and running. Simply assigning everyone full control of the data may be problematic, therefore having access to professional IT personnel who have the experience to assign and manage permissions effectively is extremely important.

Managing permissions is also a key point that reveals why having an application server in an "N-Tier" architecture is so important. Without this type of architecture, all users must be assigned permissions to the file storage, and relational databases. With an "N-Tier" architecture, access to these resources may use a proxy account from the application server. This greatly simplifies deploying and managing the system, especially when trying to share data across multiple sites or different domains.

Backing up the data is another area where having professional IT support is very valuable. As the data volume grows, creating, and maintaining complete backups is a difficult task. The key feature that a successful HCS backup strategy has is preventing the volume of data that needs to be backed up from growing beyond the manageable range of the backup solution.

One of the best approaches to achieve this is to store the HCS data in different locations based on time. A location's time may then be used to determine whether the data has already been backed up. Once a particular location is no longer having data added, a final backup of this location may be completed. This location may then be removed from the periodic backup regimen.

IT professionals can also help with the archiving of data. This is especially true if the data may be archived based on metadata criteria such as creation date, storage location, or creating user. However, if biological metadata like projects, compounds, or hits drive the archive process, then scientists will need the ability to archive data. Regardless of who actually performs the archiving, coordination among users, and IT staff is vitally important to effectively manage HCS data (*see also* Chapter 21).

7. Visualization

Visualization tools are one type of client application mentioned earlier that provide a quick and effective means to interrogate HCS data and images stored in a secure repository. Users want to view the data, share it with colleagues, and compare results. Visualization software should provide powerful search and navigation tools to rapidly locate plate, well, cell, and image data. Rich search functions should be available to find data based on various metadata and derived data parameters (e.g., user name, dates/times, assay type, features, and so on) (*see also* Chapter 22).

The most basic form of any HCS data visualization tool should provide interactive tools for reviewing data with drill-down capabilities from the plate, well, and cell level together with links to images, and any graphical image overlays. Various forms of viewing the data should be provided including tables/spreadsheets and graphs (bar charts, scatter plots, and so on, *see* **Fig. 2**). Various views should also be provided for different types of users (e.g., managers, scientists, operators, IT personnel, and so on).

Capabilities should be provided for comparing data within a plate, across plates, and so on. Additional capabilities should also be provided for generating statistics on groups of data (e.g., groups of wells, cells, and so on). The data should be displayed in ways that allow the user to explore patterns and recognize patterns and outliers. Users want to be able to save their analyses and visualizations as well as build reports and save these. Making annotations on the data is also very important.

Common uses for visualization in HCS include assessing the quality of the dataset (e.g., identifying outliers and false positives), and identifying hits. There are many possibilities for visualization of HCS data using commercially available tools (e.g., Spotfire (www.spotfire.com), OmniViz (www.omniviz.com), and so on) (*see also* Chapters 13 and 23).

8. Data Mining

The large amount of multiparameter data inherent in HCS provides opportunities to reveal patterns or trends in the data using data mining tools *(7,8)*. Data mining tools are another type of client application mentioned earlier. These tools can include pattern recognition techniques, self-organizing maps, fuzzy logic, statistical methods, and machine learning methods. In addition to identifying patterns and trends from the data, data mining technologies can be used in making predictions and simulations of future events.

Used together with visualization tools, data mining can be used to discover knowledge in HCS data sets in a form that is more easily understood. The goal is to reduce complexity and extract relevant and useful information from large HCS data sets in an intuitive and efficient manner so that better decisions can be made (*see also* Chapter 23).

Although data mining tools can be a very powerful aid to making important decisions, they are not self-sufficient. To be successful, data mining requires skilled technical and domain specialists who can structure the analysis and interpret the output that is created. For example, data mining can help identify patterns and relationships, but it does not tell the user the value or significance of these patterns. These types of determinations need to be made by the user. Similarly, the validity of the patterns discovered is dependent on how they compare to real world circumstances. Nevertheless, data mining holds great promise as a critical tool for HCS analysis and we expect that data mining will therefore have a significant impact, much as it has had in other industries that have large quantities of data.

9. Integrating HCS Data With Other Data and Systems

With the widespread adoption of HCS throughout the drug discovery and academic research domain, the need to integrate HCS data with other discovery data and external systems has arisen. Indeed, integration has become a key issue as HCS data is used to make decisions that require multiple data sources, from target validation data through to ADME/Tox and preclinical domains. HCS data cannot be a critical part of the drug discovery decision process unless it is effectively integrated. Integration can take many forms, but can be categorized as data-level integration, database integration/federation, and application/software integration.

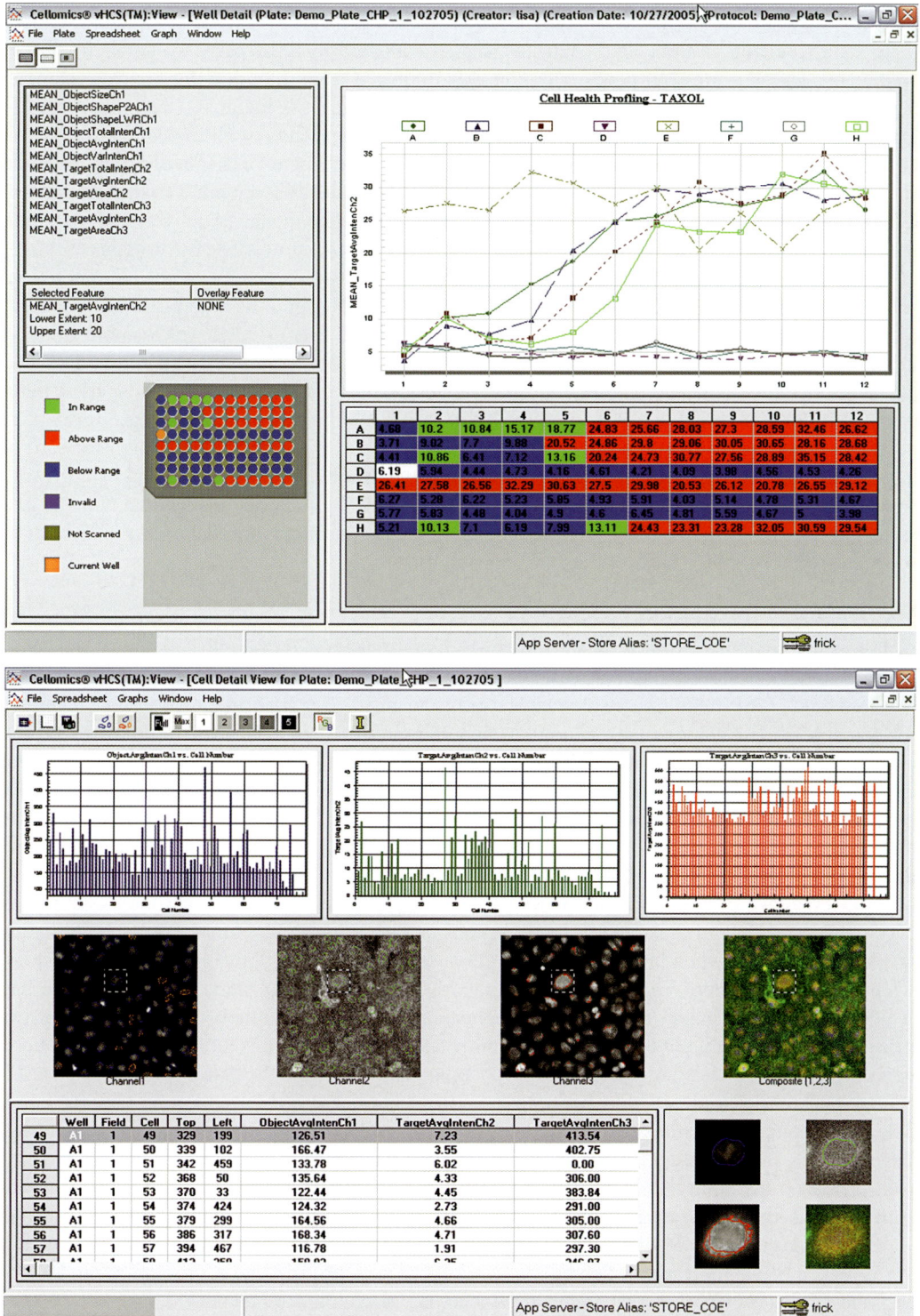

Fig. 2. Example HCS data visualization with interactive tools for reviewing data with drill-down capabilities from the plate and well level (top display) to cell level together with links to images and graphical image overlays (bottom display) (Courtesy of Cellomics, Inc.).

9.1. Data-Level Integration

In data-level integration, HCS data is usually exported to third party systems, for either archive or "warehousing" purposes. Often data is imported from the same third party system so that HCS data can be annotated appropriately so that links may be made. An example of this is linking HCS data to sample information (e.g., chemical compounds or biological test samples). Often the user has centralized systems for collating all instrument or discovery data together, such as IDBS ActivityBase (www.idbs.com) or other laboratory information management system software. Linkage at the data level via an export is a simple means to deliver HCS data into the enterprise as well as integrate HCS data into laboratory workflows. The informatics architecture therefore needs to support both the necessary relational data structures to permit annotation, such as sample identifiers for compounds. In order to push data into the enterprise and link it in, requires flexible, format neutral export tools. Over the past 2–3 yr XML (eXtensible Markup Language) *(9)* has arisen as the format of choice for data export, as it is self-describing text (i.e., not only does it contain the data to export but a description of the data in the same file [metadata]). Virtually any software can interpret XML and it can be translated into other formats, if necessary. Data-level integration has certain advantages in that it is relatively straightforward to implement, almost any data can be integrated, and few changes, if any, are required by either the source or target applications. Disadvantages are that an additional copy of the data is made and there may not be a way to actively link content (e.g., see an interesting data point and wish to see the associated image without further programming).

9.2. Database Integration/Federation

In this integration type, HCS data is either (1) directly integrated or published into a data warehouse with other discovery data sources, loader scripts or database views are used, and data is often cleansed or (2) some middleware software is used as an abstraction layer to more loosely "federate" for example HCS databases with genomics, and cheminformatics databases. Middleware layers, often called metalayers, provide consumers of data with a single "view" on the data, independent of the native data format or schema, so that a user application can query and work with data across perhaps dozens of data sources, be they relational databases or unstructured data such as text files and images. The integrated data warehouse approach to database integration does have some advantages in that it is relatively simple to implement and there are now sophisticated data warehousing tools for carrying this out, however, as the desire to integrate more data sources grows, the system has to scale and this requires hands on effort. The volume and complexity of HCS data is also a consideration when building a data warehouse/integrated data integration. Using a middleware or metalayer approach to federating databases became popular particularly during the late 1990s in the bioinformatics revolution, as sophisticated data analysis tools needed to look across many data sources. Several life science and informatics vendors use this kind of technology. Such approaches have more of merit as no data gets copied anywhere and the data sources stay intact. It is also much easier to make links between data. In addition, the metalayer can be "smart," being able to semantically interpret data queries, for example. HCS data can certainly be federated using this approach, providing the advantage of best in class management of the large volumes of complex data with the ability to more actively link this data with other key discovery data sources. However, this kind of integration comes at a price, adapters have to be written for every data source, which demands an intimate knowledge of the database schema and business logic of the source data. Performance of the metalayer when querying across dozens of disparate data sources can also be an issue. If the schema of the source changes then the adapter has to be updated.

9.3. Application/Software Integration

The third category of integration focuses on more of a programmatic integration (i.e., an application programming interface [API] rather than a pure data integration). Data might well be a result of the integration, but the primary point is that some third party application requests either data or a function to be performed by the source software. For example, a third party application might ask for an image to be analyzed or for an HCS experiment to be statistically evaluated, sending the data to a visualization application such as Spotfire. From a user perspective, the user is working with an application that perhaps spans several functions (e.g., gene sequencing, proteomics with an HCS analysis being just another choice). From an IT perspective, applications, and workflows involving HCS data can be built as need dictates. No special database is needed, no metalayer adapter, no knowledge of the underlying schema is required and no copy of the data needs to be made.

From an informatics architecture perspective and in terms of integrating HCS data, workflow, and business logic into the life science enterprise, the API integration has considerable merit. However, traditional APIs are often compile time code and so changes by the API vendor force a change to the calling application, in addition the API might only work with a limited set of programming languages or tools. Recent advances in web services *(4)* overcome many of the disadvantages of using traditional APIs. Web services are part of a more distributed, federated approach to data/application integration that does not require programmatic integration in the traditional sense. External applications are seen as services (irrespective of location or hardware), which are "consumed" by other applications. System architects can then build very powerful systems based on a loose coupling of web services in a so called "service oriented architecture" *(3)*. Given the data volumes and the emerging business, and workflow of HCS, exposing this functionality through a web service is considered as the best practice for integrating HCS into life science workflow. Furthermore, this approach fits very well with "workflow" software such as that provided by Scitegic's PipelinePilot (www.scitegic.com). Workflow software allows large scale integration of business functions (rather than data) to achieve an end point, one could envisage such an application, looking at genes of interest for a particular group of targets, analyzing the literature, finding the appropriate RNAi, interpreting the subsequent RNAi experiments using HCS, evaluating the proteins of pathways knocked out and suggesting compounds likely to have an effect. All this could be achieved if all the functions were available as services that can be coupled as needed.

10. Summary

Advances in various HCS technologies, including new cell-based assays, imaging algorithms, and higher throughput instrumentation, have created an explosive growth of available HCS data. The massive volumes of information-rich data being generated by HCS systems and the effective management and use of this data has evolved into one of the most pressing issues organizations face today. The success of HCS is now more heavily linked to informatics capabilities than ever before. In the same way that the impact of genomics was greatly enhanced by bioinformatics, so HCS requires its own unique informatics infrastructure and tools. Getting access to the huge volumes of information rich HCS data, managing it, sharing it, analyzing it, and, in general, effectively using it, is critical to the overall success of the field of HCS as a whole. Informatics is and will continue to be a key critical technology for the ongoing success and widespread acceptance of HCS.

11. Notes

1. *Example data volume calculations:* in a typical drug discovery screening scenario, the image data generated for a plate with 96 wells, with four fields per well, three images per field, and one timepoint, and an image size of 0.5 MB ($512 \times 512 \times 2$ bytes) would be about 1152 (96 wells × four fields/well × 3 channels/field) images requiring about 576 MB (1152×0.5 MB) of storage (uncompressed). If 100 cells per field are selected with 10 features per cell calculated, then 384,000 (96 wells × four fields/well × 100 cells/field × 10 cell features/cell) cell feature records would be required. If 50 well features are calculated per well, then 4800 (96×50) well feature records would be required.

Depending on the speed of the HCS reader, such a plate could typically be analyzed in 10–30 min. If we use 20 min as an example scan time, and 48 plates being analyzed in a 16-h period per day, this results in a need for 55,276 (48 × 1152) images requiring about 27.64 GB (48 × 0.576 GB) of storage and about 18,662,400 [(48 plates × 384,000 cell features/plate) + (48 plates × 4800 well features/plate)] cell and well feature records requiring about 600 MB (18,662,400 × 32 bytes per record) per day. (The record size for storing features in a database depends on the database technology used and the specific implementation—for this example we use 32 bytes per feature record). In a 5-d period, this results in 276,380 images requiring about 138 GB of space and 93 million features requiring about 3 GB of storage. In 1 yr this translates into over 13 million images and over four billion cell and well feature records with combined storage requirements of over 7 TB. In a production screening operation, where multiple plates could be scanned in parallel on three or four HCS instruments operating continuously, the storage requirements could easily exceed 25 TB per year.

References

1. Giuliano, K. A., Haskins, J. R., and Taylor, D. L. (2003) Advances in high content screening for drug discovery. *Assay Drug Dev. Technol.* **1,** 565–577.
2. Comley, J. (2005) High content screening—emerging importance of novel reagents/probes and pathway analysis. *Drug Discov. World,* **6,** 31–53.
3. Erl, T. (2004) *Service-Oriented Architecture: A Field Guide to Integrating XML and Web Services,* Prentice Hall, Upper Saddle River, NJ.
4. Newcomer, E. (2002) *Understanding Web Services: XML, WSDL, SOAP, and UDDI, First ed.,* Addison-Wesley Professional, Boston, MA.
5. McConnell, S. (2004) *Code Complete,* Second ed., Microsoft Press, Redmond, Washington.
6. Bass, L., Clements, P., and Kazman, R. (2003) *Software Architecture in Practice, Second ed.,* Addison-Wesley Professional, Boston, MA.
7. Han, J. and Kamber, M. (2001) *Data Mining: Concepts and Techniques,* Morgan Kaufmann Publishers, New York.
8. Adriaans, P. and Zantinge, D. (1996) *Data Mining,* Addison Wesley, New York.
9. Harold, E. and Means, W. S. (2004) *XML in a Nutshell, Third ed.,* O'Reilly, Sebastopol, CA.

21

Large-Scale Data Management for High Content Screening

Leon S. Garfinkel

Summary

High content screening (HCS) plays an important role in target selection in primary and secondary screening, but further developments in informatics and data management are needed for strategic implementation of HCS in the drug discovery process. An organization charter for the Research Informatics and Infrastructure Organization is described and is consists of four basic parts: Partner, Build Trust, Champion, and Core vs Noncore. The successful evolution of the charter over the last 5 yr is mapped using high-throughput screening and HCS data as an example. A future view of large-scale data management for the drug discovery process will incorporate all scientific information into multiple parameter type runs for many aspects of the science. This information will subsequently be aligned into a subset that an individual can digest and more easily choose the next appropriate steps.

Key Words: Bioinformatics; data integration; information technology; large-scale cell-based assays; platform independent.

1. Introduction

The importance of high content screening (HCS) in the drug discovery process is target selection in primary and secondary screening but it is only in infancy from the standpoint of informatics and data management. As the imaging equipment in the laboratories becomes more sophisticated; produces better images, more graphics with higher resolution, and allows for faster collection of data, the infrastructure specialists working in the background will not only have a hard time keeping up in some organizations, but might at times ask the scientists to stop while they catch up. Where does one want your HCS informatics organization to be? Ideally, one wishes to be a step ahead of the drug discovery process and taking a strategic view of this area as opposed to being in a purely operational tactical solution oriented mode, which will need daily management. How does one get to this point, what are some of the options, methods, and proven answers that will get one ahead of the curve? The author will try and address as many of these issues as possible and convey the real hands on information to make it happen for your organization.

The key theme and piece of information repeated throughout this chapter is "partnering." Scientific research and informatics must work together for the mutual benefit of the drug discovery process. To really be part of the winning team in any organization, all areas must bring their collective expertise together and make the extra effort to understand one another and defer where there is lack of knowledge to those on the team with the experience and expertise or to seek external advise. It is necessary to start off by setting the stage concerning where laboratory computing, which includes the data management (we will discuss a bit later in the chapter), has progressed in order to gain the necessary understanding of where it currently is and where we anticipate it will be going in the HCS area in the future.

First let us examine a brief bit of history and background that will first put in place the foundation for computing in a Scientific Research environment as opposed to general computing in business. In 2000, out of all the departments in the Roche Nutley Research organization, (chemistry, biology, four therapeutic areas, and screening), there were two items that stood out which needed immediate attention. First, there was no single primary user of computer systems and, second, all of the departments combined had accumulated almost two terabytes of storage. In 2000, we were aware of the inefficiencies of trying to maintain multi-vendor computer systems and we anticipated a huge upswing in the amount of data storage space needed by the servers to handle the increases coming from the laboratory equipment. Although most of the information contained here might be specific to our organizational history, the author will certainly include examples and relevant comments from peer organizations in which the same situations might have occurred being better or worse. Some of the typical environmental situations that were contained within the labs:

- Computers were stored under laboratory work benches.
- Add-on disk drives were stacked up on cardboard boxes.
- Network connections went from some slow speed LAN to internal dial-ups.
- Servers were stored in closets in which space was available.
- Backups were done sporadically or not at all.
- There were published incidences in the industry in which tapes or other valuable/confidential backup computer documents were tossed into whatever available drawer space there was or even tossed out into dumpsters by accident.
- A mixture of Macintosh and PC type computers were in use, to the delight of scientists.
- Highly specialized, nonstandard computers that required extensive hands-on maintenance were in use throughout the facility.
- Large CRT monitors took-up valuable laboratory counter space.
- Scientists were doing science as well as Informatics work.

The first step in altering the situation was the realization that the research department had very specialized computing needs, which were very different from every other division of the company. Sales and marketing, finance, human resources, and even to some degree manufacturing departments were able to make use of off-the-shelf computer software and hardware. Scientific research on the other hand was not able to use of off-the-shelf items in most cases; there were requirements for special hardware to connect to instruments, special software for collecting, manipulating, synthesizing and managing the data coming from those instruments. Once this realization took place, an infrastructure and informatics organization was formed inside of the research department, reporting to Research to handle all of these issues.

2. Organizational Structure

There are four basic parts to the charter for the research informatics and infrastructure organization, they are: partner, build trust, champion, and core vs noncore.

2.1. Partner

Partner with the scientists to allow IT and IM professionals to take over responsibility for informatics tasks, thus, proving to the scientists that the informatics staff could successfully handle all of the informatics duties for the scientists. This would not only allow everyone's expertise to shine but would also return a large amount of time to the scientists, whom in the past were making extensive use of their time for informatics work. The side-by-side work would allow the informatics organization to learn about the workflows of the scientists, so the computer processes could be placed inline with the workflow and not be forced to add extra steps to the workflow.

2.2. Build Trust

Build the necessary level of trust with research management and scientists to ensure that the desired partnership "bears fruit."

- Do the installations of computer equipment for the scientists on time.
- Write the necessary utilities and programs to meet the needs of the scientists.
- Be diligent in purchases because whatever money is spent on IT cannot be spent for science.
- Require perfection from the informatics staff.
- Have the scientists validate the informatics work and sign-off that it meets the specified requirements.

2.3. Champion

Champion the research department's needs with corporate IT and IM and take full responsibility as the intermediary between these parts of the organization. Own the responsibilities of being in the research organization, yet being an IT professional.

- That is, system maintenance on databases can only be done twice a year rather than four times a year as the rest of the organization does, because of the nature of programs in the research department and the fact that programs might be executing for weeks at a time.

2.4. Core vs Noncore

Defining Core vs Noncore activities for the Research Informatics area. Does it make sense for Research to own, operate, and manage a data center? It makes sense because Corporate IT, already has the market on this, can provide the service for the RO also. It allows Research to take advantage of economies of scale. When an organization decides to share commodity services, there are security polices, practices, and standards available that are good enough for the entire company? Do they work for the research department also? In 99% of the cases the answer is "yes." We can make use of these governances; however, where it is not appropriate, research IT must have the latitude to break the rules.

2.4.1. Core Activities

- What type of desktop and servers are required for scientific computing? (brand, type, speed, and memory)
- How do the computer systems interface with the necessary data collection instrumentation and "talk" to the network and servers at the same time?
- How is the data stored, managed, and protected for short-, medium-, and long-term use?
- Are there government regulatory requirements that need to be met?
- Is the required software available off-the-shelf or must it be written in-house? This decision is collaboration between IT and scientists based on the defined requirements.
- Can allowances and accommodations be made for external collaborations and programs shared among scientists? Will we in the corporate world make the necessary adjustments to the IT systems so the scientists can freely collaborate with their peers in the academic world, whose computer systems might not be up to the standard we have set for security and virus protection? Can we create a safety buffered zone outside of our firewalls to allow this external data exchange?

Answering the earlier questions tells one exactly which business you want to be in and what type of work you want to be doing. This obviously needs the cooperation of the entire organization and all aspects of it. It re-enforces that there are experts in different areas and although scientists are highly skilled and learned, they do not have the specific professional knowledge that people in the computer industry have about making the correct business decisions related to IT. In today's day, in which just about everyone has a computer at work and at home, we tend to think of this not as a profession but as a hobby. If we bring our hobby to work with us, we might even save some dollars by not calling on the professionals. Wrong attitude and thought process, going down this path gets everyone in trouble. There is a world of expertise beyond the small domain of home computing that gives professionals that edge to do strategic business planning as well as tactical operations for the benefit of the organization. It is end-to-end thinking rather than dealing with a single isolated situation.

2.4.2. Noncore Activities

- Physical data center along with facilities management (electricity, air conditioning, fire protection, physical security).
- Networking.
- Security.
- Backup and recovery.
- Standards for PCs and peripherals, servers, desktop applications, middleware applications, web standards.

2.5. Organizational Summary

The accomplishments of RO/IT partnering actually amazed many people on both sides of the organization. The Informatics people never thought it would be possible to gain the acceptance, respect and ultimate responsibility to take computing away from the scientists. On the other hand, the scientists did not believe that anyone could do as good a job as they had been doing, no matter what level of expertise people had. Additionally, this opportunity ultimately allowed a number of scientists, who no longer wanted to work in labs to take their expertise and become computer experts and start new careers. There are a number of people in the organization that were PhD chemists and are now UNIX system administrators, Visual Basic programmers, project managers and workflow experts. Below are a number of outcomes that this effort was able to "bear fruit on":

- Standardized desktop systems (hardware and software) were in place throughout Research including the same application suite for all (i.e., Microsoft Windows XP, Microsoft Office, antivirus software, and Web access and portals).
    - o All users defined and categorized to ensure proper equipment (super users are equipped with very high-end pc or workstations, normal users, users traveling/working from home are equipped with laptops or notebooks).
- All servers are located in the corporate data center.
- Backups are done on a scheduled "off hours" basis to minimize work disruption.
- Data storage is part of the enterprise storage program used by the entire corporation, but with the understanding that the research department is the single largest user and might have special requirements.
- Most laboratories have only the minimum amount of computer equipment located on the actual premises now (flat panel monitors, 100 megabyte or gigabyte network connections).
- Partnership between research informatics and corporate informatics with a new understanding of requirements and demands in both parts of the organization.
- Scientific hardware/software vendors are now partnering with computer hardware/software vendors as a result of us sitting in the middle and demanding the best of all possible solutions. This type of partnering is something we in the IT field have been doing for years. The author was told by his colleagues on the science side that this partnering is something new and they are not used to working with Vendors in this type of fashion. The bottom line here is we challenged all of our "business partners" not vendors to come work with us on what for us was a unique set of initial requirements—a solution for data archiving of HCS imaged data. However, it turned out once the solution was tested and put in place that many other organizations had the same need for this solution.

The accomplishments described above took approx 3 yr to put in place. The infrastructure (hardware—desktops, servers) staff was six people and Informatics (software—packages, code writing) staff was approx 11 people. A basic operating expense budget for this type of area was approx $5–8 million annually and a capital budget of approx $2–5 million annually. This was to support approx 250–300 scientists and all of their related laboratory computer equipment and data collection. This work set the foundation for all of the integration of laboratory equipment and computer systems to follow. It allowed almost a cookie-cutter approach to future solutions that would be required.

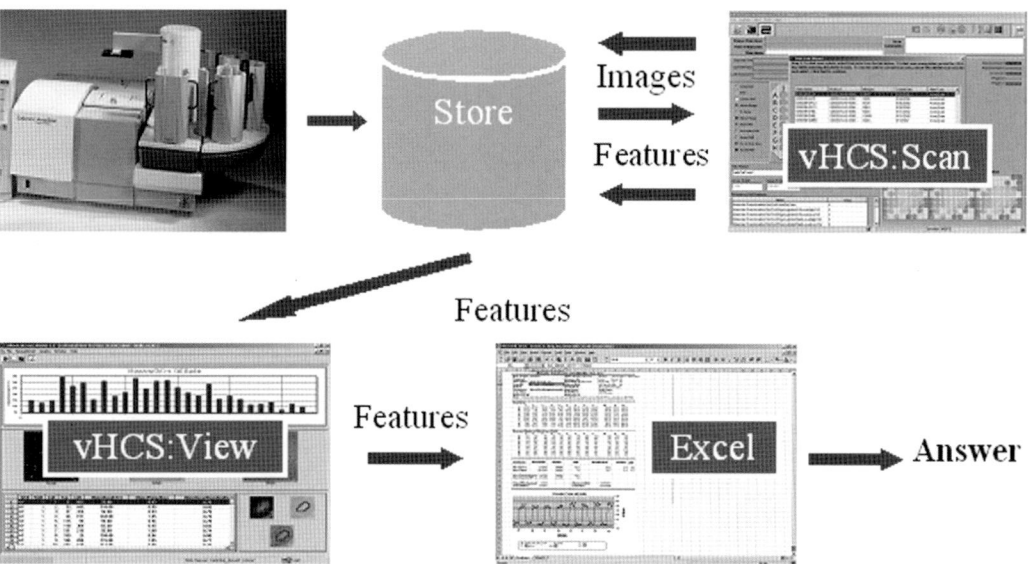

Fig. 3. Current features of Cellomics database management system. (Please *see* the companion CD for the color version of this figure.)

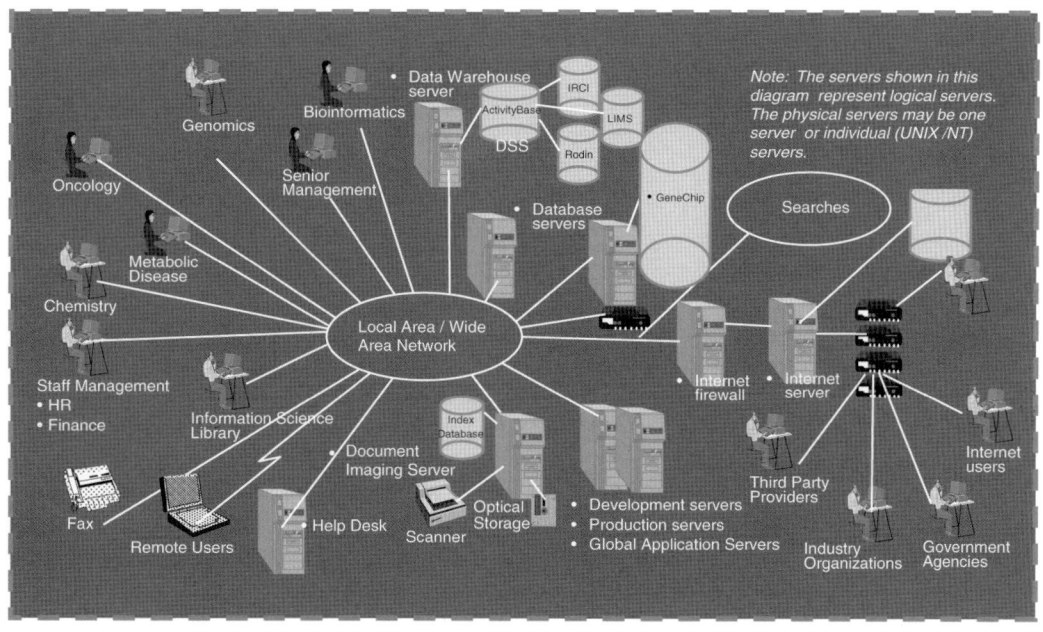

Fig. 4. High level logical view of the information technology architecture. (Please *see* the companion CD for the color version of this figure.)

The solutions depicted here were engineered with some of the following concerns in mind:

- As little daily operator intervention as possible.
- Cost effective in the short term as well as the long-term.
- Scalability to ensure that the solution can grow as the science changes with minimal intervention.

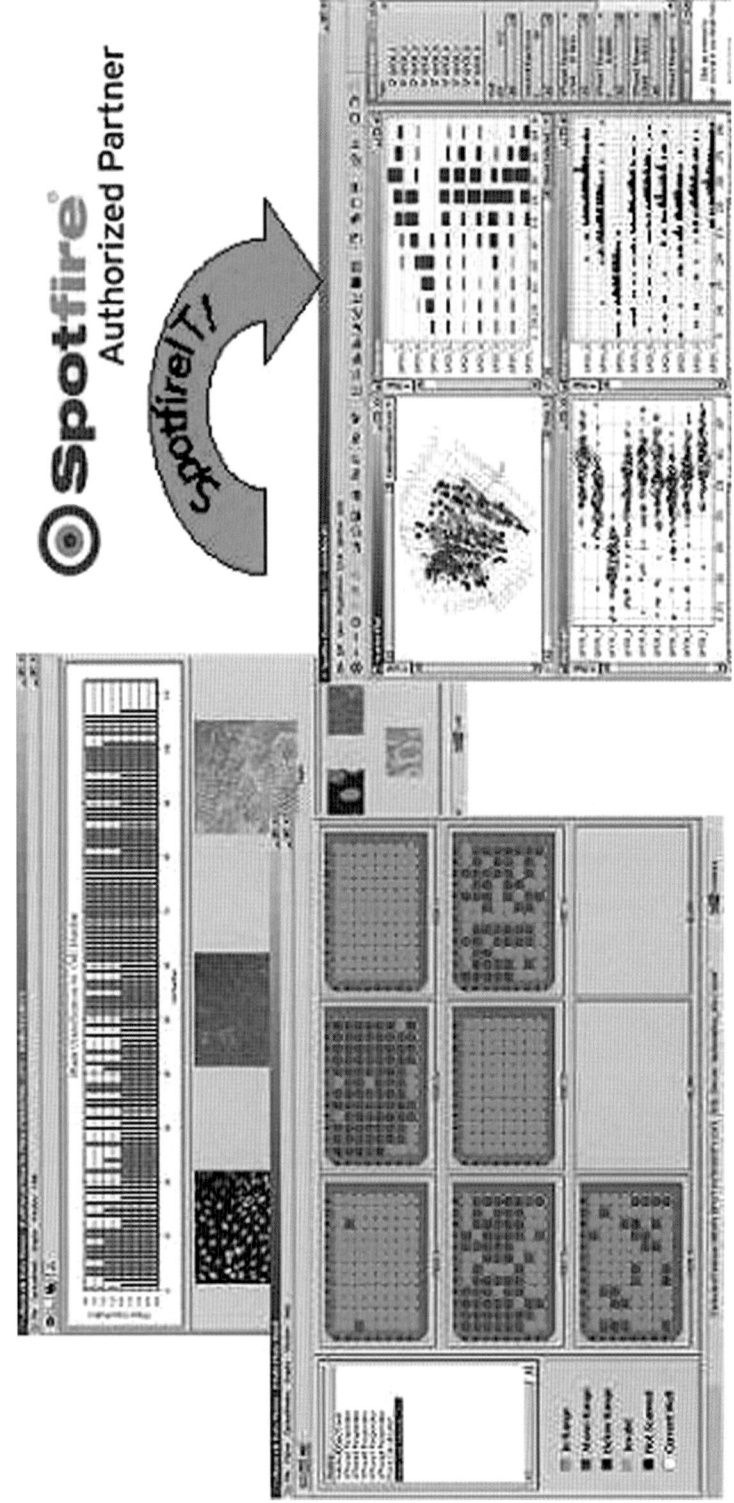

Fig. 5. Cellomics Data Handling of All Multiparameter Bioapplications. (Please *see* the companion CD for the color version of this figure.)

- Flexibility to ensure that additional scientific instrumentation could be integrated with little or no involvement from the Informatics area. This solution has allowed almost any device to be brought into the organization for testing and or permanent placement with minimal intervention by the Informatics organization. All of the information below is from projects already undertaken and proven within the environment.

5. Summary and Future Consideration

Where are we today, about 5 yr later? Scientific research is now consuming approx 12–16 terabytes of data. We seem to be adding around 4 TB each year. The single largest user, group amassing data is the HTS area. This is not only because of their increasing the number of wells on a plate, but as you can see by the instruments below the quality of the data is becoming much more robust. For the images being stored, studied and passed on to other areas, the data sets are getting larger and larger. One additional piece of information that can easily facilitate the entire process of implementing good Informatics practices and having a proactive organization to ensure the proper strategic planning for growth and technology is a user advisory committee or steering committee. This is a low cost option and should have full management support to ensure the success.

The mission of this steering committee is to:

- Assure alignment of Informatics activities with local site and global priorities.
- Recommend prioritizations of activities to senior management.

The members of this group:

- Have a strategic view of the interactions between scientists and Informatics.
- Represent the overall needs and concerns of their constituents as well as the site.
- Membership should include—biology, chemistry, HTS, genetics, genomics, bioinformatics, therapeutic areas (3), pharmacology, safety, animal resources, informatics, finance, senior management.
- Be able to make decisions on operational issues and not for the sole benefit of individuals.
- Improve communication regarding requirements and deliverables.
- Pro-actively provide management with information that will have an effect on deliverables and on current projects.
- Provide transparency around resource allocation as well as costs.
- Dedicate the time to the meeting and bring information back to your department as well as to the next committee meeting. Once a month was sufficient for our organization but this can be adjusted based on topics and needs.

Attempting to look into the future, we hope to be able to see that the technology used in science will not only continue to improve, from the standpoint of what the scientists are doing on a single biological or chemical level, but will incorporate all of the information into multiple parameter type runs for many aspects of the science. In other words, while the author performs screening, can a scientist combine the biological analysis and chemical analysis and also do some mapping? include genomics information and genetics. The ultimate goal would be to align all of this information into a subset of the information that one individual can digest and relate patent information and competitor information. Armed with such information, one can more easily choose the next appropriate steps.

Acknowledgments

Special thanks to Dan Weiss, Keith Groco, Ann Hoffman, and Ralph Garippa, who knew the value of collaboration and realized that this was the team that would make the projects a success. Cellomics and EMC for not being mere vendors, but partnering with Roche to come up with a unique solution to our problem. Irene, Rebecca, and Mark for allowing me to spend all the necessary time at work and work at home when possible to ensure that these projects were a success.

22

An Integrated Biomedical Knowledge Extraction and Analysis Platform

Using Federated Search and Document Clustering Technology

Donald P. Taylor

Summary

High content screening (HCS) requires time-consuming and often complex iterative information retrieval and assessment approaches to optimally conduct drug discovery programs and biomedical research. Pre- and post-HCS experimentation both require the retrieval of information from public as well as proprietary literature in addition to structured information assets such as compound libraries and projects databases. Unfortunately, this information is typically scattered across a plethora of proprietary bioinformatics tools and databases and public domain sources. Consequently, single search requests must be presented to each information repository, forcing the results to be manually integrated for a meaningful result set. Furthermore, these bioinformatics tools and data repositories are becoming increasingly complex to use; typically they fail to allow for more natural query interfaces. Vivisimo has developed an enterprise software platform to bridge disparate silos of information. The platform automatically categorizes search results into descriptive folders without the use of taxonomies to drive the categorization. A new approach to information retrieval for HCS experimentation is proposed.

Key Words: Data mining; document clustering; federated search; information retrieval; metasearch; ontology; structured and unstructured data; taxonomy.

1. Introduction
1.1. Information Explosion

The genomics information explosion that started in the 1990s has evoked various methods to store, annotate, retrieve, and analyze the bourgeoning data aftermath. Information assets that have been created include gene expression data from microarray experiments, genomic sequences, proteomic identification, and data from high throughput SNP arrays *(1)*. In addition there has also been a growth in high-throughput and high content screening (HCS) data. These data are typically highly structured but lack cross-informatics platform capabilities.

The continued growth of biological research articles, fueling the continued information expansion and the emerging biotools that created them, lie juxtaposed with the aforementioned structured information entities. For example, Entrez PubMed, developed by the National Center for Biotechnology Information (NCBI) at the National Library of Medicine, is a web-based citation repository and search tool spanning more than 12 million citations across 4800 biomedical journals published in over 70 countries *(2)*. Staying current with this expanding research repository, in addition to the various commercial informatics tools to retrieve and analyze structured and unstructured data, becomes ever more challenging.

NCBI has extended Entrez's reach beyond journal citations to help store and retrieve the more structured information assets. These tools include: PubChem, GenBank, Entrez Protein, and Online Mendelian Inheritance in Man. In addition to these publicly available information resources, growing repositories of gene, protein, cytotoxicity, compounds, and other biological information are growing within the data-warehousing confines of the highly competitive pharmaceutical and biotechnology companies. A growing challenge has emerged for industry to blend proprietary internal data assets with the growing public assets, and to make swift analysis of these data to optimize the relatively new commitment to HCS experimentation.

Beyond NCBI, a growing legion of private corporations is developing information storage and information mining solutions to help swiftly gain insights from these growing repositories. New tools now exist to:

- Input genome sequences to identify relevant patents (Gene-IT; www.gene-it.com).
- Identify novel protein–protein interactions through canonical pathways systems (GeneGo, Ingenuity; www.genego.com).
- Rapidly create cloning vectors (Vector NTI from Invitrogen; www.invitrogen.com).

As the situation stands, not only the information, but the islands of bioinformatics tools to analyze the information remain scattered throughout the enterprise. Now the questions of where to go, how to search, and which tools are best suited to analyze the specific data continue to be difficult to answer. This chapter will explore beyond the methods by which traditional high throughput and high content data have been stored and analyzed; it will propose an information architecture to blend the structured with the unstructured knowledge that have fueled the information explosion and knowledge discovery. This new information infrastructure brokers output from various bioinformatics tools—public and proprietary—as inputs to other tools. This new platform will help promote information integration across primary functional groups in industry—from preclinical to marketing and competitive intelligence—as well as academic research enterprises, by means of a unified, single-point research architecture that transcends complex query syntax.

1.2. Current Information Retrieval Challenges

The traditional information mining and retrieval challenges have been:

- Disparate information silos (information residing in multiple physical locations).
- Information overload (too much information being returned to the user).
- Information overlook (missing critical information that has been clouded by information overload).

Researchers in academia and industry may have a wealth of information sources and bioinformatics tools at their disposal. These information sources include public, private/licensed, and internal proprietary content residing in multiple data repositories. For example, a researcher might be interested in all published and proprietary documents related to the interaction between the proteins p53 and mdm2. Relevant information research sources may include PubMed, Science Direct™, Genbank, and an internal Oracle database that stores drug discovery experimental results. The traditional method to search for information such as the pathway components p53 and mdm2 is to access each source in series. The researcher must then know where to go, how to search once they get there, and how to synthesize the results into a manageable result set for further review and study.

1.3. Current Functional Workflow—From High Content Results to New Discoveries and Insights From the Literature

HCS was originally implemented as a smaller vertical within secondary screening in drug discovery and has begun to penetrate primary screening. Given early HCS successes, HCS has

An Integrated Biomedical Knowledge Extraction and Analysis Platform

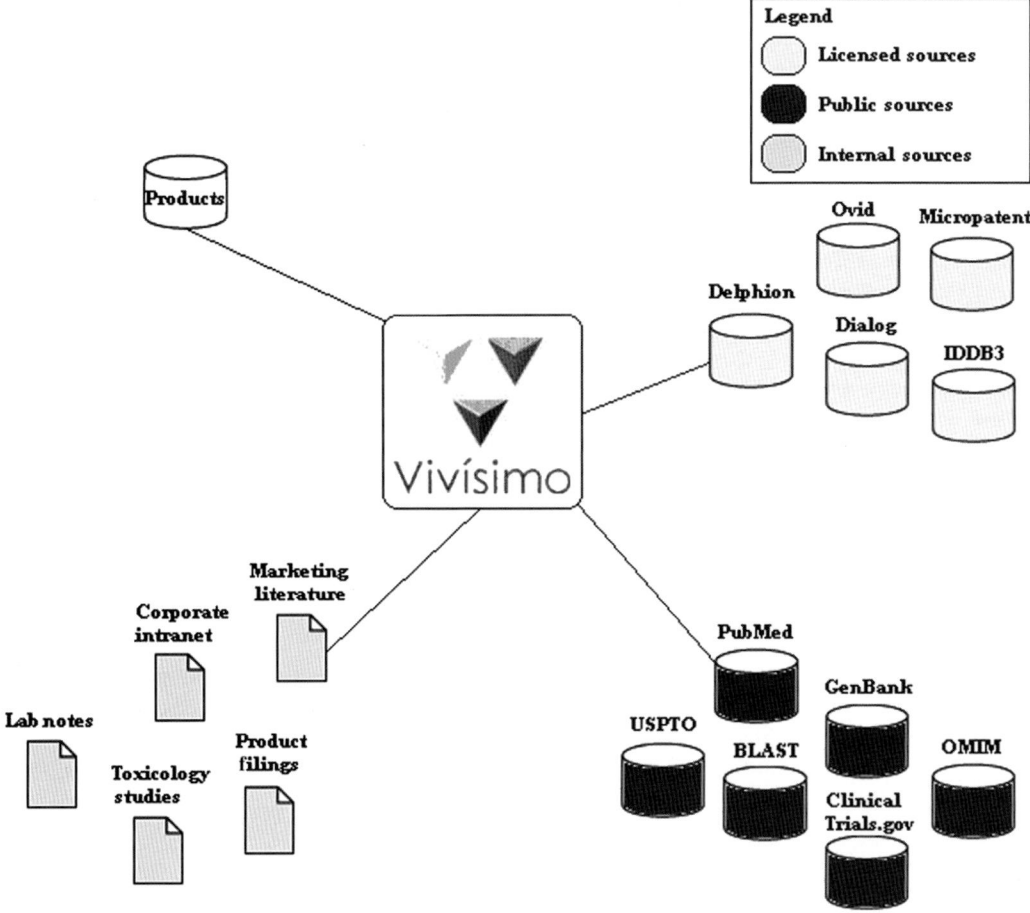

Fig. 1. Biomedical information landscape.

expanded its reach upstream and downstream of the second and primary screening continuum. HCS is now applied to target identification/validation, to lead optimization, and to toxicology *(3)*. Consequently, HCS assay design, the information collected, and the informatics tools to help derive knowledge have a greatly expanded scope. Additionally, interpreted results from an HCS experiment when combined with additional information from the literature and other information sources can lead to the next level of HCS investigations. The domain of scientific literature serves as a potential validator for HCS experimental design and results alike *(3)*. Imagine the bottleneck created when critical pre- or postexperimental knowledge is buried within over 12 million citations hosted by the PubMed search engine—in addition to the internal data stores scattered throughout the enterprise.

2. A New Approach—Integrated Research Portal Bridging Both Data and Information Tools

Vivisimo has developed a webserver-based software application (Velocity for Life Sciences™ [VVLS]) to bridge the information retrieval and information brokering challenges (*see* **Fig. 1**). With this application, researchers are able to simultaneously search across multiple information sources by means of a single query. The Vivisimo application then organizes—spontaneously—search results into descriptive, hierarchical folders allowing for the consolidation of similar documents into logical groupings. Moreover, the Vivisimo software intelligently routes queries

2.1. Content Integration

Vivisimo technology provides a single search interface to access and retrieve documents from any number of internal or external sources. This capability is called metasearch, or federated search; the application accesses host search engines remotely, and the content is maintained by each host source. In this way, federated search enables users to access the most up-to-date documents across multiple information domains simultaneously. Examples of consumer-based, public metasearch engines include Clusty.com and Ixquick.com. Clusty.com federates search results from host search engines such as Yahoo, and Ask Jeeves. Clusty routes each search query to multiple search engines; each host search engine then returns results from their current document index.

The Vivisimo technology includes a federated search component. Unlike public metasearch engines such as Ixquick.com, the VVLS technology may be configured to access any number of electronic sources internal and external to the organization (*see* **Table 1** for sample external and internal sources).

The VVLS technology will allow the end user to choose any number of sources to federate per query. VVLS will access each source's search interface and will return a preconfigured number of results per source. VVLS also includes advanced duplication-prevention algorithms. VVLS presents the search results back to the user in a ranked result set, allowing the user to hyperlink directly to documents of interest. Although federated technology has overcome the challenge of disparate information repositories, hundreds (if not thousands) of search results may be returned per query. Although the host search engines' rankings have been preserved, end users may wish to delve more deeply into the search results. Thus, some mechanism must be implemented to organize the search results for greater knowledge discovery.

2.2. Document Clustering

VVLS includes a document clustering module that dynamically transforms search results into "crisp, hierarchical folders" *(4)*. This clustering is performed on the fly without the use of taxonomy. The VVLS software is preconfigured to use specific outputs from each of the sources as inputs for the Clustering Engine. For example, one clustering method may be to use the title along with the abstract (or snippet) from each document as inputs to the Clustering Engine. The Clustering Engine will then group the documents based on their similarities; from the grouping, VVLS generates meaningful folder headings. The Clustering Engine may be configured to incorporate metadata as inputs. Metadata may also be used as clustering inputs. For example, documents from sources that share metadata (including author and date of publication) may be organized accordingly.

3. Methods—An Example Application
3.1. Intelligent Query Routing

The Vivisimo technology is capable not only of remotely administrating host search engines from information sources, but is also capable of routing queries to bioinformatics and other software tools. For example, consider the following hypothetical scenario wherein a researcher is broadly interested in the interaction between proteins p53 and mdm2 (*see* **Fig. 2**). In this example application, the query will be presented to the Vivisimo technology and will then be brokered to public information sources, iPath (systems biology tools created by GeneGo and hosted by Invitrogen), and Invitrogen products and services relevant to the query. This is only one example on how the Vivisimo tools can be applied to multiple databases and biotools at the same time.

Table 1
Sample Internal and External Information Sources

Government scientific sources
- Biomedical literature (NLM)
- Multidimensional drug screening results (NCI-DTP)
- Genomics databases (NIH-NCBI)
- Protein data bank (NSF-NIH)
- Clinical trials (NLM)
- Regulatory issues (FDA)
- Scientific project information (NIH-CRISP)

Commercial scientific sources
- Genomics (e.g., Gene cards, XenneX)
    - Gene function
    - Role in disease
- Proteomics (e.g., Proteome bioknowledge library, Incyte)
    - Organismal distribution
    - Protein structure
    - Enzyme activity
- Chemiformatics (e.g., iResearch library, ChemNavigator)
    - Compound structure–function
    - Biological activity
    - Business intelligence
    - Biomedical literature

Competitive sources
- News and press releases
- Investor information
- Patent application information

Internal sources
- Cheminformatics
    - Compound structure–activity relationship data
    - Biological activity
- High-throughput and HCS results
    - Compound structure–activity relationship data
    - Target validation
    - Dose–responses
    - Phenotype effects
- Toxicology studies
    - Drug–drug interactions
    - Metabolism
    - Animal toxicity
- Clinical development
    - Patient responses
    - Cumulative reports

Internal source continued
- Regulatory correspondence
    - Product filings
    - FDA responses
    - Warning letters
- Manufacturing
    - Formulations
    - Production specifications
    - Packaging requirements
- Institutional intranet
    - Reports
    - Memos
    - Presentations
    - Marketing materials

1. The researcher begins by presenting the query, "p53 and mdm2" to the Vivisimo software. The demonstration URL is: http://vivisimo.com/metademoHCS. The software will route the query to the aforementioned sources and return information results sets in the form of citations, products, and pathways. Initially, the researcher may be interested in the overall landscape of citations that are related to p53 and mdm2 to understand the scope. The left hand pane will dynamically cluster the returned citations to accommodate this exploration. The researcher may then discover a clustered folder titled, "serine, phosphorylation p53."
2. Each folder may be expended into subfolders by clicking on the "+" sign captioning each cluster. So for example, out of 500 returned results for "p53 and mdm2," the cluster titled, "serine, phosphorylation p53" contains 34 related citations. Further exploration within this folder illuminates the process by which p53 is phosphorylated causing the disassociation from mdm2 and the subsequent transcription of p21. This knowledge is easily derived from scanning the abstracts wherein p53 and mdm2 are indicated in bold font.

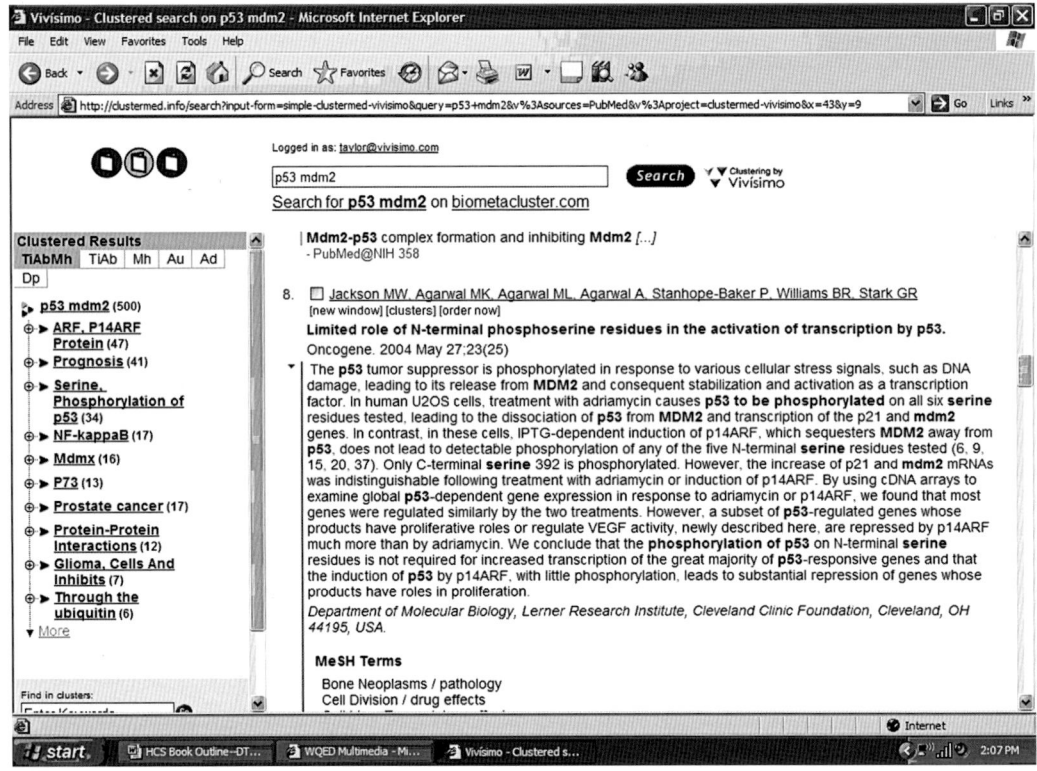

Fig. 2. Velocity for Life Sciences™ screen shot.

3. Now let us assume that p21 is not well understood by the researcher. The Vivisimo software can be requeried to append p21 to the original query (query expansion), or a new query with "p21" as the sole search string may be presented. The latter is a more traditional approach to further exploring "p21" as a topic of interest. However, let us assume that the researcher may wish to implement a systems biology search using p21 as the search string in order to identify the pathways in which p21 is implicated (as well as the relevant literature citations). Without the Vivisimo technology, the researcher must know where and how to access their systems biology platform (e.g., GeneGo, Ingenuity Pathways, and so on).
4. With the Vivisimo technology the researcher simply enters p21 and performs another search. Now the query is routed directly, in this example, to iPath by Invitrogen that is built on the GeneGo platform. Selecting the iPath tab at the top of the Vivisimo search screen will take the research directly into the list of Genes and associated pathways pertaining to p21 (*see* **Fig. 3**). Now the researcher is able to link directly to the following pathway maps:

- ATM/ATR regulation of G1/S checkpoint.
- IFN-γ signaling pathway.
- Role of AP-1 in regulation of cellular metabolism.
- Role of HDAC and calcium/calmodulin-dependent kinase in control of skeletal myogenesis.
- ATM/ATR regulation of G2/M checkpoint.
- Putative integrins pathway. Part 1.
- Putative integrins pathway. Part 2.
- TPO signaling through JAK-STAT pathway.
- DNA damage response in inhibition of CDK during G2.
- TGF-β receptor signaling.
- Cell cycle regulation by Brca1.
- Angiopoietin—Tie2 signaling.
- AKT signaling.
- PTEN pathway.

Visualization of HCS Data

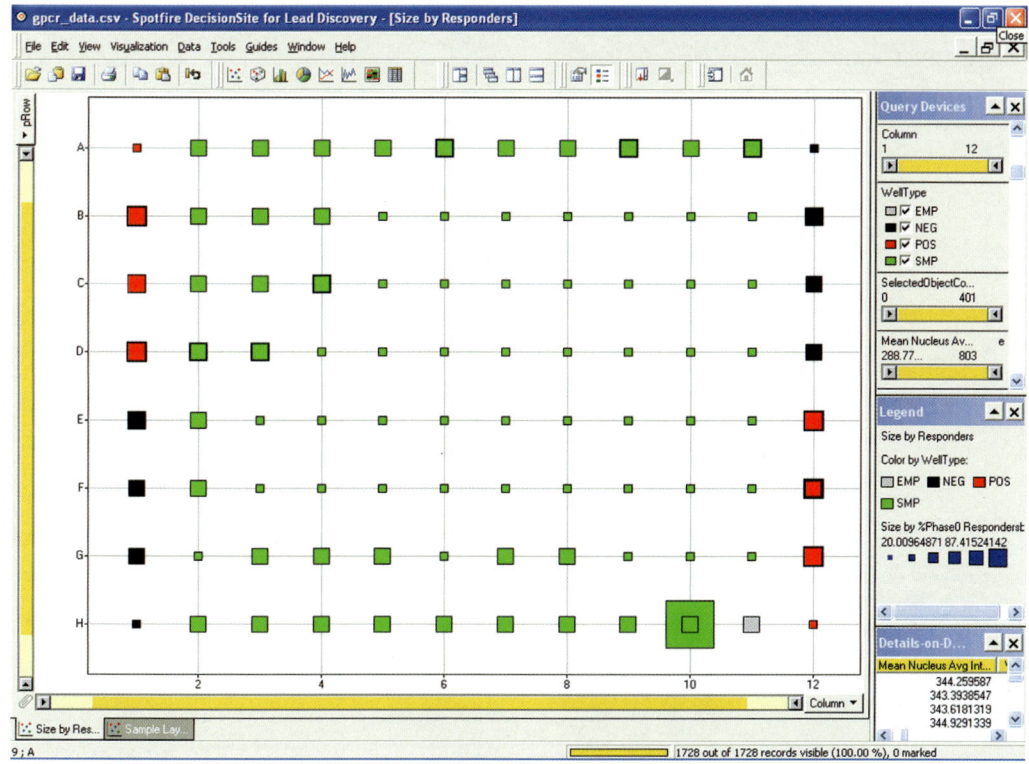

Fig. 2. Spatial view of bias: in this plot color is used as in **Fig. 1**, but the added dimension of size reflects the experimental result of "%Phase0 Responders." Here one notes that the controls and sample wells that surround the plate are larger in size, reflecting higher values for these wells and thus, potential edge effect bias in control wells.

the experiment represented as pixel intensity. Algorithms have been developed that use intensity information to generate quantitative measurements or derived data representing, for example, nuclear volume for every cell in a single test well within a 96-well plate. The primary or raw data is often stored as large image files. Derived data can be described as the data that is refined and presented at a higher level in which it is aggregated, characterized statistically, and interpreted to drive decision making *(3)*. Images and derived data might be stored in files or database schemas. As HCS technology advances this data is approaching the Terabyte size range and beyond thus, effective storage and access becomes essential. For example, access to the derived data combined with the ability to reference back to original images can be useful in confirming a particular finding and identifying problems with images that might not be evident from the derived data. Metadata is also captured which contains summary information, plate name, assay protocol, information about controls, and experimental design. We will describe how visualizations of primary data, derived data and metadata can be constructed to reveal systematic bias and improve the identification and understanding of features that drive biological events.

2.2. Transformation

Data transformation might be necessary in order to visualize a particular trend, normalize to a control, for example, by subtracting background noise, divide by a baseline value to view fold change, remove systematic bias, or summarize multiple measurements. A view of the metadata as a plate layout can be useful before performing normalization. Selected controls might be

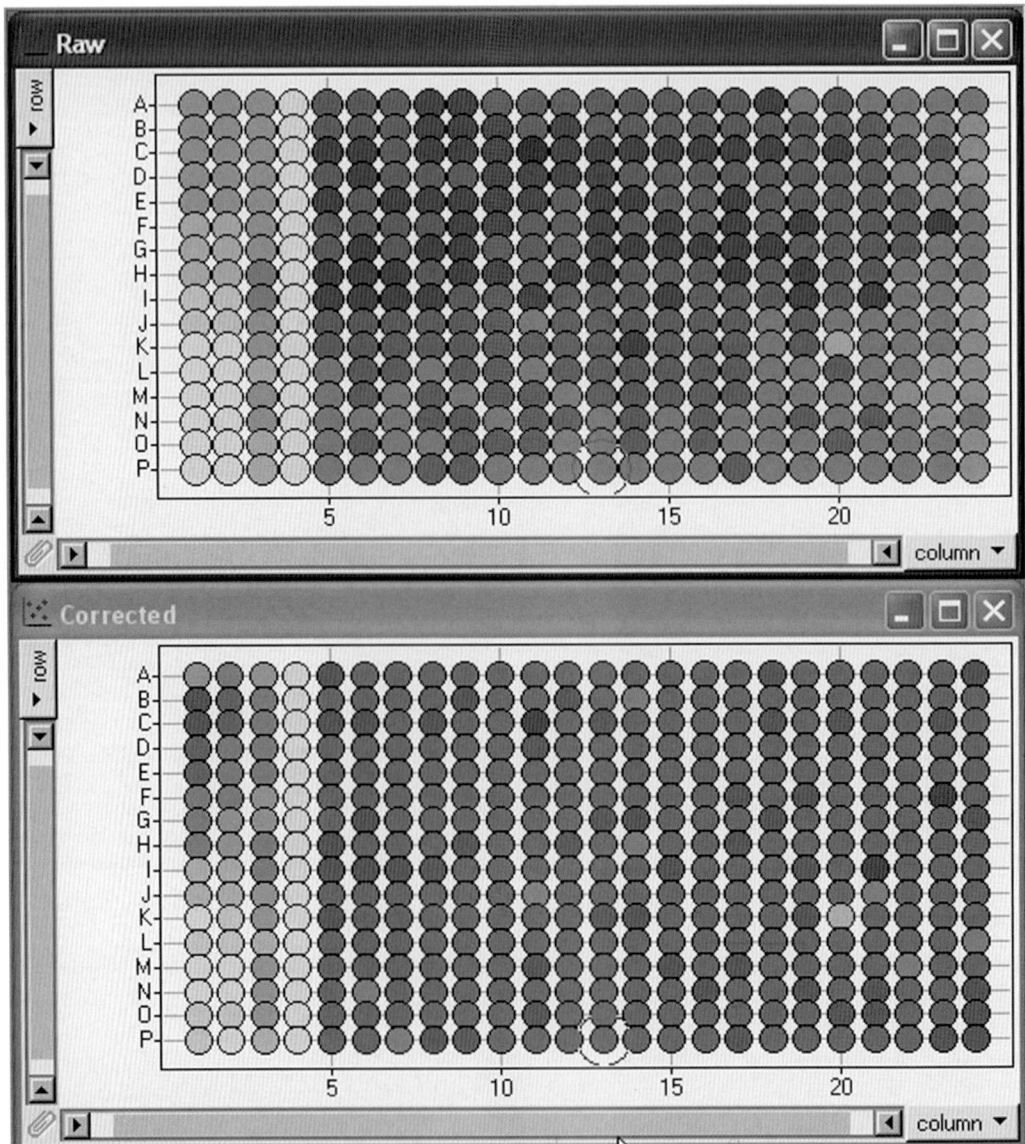

Fig. 3. Raw and corrected data: a plate view, shaded by raw and corrected experimental results removes bias that masks true differences in wells which are more clearly visible in the corrected view. (Please *see* the companion CD for the color version of this figure.)

grouped and used to normalize the data. Coloring by sample type in a virtual plate view is often the first visualization that is used in beginning an analysis (**Fig. 1**).

Visualization of the results can be added to the virtual plate map by sizing markers by an experimental result while retaining color, which denotes sample type in each well. In doing so, one might notice systematic bias, such as edge effect that might occur when stacking plates in a cell culture incubator. The response of the cell culture is reflected in a measured result, which is recognized visually as a bias to large values at the plate edge (**Fig. 2**).

Methods for reducing systematic bias in such experiments have recently been developed *(5,6)*. Applying these data correction methods can reduce bias that masks true results. For example, systematic bias that yields higher values in a pattern across a plate can mask the real differences

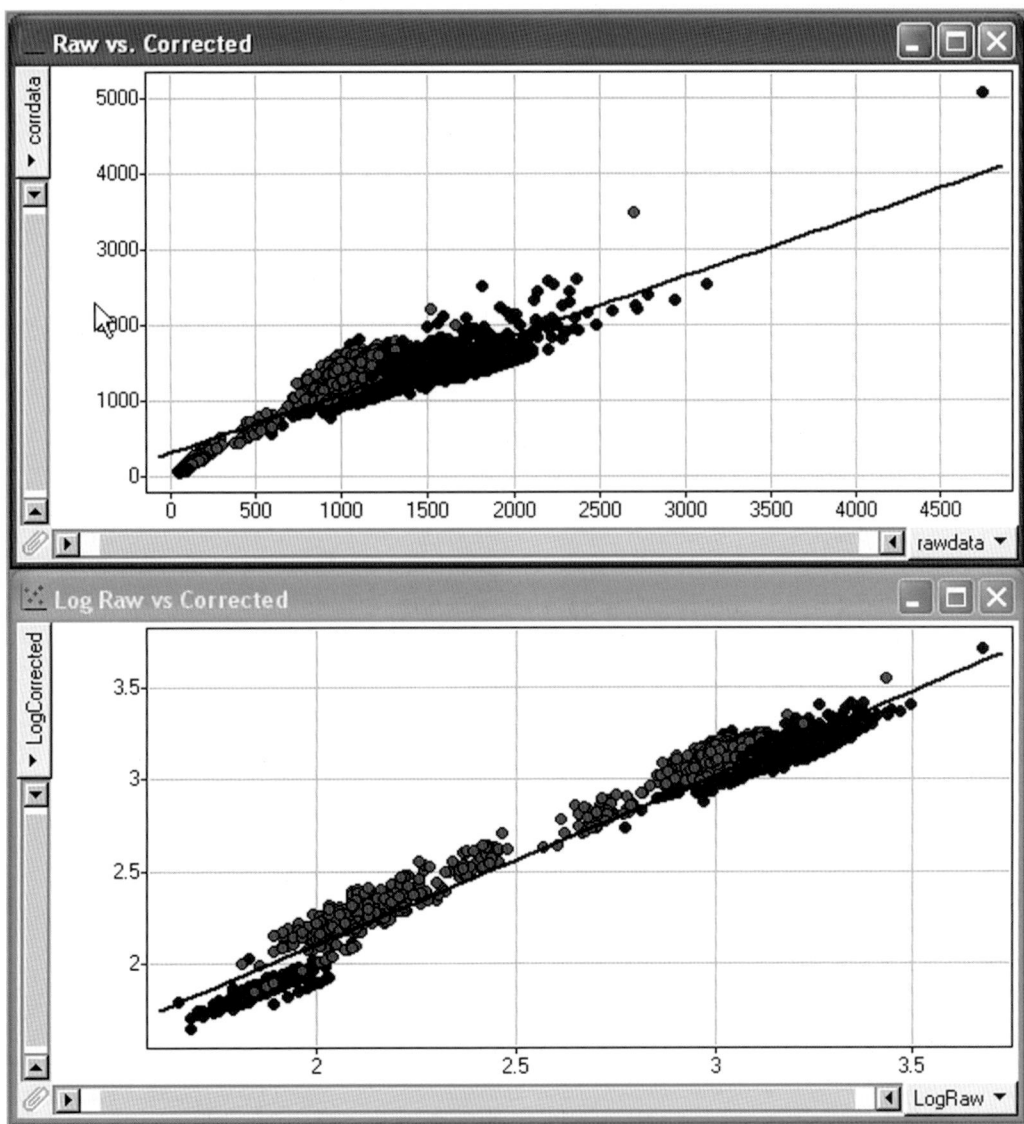

Fig. 4. Log transformation: scatter plots show the correlation of raw and corrected data on the x- and y-axis, respectively, of both original and log transformed data (upper and lower plots). Logarithmic transformation provides a more detailed visualization of data clusters within the lower value ranges. (Please *see* the companion CD for the color version of this figure.)

between wells, thus hiding important information. Visualization of raw and corrected data together, confirms the effect of employing such methods to cleanse data before analysis (**Fig. 3**).

Log transformation is a transformation method used before visualizing data that exists across a wide range of values. Large outliers can skew the visualization scale creating a condensed cloud of data points at lower ranges. By applying log transformation, the data is spread more evenly across the visualization, providing more information about data within the lower range (**Fig. 4**).

Another data transformation method often used normalizes data based on controls or experimental condition such as time. For example, based on experimental design, samples might be normalized to a zero time-point to identify significant changes over the experimental conditions. This can be done by taking the signed log ratio of each sample to a specific baseline value-typically

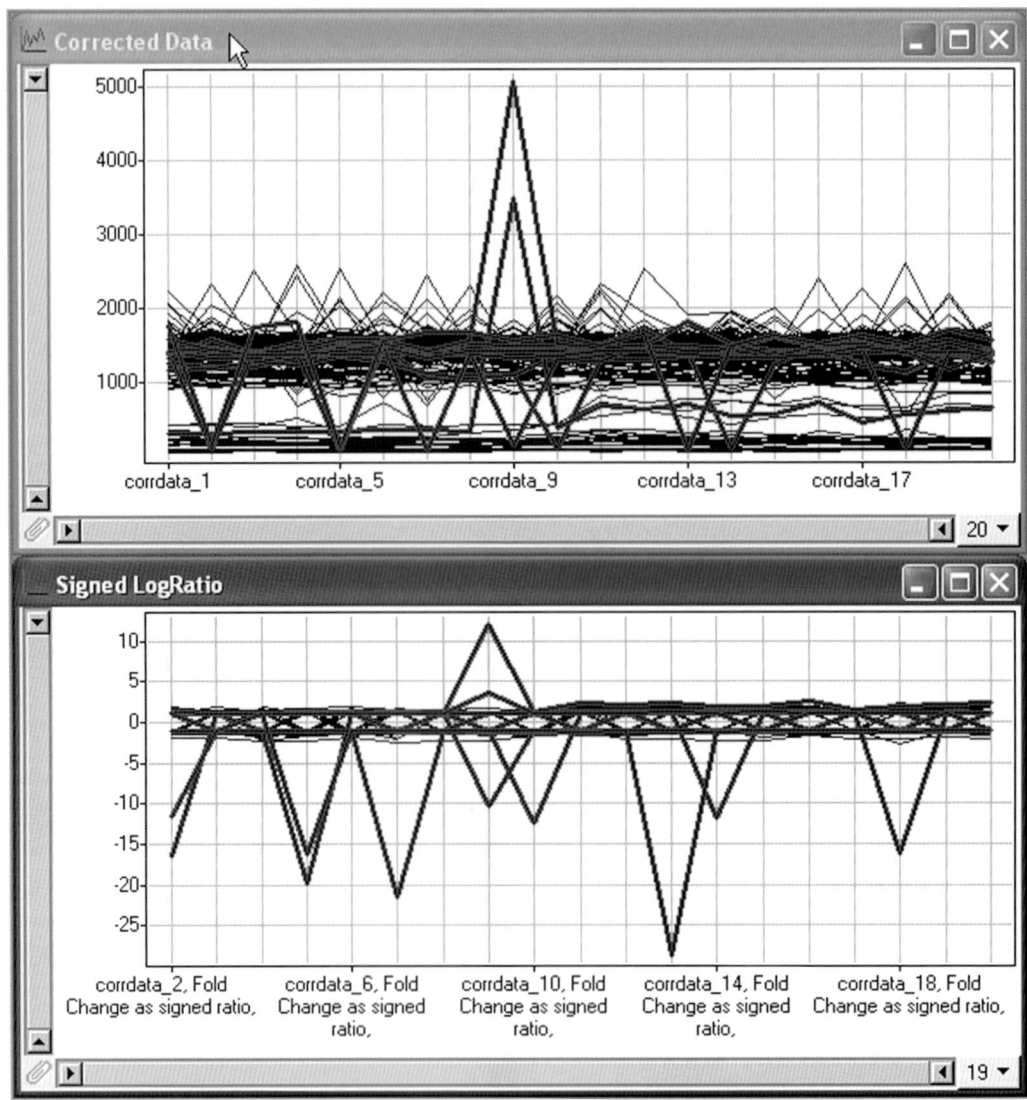

Fig. 5. Fold change: samples might be compared to a baseline value, for example time zero before treatment. By taking the signed log ratio to a baseline, shown in the lower profile plot, the largest changes in a positive or negative direction from baseline across all experiments can be easily identified. Corrected data before fold change calculation is shown in upper plot. (Please *see* the companion CD for the color version of this figure.)

time zero, before addition of drug or start of experimental condition. Visualization of data before and after this normalization allows one to isolate those samples that are most changed from baseline (**Fig. 5**). Normalization might also involve rigorous statistical methods that fit the data most appropriately based on assumptions about the type of data or instrument bias. These and additional quality control methods can provide an assessment of the assay quality, improvement in hit selection and overall data quality *(7–9)*.

Scaling data is another method that affects visualization. For example, z-score normalization sets the mean of the data to zero and presents data in standard deviation units. This is another useful way to see relationships between samples that can be masked by large differences in scale (**Fig. 6**).

Visualization of HCS Data

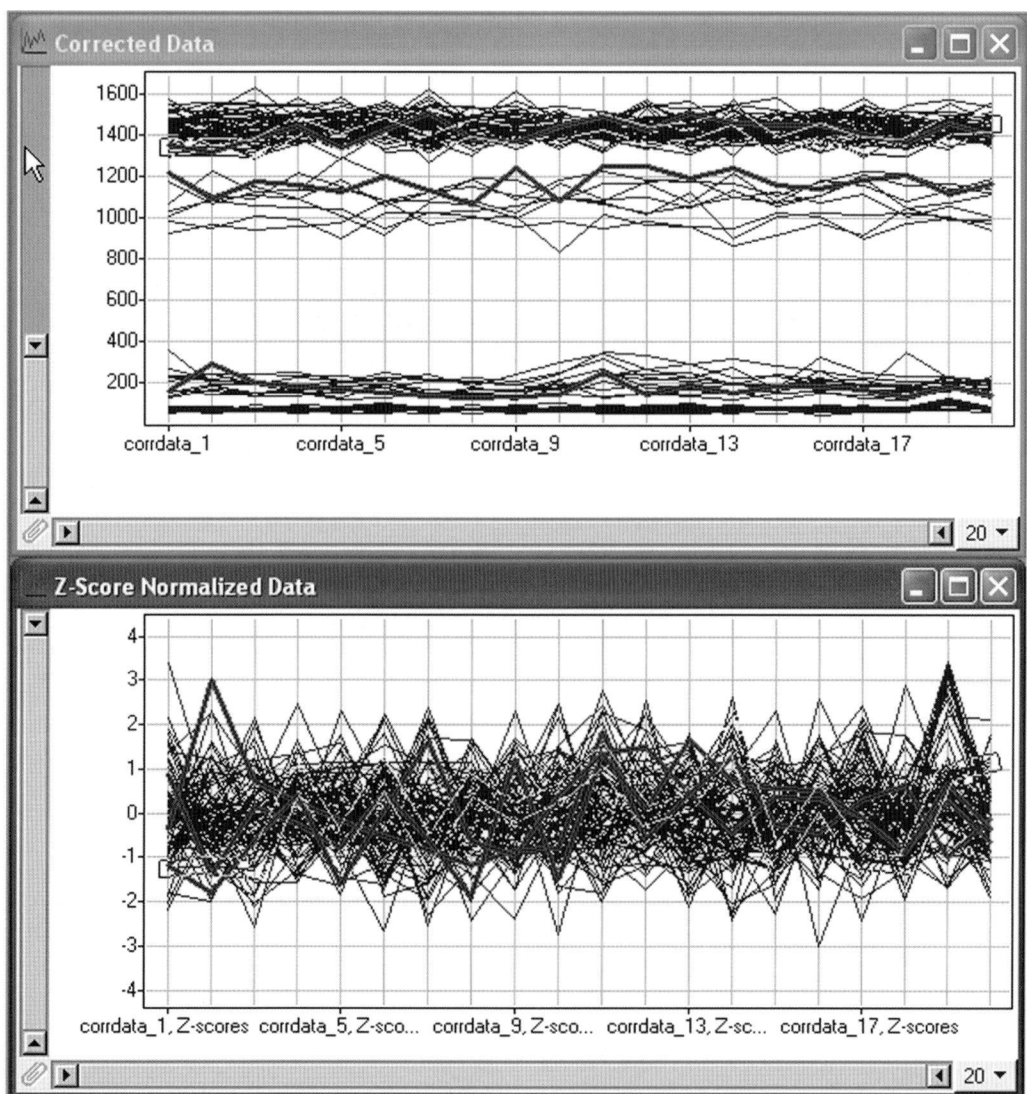

Fig. 6. z-Score normalization: data is normalized by setting the mean to zero and displaying the data in standard deviation units. In these profile plots the visual comparison of patterns is facilitated by normalization even though scales differ widely as shown in the upper, nonnormalized plot. (Please *see* the companion CD for the color version of this figure.)

Data transformations can be used for a variety of purposes to display and analyze data more effectively. Some of these methods, which are briefly described here, help to remove bias, and provide a better representation of values over a broad range within the same plot. They might also be used to display the response of cells within an experiment in relation to controls or baseline values, and transform data so that similar patterns can be more easily identified regardless of scale.

2.3. Combining Data

Data from different screening runs might be combined to view trends over time. Incorporating new information about each run is useful in detecting problems with a particular run. For example, screens run on different days might have bias introduced by different instruments, a miscalibrated pipet, varying lot numbers of reagent, temperature variations, or even culture conditions

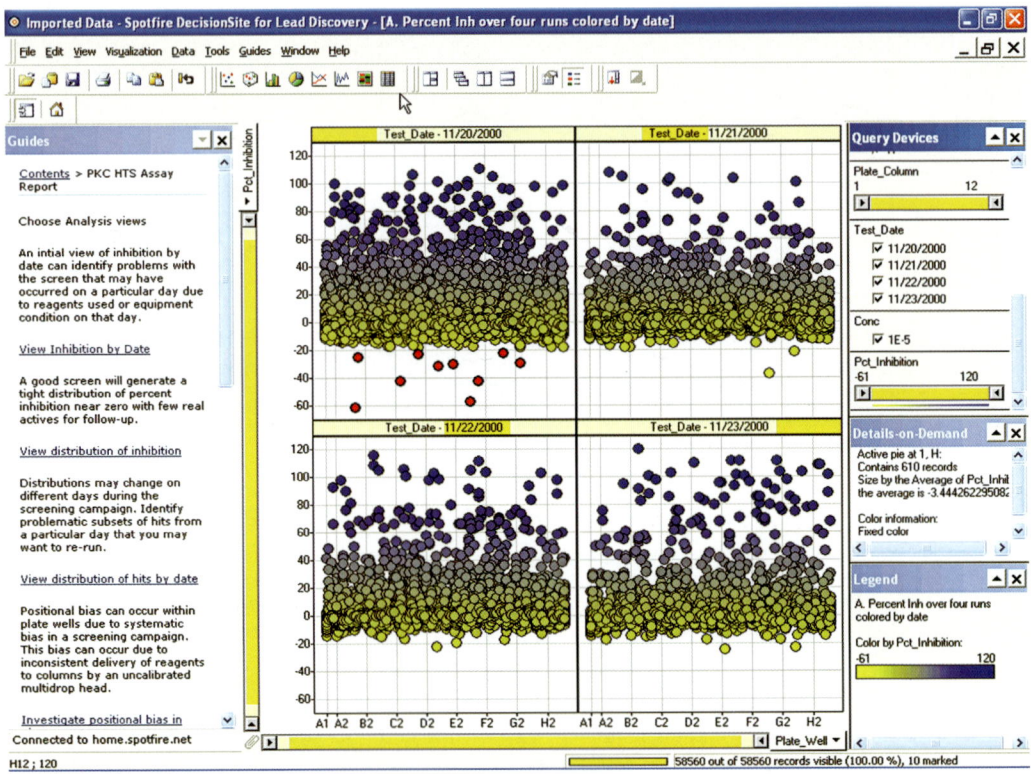

Fig. 7. Screening runs by date: scatterplots of experimental results are displayed in a trellis view by the date the experiment was run. This can identify outliers that might be specific to the instruments or reagents used on that date. The red markers show a number of negative results from experiments run on a particular date that might have resulted from factors or conditions on that day.

(**Figs. 7** and **8**). By including this experimental information researchers can be aware of any extraneous factors that might affect the results of their experiments.

Derived HCS data can be displayed and analyzed within visualizations that combine spatial information, such as well position, with data from specific cells within that well. It might also be valuable to incorporate original image information from which the quantitative data was derived. This allows one to spot specific problems with the image or better understand the relationships within the wells from which derived data is acquired (**Fig. 9**).

Associated data might also include data from HTS, proteomics, genomics, and the chemical structures of the compounds tested within each well. In large-scale experiments, many compounds might be tested and the responses identified within visualizations can be correlated to structural motifs that influence drug design (**Fig. 10**).

Combining derived HCS data with metadata such as dates of experiment or lot number of reagents can help to identify the causes of bias in results because of extraneous variables. Merging original images can also help identify bias and provide additional information such as cell shape that might be lost in quantitation. Incorporating other data such as chemical structures in the analysis of HCS data, assists in identifying structural motifs that might be responsible for cellular response and can drive the direction of synthesis of the most promising drugs.

These examples demonstrate the power of visualization in the analysis of HCS data and related information in order to better understand the complex interactions within biological systems.

Visualization of HCS Data

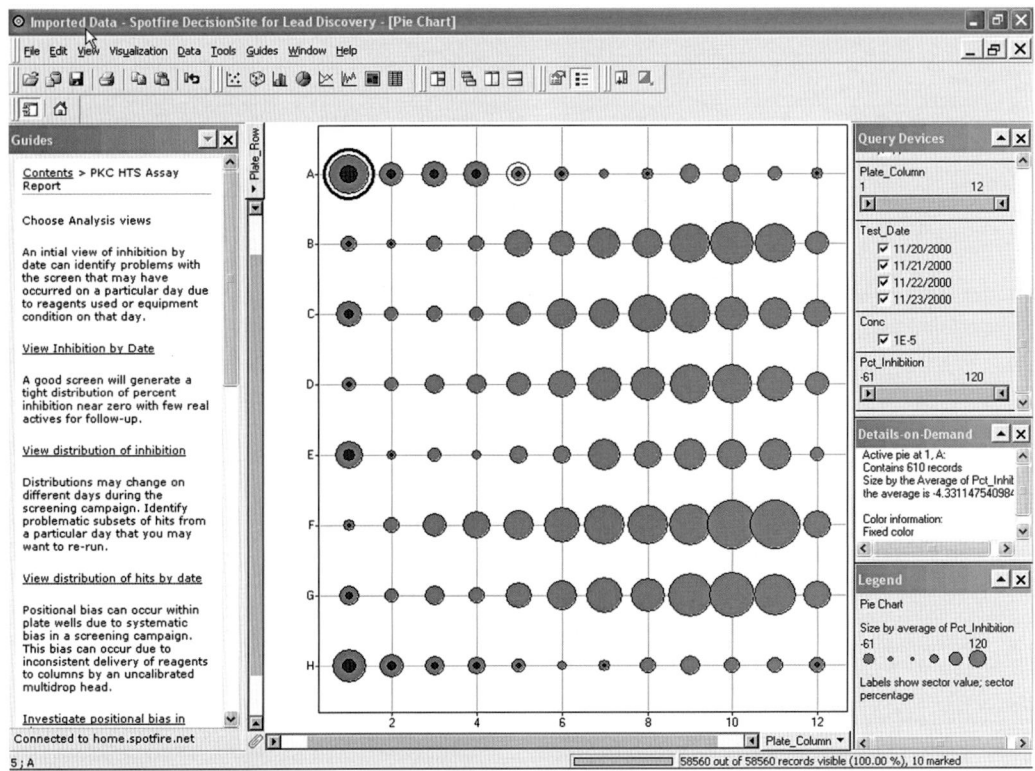

Fig. 8. Investigating positional bias: pie charts with positional information are used to display and summarize all 96-well plates from a screening run. Pie size reflects the average results for each well across all plates and larger pies on the right of the plate indicate potential pipetting bias for instruments moving from left to right in the plate. (Please *see* the companion CD for the color version of this figure.)

3. Visualization

3.1. Spatial and Temporal Views

Data gathered from HCS often includes information on cell density and morphology, for example, cellular volume or nuclear volume. Spatial relationships within a cell culture might reflect induction of certain pathways by an agonist or repression by antagonists.

Temporal views are often presented as profile charts. This type of view describes relationships between time-points for compounds or cultures. Profiles might represent compounds over time or cell culture over time in the presence of compounds. Together with clustering techniques such as k-means clustering, one can isolate the overall pattern of cellular response as a group of similarly shaped profiles (**Fig. 11**). Temporal views provide dynamic information that can reflect cyclic responses or complex upstream and downstream induction events. These views can also include spatial relationships such as cellular migration over time (**Fig. 12**). This visualization of cell tracking data provides information on both displacement and velocity of motile cells *(2)*.

3.2. Well- and Plate-Level Information

HCS captures many values for each cell within individual wells. This well level data is displayed and evaluated along with well summary information yielding both the cell level and summary values for each well (**Fig. 13**). With these types of views one can inspect the background intensity across all wells in the bar chart, evaluate correlation between nuclear elongation and cell area, and

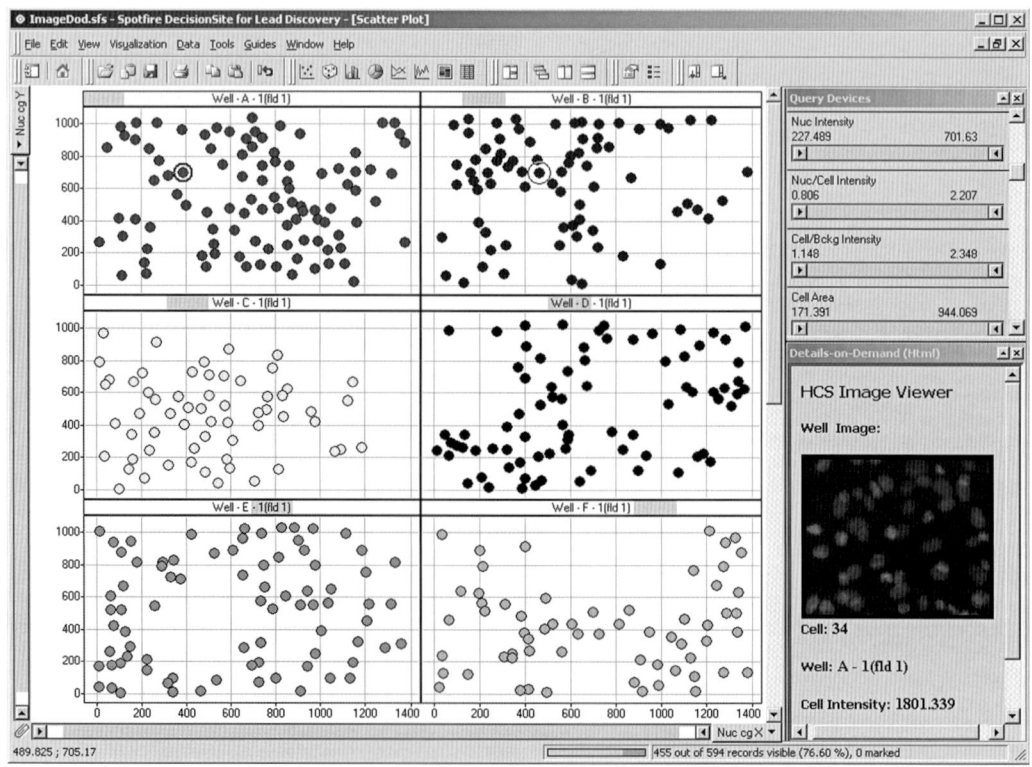

Fig. 9. Combine primary image data: scatterplots can be used to display the *x* and *y* coordinates of data captured from an image. In this visualization, the scatterplots are trellised by well. By clicking on markers representing cellular measurements within a well, the original well image is retrieved and can be reviewed along with the derived data. (Please *see* the companion CD for the color version of this figure.)

isolate outliers that can affect the summary information. Using these techniques, cell populations can be identified and explored within and across wells both at the cell and well level.

HCS data can also be viewed at the plate level to look for trends across a series of test samples or concentrations. This type of view also gives an indication of overall performance, and in which plate biases occur resulting from, for example, edge effects caused by stacking of plates in a cell culture incubator (**Fig. 2**).

3.3. Visualizing Related Data

Bringing together the large number of variables measured in HCS can be difficult. Visualization of more than three to four dimensions within a single plot pushes the limits of perception. By plotting key variables in multiple linked views it is possible to select a cluster of data points in one view and identify where these same data points lie in a second visualization of additional variables. Analysis can then proceed stepwise from one view to another. For example, viewing the results of principle component analysis (PCA) along with well level and correlation plots can help to assess a cluster of cells having a similar response that group together in PCA space. From a selected cluster in this view, one might then explore any spatial relationship within the wells in a second view and lastly, how these cells correlate in terms of nuclear and cytoplasmic intensities in a scatterplot (**Fig. 14**). Using color and size to add phase classification and additional information like spot average intensity, adds information to drive the selection process. Thus, complex interactions can be explored using statistical methods such as PCA within multiple visualizations in order to identify features that drive biological responses.

Visualization of HCS Data

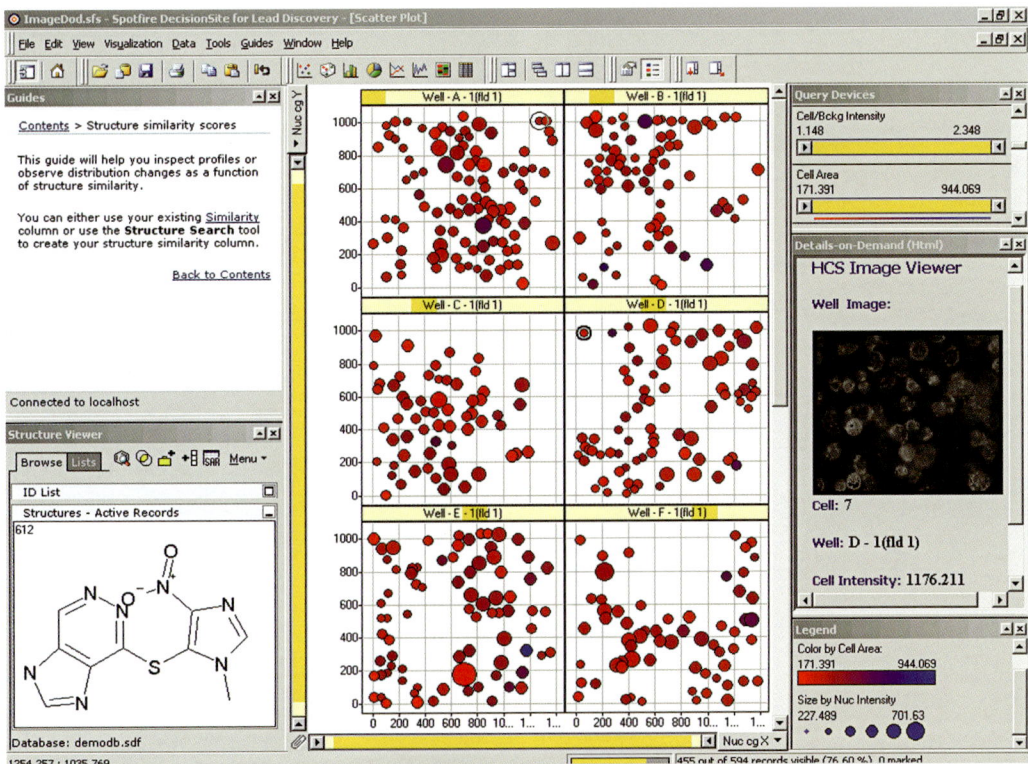

Fig. 10. Adding chemical structure information: well images and chemical structures are retrieved from chemistry and image databases and displayed simultaneously when clicking on data points within scatterplot visualizations. Scatterplots are split by well number. Markers in the scatterplots are colored by cell area, sized by nuclear intensity, and, as expected, reflect an inverse relationship.

3.4. Interactive Visualization

Although exploring the large number of measurements in an HCS data it is often valuable to remove subsets of data that might obscure patterns and trends. This type of interactive analysis requires fast and intuitive methods that facilitate understanding of the data. By making selections within linked views, patterns can emerge that are obscured by the density of data. By selecting a grouping of cellular measurements within the PCA results that cluster together and are from the same phase classification (blue color) of **Fig. 14**, one can isolate three cellular measurements from the entire dataset and observe their spatial and morphological relationships in linked views (**Fig. 15**). The ability to rapidly filter information using visual filtering devices facilitates inquiry and analysis of complex relationships found in HCS data (**Fig. 16**).

4. Conclusion

The analysis and interpretation of HCS data poses new and sometimes difficult issues for researchers. Although this technology provides rich biological response data, it is often difficult to access and visualize in an optimal manner to address the questions that it is designed to answer. A key aspect to understanding the information is to design the way in which metadata, primary data, and derived data should be visually presented in order to answer specific questions. Too often, data integration efforts place too much emphasis on data access and transformation, losing sight of the hypothesis. Beginning with the hypothesis one can construct the appropriate questions, define the visualizations and analyses that answer these questions, and lastly, define the data

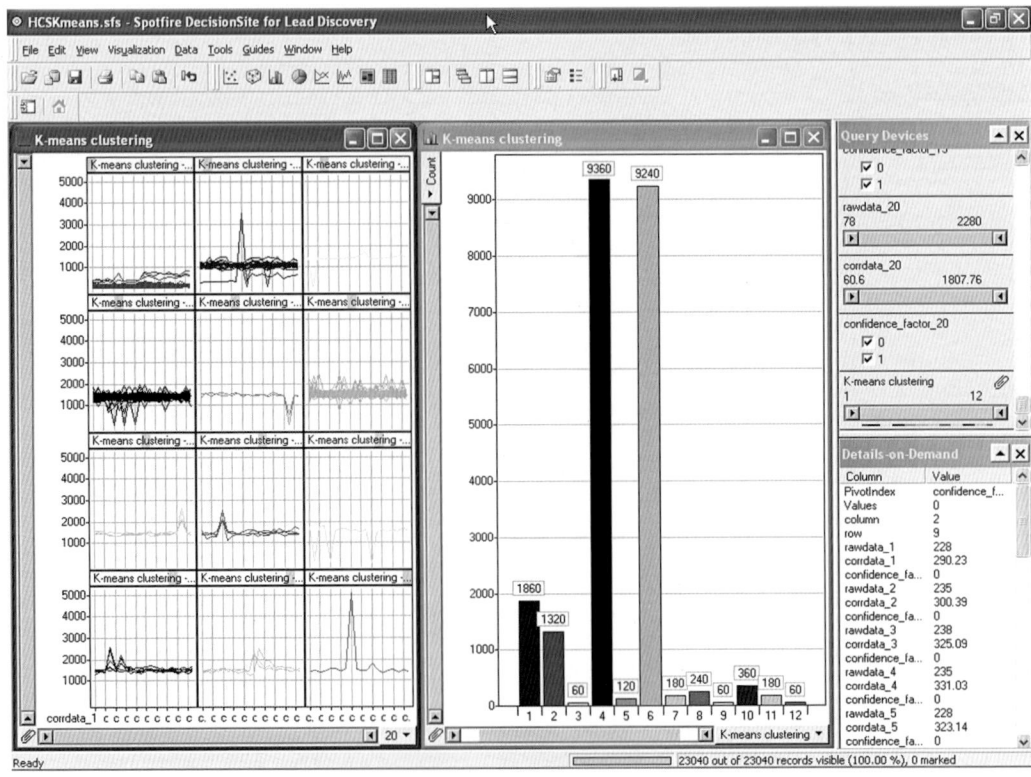

Fig. 11. Cellular response: a temporal view of a cellular response or compound performance over time is captured in profile charts which are trellised by similar pattern. A linked histogram displays the distribution of members in each cluster showing that most fall into two clusters. (Please *see* the companion CD for the color version of this figure.)

access and transformation steps needed to create the appropriate views and analyses. Additional hypotheses can also be explored using an interactive visual environment that immediately responds to repeated inquiries necessary when working with such highly dimensional information.

Acknowledgments

The author would like to thank Cellomics, Inc., GE Healthcare, and Cybio AG for providing data used to create some of the visualizations in this chapter.

Note

1. For more information on interactive visualization and data analytics contact www.spotfire.com.

References

1. Giuliano, K. A., Haskins, J. R., and Taylor, D. L. (2003) Advances in high content screening for drug discovery. *Assay Drug Dev. Technol.* **1,** 565–577.
2. Abraham, V. C., Taylor, D. L., and Haskins, J. R. (2004) High content screening applied to large scale cell biology. *Trends Biotechnol.* **22,** 15–22.
3. Searls, D. B. (2005) Data integration: challenges for drug discovery. *Nat. Rev. Drug Discov.* **4,** 45–49.
4. Gosh, G. N. and Haskins, J. R. (2004) A flexible large-scale biology software module for automated quantitative analysis of cell morphology. *Business Briefing Future Drug Discov.* 1–4.
5. Kevorkov, D., Makarenkov, V., and Zentilli, P. (2005) New methods for statistical analysis and data correction in HTS. *Cheminformatics, Library Design and Virtual HTS: Poster Session, SBS 11th*

Visualization of HCS Data

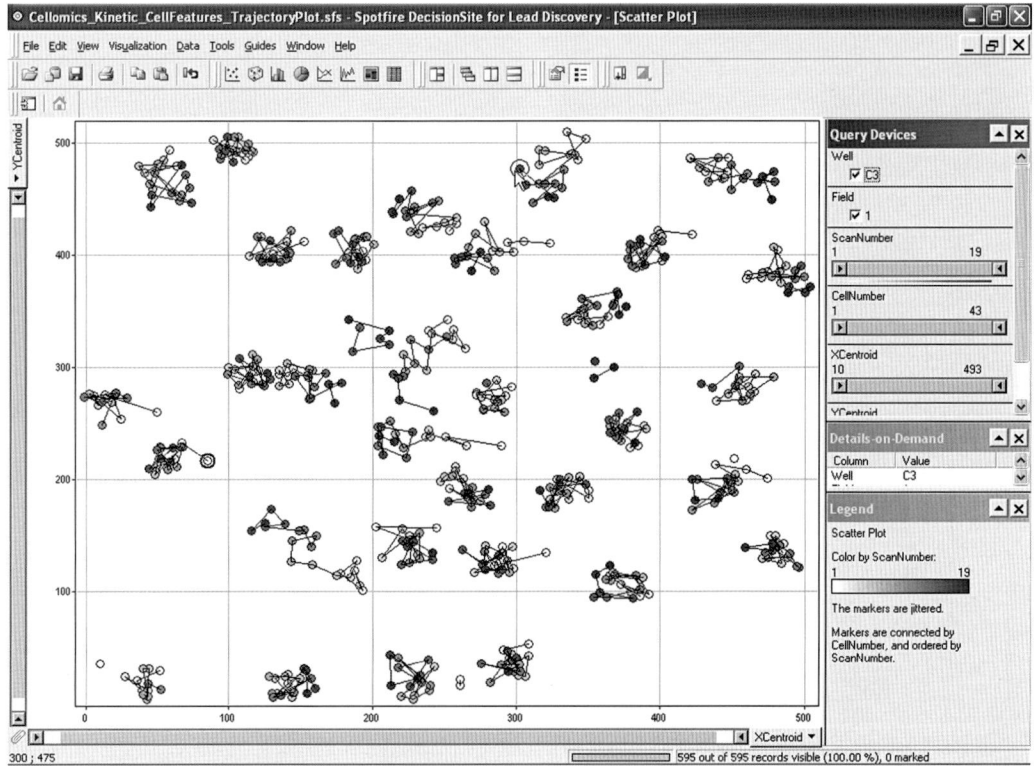

Fig. 12. Cellular migration: spatial coordinates for cells in culture are captured over time and connected by lines with arrows to indicate the migration of the cells in culture. (Please *see* the companion CD for the color version of this figure.)

> *Annual Conference and Exhibition Drug Discovery: From Targets to Candidates*, 11–15 September 2005 Geneva, Switzerland.
> 6. Heuer, C., Haenel, T., and Prause, B. (2002) A novel approach for quality control and correction of HTS data based on artificial intelligence. *The Pharmaceutical Discovery and Development Report, 2003/03, PharmaVentures Ltd.*, [Online] Retrieved from http://www.worldpharmaweb.com/pdd/new/overview5.pdf. Last accessed March 1, 2006.
> 7. Zhang, J. H., Chung, T. D. Y., and Oldenburg, K. R. (1999) A simple statistic parameter for use in evaluation and validation of high-throughput screening assays, *J. Biomol. Screen.* **4,** 67–73.
> 8. Brideau, C., Gunter, B., Pikounis, B., and Liaw, A. (2003) Improved statistical methods for hit selection in high-throughput screening. *J. Biomol. Screen.* **8(6),** 634–647.
> 9. Gunter, B., Brideau, C., Pikounis, B., and Liaw, A. (2003) Statistical and graphical methods for quality control determination of high-throughput screening data. *J. Biomol. Screen.* **8(6),** 624–633.

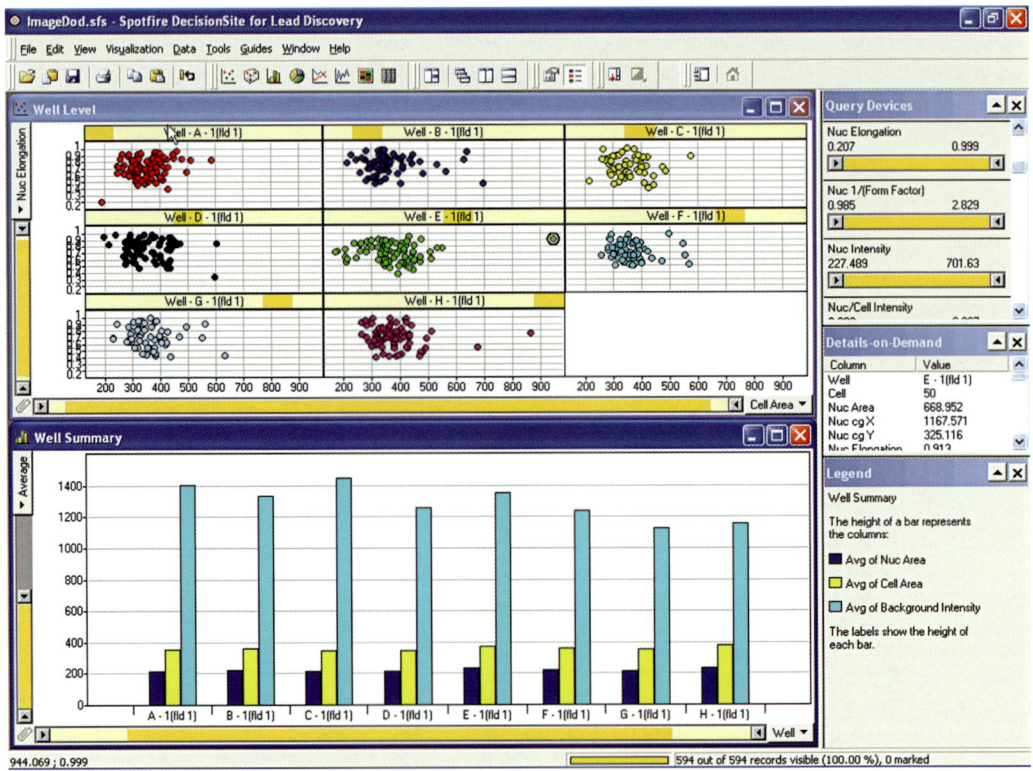

Fig. 13. Well and plate level data: each trellised pane in the upper scatterplot represents a single well with the cells' nuclear elongation on the *y*-axis and total cell area on the *x*-axis. The bar charts below summarize information from each well as average nuclear area, cell area and background intensity.

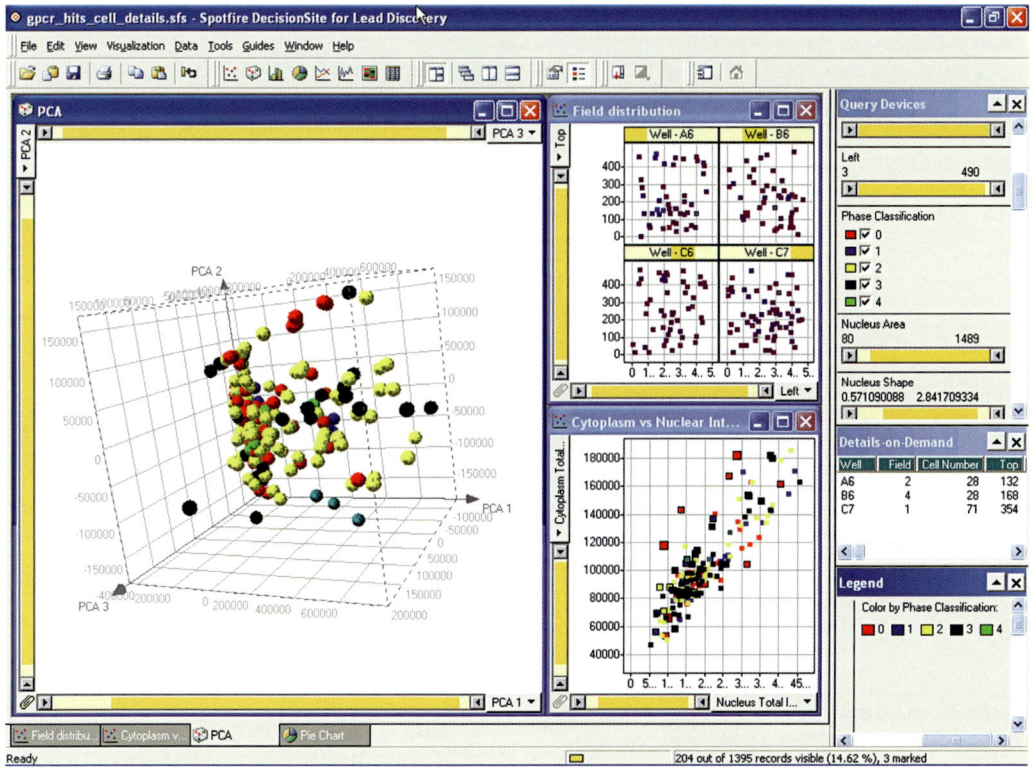

Fig. 14. Linked visualizations: results of principle component analysis groups cells in a new three-dimensional (3D) space designed to capture and summarize the variability of a large number of experimental measurements. Selecting a cluster of these cells in the 3D plot automatically selects the corresponding points in all visualizations. Thus, their location within wells and correlation between cytoplasmic and nuclear intensity can be assessed together in a stepwise fashion from one view to another.

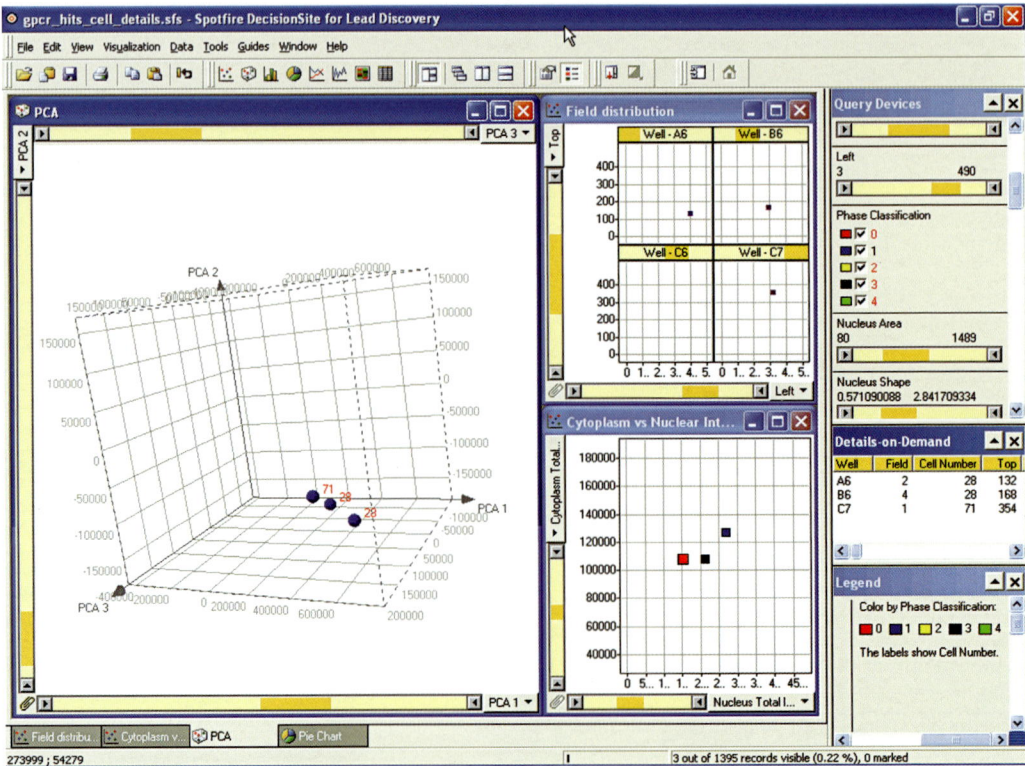

Fig. 15. Interactive visualization: by selecting a subset of cellular measurements clustering in 3D space, the spatial and morphological relationships can be assessed in well level and cytoplasmic vs nuclear intensity correlation plots.

Pathway Mapping Tools for Analysis of High Content Data

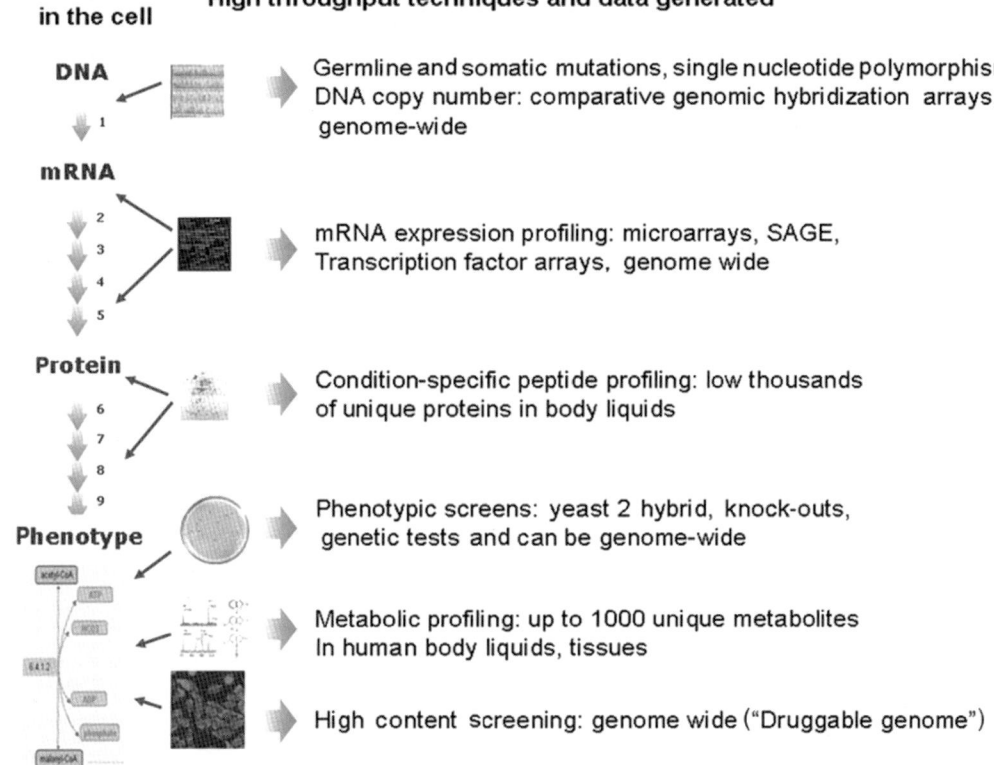

Fig. 1. This is the general schema for network analysis from different data types. Various high-throughput or high content data can be linked to the tables of human protein interactions. Nine levels of regulation of protein activity in a human cell can be summarized: (1) gene transcription, (2) mRNA processing and editing, (3) mRNA transport from nucleus, (4) mRNA stabilization, (5) protein translation, (6) protein transport, (7) folding and protein stabilization, (8) allosteric modulation, and (9) covalent modification. The types of data generated corresponding to these levels are also shown. (Please *see* the companion CD for the color version of this figure.)

include EMP/MPW, BRENDA, ERGO, and KEGG, which have been described previously *(19)*. KEGG is probably the database most frequently referred to for metabolic pathways in multiple species. Unlike bacteria with more than 80% of genome-content encoding core metabolic genes, the majority of eukaryotic biology is in signaling: membrane receptors, signal transduction enzymes, transcription factors, and modulators. With the progress of eukaryotic biology in the last decade, signaling pathways have also been collected in multiple databases. Some of examples of these signaling databases are listed later.

Biocarta is a commercial collection of about 350 maps on human biology representing canonical, mostly signaling pathways. Biocarta does not support mapping of experimental data on the pathways.

Gene MicroArray Pathway Profiler is a database of gene ontology (GO) derived diagrams designed for viewing and analyzing gene lists from experimental data *(5)*. A free tool, Gene MicroArray Pathway Profiler became popular among academics as the first pass functional analysis of microarray expression data.

Protein Lounge is a commercial package with about 600 mixed-species metabolic and signaling pathway maps. These pathways can only be browsed, as there are no tools for mapping experimental data.

MetaCore is a commercial package that contains more than 400 human signaling and metabolic pathway maps available for mapping gene expression, proteomics, metabolic, and HCS data. The data generated can be exported from individual maps and clusters of maps and analyzed further with networks. MetaCore will be described in considerably more detail later in this chapter (**Subheading 3.**).

iPath (Invitrogen; http://escience.invitrogen.com/ipath/) 225 detailed, interactive maps of interconnected biological signaling and metabolic pathways that are freely available. These maps have been created for Invitrogen by GeneGo and are linked to reagents commercially available from Invitrogen.

Databases of high-confidence human protein–protein interactions and associated tools for network generation, analysis, and mapping of experimental high-throughput data have also been developed *(19,20)*. There are multiple methods to elucidate protein–protein interactions. One approach is to screen the experimental literature, using text-mining algorithms, for cooccurrence and, therefore, associations between gene/protein symbols and names in the same text. Typically, natural language processing (NLP) and other text-mining algorithms are used for this automated mining of abstracts and titles from PubMed articles *(21–23)*. It is important to note that comparative studies have shown that up to one half of NLP associations do not correspond to experimentally verified protein interactions *(24)*, although more than 60% of interactions can be elucidated by automated text mining *(25)*. The reliability of NLP-derived associations can be enhanced further by the compilation of field-specific synonym dictionaries, using longer word strings for searching and full-text articles to query against, and statistical methods *(26)*. In a recent study, the NLP engine MedScan was used to extract hundreds of thousands of functional relations including 20,000 protein interaction facts between human proteins from full text articles with a precision of 91% for 361 randomly extracted protein interactions *(27)*.

The protein–protein interactions can also be derived from high-throughput experimentation. For example, the yeast 2-hybrid (Y2H) screen test identifies protein interactions in yeast cells *(28)*. A widely used wet lab technique, Y2H was scaled-up for global mapping of protein interactions in yeast *(29)*, fly *Drosophila melanogaster (30)*, and worm *Caenorhabditis elegans (31)*; however, Y2H-derived interactions are known for a high (more than 50%) level of false positives and false negative interactions *(32)*. In one study, over 70,000 interactions for 6231 human proteins were predicted assuming the interactions between these proteins' orthologs in yeast, worm, and fly *(33)*. However, the accuracy of predicted interactions remains questionable. Although it was assessed computationally based on the relative correspondence of interacting protein pairs to the GO processes, the interactions were not directly compared with any high-confidence experimental data set available *(5,34)*. The interactions can also be deduced from condition-specific cooccurrence of gene expression based on the assumption that interacting proteins must be expressed in parallel *(35)*, especially when encoded by the homologous genes *(34)*. However, the overall confidence in coexpression-derived interactions in yeast is about 50% (47% anticorrelation for novel interactions) *(37,38)*. Another method, coimmunoprecipitation (CoIP) consists of the affinity precipitation of protein complexes in mild conditions using antibodies to one of the complex's subunits, followed by mass-spectrometry or Western blot analysis. CoIP enables direct and quantitative detection of interactions between active proteins, and so it is a true proteomics method. CoIP was used in simultaneously published studies of the yeast interactome *(39,40)*. The other, less commonly used experimental and computational methods include protein arrays, fusion proteins, neighboring genes in operons (for prokaryotic proteins), paralogous verification method, colocalization, synthetic lethality screens, and phage display; each method with its own particular pros and cons *(41,42)*. The overall confidence in protein interactions, defined as the intersection between interacting pairs obtained using different methods, remains poor. In general, it is believed that only manually curated, physical protein interactions extracted from the original small-scale experimental literature can be used with sufficient confidence in databases *(41,42)*.

There are a number of academic or commercially available pathway databases and network building tools available to the user. We have previously reviewed in some detail the major protein interaction databases, pathway maps and process ontologies *(19,43)*. However, there many more databases and they generally enable the visualization of cellular components as networks of signaling, regulatory, and biochemical interactions. These databases include at the time of writing: human protein reference database (HPRD *[44]*) is the most advanced public source of curated human protein–protein interactions; a joint project between Johns Hopkins University and the Indian Bioinformatics Institute. HPRD currently contains more than 25,000 interactions for 15,000 human proteins, protein domain information, and seven signaling maps. Drawbacks of HPRD include there being no network building tools and this database can only be browsed at present. Second, most of the cited interactions are indirect and only two interaction mechanisms are supported.

Database of interacting proteins is a database of experimentally determined protein–protein interactions, mostly from yeast. About 10% of database of interacting proteins interactions are derived from high-confidence small-scale experiments *(41)*.

Biomolecular interaction network database is a curated database of interactions, derived both from the literature and experimental data sets. Currently more than 8500 interactions are deduced from high-confidence small-scale experiments from multiple species. Biomolecular interaction network database can be used for querying and browsing *(45)*.

Molecular interaction database is a searchable interaction database with more than 40,000 total interactions, mostly from yeast and fly. Seventy percent of these interactions are from lower-confidence Y2H screens. Only 3800 interactions include human proteins *(46)*.

PathArt (Jubilant Biosystems; http://www.jubilantbiosys.com/pd.htm) is a manually curated database of about 7500 protein–protein and protein–compound interactions and pathways.

Pathways knowledgebase (Ingenuity, Inc.; http://www.ingenuity.com) represents mammalian interaction content for which the number of interactions has not been described.

ResNet database (Ariadne Genomics; http://www.anadnegenomics.com) is an automatically extracted and manually validated database of human protein interactions (more than 30,000), with transcriptional regulation (10,000), protein modifications (10,000), and functional regulations (350,000) *(27)*.

Search tool for interacting genes and proteins is a database of known and predicted protein interactions deduced from more than 110 genomes using high-throughput experiments and gene coexpression data *(47)*.

Other categories for the functional classification of proteins are gene, protein, and processes ontologies. Unlike pathways in which proteins are connected through consecutive single-step reactions and direct interactions, in ontologies proteins are assigned to particular categories because of functional or sequence similarity. They might or might not be physically connected in the cell. The best known among these is GO, a publicly available protein classification based on cellular processes developed by the GO Consortium *(48)* and curated by the European Bioinformatics Institute. Another popular process classification, PANTHER (protein analysis through evolutionary relationships) classification system is currently freely available from Applied Biosystems and classifies proteins according to families (>6683) and subfamilies (31,705), molecular functions and biological processes.

2.3. Network Theory and Tools

In recent years, it has become apparent that biological networks are ubiquitous *(26)* and therefore network analysis represents a powerful tool for the functional mining of large, inherently noisy experimental data sets. Proteins are represented as the nodes and physical protein–protein and protein–DNA interactions are represented as edges on these networks *(4,49)*. One-step binary interactions between proteins can be extracted from the experimental literature and when combined they form multistep modules and pathways. These inturn are connected into higher level clusters of multistep pathways using all proteins of known function *(50)*. Such protein interactions described for one cell type and condition are also possible in other cells and tissues, resulting in billions of possible multistep network combinations *(19)*, although only a fraction might be in use at any one time. A subset of the functioning or activated cellular machinery can be captured by high-throughput experiments, which in turn can be visualized on networks using unique algorithms.

There is an important distinction to be made between networks and other functional analysis tools. Unlike the preset, somewhat arbitrary groupings of objects into categories like pathways and processes, network edges bear the primary experimental information on connectivity between proteins, their subunits, DNA sequences, and compounds. The complete set of interactions therefore defines the potential of the core cellular machinery with potentially billions of physically possible multistep combinations *(19)*. Obviously, only a fraction of all possible interactions are activated with any given condition, as only some (10–30%) of the genes are expressed at a given time in a tissue and only a fraction of the cellular protein pool is therefore active. The subset of activated (or repressed) genes and proteins are unique for the experiment and are captured as snapshots by high-throughput data. Because each interaction represents a binary connection between individual proteins, the network gives the highest possible resolution of the resulting data patterns. The use of such networks can, therefore, overcome some of the drawbacks of high-throughput analysis. First, the interaction modules graphically represent biological mechanisms connecting the data. Second, interaction sets are comprehensive and eventually cover the majority of an organism's genes and proteins. Third (and most importantly), networks are dynamic and unique for each data set. Once generated, the networks can be interpreted in terms of these higher level processes, and the mechanism of an effect can be elucidated. This is achieved by linking the network objects to GO *(43)* and other process ontologies, as well as metabolic and signaling maps, and statistical analysis.

Biological networks are currently generated and analyzed by the methods of modern graph theory *(26)*. The default random network theory states that pairs of nodes are connected with equal probability and follows a Poisson distribution. This implies that it is very unlikely for any node to have significantly more edges than average *(51)*. As the field of network analysis has developed biological networks in yeast and elsewhere have been shown to be nonrandom *(28,29,50)*. The distribution of edges is very heterogeneous in these networks, with a few highly connected nodes (hubs) and the majority of nodes possess very few edges. Such a topology is defined as scale free, meaning that the node connectivity obeys the power law: $P(k) \sim k^{-\gamma}$, where $P(k)$ is the fraction of nodes in the network with exactly k links *(52)*. The hubs of such networks are predominantly connected to low-degree nodes, a feature that gives biological networks the property of robustness. Hence, the removal of even a substantial fraction of nodes still leaves the network connected *(53)*. The overall possible network topology correlates with the biological properties of the constituent node proteins *(31,54,55)* as highly connected hubs are conserved by evolution *(28,30,31)*. These essential proteins also tend to be more closely connected to each other. Furthermore, essential proteins are frequently the more promiscuous transcription factors and target genes that are in turn regulated by fewer transcription factors. Many of these targets are known as house-keeping genes with high-expression levels and demonstrate less fluctuation in expression *(56)*. The use of such network visualizations suggests an organized modularity in complex systems *(57)*, which has also been applied to interpret the connectivity of small molecules and their interaction with proteins *(58–61)*. Combined, these findings might have substantial implications for the practice of drug discovery in terms of target prioritization and identification of multigene/multiprotein biomarkers.

Network organization therefore has a characteristic importance for all levels of information flow in the cell. The networks can be generated at each level directly from available high-throughput data and assembled from protein–protein interaction databases. Although, traditionally the differentiation of all cellular processes into metabolic and signaling components is the norm, in reality, in a living cell both cascades work together and are codependent. In this case many endogenous compounds act as ligands for signaling cascades and almost all transcription factors, ultimately regulating metabolic enzymes and transporters. Some of these networks at different levels of the cell will now be described.

Genetic association networks visualize interconnections between the gene variants associated with a certain phenotype or disease, which are typically associated with mutations, single nucleotide polymorphism (SNPs), and in some cases, chromosomal rearrangements on the order of dozens to hundreds (in the case of cancers) of genes. Disease-associated genes vary greatly by their impact in particular diseases and are interconnected by complex epistatic relationships *(62,63)*. Although, more than 3000 disease associated genes are described and stored in the National Center for Biotechnology Information's (NCBI) online Mendelian inheritance in man (OMIM) database, disease-specific epistatic clusters (or networks) are poorly studied because of the inherent complexity of disease genetics and to the lack of tools with which researchers can tackle the problem. Recently, some of the first network studies were conducted for Alzheimer's disease *(64)* and glaucoma *(19)*. In both cases, large sets of protein–protein interactions were used to connect approx 60 genes associated with the diseases and provided new insights into potential therapeutic target genes. This approach is likely to be repeated for many other diseases.

Genome-wide transcription data are ubiquitous in disease research, thanks to the relative robustness and reliability of mRNA microarray technology. Based on the assumption that functionally related genes should be cotranscribed at the same time under the same conditions, several computational methods have been applied for the generation of gene coexpression networks in human and model organisms *(65)*. In one study, more than 3000 individual microarrays from human, fly, worm, and yeast were tested for coexpression of orthologs in multiple organisms (metagenes) and 3400 such orthologous metagenes appeared to be connected through 22,000 interactions *(66)*. By virtue of the underlying experimental data, coexpression networks mostly describe transcriptional regulation and at a basic level includes transcriptional factors and their downstream targets. Regulatory networks are however topologically complex and multilayered, with basic elements organized in; small one-step connected motifs which repeat frequently in the networks; semi-independent larger modules consisting of several motifs and, finally, the whole network as an interconnected set of modules *(65)*.

The level of active proteins impacts the information flow in the cell, as proteins are the main building blocks for biological function. In recent years it was realized that most proteins function as physically connected complexes best described as combinations of physical binary interactions *(55)*. Protein interaction networks show properties similar to other networks, with a few highly connected proteins and domains defining the network topology *(67)*. These hubs can be separated into two types: "party hubs" of simultaneously connected partners and "date hubs" with different partners at different times and conditions *(57)*. An important new direction consists of deducing protein interactions based on structural domain information *(68)*. For instance, proteins are considered to be interacting if their domain sequences are compatible with the X-ray structure of heterodimers *(69)* or with domains that have been observed among interacting proteins *(70)*.

A study on 43 organisms including human showed that similar to other types of biological networks, endogenous metabolic networks follow the power-law distribution with a few highly connected major metabolites, and that any two metabolites can be connected by at most three steps *(71)*. Metabolic networks allow a semiquantitative evaluation of the balance of major metabolites, known as metabolic flux analysis and constraint-based modeling *(72)*.

2.4. Other Resources for Network Analysis

The 2005 version of Molecular Biology Database Collection, the benchmark resource compiled at NCBI, contains more than 700 mostly public databases of variable quality and utility. However, several other biology resources are available.

- NCBI includes many useful subsections; Gene, Nucleotide, OMIM, UniGene, SNP, and PubMed. These all provide additional information about genes and proteins. PubMed is especially useful because it is a repository for published papers in the form of abstracts.
- The human genome database contains annotations for the Human Genome.

- The Human Gene Mutation Database lists any mutations for a given gene; it also lists the information in which mutations were first reported.
- European Molecular Biology Laboratory Contains the sequence for a given gene or protein along with PubMed references.
- SwissProt (Swiss Protein) lists additional information about proteins. It contains information such as synonyms, accession numbers, and links to related Medline and PubMed articles.
- The protein information resource supports genomic and proteomic research. The protein information resource maintains the Protein Sequence Database, which is a protein database containing more than 283,000 sequences.
- The Protein Data Bank is a worldwide repository for the processing and distribution of 3-D structure data of large molecules, for example, proteins and nucleic acids.

3. Integrated Network Data-Mining Suites

3.1. MetaCore

MetaCore is a web-based computational platform for multiple applications in systems biology. It is primarily designed for the analysis of high-throughput molecular data (microarray-based and serial analysis of gene expression (SAGE) gene expression, array-comparative genomic-hybridization DNA arrays, proteomics data, metabolic profiles, and so on) in the context of human and mammalian networks, canonical pathways, diseases, and cellular processes. MetaCore is an integrated system, which consists of (1) a curated database of mammalian biology, (2) a suite of tools for querying, visualization, and statistical analysis including pathways maps, network algorithms, and filters, (3) a toolkit (pathway editor) for custom assembly of functional networks, and (4) a set of parsers for uploading and manipulating of different types of high-throughput molecular data *(19,43)*.

3.1.1. Content

As a foundation, MetaCore has a database of protein–protein, protein–DNA, and protein–compound interactions, metabolic reactions, pathway maps, bioactive compounds (metabolites, drugs, and ligands), and diseases. Human pathways have been manually collected from the experimental literature for more than 5 yr. This represents one of the most comprehensive databases in the field, the core of MetaCore consists of more than 4.5 million individual findings resulting in about 50,000 signaling interactions and 20,000 human metabolic transformations (covering both endogenous and xenobiotic metabolism). The database has interaction information for more than 90% of known human proteins, including 1720 transcription factors and 650 GPCRs. This content is linked to 3200 human diseases and conditions. The bioactive chemistry component includes more than 7000 known drugs with protein targets and 5000 endogenous metabolites. The pathway information is organized in more than 400 signaling and metabolic maps with more than 3000 canonical pathways represented. This information is organized in an Oracle database.

3.1.2. MetaCore Database Architecture

The software currently runs on an Intel-based 32-bit server running RedHat Linux Enterprise 3 AS (RedHat, Raleigh, NC) and the web server runs Apache 1.3.x/mod_perl. Software on the server side is written in Perl, whereas the client side requires HTML/JavaScript and the Macromedia Flash Player Plug-in (Macromedia Inc, San Francisco, CA). The MetaCore database is generated from manual annotation of full text articles as well as disease relevant information from OMIM and EntrezGene. This database has functional processes as the core objects, which can be unique and have different relationships with molecular entities. We have used three major types of functional processes: effects, transformations, and blocks. In addition, we introduce the notion of a component that describes molecular species or functional groups of molecules in their biological context described as follows.

3.1.2.1. COMPONENT

This represents major functional objects in the context of a biological system. For example, a component might represent a single gene and its protein product, protein complex, a family of related proteins, small molecules, such as drugs, and metabolites. In the visual representation scheme of MetaCore, component corresponds to objects represented on the network and pathway maps. A component is related to a molecular entity, localizations, cells/tissues, and/or organisms. Thus, network classes represent biological molecules within their biological context. The molecular entity is treated in a broader sense than just being a specific chemical compound. In our current representation a component could also be a group of molecules (e.g., a protein family or class of chemical compounds) or a molecular complex. This is particularly useful for representing cellular processes, or when the exact chemical composition or a particular isoform of a protein participating in a pathway is unknown or ambiguous (e.g., EC numbers). Essentially the component category unites proteins, and compounds (small molecule ligands, endogenous metabolites, xenobiotics), and proteins. For example, p53 and ATP localized in the nucleus would both be components. Similarly, ATP in the cytoplasm will present itself as another component. In the nucleus, ATP is needed for RNA synthesis and in the cytoplasm as an energy source. Activated (phosphorylated) p53 in the nucleus is a potent transcriptional factor and a different component than the inactive p53 in the cytoplasm. At the same time, the family of integrins can be considered as one component in certain conditions.

3.1.2.2. TRANSFORMATION

This is an entity that is used to store information on biochemical reactions, transport, transcription, and translation, or on any biological process whose primary function is to change the state of a molecule (e.g., a reaction, in a broad sense), which is considered in its particular environment as linked to a subcellular compartment, tissue and organism. Transformation defines the morphing of components into each other. One example of a transformation is a one-step metabolic reaction, such as the synthesis of ATP from ADP and phosphate during oxidative phosporylation in the mitochondria. The transfer of ATP from the mitochondria to the cytoplasm is also a transformation. Another example is of protein phosphorylation in signal-transduction cascades. In all these cases, transformations share the property of modification of one component into another.

3.1.2.3. EFFECT

This is an entity that represents the influence that molecules exert either on transformations or on each other. Each effect has an agent (a component, which corresponds to the molecule[s] involved), a target (transformation, another component, or entire block [*see* **Subheading 3.1.2.4.**]), type, and a set of numerical values that could be associated (e.g., a kinetic end point). The notion of an effect is convenient for the description of biological activity whether or not the exact mechanism is known, as incomplete information can be stored allowing for the later reconstruction of cellular networks. For example, H-dependent ATP-synthase catalyzes ATP synthesis. This effect is presented as a link between the protein and the corresponding reaction. Another example of an effect is the phosphorylation of p53 by CHLK1 kinase. In turn, p53 is a transcription factor whose effect is modulation of transcription of multiple genes. This effect is the first step in a long chain of chemical and transport transformations induced by activation of the transcription of multiple genes.

3.1.2.4. BLOCK

This is used to describe functional units, be it a particular category of metabolism, or any other functional process. Blocks link together components, effects, and transformations that are themselves functionally related. Blocks are hierarchical as they might contain other blocks as elements. On the other hand every element might be a part of more than one block. Blocks are linked to each other by shared elements. Assembling different entities within functional blocks therefore enables the rapid searching of functional links and the function-centered analysis of expression and other high-throughput molecular data. A sequence of chemical reactions and regulatory protein interactions might have its own biological effect, and therefore can be identified as a block. An example of metabolic block is the oxidation of pyruvate followed by CoA attachment. This process consists of six discrete metabolic steps connected in three cycles and catalyzed by three enzymes. These enzymes form a pyruvate-dehydrogenase complex needed for coregulation kinetics of all reactions in such a way that the products of a downstream reaction are timely and directly assessable to the active site of an upstream enzyme. As a result, the coregulated and spatially unified set of six reactions can be considered as one functional block. At a higher level, The Krebs Cycle can be summarized as a unified block of reactions, which transform pyruvate into carbon dioxide, a process accompanied by reduction of NAD, ubiquinone, and GTP synthesis from GDP. On the one hand, a block can be divided into individual reactions, linked by functional (kinetic or regulatory) connections. On the other hand, it can also be considered as a united and separate entity interacting with other blocks.

This overall organizational structure has been described as a graphic diagram previously *(19,73)*. To summarize, functional processes and components serve as the core information space-holders in our database with many-to-many relationships between them. The corresponding molecular and mechanistic data are then linked to these space-holders as they become available. Functions serve as the "linking portals" for heterogeneous data. Once linked, the heterogeneous types of high-throughput data become a part of a larger system-level picture, in which functional relations among them can be more easily established and elucidated (e.g., all proteins in a pathway and their genes with expression patterns). Every pathway and its elements (interactions, reactions, enzymatic functions) are linked to available molecular data (genes, proteins, compounds, expression data, SNPs, and so on) annotated with relevant information about their involvement and importance in a number of common human diseases. This software allows the superimposition of relevant biological data such as microarray, SAGE expression data, metabolic profiles, and protein interactions on the pathways networks. Currently, by manual annotation of full text articles we have established more than 32,000 links between pathways from our collection and more than 3200 disease states, classified into six major categories:

- Cause—The highest level of verification. Cause means that it was clearly established experimentally that a deviation in a pathway directly causes the disease.
- Manifestations—A clinically confirmed strong correlation between the disease and a deviation in the pathway, but without the direct evidence for cause.
- Hypothesis—A correlation between the disease and the pathway has been demonstrated in some cases, but not in the others.
- Animal models—A correlation is described for one or more model organisms.
- Treatment—The changes in the pathways observed during or after therapeutic intervention.
- No relation—No links have been found between the disease and particular pathway or its elements.

3.1.3. Network Algorithms and Filters

Within MetaCore the networks are generated as a combination of binary single-step interactions (edges) which connect proteins and genes (nodes). The nodes and edges derive from the corresponding interaction tables in the MetaCore database and are visualized as clusters of interconnected nodes with the Macromedia Flash Player Plug-in. The end nodes on the networks have only one edge; the internal nodes might have anywhere from two to several hundred edges depending on connectivity with other nodes. The networks can be built from any input list of genes, proteins, and compounds corresponding to the components (network classes) in the database. The nodes in the input list are therefore considered as root nodes. The input list can be generated in several ways. Gene and protein names can be input and recognized with a built-in

Ingenuity pathways analysis (ingenuity)

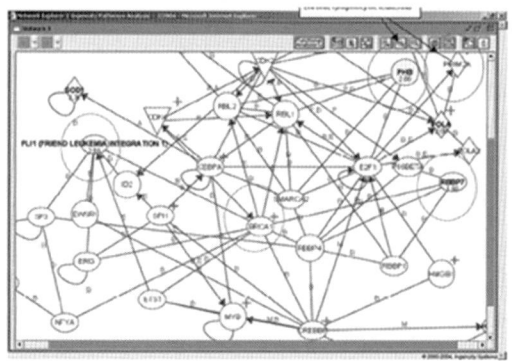

Pathway assist (Ariadne Genomics)

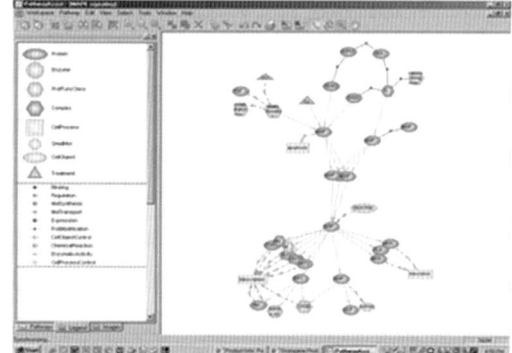

Metacore pathways analysis (GeneGO)

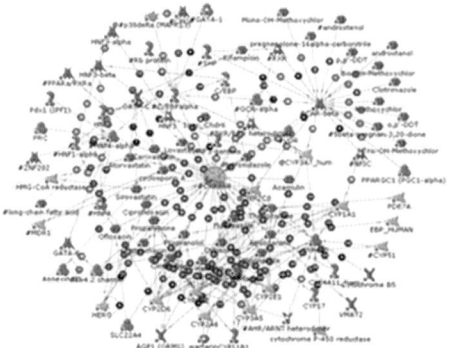

Pathart-pathway articulator (Jubilant BioSys)

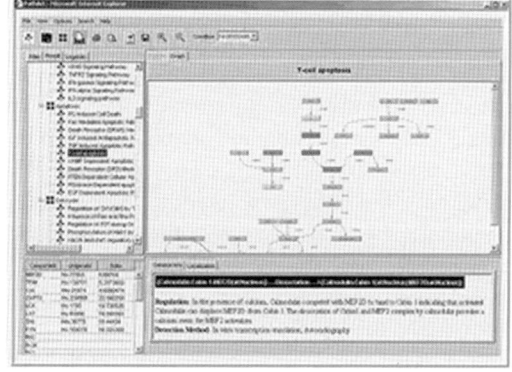

Fig. 3. Screenshots of the four major commercial software suites for network and pathway analysis. (Courtesy of Jack Collins, National Cancer Institute.) Note that Pathway assist in now known as Pathway Studio. (Please *see* the companion CD for the color version of this figure.)

$$pVal(r,n,R,N) = \sum_{i=\max(r,R+n-N)}^{\min(n,R)} P(i,n,R,N) = \frac{R! \cdot n! \cdot (N-R)! \cdot (N-n)!}{N!}$$

$$\sum_{i=\max(r,R+n-N)}^{\min(n,R)} \frac{1}{i! \cdot (R-i)! \cdot (n-i)! \cdot (N-R-n+i)!}.$$

3.2. Other Pathway Suites

Currently there are at least three other commercially available pathway analysis software suites that are widely used by the pharmaceutical industry and major research institutes (**Fig. 3**). They all have slightly different data and visualization capabilities.

3.2.1. PathwayAnalysis (Ingenuity, Inc.)

An integrated analytical suite based on a manually curated database of literature-derived mammalian protein–protein interactions used for visualization of data on networks and analysis. Networks are connected to GO processes, ~60 KEGG metabolic maps and Cell Signaling Inc.'s signaling maps. Web access and an enterprise solution is available.

3.2.2. PathArt

A curated database of generic protein interactions described earlier, pathways and bioactive molecules supported by high-throughput data parsers and visualization tools. This tool has connectivity with ligand databases and GO categories, while web-based access is available.

3.2.3. Pathway Studio (Formerly Known as Pathway Assist) (Ariadne Genomics)

A desktop software tool for mapping the high throughput data on networks, maps, and pathways. The source of the interaction data is NLP mining of PubMed abstracts. PathwayAssist is bundled with Jubilant and Integrated Genomics pathways content.

3.3. MetaDrug

The parallel development of different high-throughput methods, databases, absorption, distribution, metabolism, and toxicology (ADME/Tox) modeling, and systems modeling is currently ongoing *(74)* and will result in systems-ADME/Tox as a new area for research. We have used MetaCore as a foundation for building a software suite for ADME/Tox, called MetaDrug™. The ultimate goal of this platform is to predict from an input structure the major xenobiotic metabolites in humans and their predicted binding interactions with enzymes and other key ADME/Tox proteins in humans. MetaDrug includes more than 10,000 xenobiotic reactions, more than 1500 enzyme substrates, and 1000 enzyme inhibitors with kinetic data. MetaDrug has been used to derive some of the major metabolic pathways and determine the involvement of particular cytochrome P450s for compounds *(75,76)*. This database has also enabled us to generate more than 85 key metabolic pathways for predicting likely metabolic reactions coded in the software. A molecular structure can be parsed to rapidly create possible metabolites, which are prioritized using a further algorithm. In addition, the molecules can be scored using more than 40 integrated quantitative structure activity relationship (QSAR) models covering a wide range of ADME/Tox properties. Alternatively, the user can generate and use their own QSAR models with the software. Likely reactive metabolites for the input molecule/s are readily highlighted using 89 rules. Ultimately the predicted molecules and their interactions might be visualized as temporary objects with connections on a network diagram derived using one of two network algorithms.

To our knowledge MetaDrug is presently the only commercial product that combines all of the key properties of a human drug metabolism database, QSAR, rule-based methods for metabolism and reactive metabolite formation, and systems-biology approaches. The total effect of combining these different functions represents a significant step toward developing a Systems-ADME/Tox platform approach integrating computational predictions with data from all experiments to provide an understanding of the effect of xenobiotic and endobiotic molecules on ADME/Tox properties in humans *(76)*. The software also has additional valuable roles of providing a means to visualize predicted data in the context of empirical information on complex networks *(74)* and identify gene-signature networks *(77)*.

Future developments for MetaDrug include the integration with pipelining software such as Pipeline Pilot (SciTegic; www.scitegic.com) to allow MetaDrug to be used seamlessly as part of a larger data generation protocol such as, for large virtual library screening. Second, we will produce a version of MetaDrug with rat and mouse metabolism data in the underlying database, to enable predictions for these species and ultimately enable comparisons with the human predictions. Third, we are developing more sophisticated machine learning algorithms for metabolite prioritization to enable increased accuracy of predictions.

3.3.1. Applications of MetaDrug

Previously we have used MetaDrug to generate networks around nuclear hormone receptors (NHR) as well as analyze high-throughput microarray data *(20)*. MetaDrug was applied to analyze NHR, transcriptional factors and their associated interactions with other proteins, and small molecules relevant to drug disposition and toxicology to result in a very complex network using

Pathway Mapping Tools for Analysis of High Content Data 335

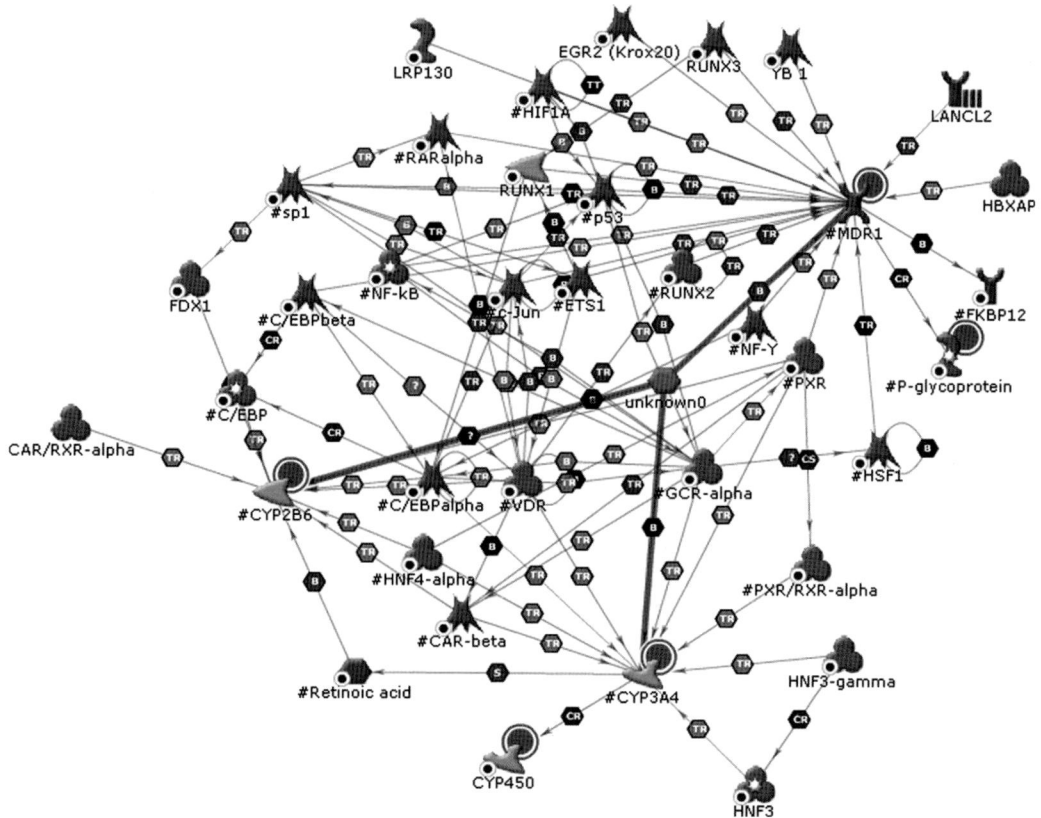

Fig. 4. Network visualization for Artmisinin (pink hexagon) using PCR data from human hepatocytes (red circles) *(98)*. Small molecules are hidden from this network for clarity. Ligands (purple hexagons) linked to transfactors (red), enzymes (yellow arrows), and transporters (blue) from the MetaDrug database. Highlighted lines show predicted interactions. Small colored hexagons indicate the type of interaction between two nodes, for example, Tr, transcriptional regulation; P, phosphorylation; B, binding. (Please *see* the companion CD for the color version of this figure.)

the autoexpand algorithm *(20)*. This visualization represented the current literature around NHR in terms of a network focused on the proteins of importance to drug disposition. Microarray data was also transposed on metabolic and signaling networks generated with this database using data from published experiments, in which MCF-7 breast cancer cells had been treated with 4-hydroxytamoxifen (OHT) for 24 h *(78)*. A network was generated around the enzymes of interest relating to the metabolism of OHT and the microarray data was visualized on this. We have recently provided further test cases using MetaDrug that: enabled the prediction of metabolites for molecules based on their chemical structure; predicted the activity of the original compound, and its metabolites with various ADME/Tox models; incorporated the predictions with human cell signaling, metabolic pathways, and networks; and integrated networks and metabolites with relevant toxicogenomic, or other high-throughput data. We have demonstrated the utility of such an approach using recently published data from in vitro metabolism and microarray studies for Aprepitant, L-742694, Trovofloxacin, and artemisinin (**Fig. 4**), and other artemisinin analogs as well as OHT. This enabled us to show the predicted interactions with cytochrome P450s, pregnane X-receptor and P-glycoprotein, the metabolites and the networks of genes that are affected (Ekins. 2006). These examples represent how MetaDrug could be used as a novel method for analysis of computational predictions and microarray data on networks of interacting genes in order to visualize data in the context of the complete biological system. This provides insights for the up or down regulation of particular genes involved in a phenotypic response and also highlights genes not on the microarray but central to a network.

4. Examples Using MetaCore With Different Data
4.1. Mapping HCS Data on Networks

MetaCore is well suited for mapping phenotypic data such as HCS and siRNA as long as the data points are linked to either genes, or protein, or metabolic IDs. The mapping, visualization, and analytical procedures are virtually the same as mapping of molecular data such as gene expression or protein abundance. It is important that HCS data can be compared and cross-validated with molecular data on the same pathways and networks. One such analysis is currently in progress at the Translation Genomic Research Institute (Jeff Kiefer, personal communication). In this study, 162 genes/proteins were identified as hits from a high-throughput siRNA screen of 5000 genes constituting the "druggable genome" *(79)*. The shortest paths network was built from this data (**Fig. 5A**). When targeted by siRNA these genes were able to increase the sensitivity of the cancer cell line used to the effects of a low dose of a chemotheraputic compound. The cell proliferation GO process was selected and traced on the same network (**Fig. 5B**) indicating an agreement with the observed data.

A recent study by Cellomics, Inc. (www.cellomics.com) describes the prototype FluoroTox system used for detection and classification of chemical and biological agents using HCS *(14)*. This focuses on one cell type and multiple parameters that were measured including p38 activation, NF-κB activation, NF-κB inhibition, CREB activation, ERK activation, and cytotoxicity. Apart from the latter general assay, the rest relate to four specific proteins, which can be mapped in MetaCore using the Analyze Network algorithm (**Fig. 6A**). A second HCS study studied a collection of 720 natural compounds to find inhibitors of the mitogen-activated protein kinase phosphatase-1 (MKP-1), a dual specificity phosphatase overexpressed in many cancers *(12)*. An alkaloid sanguinarine was found to inhibit MKP-1 and induce phosphorylation of *ERK* and *JNK*. Using MetaCore we can visualize the linkage between MKP-1, *ERK* and *JNK* with the Analyze Network algorithm (**Fig. 6B**). These MetaCore networks could therefore be used to visualize HCS data following treatment with different compounds to visualize the extent of protein deactivation or activation. This data can also be combined with any of the other data types described later.

4.2. Mapping Metabonomics Data

MetaCore and MetaDrug have the capability to upload metabonomics data either as a list of molecule names or molecular formulas at present. We have previously illustrated the frequency distribution of molecules and their molecular formula in MetaDrug *(76)*, which indicated the majority of molecular formulas corresponded to one to two metabolites. At present, we can visualize all the metabolites suggested for each unique formula or name, and after highlighting a molecule of interest further information can be retrieved including the molecule structure, synonyms, and reactions. We can also visualize these metabolites on maps or networks in MetaCore. Using a data set from a recent publication *(80)*, which determined the differences in hydrazine toxicity between rat and mouse by collecting urine and analysis using ^{1}H NMR, we are able to demonstrate this utility. We have used the endogenous urinary metabolites observed in rat and parsed them with our software to visualize the seven of the 17 metabolites alongside proteins on networks, after using the Analyze Network algorithm (G-score = 46.31, $p = 9.30$ e^{-18}, **Fig. 7**). These networks might be useful for indicating the type of toxicity that could be observed following compound treatment from metabolite data alone and lead to the generation of signature metabolite networks.

4.3. Mapping Genomics Data
4.3.1. Tat-Upregulated Genes at the G_1/S Phase

MetaCore has been primarily used to date for the analysis of microarray data *(19,73,77,81–87)*. To further illustrate the utility for analysis of genomics data we have taken a recently published

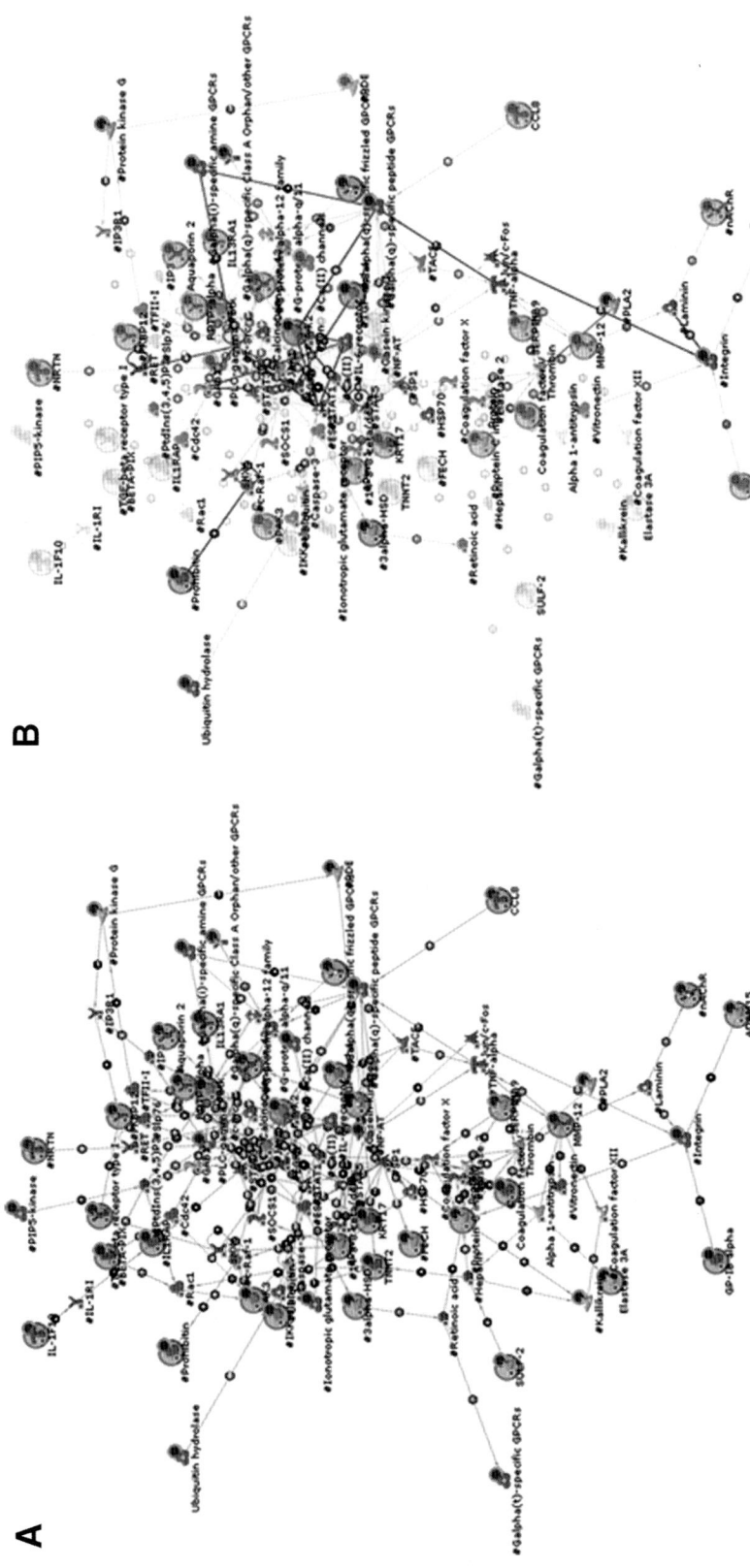

Fig. 5. Mapping of siRNA data on networks. High-throughput siRNA cell assays were conducted with 5000 genes and networks were built in MetaCore. (**A**) The network of 162 initial nodes was built run using Shortest paths algorithm. Tricolored circles mark the genes from input list; solid blue circles mark the nodes with numerical data. (**B**) GO process for cell proliferation is traced on the same network as a blue line. (Courtesy of Jeff Keifer, Translation Genomic Research Institute.) (Please *see* the companion CD for the color version of this figure.)

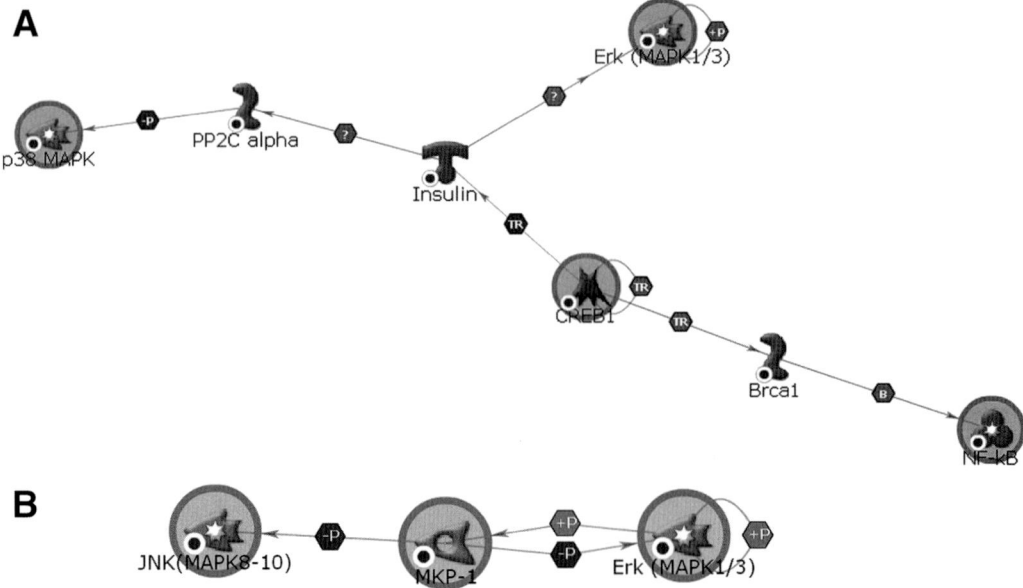

Fig. 6. Visualizing pathways from cellomics data. (**A**) Assays used in the FluoroTox system *(14)* visualized with MetaCore using the Analyze Network algorithm (G-score = 91.2, $p = 1.70\ e^{-14}$), (**B**) Assays used with the alkaloid sanguinarine *(12)* visualized with MetaCore using the Analyze Network algorithm (G-score = 121.19, $p = 1.81\ e^{-12}$). Nodes surrounded by a blue circle indicate those from the input list corresponding to therapeutic targets. Small colored hexagons indicate the type of interaction between two nodes, for example, Tr, transcriptional regulation; P, phosphorylation; B, binding. (Please *see* the companion CD for the color version of this figure.)

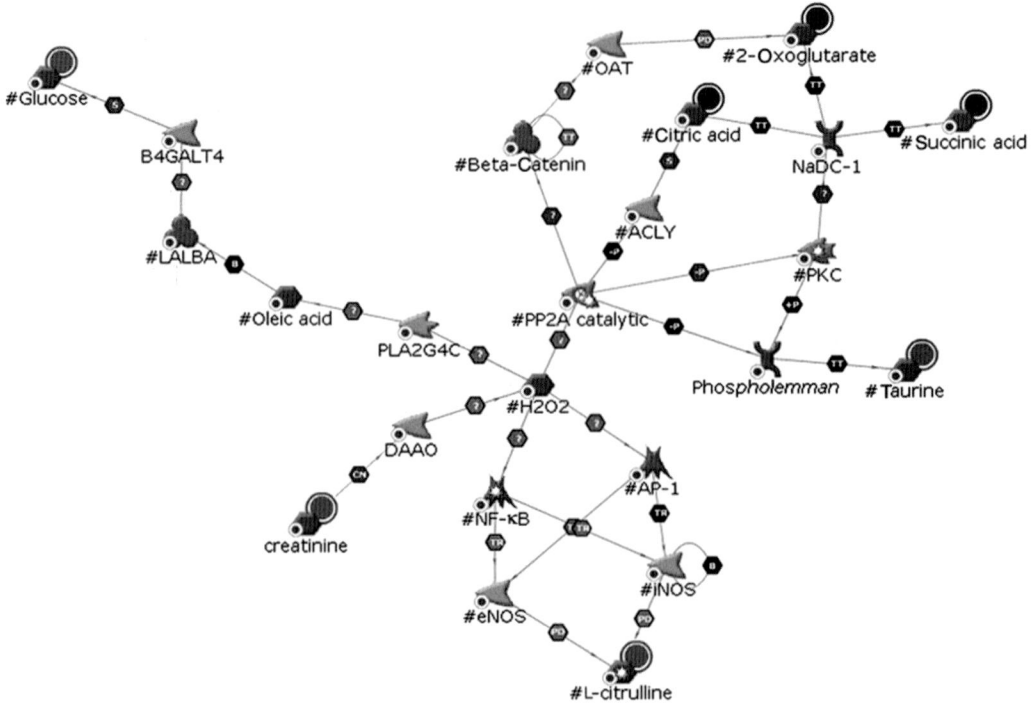

Fig. 7.

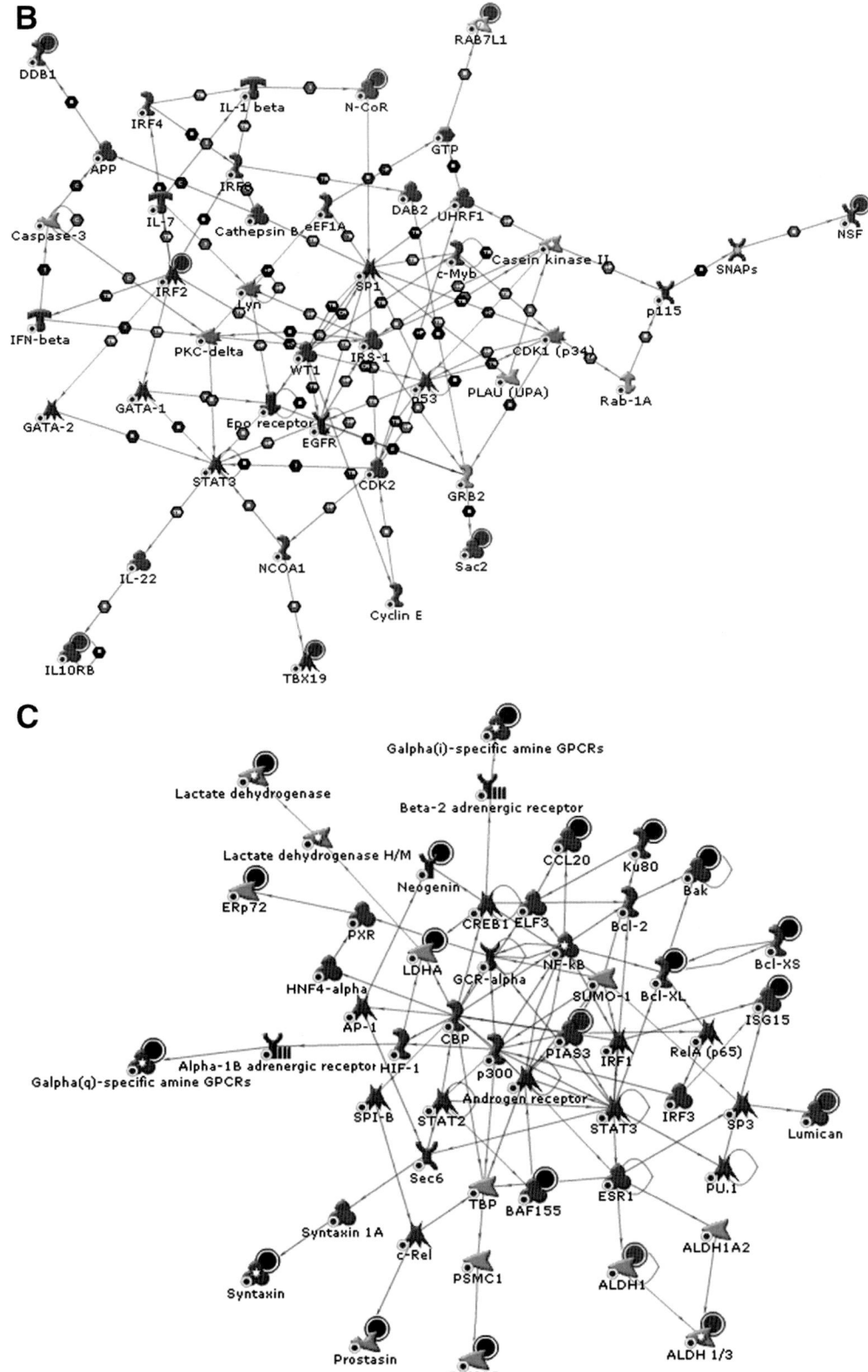

Fig. 9. *(Continued)*

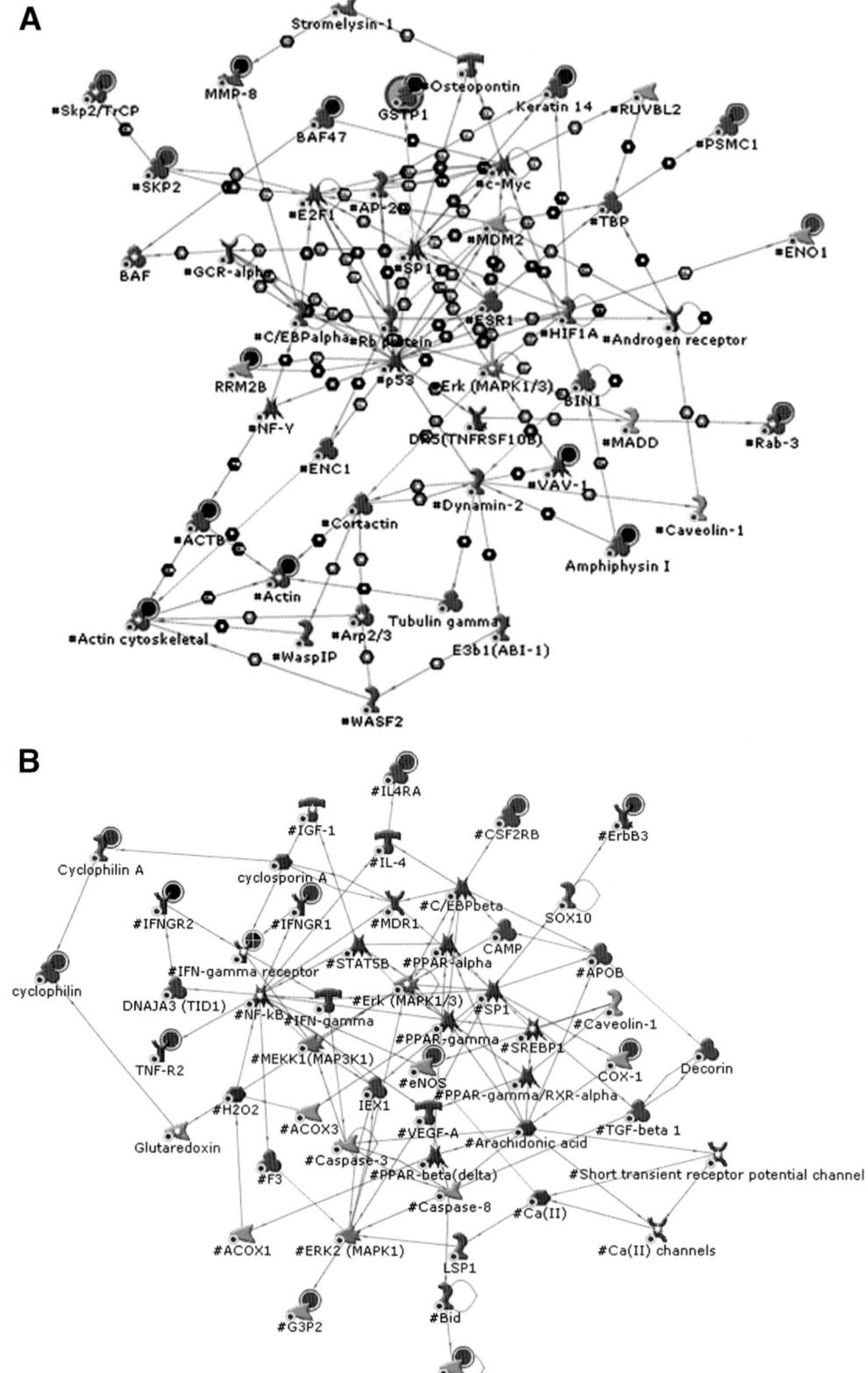

Fig. 10.

25

Systems Biology in Cancer Research

Genomics to Cellomics

Jackie L. Stilwell, Yinghui Guan, Richard M. Neve, and Joe W. Gray

Summary

Cancers result from large-scale deregulation of genes that lead to cancer pathophysiologies such as increase proliferation, decreased apoptosis, increased motility, increased angiogenesis, and others. Genes that influence proliferation and apoptosis are particularly attractive as therapeutic targets. To identify genes that influence these phenotypes, we have developed simple and rapid methods to measure apoptosis and cell proliferation using high content screening with YO-PRO®-1 and anti-BrdU staining of BrdU pulsed cells, respectively.

Key Words: Apoptosis; BrdU; cell cycle; cellomics; high content analysis; high content screening; image analysis; propidium iodide; siRNA; YO-PRO-1.

1. Introduction

The genomic revolution has spawned a variety of global analysis methods to assess abnormalities in the genome, epigenome, and transcriptome that occur during carcinogenesis and cancer progression. Functional assessment of the genes revealed by these analyses is important both to identify potential therapeutic targets and/or to suggest molecular markers that might predict patient outcome. This requires efficient methods to measure cancer-related changes in cellular phenotypes that result from manipulation of the genes revealed by "-omic" analyses. Relevant phenotypic changes in cells involve cancer hallmarks such as genome instability, cell proliferation, apoptosis, angiogenesis, invasion, and metastasis *(1)*.

Functional prioritization of aberrant genes is challenging because literally thousands of genes already have been implicated by profiling experiments *(2)*. Our approach has been to use high content imaging to assess changes in two important cancer related end points—apoptosis and proliferation—during manipulation of candidate genes using siRNAs and/or gene-specific small molecule inhibitors *(3,4)*. We have used this strategy in multiple cancer cell lines to evaluate genes that have been identified as potential therapeutic or diagnostic targets. Two high content screening methods employed by our lab are YO-PRO®-1 staining of cells to measure apoptosis and BrdU incorporation to measure proliferation *(5,6)*. An example, using these techniques to evaluate a gene amplified and overexpressed in some ovarian cancers, will be used to demonstrate the evaluation of cellular phenotype after treatment with siRNA to this gene.

2. Materials

1. Cell lines testing positive for amplification and overexpression.
2. Cell lines negative for amplification and overexpression.

3. Tissue culture media selected for optimal growth of cells.
4. Lipofectamine 2000 (Invitrogen, Carlsbad, CA).
5. Optimem.
6. 1X Sterile PBS (phosphate-buffered saline).
7. Multichannel pipet (20–200 µL).
8. Aspirator.
9. Coulter Counter® (or hemocytometer—*see* **Note 1**).
10. Cuvet for Coulter Counter.
11. Isoton II Diluent (Beckman Coulter; Fullerton, CA).
12. Sterile disposable multichannel pipet basins.
13. YO-PRO-1 iodide 491/509 (Molecular Probes, Eugene, OR).
14. Hoechst 33342.
15. BrdU (bromodeoxyuridine).
16. Anti-BrdU monoclonal antibody (BD Biosciences, San Diego, CA).
17. Rabbit antimouse Alexa Fluor® 488 (Invitrogen).
18. PI (propidium iodide).
19. NaOH.
20. 70% ethanol.
21. Tween-20.
22. KineticScan HCS Reader (KSR; Cellomics, Inc., Pittsburg, PA).

3. Methods

The methods described next outline (1) seeding and treatment of cell lines, (2) staining cells with YO-PRO-1 and Hoechst 33342, (3) incorporation of BrdU and antibody staining in proliferating cells, (4) scanning the plates using a high content screening system, and (5) analyzing the data.

3.1. Seeding and Treating Cells

The seeding and treating of cells with siRNA is described in **Subheadings 3.1.1.–3.1.4.** This includes (1) a description of criteria to select cell lines to test, (2) counting and seeding the cells onto a 96-well plate, and (3) treating cells in 96-well plate.

3.1.1. Selection of Cell Lines

In the example presented here we identified target genes by finding regions of the genome that are amplified (or lost) in ovarian cancer cell lines and tumors *(7–9)*. These regions were identified based on DNA copy number changes of many different cell lines measured by comparative genomic hybridization arrays *(10,11)*. Once amplified regions were identified, the genes in these regions were examined for overexpression. One of the amplified regions identified was chromosome 8q24 and is associated with poor outcome in ovarian cancers *(9,12,13)*. Several genes were selected for testing using high-content analysis and one example is presented here. Once a gene was selected for testing in this system, cell lines were then selected based on either being positive or negative for the amplification and over expression of this gene. Four representative cell lines were selected for each category (*see* **Note 2**). For other applications cell lines would be selected based on the information required.

3.1.2. Cell Lines

The cell lines with the 8q24 amplification used for this example were CAOV4, HEY, OVCA432, and OVCAR-8. Four others, A2780, SKOV3, OV-90, and CAOV3 were also selected because they do not contain this amplification. CAOV3, CAOV4, OVCAR-8, and OV90 were purchased from American Type Culture Collection and the remaining cell lines were kindly donated by Gordon Mills, MD Anderson, Dallas, TX. The cell lines were maintained in L-15 (CAOV4), RPMI 1640 (HEY, OVCAR8, and A2780), MEM (OVCA432 and CAOV3), and M199/MCDB (OV90), supplemented with 10% fetal bovine serum (FBS). The media selected should be optimal for propagation of the cell line used.

The treatment used for the cells depends on the experimental goals, but this example, treatment with siRNA will be described. One of the important considerations for any treatment is to include a group of mock-treated cells. In our experiments, each condition is repeated in one of four wells to ensure reproducibility. If it is possible and convenient, randomizing the position in which, repeat wells are placed is a good practice, because of potential variability in the plate, edge effects, or any other unforeseen effects that might be dependent on the position of the cells in the plate.

3.1.3. Seeding Cells on a 96-Well Plate

The appropriate number of cells per well on a 96-well plate varies with the size of cells and the length of time from seeding to analysis. In this example, 5000 cells per well are plated in a 96-well plate (*see* **Note 3**). We have also used similar protocols in a 24-well plate and have found that 50,000–100,000 cells per well is optimal. Typically, the assays are performed within 48 h after seeding.

Adherent cells are dissociated from the culture dish surface with 0.25% trypsin EDTA and then the trypsin EDTA is inactivated with media containing 10% FBS. If the cells are sensitive to trypsin they can be spun down at this point and resuspended in media plus 10% FBS. Spin trypsin-sensitive cells down gently in a conical tube in a table-top centrifuge at approx 1200 RCF, remove media, and resuspend cell pellets into new media.

Approximately 100 µL of the cell mixture is then transferred to a Coulter Counter cuvet containing 10 mL isotone. A dilution of 101 is set on the Coulter Counter display. After the system is primed according to manufacturer's instructions, cells are counted and a calculation is made to determine the dilution of cells to obtain media containing 50,000 cells/mL. The Coulter Counter is then either put through a priming cycle so that another group of cells can be counted or, if counting is completed, the cleaning solution is put into the Coulter Counter cuvet and the system is cleaned to prevent future clogging. The cells are then diluted in the same media used for propagation. Before each step the cells should be gently, but thoroughly mixed to ensure accurate counting and even spreading in the wells of the plate. Approximately 10 mL of diluted cells are required per 96-well plate.

Once the dilution has been established the appropriate volume of diluted cells, depending on how many 96-well plates are required, is added to a sterile trough. The maximum volume for these receptacles is 55 mL. The multichannel pipet can be used to add 100 µL of diluted cells per well. The plates are then immediately placed at 37°C in 5% CO_2 for incubation overnight. This will allow the cells to adhere to the plates and recover from the trypsin treatment.

3.1.4. siRNA Transfection

1. Label two sterile microcentrifuge tubes with the designation A or B.
2. In tube A add 10 µL of Opti-MEM+ X µL of 20 µ*M* of siRNA stock (*see* **Table 1** for amount of siRNA oligonucleotide to add).
3. In tube B add 10 µL of Opti-MEM+ X µL of 1 µg/µL Lipofectamine 2000.
4. Incubate each tube at room temperature for 5 min.
5. Add the contents of tube B to tube A. Incubate the mixture at room temperature for 20 min.
6. Approximately 5 min before the 20 min incubation is over, replace culture media with 100 µL of Opti-MEM in each well of a 96-well plate and with 500 µL in each well of a 24-well plate.
7. Add the tube A + B siRNA-Lipofectamine 2000 mixture gradually and gently to wells containing Opti-MEM. For each well of a 96-well plate, add 20 µL of the mixture. For each well of a 24-well plate, add 100 µL of the mixture.
8. Incubate plate at 37°C in a CO_2 incubator for 2–4 h.
9. Replace Opti-MEM with complete culture media.

A positive control should also be added to ensure the assay is performed properly. In this example, cells were treated with paclitaxel, causing both apoptosis and proliferation effects as expected. After treatment, cells are incubated for an appropriate period of time to ensure that

Table 1
Recommended Volumes of 20 μ*M* Stock siRNA and 1 μg/μL Lipofectamine 2000 for Different Final Concentrations of siRNA

Culture format	siRNA final concentration		
	40 (n*M*)	80 (n*M*)	120 (n*M*)
24-well (μL/well)	1.25	2.5	3.75
96-well (μL/well)	0.25	0.5	0.75

both the siRNA has had time to downregulate the gene of interest and any resulting downstream effects have occurred. Initial testing is generally required to establish a range of concentrations of siRNA or times relevant for the analysis. Additional cells are treated with the siRNA so that some can be processed for RT-PCR and Western blots to verify knockdown of the transcript and protein of interest. After cells are incubated sufficiently long, the protocol for identifying changes in apoptosis and/or proliferation can begin.

3.2. Staining Cells With YO-PRO-1 and Hoechst

YO-PRO-1 iodide permits analysis of apoptotic cells without interfering with cell viability *(14)*. Cells stained with YO-PRO-1 are counter stained with Hoechst 33342, a nucleic acid stain permeant in both live and dead cells, whereas YO-PRO-1 is only permeant in cells that are beginning to undergo apoptosis *(15–17)*. Therefore, apoptotic cells fluoresce both green and blue, whereas live cells fluoresce only blue (*see* **Fig. 1A,B**). This section describes staining of cells with YO-PRO-1 iodide to measure apoptosis and imaging the plates using the KSR.

3.2.1. Staining Cells With YO-PRO®-1

After incubation with the siRNA or other reagents is completed the cells are then stained with YO-PRO-1 and Hoechst 33342 (*see* **Note 4**). Approximately 10 mL of the appropriate media containing these DNA stains is required per 96-well plate. The following steps are performed to complete the staining of the cells.

1. Warm the appropriate media to 37°C in a water bath. Use the same media that was used to propagate the cells.
2. Thaw YO-PRO-1 and Hoechst.
3. Transfer 10 mL of prewarmed media per 96-well plate to a 50-mL conical tube.
4. Add 20 μL of 10 mg/mL YO-PRO-1 per 10 mL media.
5. Add 20 μL of 10mg/mL Hoechst per 10 mL media. This will make a 2X solution for staining.
6. In a laminar flow hood transfer the media containing YO-PRO-1 and Hoechst into a solution basin for transfer into 96-well plates.
7. Using a multichannel pipet, gently transfer 100 μL of the dye plus media mixture into each well.
8. Rock the plate gently to mix and then transfer back into a 37°C incubator with 5% CO_2 for 30 min (*see* **Note 5**).
9. After 30 min the plate is immediately imaged using the Cellomic's KSR or the ArrayScan.

3.2.2. Imaging 96-Well Plates

Plates are immediately imaged using the KSR system. This high content analysis system scans plates by imaging a designated number of fields in each well of a 96-well plate with user designated filter sets. The XF100 filter set is used for this protocol, although any filter set that allows the visualization of FITC and Hoechst can be used. The user also designates the wells to be imaged. In this assay, two images are collected per field, one in the blue channel image (Hoechst) and one in the green channel image (YO-PRO-1). Before beginning the scan, plates should be examined using an inverted light microscope to assess the density of cells. The density of the cells will help determine how many fields per well should be imaged with a selected

Systems Biology in Cancer Research

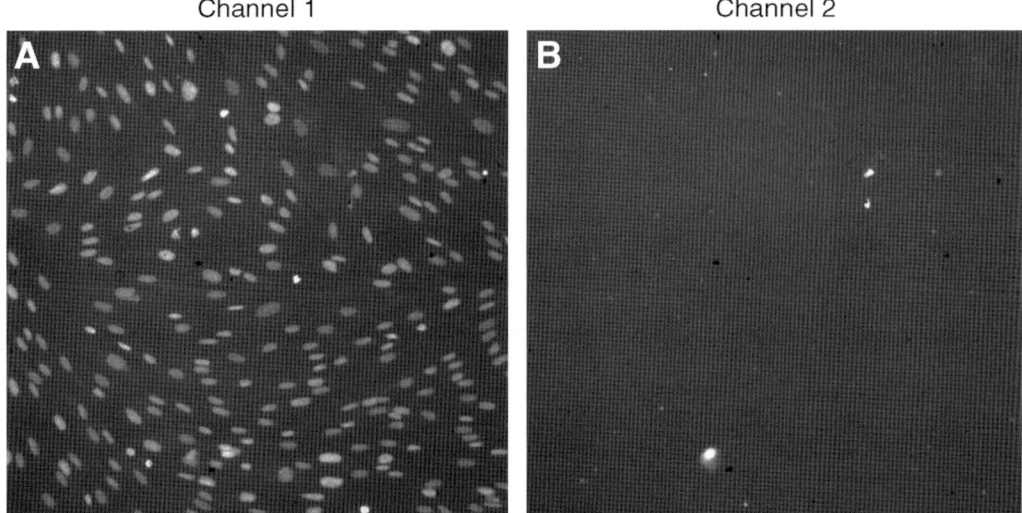

Fig. 1. Images generated by the KineticScan® HCS Reader. Cells were plated on 96-well plates, transfected with siRNA, incubated for 48 h, and then stained with Hoechst and YO-PRO®-1. Plates were transported to the KineticScan for image generation, then automated analysis was performed on the collected images. Each nuclei imaged by the KSR is identified with the Cell Health Profiling BioApplication software in (**A**) the blue channel by Hoechst staining or (**B**) the green channel by YO-PRO-1 staining. Measurements of intensity and area, among others, are then made using the Cell Health Profiling algorithm. (Please *see* the companion CD for the color version of this figure.)

objective. Typically, the ×10 objective is used and 5–10 fields per well are imaged. If the cell density is low up to 30 fields can be imaged in each channel.

Images are acquired by the KSR and then analyzed using the Cellomics, Inc. BioApplication® software, in two separate steps. This is in contrast to the ArrayScan, in which these two processes are performed simultaneously. Image acquisition is executed as specified by the manufacturer of the instrument (Cellomics, Inc.). Briefly, the instrument and light source are turned on 10 min before the end of the last incubation of when using the YO-PRO1 staining protocol. The user then double clicks on the Cellomics, Inc. KineticScan icon and types in a username and password. The Cell Health Profiling® protocol is selected for this assay and modified, as needed, according the manufacturer's instructions. The main modifications required are (1) the length of exposure for each channel, dependent on the intensity of staining, and (2) the wells to be imaged. Because the images are analyzed after they are collected, it is important to ensure they are saved to disk in the final dialog box of the scanning software. After the scan is completed the plate is removed. The images are then analyzed as discussed in **Subheading 3.4.** Results from a typical experiment are graphed (*see* **Fig. 2**).

3.3. Staining Cells With Incorporated BrdU

The protocol used for examining cell proliferation is based on protocols typically used for flow cytometry *(18)*. Cells are pulsed with a 1 m*M* solution of BrdU before they are fixed and stained for BrdU and DNA content. Only cells that are actively replicating DNA incorporate BrdU into the newly synthesized DNA, providing a measurement of the number of cells in S-phase during the pulse. Because BrdU is a derivative of uridine that replaces thymidine in replicating DNA, it provides an indication of the rate of cell proliferation. The duration of the pulse with BrdU can vary depending on the proliferation rate of the cells. The typical time for incubation with media containing BrdU is 30 min, however if the cells proliferate rapidly less time is required. In this protocol a mouse monoclonal anti-BrdU antibody is used and its location is visualized with an

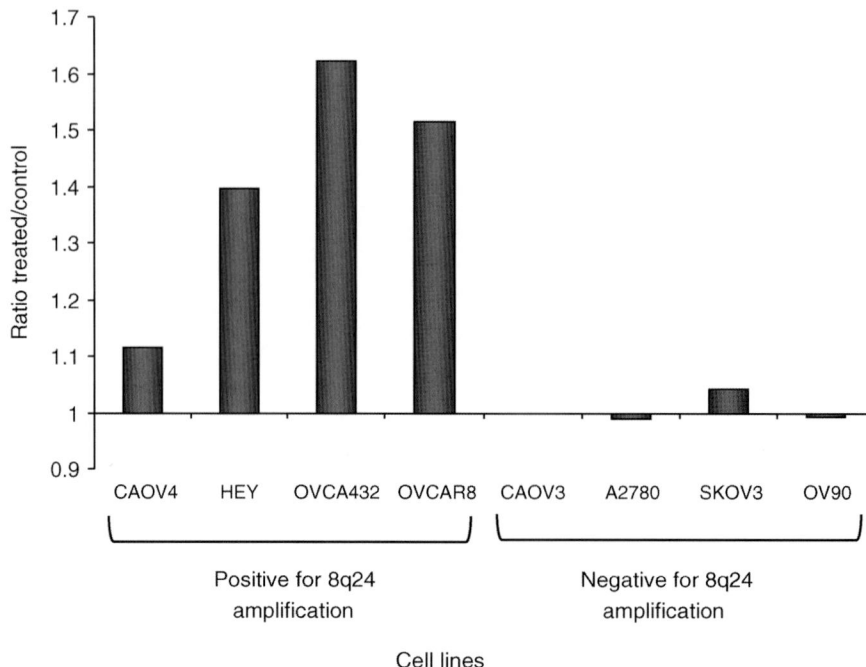

Fig. 2. Results of YO-PRO®-1 staining in ovarian cancer cells. Amplification of 8q24 (amplified in 50% of ovarian cancers) is associated with poor survival. siRNA was developed against a gene in this region and used to transfect ovarian cancer cells, both negative and positive for amplification of this region. Staining with YO-PRO-1 shows that treatment with siRNA against this selected gene was demonstrated to induce apoptosis in four overexpressing cell lines compared to lipofectamine treated cells, whereas not inducing significant apoptosis in three of the four negative cell lines. (Please *see* the companion CD for the color version of this figure.)

antimouse AlexaFluor®488 antibody (*see* **Fig. 3B**). Cells are counterstained with propidium iodide to provide an indication of total DNA content and for object identification (*see* **Fig. 3A**).

3.3.1. Pulsing Cells With BrdU and Fixing

1. Add BrdU directly to cell media to achieve a final concentration of 10 µM (*see* **Note 6**).
2. Pulse for 30 min. If the cells proliferate more rapidly or slowly than typical tissue culture cells the length of pulse can be determined empirically and adjusted accordingly. For example, when using rapidly proliferating cells the pulse can be as short as 5 min to obtain labeling of a significant percentage of the cells. However, for most cells 30 min is optimal. Place cells at 37°C in 5% CO_2 during the pulse.
3. Remove media. It is preferable to use a multichannel pipet for this step, because it is gentler than an aspirator and fewer cells are lost.
4. Add enough cold 70% ETOH to cover cells. With a 96-well plate 100 µL is sufficient. The 70% ethanol should be stored at –20°C.
5. Place cells covered, at 4°C until ready to stain. The incubation should be at least 1–2 h, but we have stored cells in 70% ethanol for several days and have obtained adequate results.

3.3.2. Staining Cells With Incorporated BrdU

1. Remove 70% ETOH and allow cells to air-dry for 1–2 min. Cells should not be allowed to dry completely.
2. Add 0.07 N NaOH to denature the DNA and incubate at RT for 3 min. This will allow the incorporated BrdU to be accessible to an anti-BrdU antibody.
3. Remove NaOH and add 100–200 µL of PBS to neutralize the base.
4. Mix anti-BrdU with PBS/Tween at a 1/100 dilution. For one 96-well plate, at least 5 mL of PBS/Tween is required, containing 50 µL of the mouse anti-BrdU antibody.

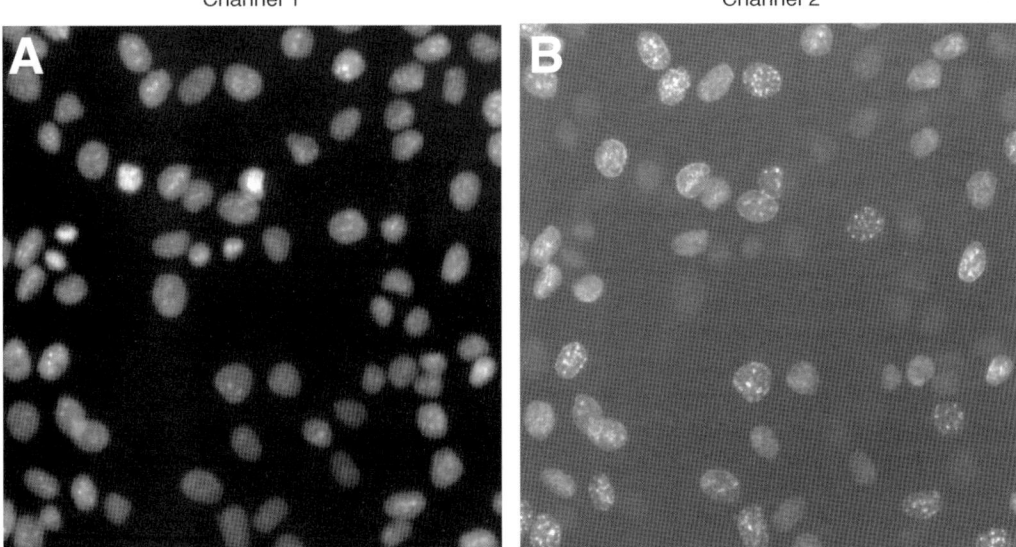

Fig. 3. Images generated by the KineticScan® HCS Reader. Cells were plated on 96-well plates, transfected with siRNA, incubated for 48 h, and then fixed. They were then stained with propidium iodide and the monoclonal antibody for BrdU followed by an antimouse AlexaFluor®488. Plates were transported to the KineticScan for image generation, then automated analysis was performed on the collected images as described in the text. Each nuclei imaged by the KSR is identified with the Cell Health Profiling Bioapplication Software in (**A**) the red channel by PI staining or (**B**) the green channel by AlexaFluor 488 staining. This staining indicates the cells that underwent DNA replication during the pulse with BrdU.

5. Gently add 50 µL of anti-BrdU dilution per well to cells (*see* **Note 7**). Incubate 45 min at room temperature.
6. Wash twice with PBS/Tween. Do this step gently so that the cells remain attached to the plate.
7. Mix the antimouse AlexaFluor 488 antibody in PBS/Tween at 1/500. For a 96-well plate use 5 mL PBS/Tween and then add 10 µL of the antimouse antibody AlexaFluor 488. Protect the mixture from light.
8. Add 50 µL to each well, and incubate for 30–60 min, in the dark, at room temperature.
9. Wash three times with PBS/Tween.
10. Incubate cells for 5 min in 0.5 µg/mL PI in PBS.
11. Wash cells 1X with PBS. Add 200 µL PBS, seal with clear and adhesive plate sealing film, and either scan immediately or place the plate at 4°C until ready to scan.

3.3.3. Imaging 96-Well Plates

The KSR® system scans plates by imaging a designated number of fields in each well of the plate, with a user designated objective and filter set, in a similar manner to the YO-PRO-1 staining previously described in **Subheading 3.2.2.** Because the cells in this assay are fixed the plate can be stored at 4°C for several days or imaged immediately. In this particular assay, two images will be collected per field, one red channel image (PI) and one green channel image (AlexaFluor 488). Before beginning the scan, as described in **Subheading 3.2.2.**, the plates should be examined using an inverted light microscope to assess the density of cells to determine the number of fields imaged per well. In this protocol, the XF93 filter set and ×10 objective are typically used and 5–10 fields are imaged unless the density is low. With lower cell densities, a greater number of fields can be imaged. The software should be set so that the first channel is set for Texas Red and used for object identification. This is because PI stained nuclei can be visualized in this channel and all cells are stained with PI.

As described in **Subheading 3.2.2.** images are acquired by the KSR and then analyzed using a Cellomics Bioapplication as a separate step. Cell health profiling is the bioapplication used for this assay and also, modified as needed according the manufacturer's instructions as in the previous protocol. Typical images acquired using this protocol are shown (*see* **Fig. 3A,B**). After the scan is complete the software is used to eject the plate. The images are then analyzed as described in the next section.

3.4. Processing of the Acquired Images

Described below are the two different methods we use for (1) analyzing the plate stained using YO-PRO-1 and (2) the plate stained using the anti-BrdU and secondary antibodies. When staining with YO-PRO-1, the essential information is the percentage of the total number of cells stained with YO-PRO-1 over a certain threshold of staining intensity. In addition, the intensity of the Hoechst staining in this assay can also be used as an indicator of apoptosis, because nuclei condense while undergoing apoptosis. This causes a greater average intensity of fluorescence within the apoptotic nuclei.

When cells are stained with PI and anti-BrdU, the fluorescence intensity for both of the dyes is important. This is because the total DNA content and the amount of DNA that has incorporated BrdU in each nucleus are both important pieces of information for the final analysis.

3.4.1. Image Analysis of Hoechst and YO-PRO 1 Stained Cells

The Cell Health Profiling Bioapplication (Cellomics, Inc.) is used to analyze the images obtained with the KSR, using manufacturer's instructions, with a few modifications. It is important to have enough cells to obtain statistically significant results, but not so many cells that they become clumped and do not allow the image analysis software to adequately identify individual nuclei (*see* **Note 8**). For initial optimization of the image analysis software, the Hoechst stained nuclei images from channel 1 are used to define individual nuclei.

In this part of the optimization it is important to identify all of the nuclei that are stained with Hoechst and to distinguish individual nuclei. In order to start this process a single typical image of Hoechst stained nuclei is selected from the collection of images in an experiment. In order to optimize both object selection and discrimination, the user can alter several settings within the software. There are many examples of this type of optimization in the documentation provided by the developer of the software (Cellomics, Inc.).

Briefly, the first parameter optimized in the image analysis software, allows all the present nuclei to be identified as an object for analyses in addition to the area of the object in which the algorithm will be applied. The algorithm we use the most frequently for this purpose is the Isodata threshold object identification method, which selects a certain percentage of the brightest pixels in an image. If nuclei are not brightly stained the fixed threshold object identification method can be used. In optimizing the Isodata threshold algorithm, we apply numbers between -0.2 and -0.8, depending on the intensity and uniformity of the Hoechst staining of the nuclei. Larger values result in the rejection of dimmer pixels in the image. The optimal setting for this algorithm is ultimately determined empirically by using several different numbers, and each time a new number is chosen, the algorithm is applied to the selected image. Once a value is chosen, several images should be tested with this algorithm to insure uniform results. More details are found in manufacturer's protocol entitled "ArrayScan HCS Reader: Cell Health Profiling BioApplication Guide." This is either provided with the software or can be obtained online at www.clubhcs.com after registering.

In order to select individual nuclei, optimizing the image processing parameters of background correction, object smooth factor, and object segmentation in this software can minimize the effect of clumping on the final analysis. Instructions and examples of how changing these parameters can affect the image analysis can also be found in the manufacturer's protocol. In general these parameters are optimized using images from both a negative and positive control

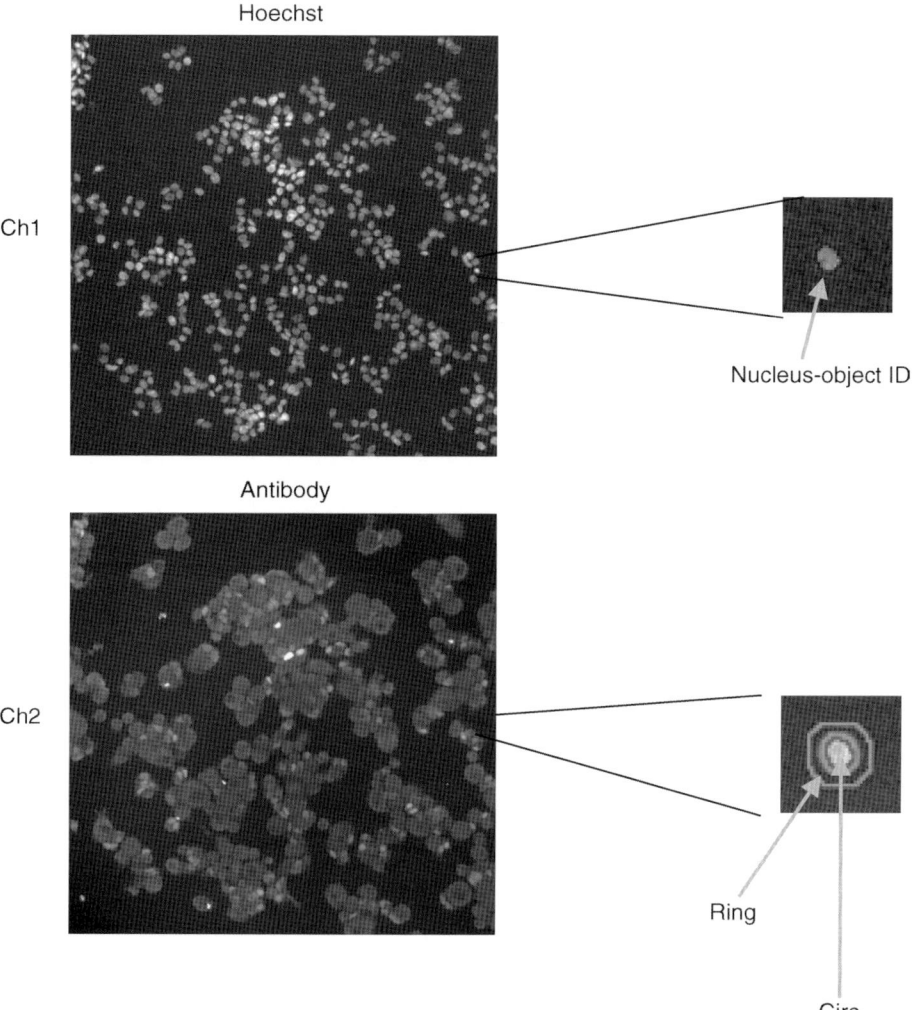

Fig. 4. Example of image analysis performed by the Cell Health Profiling BioApplication. The image is identified in channel 1 and a circle is placed around the nucleus. The area within this circle is termed *circ*. Using the image in channel 2, a second circle is drawn a specified number of pixels from the first circle (*circ*) and this is termed *ring*. Separate measurements of fluorescent intensity can be obtained for both of these regions of the cell. Because the stains that are used in this protocol are nucleic acid specific, the analysis of the region termed *ring* is not necessary.

and once parameters are optimized they are tested using images from a subset of the experimental wells to ensure that visual observations are reflected in the results of the image analysis.

For measurements of apoptosis, the desired information is the percentage of cells stained with YO-PRO-1, because this provides a measurement of the portion of cells that have become permeable to the nucleic acid stain. The cell health-profiling algorithm allows measurements of staining intensity in both the nucleus (circ) and the cytoplasm (ring) (*see* **Fig. 4**). Because YO-PRO-1 is a nucleic acid stain, measurements of staining intensity are only required for the nucleus, so the algorithm is set with a zero ring width. Once acceptable parameters are selected for the image analysis software that adequately identifies individual nuclei using channel 1, then image selection parameters in channel 2 are modified so that only stained nuclei are selected for analysis. In general, the only parameter that needs to be modified here is the average intensity threshold. The threshold should be set using both negative and positive controls so that only clearly stained nuclei are selected. This number will be reflected in the final analysis as the percentage selected. It is this number that is compared between control and experimental wells.

3.4.2. Image Analysis of PI and BrdU Antibody Stained Cells

The image analysis software that we use to analyze these images is also the Cell Health Profiling Bioapplication. The same basic tenets outlined in **Subheading 3.4.1.** for identifying individual nuclei apply in this section. The only difference is that total intensity measurements are important, so it is necessary to ensure that the entire nucleus is identified and circled. In this case initial optimization is done using the PI stained nuclei from channel 1. If some cells are clumped and the software is unable to identify them as separate objects, they can be removed from the final analysis by altering the object selection parameters in channel 1. Parameters that can be changed to eliminate these include total intensity, area, and shape. Nucleus area is the most reliable parameter to vary in our hands, because this is relatively constant from individual cell to individual cell and if the cells are clumped this number increases as a function of the number of nuclei in the clump.

In this particular protocol, all of the objects identified and selected in channel 1 should be selected and identified in channel 2. This is because the total intensity of anti-BrdU antibody staining in channel 2 is the important measurement, not the presence or absence of staining. The negative controls for anti-BrdU antibody staining should include wells that have been stained with only the secondary antibody to access the intensity of the nonspecific staining. The total intensity measurement of these nuclei should be extremely low and significantly different from positively stained nuclei. Total intensity measurements are obtained for both the PI and the anti-BrdU antibody staining, and these are plotted to obtain information about cell proliferation (*see* **Fig. 5**). As an alternative, the percentage of cells in S-phase, during the BrdU pulse can be used as an indication of cellular proliferation. This would require object selection in channel 2 as in **Subheading 3.4.1.**

3.5. Data Analysis

Analysis of data is different for each type of staining and will be described separately.

3.5.1. Analysis of YO PRO 1 Stained Cells

As stated in **Subheading 3.4.1.** the important measurement is the percentage of cells stained by YO-PRO-1. These numbers are obtained by opening the files using Cellomics, Inc. vHCS View software. For this protocol, the parameters that are typically downloaded into an Excel spreadsheet include the total number of Hoechst stained nuclei, the average intensity of the Hoechst staining, and the percentage of nuclei selected in channel 2. Once these numbers are downloaded into Excel they can be graphed using a program such as GraphPad Prism® to allow curve fitting and basic statistical analysis. Standard deviations are determined for the experimental results using the averaged values from four independent wells. High standard deviations for an experiment might indicate that the image analysis software was not optimized for this application, high background, or uneven staining. If the variability is high, individual images should also be examined for evidence of software or hardware failure.

3.5.2. Analysis of Anti-BrdU Antibody Stained Cells

As stated in **Subheading 3.4.2.** it is important to obtain fluorescent intensity measurements for both PI staining and the AlexaFluor 488 staining of individual nuclei. The data is accessed using Cellomics, Inc. vVHS view software and instead of downloading average values for each well, intensity measurement both channels in each nucleus are downloaded. From these measurements, a bivariate distribution of DNA content (PI staining) Vs BrdU content (AlexaFluor 488 staining) can be generated on a scatter plot and analyzed for the proportion of cells in G1, S, and

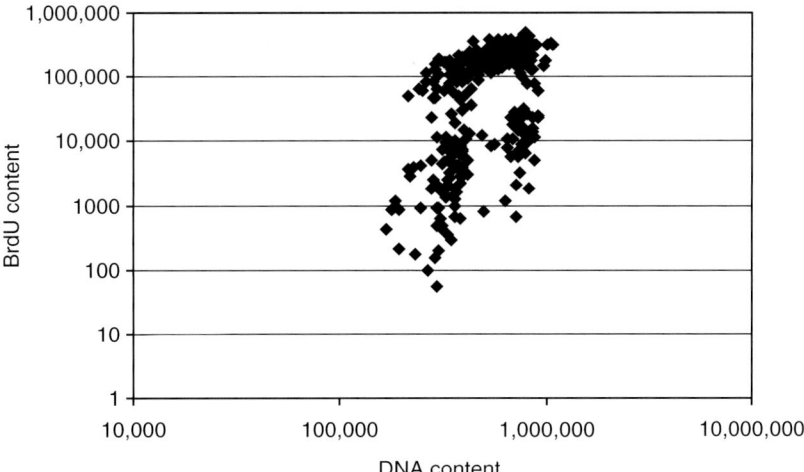

Fig. 5. Measurement of cell proliferation after treatment with siRNA. Cells were plated on 96-well plates, treated, pulsed with BrdU, fixed, and then stained with anti-BrdU and PI. Plates were transported to the KSR for image collection and then automated analysis was performed on the collected images. This is a typical scatter plot of BrdU staining intensity vs PI intensity. This is used for calculating the number of cells in G0/G1, S, and G2/M phases.

G2/M phases of the cell cycle (*see* **Fig. 5**). This is performed as described in **ref. *19***. A simple Student *t*-test can also be used to assess if there is a significant difference between the BrdU staining in the control vs the experimental.

4. Notes

1. A hemocytometer can also be used to count cells, but the process is longer and subject to greater variability than the results obtained with the Coulter Counter.
2. Cell lines must be selected based on the question that is being answered. For this application the question was whether or not the presence or absence of an amplification of a specific chromosomal amplification had an effect on cell proliferation and apoptosis. Cell lines were chosen that contain this amplification (positive) and cell lines that do not contain this amplification (negative). The experiment was then to knock down expression of a particular gene in this region with siRNA to determine if it plays an important role in cell proliferation or apoptosis. Because cancer is associated with the deregulation of these processes, it is important to understand the effect of overexpression of specific genes in this region.
3. When using larger cells, such as lung or skin fibroblasts, we typically reduce the number to 2500–3000 of cells per well in a 96-well plate. With this particular protocol, only the nucleus is used in the image analysis, therefore, 3000 cells should be a good starting point when evaluating larger cells in this protocol. Before doing this protocol on a large scale with a particular cell line it is a good practice to plate various numbers of cells in the wells of a 96-well plate, spanning 3000, and then check cell growth at the time-point they will be imaged in the final assay. It is important to avoid clumping of cells as much as possible, and this is achieved by mixing well before plating and keeping the density low enough so individual cells can be clearly differentiated from each other. This becomes important when performing image analysis, because the program analyzes individual nuclei. When cells are very close together the image analysis software might recognize several nuclei as one nucleus, impacting the final results.
4. Because the Hoechst is visualized in the blue channel and the YO-PRO-1 is visualized in the green channel, PI can also be added as a nucleic acid stain to identify necrotic cells. This would be visualized in yet a third channel. The only modification to the protocol is to add PI at a final concentration of 1 µg/mL to the staining mixture containing YO-PRO-1 and Hoechst. The filter set used would then be XF93 and the Texas Red channel would be added to the analysis.
5. Staining of the nuclei with YO-PRO-1 occurs almost immediately, but a short amount of time is required to obtain the desired intensity of staining with Hoechst.

6. During this step the media should not be changed because this could affect the growth rate of the cells. Growth media becomes "conditioned" after cells have been grown in it for a period of time, because it contains extracellular factors and metabolites secreted from the cells in the media, in addition to the original components of the media. These extracellular factors might include growth factors or other molecules that effect the proliferation of cells. We typically make our standard solution as 1 mM BrdU in PBS and add 1 μL to the 100 μL of media in each well of a 96-well plate.
7. If a 24-well plate is used add 100 μL and rock the plate gently back and forth on a rocker so that all cells are exposed to the antibody mixture. This can also be applied to the secondary antibody.
8. This is the step in which the even distribution of the cells makes the most difference. It is important, especially for the YO-PRO-1 staining, that individual nuclei can be distinguished. This is because have an accurate count of the total number of nuclei affects the accuracy of the results. The most important information obtained from this analysis is the percentage of nuclei stained with YO-PRO-1, indicating the number of cells undergoing apoptosis. The percentage obtained might be artificially high, if clumps of cells are not distinguished from one another. The image analysis Cell Health Profiling Bioapplication Software has some features that allow it to distinguish individual nuclei in clumps, but it limited.

Acknowledgments

The authors thank Dr. Gordon Mills for supplying many of the cell lines used in these studies.

References

1. Hanahan, D. and Weinberg, R. A. (2000) The hallmarks of cancer. *Cell* **100,** 57–70.
2. Albertson, D. G., Collins, C., McCormick, F., and Gray, J. W. (2003) Chromosome aberrations in solid tumors. *Nat. Genet.* **34,** 369–376.
3. Silva, J., Chang, K., Hannon, G. J., and Rivas, F. V. (2004) RNA-interference-based functional genomics in mammalian cells: reverse genetics coming of age. *Oncogene* **23,** 8401–8409.
4. Singer, O., Yanai, A., and Verma, I. M. (2004) Silence of the genes. *Proc. Natl Acad. Sci. USA* **101,** 5313, 5314.
5. Dolbeare, F., Gratzner, H., Pallavicini, M. G., and Gray, J. W. (1983) Flow cytometric measurement of total DNA content and incorporated bromodeoxyuridine. *Proc. Natl Acad. Sci. USA* **80,** 5573–5577.
6. Wronski, R., Golob, N., Grygar, E., and Windisch, M. (2002) Two-color, fluorescence-based microplate assay for apoptosis detection. *Biotechniques* **32,** 666–668.
7. Kiechle, M., Jacobsen, A., Schwarz-Boeger, U., Hedderich, J., Pfisterer, J., and Arnold, N. (2001) Comparative genomic hybridization detects genetic imbalances in primary ovarian carcinomas as correlated with grade of differentiation. *Cancer* **91,** 534–540.
8. Patael-Karasik, Y., Daniely, M., Gotlieb, W. H., et al. (2000) Comparative genomic hybridization in inherited and sporadic ovarian tumors in Israel. *Cancer Genet. Cytogenet.* **121,** 26–32.
9. Suzuki, S., Moore, D. H., 2nd., Ginzinger, D. G., et al. (2000) An approach to analysis of large-scale correlations between genome changes and clinical endpoints in ovarian cancer. *Cancer Res.* **60,** 5382–5385.
10. Collins, C., Volik, S., Kowbel, D., et al. (2001) Comprehensive genome sequence analysis of a breast cancer amplicon. *Genome Res.* **11,** 1034–1042.
11. Pinkel, D., Segraves, R., Sudar, D., et al. (1998) High resolution analysis of DNA copy number variation using comparative genomic hybridization to microarrays. *Nat. Genet.* **20,** 207–211.
12. Iwabuchi, H., Sakamoto, M., Sakunaga, H., et al. (1995) Genetic analysis of benign, low-grade, and high-grade ovarian tumors. *Cancer Res.* **55,** 6172–6180.
13. Lapuk, A., Volik, S., Vincent, R., et al. (2004) Computational BAC clone contig assembly for comprehensive genome analysis. *Genes Chromosomes Cancer* **40,** 66–71.
14. Idziorek, T., Estaquier, J., De Bels, F., and Ameisen, J. C. (1995) YOPRO-1 permits cytofluorometric analysis of programmed cell death (apoptosis) without interfering with cell viability. *J. Immunol. Methods* **185,** 249–258.
15. Daly, J. M., Jannot, C. B., Beerli, R. R., Graus-Porta, D., Maurer, F. G., and Hynes, N. E. (1997) Neu differentiation factor induces ErbB2 down-regulation and apoptosis of ErbB2-overexpressing breast tumor cells. *Cancer Res.* **57,** 3804–3811.

16. Estaquier, J., Idziorek, T., Zou, W., et al. (1995) T helper type 1/T helper type 2 cytokines and T cell death: preventive effect of interleukin 12 on activation-induced and CD95 (FAS/APO-1)-mediated apoptosis of CD4+ T cells from human immunodeficiency virus-infected persons. *J. Exp. Med.* **182,** 1759–1767.
17. Estaquier, J., Tanaka, M., Suda, T., Nagata, S., Golstein, P., and Ameisen, J. C. (1996) Fas-mediated apoptosis of CD4+ and CD8+ T cells from human immunodeficiency virus-infected persons: differential in vitro preventive effect of cytokines and protease antagonists. *Blood* **87,** 4959–4966.
18. Gray, J. W., Dolbeare, F., Pallavicini, M. G., Beisker, W., and Waldman, F. (1986) Cell cycle analysis using flow cytometry. *Int. J. Radiat. Biol. Relat. Stud. Phys. Chem. Med.* **49,** 237–255.
19. Dolbeare, F. and Selden, J. R. (1994) Immunochemical quantitation of bromodeoxyuridine: application to cell-cycle kinetics. *Methods Cell Biol.* **41,** 297–316.

26

Target Validation in Drug Discovery

Robert A. Blake

Summary

The process of target validation identifies and assesses whether a molecular target merits the development of pharmaceuticals for therapeutic application. The most valuable application of high content screening to target validation is at the early stages of the process when genetic methods (including RNA interference—RNAi) are being applied to many potential targets. At this stage both throughput and indepth analysis are required. This process is illustrated using various examples from the area of oncology target validation. The Akt signal transduction pathway is used to illustrate an efficient way of identifying HCS compatible reagents for use in assay development. RNAi transfection methods are discussed. A description is given of an HCS assay that simultaneously measures two nodes of the Akt pathway: Akt substrate phosphorylation and RPS6 phosphorylation. Another example of an assay measuring proliferation (DNA synthesis) and apoptosis (Histone H2B phosphorylation) within the same cell population is used to illustrate the combination of typical phenotypic assays.

Key Words: Akt; apoptosis; BrdU; DNA synthesis; high content screening; histone H2B; oncology; phospho-histone H2B; phospho-RPS6; PRAS40; phospho-PRAS40; 4E-BP1; phospho-4E-BP1; RNAi, RPS6; siRNA, S-phase; target validation.

1. Introduction

The term "target validation" describes the process of demonstrating that a molecular target is a therapeutically relevant pharmacological target. In reality a target is not truly "validated" until late stage clinical trials are complete and the mechanism of action understood. "Target validation" is also used to describe the very early stages of this process that precede the development of pharmacological reagents—a stage that is very reliant on genetic methods of modulation of the target. These genetic methods include RNAi or small interfering RNA (siRNA; a gene silencing technology that is finding a major application in target validation; *see* Chapter 18 on RNAi and **Note 1**).

It is this earliest stage of target validation, when multiple targets are being processed in parallel, that high content screening is most valuable. HCS is able to merge the detailed biological measurements often required for target validation with the throughput required for processing multiple targets. The focus of this chapter will be the application of HCS to the early phase of the target validation process using RNAi techniques for decreasing gene expression. This process will be illustrated by a description of methods used to identify and validate oncology targets that are associated with the Akt signal transduction pathway and affect cellular proliferation or apoptosis. The general principles behind these examples can be extended to other areas of biology and target validation projects.

2. Materials

1. HCS cell imaging equipment: Cellomics, Inc. Arrayscan with Compartmental Analysis Bioapplication.
2. 96-well plate washer: for example, Titertek MAP C2 Quadrant.
3. Microtiter plate liquid handling equipment: for example, Titertek Multidrop.
4. Microtiter plate 96/384-well-to-96/384-well liquid handling equipment: for example, Matrix PlateMate® Plus, Caliper LifeSciences RapidPlate®.
5. Transfection reagents: siRNA, positive and negative control siRNA, Oligofectamine.
6. Tissue culture reagents: DMEM growth medium. OptiMEM1 serum reduced medium. Fetal bovine serum (FBS).
7. Human lung carcinoma cell line A549, Human prostate cancer cell line PC3.
8. Insulin-Arg (Upstate Biotechnology 01-207), IGF-I (Upstate Biotechnology 01-189), LY294002 (Calbiochem, cat. no. 440202), Rapamycin (Calbiochem [EMD Biosciences, San Diego, CA] cat. no. 553210).
9. Antibodies to components of signal transduction pathway or biological process under study:
   a. Rabbit antiphospho-Ser235/236-RPS6 antibody (Cell Signaling, cat. no. 2211).
   b. Biotinylated antiphospho-PRAS40-Thr246 (biotinylated form of Biosource, cat. no. 44-1100G).
   c. Antibromodeoxyuridine, mouse IgG1, monoclonal (anti-BrdU) (Molecular Probes/Invitrogen, cat. no. A21300).
   d. Antiphospho-Histone H2B rabbit polyclonal (UBI, cat. no. 07-191).
10. Fluorophore-conjugated secondary antibodies:
    a. Alexa Fluor® 546 goat antirabbit IgG (H + L) highly cross-adsorbed (Molecular Probes [Invitrogen Corp., Carlsbad, CA] A-11035).
    b. Alexa Fluor® 488 goat antirabbit IgG (H + L) highly cross-adsorbed (Molecular Probes, cat. no. A-11034).
    c. Alexa Fluor® 546 goat antimouse IgG1 (Molecular Probes, cat. no. A-21123).
    d. Alexa Fluor® 488 goat antimouse IgG1 (Molecular Probes, cat. no. A-21121).
11. BisBenzimide (Hoechst, cat. no. 33258; Sigma, cat. no. B-1782 [St. Louis, MO]).
12. 96-well clear bottom black wall tissue culture plates (e.g., ViewPlate-96 Black, PerkinElmer, cat. no. 6005182).
13. Plate seals (e.g., ThinSeals from Excelscientific [Wrightwood, CA] cat. no. 100-THIN-PLT).
14. 4% w/v formaldehyde in PBS.
15. 0.5% v/v Triton X-100 in PBS.
16. 5-bromo-2′-deoxyuridine (BrdU; Sigma B9285).
17. 2.4 M HCl.
18. Phosphate buffered saline (PBS), pH 7.6.

3. Methods

The initial step in any target identification and validation process is defining the criteria that the prospective targets should meet. In this example the criteria are defined as follows: (1) the target must modulate components of the Akt signal transduction pathway (*see* **Fig. 1**); (2) the target must regulate either proliferation or apoptosis of cancer cells. The next step is the development of assays to survey the biology of interest. The following sections describe: (1) a process to identify suitable HCS reagents to probe the Akt signal transduction pathway; (2) a specific assay that surveys two individual nodes of the Akt signal transduction pathway; (3) an assay that simultaneously monitors effects on proliferation and apoptosis. The cell-lines selected for this study include: A549 a lung adenocarcinoma cell-line that has an upregulated Akt pathway probably resulting from a deletion of the tumor suppressor kinase LKB1 *(1,2)* and PC3 cells (a prostate cancer cell-line that has an up-regulated Akt pathway because of deletion of the tumor suppressor PTEN *(3)* (*see* also **Note 1**).

3.1. Reagent Identification

The initial limiting factor in developing HCS assays is usually the identification of suitable reagents to track the biology under study; that is, identify suitable antibodies that work sufficiently well for immunofluorescence staining to be applied to HCS. Unfortunately the manufacturers of commercial antibodies do not always accurately describe the performance of their products and

Target Validation in Drug Discovery

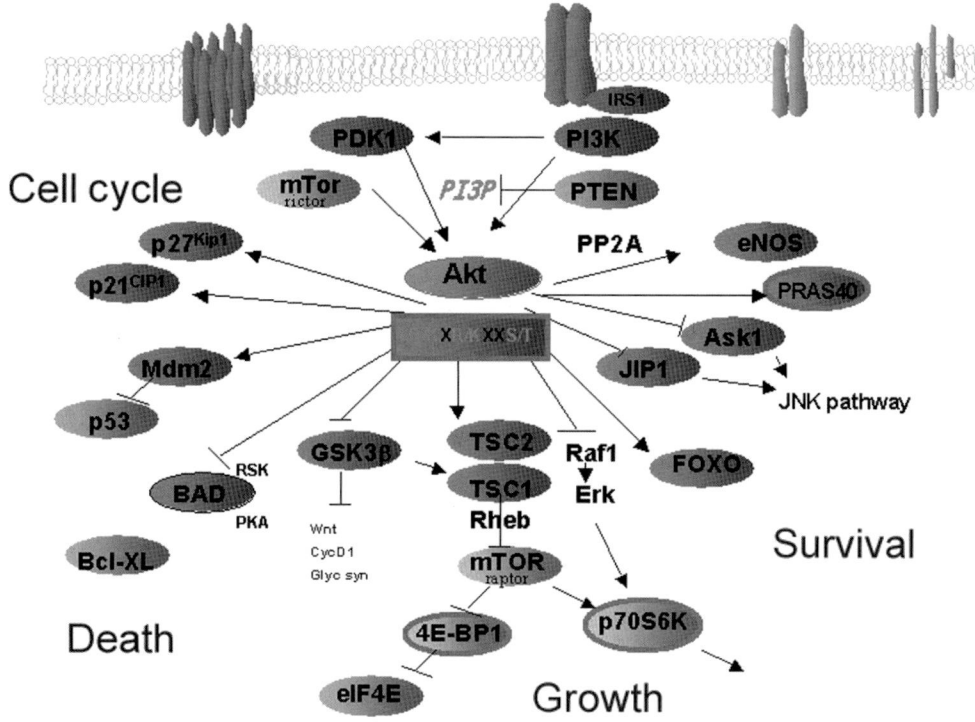

Fig. 1. Schematic representation of the Akt signal transduction pathway: The PI3Kinase/Akt/mTor pathway is generally involved in the regulation of cell survival and growth. It is has been widely implicated in many types of cancer. Akt kinases phosphorylate substrates within the generic amino-acid sequence R/K-X-R/K-X-X-S/T and often regulate binding to 14-3-3 proteins.

the choice of cell-line often affects the usefulness of an antibody owing to differences in epitope expression. Often the most efficient approach is to screen all available reagents together for direct comparison of their performance. The example used to illustrate this process is the survey of multiple reagents of components of the Akt signal transduction pathway. The method is based on testing the immunofluorescence-staining pattern for each antibody on cells, in which the Akt pathway is stimulated or inhibited (*see* **Fig. 2**). The objective is to identify antibodies whose immunofluorescence-staining pattern is modulated in a way that makes sense according to the biology of the pathway and individual protein (e.g., a decrease in phospho-staining of Akt substrates following PI3 Kinase inhibition).

1. Collect aliquots of antibodies to components of the Akt signal transduction pathway (proteins, phosphoproteins, and so on) and dispense each into a polypropylene 96-well plate, grouping antibodies according to species of origin (enables easier dispensing of the appropriate secondary antibody later).
2. Seed four 384-well black wall clear bottom tissue culture plates with A549 (4000 cells per well; 2.5% FCS); incubate overnight (*see* **Note 2** and **3**).
3. Dispense inhibitors of the Akt/mTor pathway as follows: add the PI3Kinase inhibitor LY294002 (final concentration 50 μM) to plate nos. 1 and 2 to the alternating columns 2, 4, 6, and so on. Add the mTOR inhibitor rapamycin (final concentration 100 nM) to plate nos. 3 and 4 to the alternating columns 2, 4, 6, and so on (**Note 4**).
4. Dispense activators of the Akt/mTor pathway as follows: to plate nos. 1 and 3 add insulin (concentration 500 ng/mL) to the alternating rows B, D, F, and so on. To plate nos. 2 and 3 add IGF1 (concentration 50 ng/mL) to the alternating rows B, D, F, and so on.
5. Fix the cells as follows: 40 min after addition of the insulin and IGF1 add an equal volume of 4% formaldehyde directly to the tissue culture medium to fix the cells; incubate at room temperature for 1 h (**Note 5**).

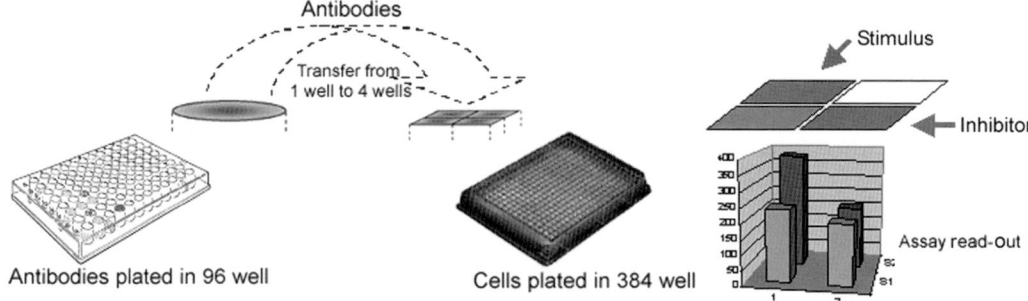

Fig. 2. Schematic depicting the process described in **Subheading 3.1** by which antibodies are screened for changes in immunofluorescence staining patterns in cells treated with activators or inhibitors of the Akt pathway.

6. Wash (four times) with 100 μL of PBS and add 0.5% Triton X-100 (in PBS) to permeabilize the cells. Incubate for 5 min and then wash (**Notes 6** and **7**).
7. Stain the cells with the various primary antibodies as follows:
   a. Transfer 1 μL of each antibody from the stock 96-well plate to a second polypropylene plate. Dilute with 200 μL of PBS.
   b. Remove the PBS from each of the 384-well tissue culture plates (by inverting and flicking over a sink). Transfer 40 μL of each of the diluted antibody from the 96-well plate to the cluster of four wells that correspond to the same position in the 384-well plate. Incubate overnight at 4°C. Wash (four times) with 100 μL of PBS.
8. Stain the cells with appropriate secondary antibodies as follows (**Note 2**).
   a. Prepare the appropriate fluorophore labeled secondary antibodies Alexa-fluor 488 conjugated antimouse and/or antirabbit IgG (Molecular Probes) (depending on the species of origin of the primary antibodies at a 1/1000 dilution in PBS containing 1 μg/mL Hoechst dye.
   b. Remove the PBS from each of the 384-well tissue culture plates and transfer 50 μL of each of the diluted secondary antibodies to the appropriate wells according to species of origin of the primary antibody. Incubate for 2 h at room temperature. Wash (four times) with 100 μL of PBS.
9. Seal the plates with plastic plate seals.
10. Read the plates on a Cellomics, Inc. Arrayscan plate reader using the "Compartmental Analysis" Bioapplication using XF100-Hoechst in channel 1 (focus channel) and the AlexaFluor488 in channel 2 with the XF100/FITC (1 s fixed exposure; *see* **Note 8**).
11. Data analysis: Export the following well features:
    a. MEAN_CircTotalIntCh2.
    b. MEAN_CircAvgIntenCh2.
    c. MEAN_RingTotalIntenCh2.
    d. MEAN_RingAvgIntenCh2.
    e. MEAN_CircRingAvgIntenDiff.
    f. MEAN_CircRingAvgIntenRatio.

These features will capture variations in the staining intensity in the nuclear and cytoplasmic regions (*see* **Note 9**). Compare the data from your untreated, stimulated, and inhibited samples and identify antibodies whose immunofluorescence pattern is consistently modulated by activation or inhibition of the signaling pathway. Determine whether the changes in the immunofluorescence make sense according to the biology of the individual proteins. **Figure 3** shows examples of antibodies identified using this procedure.

Fig. 3. (*Opposite page*) Thr 246 site (Biosource no. 44-1100G) shows a similar pattern. The antibody to the phospho-Ser235/236-RPS6 antibody (Cell Signaling no. 2211) strongly stains the cytoplasm in control A549 cells. Transfection with siRNA to Akt2 dramatically diminishes the cytoplasmic staining pattern. An antibody to the mTor substrate 4E-BP1's phospho-Thr37/46 site (Cell Signaling no. 9459) gives a diffuse cytoplasmic and nuclear stain in control A549 cells. Transfection with siRNA to Akt2 reduces the cytoplasmic but not the nuclear staining.

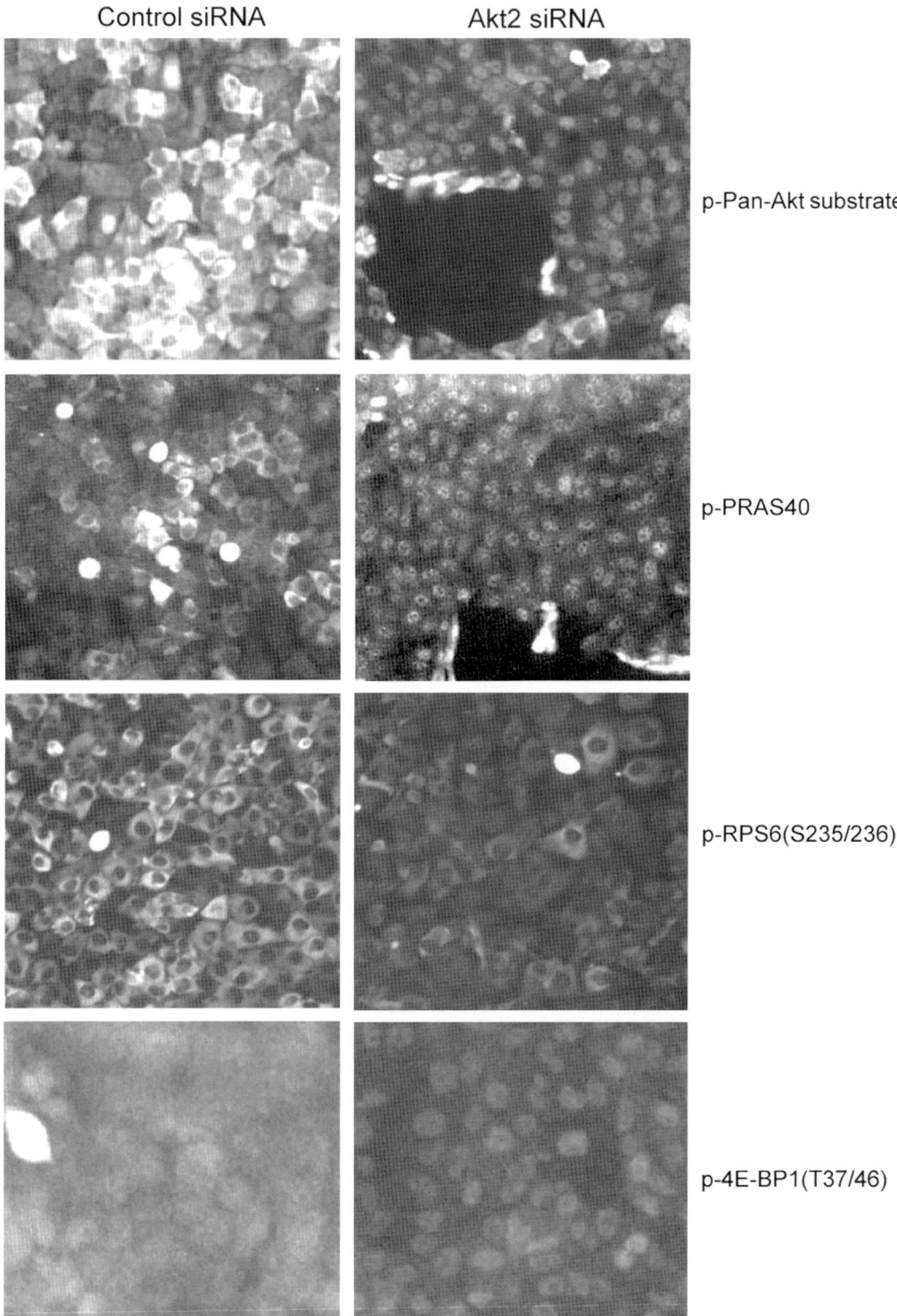

Fig. 3. Examples of antibodies identified using the process described in **Subheading 3.1.** The phospho-Serine/Threonine Akt substrate antibody (Cell Signaling no. 9611) recognizes both cytoplasmic and nuclear proteins in control A549 cells. Transfection with siRNA to Akt2 or inhibition of PI3Kinase (not shown) reduces cytoplasmic staining but not nuclear staining. The phospho-antibody to the Akt substrate PRAS40

3.2. Multiplexed Phospho-RPS6, Phospho-PRAS40 HCS Assay

Two antibodies that were identified using the method described in **Subheading 3.1.** were: antiphospho-PRAS40, which provides a measurement of Akt activity *(4)*; and antiphospho-RPS6_235/236, which provides a measurement of the p70 S6Kinase/mTor arm of the pathway *(5,6)*. This section describes an HCS assay in which both of these read-outs are combined to provide simultaneous measurements on two sections of the Akt pathway. This is an example of a multiplexed HCS assay that can be used to test siRNA for the purposes of early stage target validation (*see* **Note 10**). Other assays developed in **Subheading 3.1.** could also be applied.

1. Resuspend the siRNA according to manufacturers instructions in RNAase free H_2O at a final concentration of 20 μ*M*.
2. Seed A549 or PC3 cells at a density of 2000 cells per well (100 μL DMEM, 2.5% FBS, no antibiotics) in a 96-well. Incubate overnight at 37°C, 5% CO_2 (**Notes 8, 12,** and **13**).
3. Prepare the siRNA transfection mixes as follows in separate wells of a 96-well polypropylene plate (The volumes listed are sufficient for a single well of a 96-well plate and must be scaled up according to the number of samples being transfected. The final siRNA concentration is 25 n*M*.):
    a. Prepare Mix-A by adding 0.15 μL of 20 μ*M* siRNA to 16.85 μL of OptiMEM in a polypropylene 96-well plate (this is enough for transfecting a single well of a 96-well plate. Scale accordingly).
    b. Prepare Mix-B in bulk to add to Mix-A. For a single transfection well blend 0.6 μL of Oligofectamine with 2.4 μL of OptiMEM. Scale these volumes up to be sufficient for the total number of wells required including excess to account for dispensing dead-volume.
    c. Mix 3 μL of Mix-B with 17 μL Mix-A (Single transfection volume. Scale accordingly.) Mix gently by pipeting up and down a couple of times. Let the total mixture stand for 20 min (room temperature).
    d. Transfer 20 μL of the MixA/B blend to the each well of the tissue culture plates. Return the tissue culture plates to the incubator (**Notes 13–15**).
4. Change the cells into a no serum "starvation medium" as follows: 24 h after transfection wash the cells in DMEM by gently aspirating off the existing medium; wash with 200 μL DMEM per well; repeat the aspiration and dispense 100 μL of DMEM.
5. Fix the cells as follows: 72 h after transfection fix the cells by adding 100 μL of 4% formaldehyde to each well and incubating for 1 h at room temperature (**Note 4**).
6. Permeabilize the cells as follows: aspirate the formaldehyde from each well and add 100 μL of 0.5% v/v Triton X-100 in PBS. Incubate for 5 min at room temperature. Wash three times with 200 μL PBS (**Note 6**).
7. Stain for phospho-RPS6_S235/236 as follows:
    a. Add 50 μL of PBS containing a 1/100 dilution of rabbit antiphospho-Ser235/236-RPS6 antibody and incubate overnight at 4°C. Wash four times with 200 μL PBS (pH 7.4).
    b. Add 50 μL of PBS containing a 1/1000 dilution of Goat-antiRabbit-Alexa546 and 1 μg/mL Hoechst 33258. Incubate at room temperature for 2 h. Wash four times with 200 μL PBS (pH 7.4) (**Note 16**).
8. Stain for phospho-PRAS40 as follows:
    a. Incubate with 1/500 dilution of biotinylated antiphospho-PRAS40 antibody overnight at 4°C. Wash.
    b. Incubate with 500 μL of a 1/1000 dilution of AlexaFluor 488 conjugated Streptavidin in PBS; Incubate 2 h at room temperature. Wash leaving 200 μL PBS in the wells. Seal the plates (**Note 16**).
9. Read on the Cellomics Arrayscan using BioApplication "Compartmental Analysis" with reference and autoexposure wells activated and defined as the negative control (uninhibited) wells. Channel 1—nuclei, XF53-Hoechst, Channel 2—pRPS6, XF53—TexasRed, Channel 3—pPRAS40, XF53—FITC. Autoexposure peak target (percentile): Ch1—40%, Ch2—24%, Ch3—24%. RingAvgIntenLevel-HighCh2_CC set to 2. CircRingAvgIntenDiffLevelLowCh3_CC set to 0.1 (these are suggested starting values *see* **Note 17**).
10. Retrieve the following data:
    a. Percentage of cells positive for pRPS6 staining: HighRingAvgIntenCh2 (%).
    b. Percentage of cells positive for cytoplasmic vs nuclear pPRAS40 staining: LowCircRingAvgIntenDiffCh3 (%).
    c. Cell density: ObjectPerFieldCount (**Notes 8, 9, 17–19**).

8. Kato, M., Mochizuki, K., Kuroda, K., et al. (1991) Histone H2B as an antigen recognized by lung cancer-specific human monoclonal antibody HB4C5. *Hum. Antibodies Hybridomas* **2(2),** 94–101.
9. Kim, H. S., Cho, J. H., Park, H. W., et al. (2002) Endotoxin-neutralizing antimicrobial proteins of the human placenta. *J. Immunol.* **168(5),** 2356–2364.
10. Kunkel, M. T., Ni, Q., and Tsien, R. Y. (2005) Spatio-temporal dynamics of protein kinase B/Akt signaling revealed by a genetically encoded fluorescent reporter. *J. Biol. Chem.* **280(7),** 5581–5587.
11. van Hemert, M. J., Niemantsverdriet, M., Schmidt, T., Backendorf, C., and Spaink, H. P. (2004) Isoform-specific differences in rapid nucleocytoplasmic shuttling cause distinct subcellular distributions of 14-3-3 sigma and 14-3-3 zeta. *J. Cell Sci.* **117(Pt 8),** 1411–1420.

27

High Content Screening as a Screening Tool in Drug Discovery

Anthony Nichols

Summary

In most pharmaceutical and biotechnology companies there is a need to always improve the quality of lead candidates. This demand resulted in the use of cell-based screening as a method of choice in drug discovery *(1,2)*. High content screening (HCS) is multiplexed, functional cell-based screening *(3–6)*. HCS can be used in all aspects of drug discovery as an engine for driving lead discovery. The biological applications of HCS have been implemented in research in signaling, cell shape changes and toxicology. HCS has enabled an insight in the cellular effects of our clinical candidates in multiple cellular phenomena like dual reporter assay, subcellular target translocation and cellular morphology. Discovery of therapeutic protein and small molecule converge on diseases in therapeutic areas such neurological disorders and autoimmune diseases. HCS is used for assay development, primary, secondary screening and toxicology testing. In this chapter, the use of HCS assays in drug discovery is described and highlight the necessary step to set-up successfully these assays for screening.

Key Words: Cell-based assays; cell-membrane translocation; cell morphology assays; dual reporter assays; genotoxicity; high content screening (HCS); micronucleus; nuclear translocation.

1. Introduction

Traditional biochemical screening has not always delivered hits that can be followed up and that do not always yield the development of effective therapeutic agents. Cells are the smallest living complex entity. Biochemical screening formats do not address biological issues like cytotoxicity, complex biology or multicomponent target classes. Cell function is a complex interplay of signaling and feed back pathways lacking in isolated molecular or biochemical assays, moreover in vitro assays lack the possibility to provide information on the physicochemical properties of compound like cell permeability *(7–9)*. For these reasons cell-based assays are becoming more and more the preferred screening format to identify higher quality hits. Cell-based assays are the starting point for high information content screening (HCS) or multiple cellular event screening that provide access to major therapeutic target classes. HCS can perform multiple measurements per well or cell and allow screening for target function in a more physiological relevant setting. HCS provides more information than classical cell-free systems and can answer biological questions earlier in drug discovery. In the right format, cell-based screening can be target-based having the target in its more natural context. Cell-based screening and even more HCS for small molecule is well suited for orphan targets *(9)* or when it is not feasible to express and purify at the scale for high-throughput screening (HTS) a molecular target. Furthermore HCS may provide functional assay methodologies to identify compounds with a different mechanism of action, for example, allosteric modulator. Typically HCS can be divided into two categories

From: *Methods in Molecular Biology, vol. 356:*
High Content Screening: A Powerful Approach to Systems Cell Biology and Drug Discovery
Edited by: D. L. Taylor, J. R. Haskins, and K. Giuliano © Humana Press, Inc., Totowa, NJ

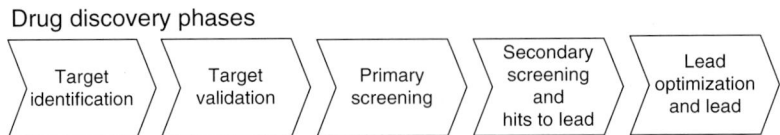

Fig. 1. Drug discovery phases.

1. Entire well measurement like dual reporter assays.
2. Multiparametric single cell measurements like cell shape, morphology, and target distribution changes within cells (*1,10*). HCS assays are used in early drug discovery process and can be described as the phase from target identification to leads. This phase can be divided as shown in **Fig. 1**. The purpose of this chapter is to review HCS methodologies used in drug discovery. Methods used for primary, secondary, and genotoxicity screening for potential therapeutic proteins and small molecules will be discussed. Drug discovery focusing on therapeutic protein discovery is well suited for the use of high content cell-based assay were the readouts are multiple phenotypic.

2. Materials
2.1. HCS Dual Reporter
1. HLR-cJun cells (Stratagene, La Jolla, CA).
2. Plasmids: pFC-MEKK, pBluescript II KS (+) (Stratagene), pRL-TK (Promega, Madison, WI).
3. Dual-Luciferase assay system (Promega), Tissue Culture Treated Microplate white 96-well clear well (PerkinElmer Life and Analytical Sciences, Boston, MA).
4. Dulbecco's modified Eagle's medium (DMEM) (high glucose, without L-glutamine, without sodium pyruvate), 2 mM L-glutamine, 10% fetal calf serum (FCS), 100 U of penicillin and streptomycin, Hygromycin B 100 µg/mL, and Gentamicin 250 µg/mL (Invitrogen, Carlsbad, CA).
5. FuGENE 6 transfection reagent (Roche Applied Science, Indianapolis, IN)
6. Microplate 2 injector luminometer Luminoskan Ascent (Labsystems, Finland).

2.2. HCS Nuclear Translocation
1. Neuro-2a (ATCC® CCL-131, Manassas, VA), HeLa cells (ATCC CCL-2, Manassas, VA), U373 (European Collection of Cell Cultures [ECACC]).
2. Minimum essential medium: with Earle's Salts, 2 mM L-glutamine, 1% nonessential amino acids, 1 mM sodium pyruvate, 10% FCS, 100 U of penicillin and streptomycin (Invitrogen).
3. c-Jun Activation HitKit HCS Reagent Kit, NF-κB Activation HitKit HCS Reagent Kit, STAT 2 Activation HitKit HCS Reagent Kits (Cellomics, Inc., Pittsburgh, PA).
4. Formaldehyde (37%), TNF-α, IL-1-α, IFN-β (Sigma, St. Louis, MO).
5. Tissue culture treated microplate black 96-well clear well (PerkinElmer Life and Analytical Sciences).
6. HCS reader ArrayScan® 3.1 and Nuclear Translocation bioapplication.

2.3. HCS Cytosol to Plasma Membrane Translocation
1. HeLa cells (ATCC CCL-2), U-2OS cells (ATCC HTB-96).
2. Weat Germ Agglutinin tetramethylrhodamine (TRITC) conjugate, TSA detection kit Alexa Fluor®488 conjugate and Hoechst 33342 dye (Molecular Probes, Eugene, OR), formaldehyde 37% and PMA (Sigma), IGF-1 (R&D Systems).
3. Goat antirabbit and antimouse coupled to Alexa Fluor 488 (Molecular Probes), PKCα Activation HitKit (Cellomics, Inc.), rabbit antiphospho-Akt (Ser473), (Cell Signaling Technology, Beverly, MA).
4. Minimum essential medium: with Earle's Salts, 2 mM L-glutamine, 1% nonessential amino acids, 1 mM sodium pyruvate, 10% FCS, 100 U of penicillin and streptomycin, McCoy's 5a medium, 2 mM L-glutamine, 10% FCS, 100 U of penicillin and streptomycin (Invitrogen).
5. HCS reader ArrayScan 3.1 and cytoplasm to cell membrane translocation bioapplication (Cellomics, Inc.).

2.4. HCS Morphology Assays

1. Neuroscreen-1 cells (NS-1) (Cellomics, Inc.).
2. RPMI-1640, 2 mM L-glutamine, 10% horse serum, 5% FCS, 100 U of penicillin and streptomycin (Invitrogen).
3. Neurite outgrowth HitKit (Cellomics, Inc.), 2.5S Nerve Growth Factor (NGF) (Promega), Tissue Culture Treated Microplate Black 96-well clear well Collagen I Coated (BD Bioscience, San Jose, CA).
4. HCS reader ArrayScan 3.1 and Neurite Outgrowth bioapplication (Cellomics, Inc.).

2.5. HCS in Genotoxicity Screening

1. CHO-K1 (ATCC).
2. F12K, 2 mM L-glutamine, 10% FCS, 100 U of penicillin and streptomycin (Invitrogen).
3. Micronuleus HitKit HCS Reagent Kit, (Cellomics, Inc.), Mytomycin C (Calbiochem), formaldehyde 37% (Sigma), tissue culture treated microplate black 96-well clear well collagen I coated (BD Bioscience).
4. HCS reader ArrayScan 3.1 and Micronucleus bioapplication (Cellomics, Inc.).

3. Methods

3.1. HCS Dual Reporter in Secondary Screening

One of the activation events of signaling pathways in cells is the phosphorylation of transcription factors, which subsequently activate transcription of genes that are dependent on the type of pathway that is stimulated. This method was used to measure the potential of c-Jun N-terminal kinase (JNK) inhibitors *(11–13)* in a cellular context. The cells used are cell line, which contain the expression plasmid for transcription factor c-Jun fused with the DNA binding domain of the yeast GAL4 and a plasmid that contains a synthetic promoter with five tandem repeats of the yeast GAL4 binding sites that control expression of the Firefly luciferase gene stably integrated into the HeLa cell. To activate the JNK pathway the upstream kinase MEKK1 *(14)* is transiently transfected with a constitutively active expression plasmid for Renilla luciferase. When c-Jun is phosphorylated by JNK, the fused DNA-binding domain binds the GAL4 binding sites and activates the transcription of the luciferase gene from the reporter plasmid. Expression (or activity) levels of Firefly luciferase reflect the activation status of JNK. The activity of Renilla luciferase reflects the transfection efficiency, the potential cytotoxicity of compounds and the presence of cells in individual wells.

In cell culture, the cells are routinely split when they reach 80–90% confluency.

3.1.1. Day 1 Morning

1. The cells from flasks of 90% confluent cultures are detached by treatment with trypsin-EDTA.
2. The cells are resuspended in culture medium and counted (*see* **Note 1**).
3. The cell suspension is diluted with medium at 3.5×10^6 cells/mL and 1 mL of cell suspension is seeded into a 10 cm culture dishes containing 9 mL of culture medium.
4. The plates are incubated at 37°C in a humidified atmosphere at 5% CO_2.

3.1.2. Day 1 Evening

The following plasmid mixtures were prepared:

1. Control: 0.2 µg pTK Renilla, 5.8 µg pBluescript KS 2. Induced: 0.1 µg pMEKK1, 0.2 µg pTK Renilla, 5.7 µg pBluescript KS (*see* **Note 2**).
2. The transfection mixture is prepared by mixing the above DNA with 18 µL of FuGENE 6 and 500 µL of OPTIMEM.
3. This mixture is added to the plated cells. The cells are incubated 12–18 h at 37°C in a humidified atmosphere of 5% CO_2.

3.1.3. Day 2

1. A 96-well plate containing 100 µL of culture medium per well is prepared. Negative control: 2 µL of 100% dimethyl sulfoxide (DMSO) is added to the 100 µL (in triplicate). Compound: 2 µL of compound in 100% DMSO is added to the 100 µL (in triplicate).

2. The transfected cells are detached by treatment with trypsin-EDTA and resuspended in 12 mL of culture medium. 100 μL of cells are added to each well of the 96-well plate.
3. The plate is incubated 12–18 h at 37°C in a humidified atmosphere of 5% CO_2.

3.1.4. Day 3

1. The medium is removed from the plate and the cells are washed two times with 100 μL PBS and the solution is removed.
2. 50 μL of 1X PLB lysis buffer is prepared according to the manufacturer's instructions and is dispensed into each well.
3. The culture plates are shacked for 15 min at room temperature on an orbital shaker to ensure complete and even coverage of the cell monolayer with the lysis buffer.
4. Luciferase Assay Reageant II and Stop & Glow Reagent are prepared according to the manufacturer's instructions; 10 mL of each solution is used per 96-well plate.
5. Transfer 20 μL of the cell lysate into a white opaque 96-well plate.
6. Load the 96-well plate into the luminometer, for reading use the following sequence: (1) Inject 100 μL of Luciferase Assay Reagent II wait 5 s, read 10 s. (2) Inject 50 μL of Stop & Glo Reagent wait 5 s, read 10 s.
7. Determine normalized results: (Firefly Luciferase light units)/(Renilla light units × 1000).

3.2. HCS Nuclear Translocation in Primary and Secondary Screens for Therapeutic Proteins and Small Molecule

Signal transduction in cell is often generated by the translocation of macromolecules (such as transcription factors or protein kinases) or smaller molecules (second messengers) from one cellular compartment to another and play a fundamental role in almost all cellular physiological processes, such as cell division, differentiation, cell motility, immune system function, neuronal transmission, and apoptosis. The regulation of transcription factors, such as c-Jun, STAT2, and NF-κB is by the translocation of the factor from the cytoplasm to the nucleus *(10,15,16)*. This method was used to measure the potential of JNK inhibitors *(11–13)* in a cellular context and NF-κB pathway modulators *(17)*. This method was also used to screen for potential therapeutic proteins that activate the JNK, the STAT2, and the NF-κB pathway. The cells are Neuro-2a mouse neuroblastoma *(18)* for JNK inhibitors, HeLa cells for NF-κB pathway modulator and U373 human astrocytoma for therapeutic proteins screen. To stimulate JNK in Neuro-2a we used the superoxide generator menadione *(19)*, JNK inhibitors were tested for their ability to inhibit phosphorylated c-Jun to translocate to the nucleus of these cells. TNF-α was used to activate the NF-κB pathway to test for compounds that inhibit the nuclear translocation of this transcription factor. IL-1 α was used as positive control for U373 for c-JUN and NF-κB *(15,16)* translocation and IFN-β was used as control inducer for STAT2.

In cell culture, the cells are routinely split when they reach 80–90% confluency.

3.2.1. Day 1

1. The cells from flasks of 90% confluent cultures are detached by treatment with trypsin-EDTA.
2. The cells are resuspended in culture medium and counted (*see* **Note 1**).
3. The cell suspension is diluted with medium at 5×10^4 cells/mL (*see* **Note 3**) and 90 μL of cell suspension is seeded into a 96-well plate.
4. The plate is incubated 12–18 h at 37°C in a humidified atmosphere of 5% CO_2.

3.2.2. Day 2

1. For a small molecule inhibitor assay, a 96-well plate containing 98 μL of culture medium, 2 μL of 100% DMSO for negative controls and 2 μL of compound in 100% DMSO is prepared. 10 μL of these solutions are transferred to the 96-well plate containing the cells and incubate 15 min at 37°C in a humidified atmosphere of 5% CO_2.
2. For a small molecule inhibitor assay, a 96-well plate is prepared containing 50 μL of medium with 5X final concentration of stimuli or 50 μL of medium for nontreated wells. 25 μL of these solutions is

High Content Screening as a Screening Tool in Drug Discovery

transferred to the 96-well plate containing the cells. For agonist protein screen a 96-well plate is prepared containing 50 µL of medium with 10X concentrated proteins and 10X concentrated protein dilution buffer for negative control in medium. 10 µL of these solutions is transferred to the 96-well plate containing the cells (*see* **Note 4**).

3. Incubate 30 min at 37°C in a humidified atmosphere of 5% CO_2 (*see* **Note 5**).
4. Add 75 µL of fixation solution (10% formaldehyde in PBS) and incubate at room temperature for 15 min in a fume hood.
5. Aspirate fixation solution and wash the plate once with 100 µL PBS per well.
6. Aspirate the PBS and add 100 µL 1X permeabilization buffer (PBS 0.5% Triton X-100) per well, and incubate for 90 s (*see* **Note 6**).
7. Aspirate permeabilization buffer and wash plate once with 100 µL PBS per well.
8. Aspirate wash buffer-M and add 50-µL primary antibody solution per well. Incubate for 1 h (*see* **Note 7**).
9. Aspirate primary antibody solution and add 100 µL 1X detergent buffer (PBS 0.01% Tween-20) per well. Incubate for 5 min (*see* **Note 7**).
10. Aspirate detergent buffer and then wash twice with 100 µL PBS per well.
11. Aspirate the PBS and then add 50 µL of staining solution (*see* **Note 8**) per well. Incubate for 1 h.
12. Aspirate staining solution and then add 100 µL 1X detergent buffer per well. Incubate for 5 min.
13. Aspirate detergent buffer and then wash twice with 100 µL PBS per well. Add 200 µL of PBS in wells.
14. Seal plate and run on ArrayScan HCS Reader using ×10 objective and the nuclear translocation bioapplication.
15. Store sealed plates in the dark at 4°C.

3.3. HCS Cytosol to Plasma Membrane Translocation in Small Molecule Secondary Screening

Stimulation of cells with growth factors initiates signal transduction cascades and subsequent intracellular activities that include recruitment to the cell membrane of macromolecules such as PKC and Akt/PKB. The cellular functions and regulation of these proteins, in most part, depend on specific subcellular localization *(20)*. Membrane targeting is mediated mainly by two conserved cysteine-rich domains for protein kinase C (PKC) and the pleckstrin domains for Akt that bind to charged phospholipids *(20)*. These cell-signaling events provide molecular targets for therapeutic intervention *(21)*. This method was used to assess the cellular activity of inhibitors of protein–phospholipid interaction.

In cell culture, The cells are routinely split when they reach 80–90% confluency.

3.3.1. Day 1

1. The cells from flasks of 85% confluent cultures are detached by treatment with trypsin-EDTA.
2. The cells are resuspended in culture medium and counted (*see* **Note 1**).
3. The cell suspension is diluted with medium at 4×10^5 cells/mL for HeLa cells and 3×10^5 cells/mL for U-2OS cells and 100 µL of cell suspension is seeded into a 96-well plate.
4. The plate is incubated 12–18 h at 37°C in a humidified atmosphere of 5% CO_2.

3.3.2. Day 2

1. For Akt, carefully aspirate 100 µL of medium with multi channel and add 200 µL of medium without serum (*see* **Note 4**). Incubate at 37°C in a humidified atmosphere of 5% CO_2.
2. Remove medium and add 40 µL of prewarmed serum free medium to all wells.
3. Add 10 µL of medium with 5% DMSO in medium for controls or medium with compound at 5% DMSO final concentration.
4. Incubated 20 min at 37°C in a humidified atmosphere of 5% CO_2.
5. Add 50 µL of inducer 2 µM of PMA for PKC translocation and 50 µL 600 ng/mL of IGF-1 for Akt, incubated 10 min for PKC and 5 min for Akt at 37°C in a humidified atmosphere of 5% CO_2.
6. Aspirate culture medium and add 100 µL of fixation solution (3.7% formaldehyde in PBS) to each well. Incubate in fume hood for 15 min at room temperature.
7. Aspirate fixation solution and wash wells once with 100 µL of PBS.
8. Aspirate PBS and add 100 µL membrane stain to each well. Incubate for 30 min (*see* **Note 9**).

9. Aspirate membrane stain and wash three times with 100 μL PBS.
10. Aspirate the PBS and fix by adding 100 μL fixation solution. Incubate for 5 min in fume hood.
11. Aspirate fixation solution and wash twice with 100 μL PBS.
12. Aspirate the PBS and add 100 μL of 0.2X permeabilization buffer (PBS 0.1% Triton X-100). Incubate for 15 min.
13. Aspirate permeabilization buffer and wash twice with 100 μL PBS.
14. For Akt aspirate the PBS and add 100 μL of blocking buffer (PBS 10% FCS, 1% BSA). Incubate 45–60 min.
15. For PKC aspirate the PBS and add 50-μL primary antibody solution (*see* **Note 10**). Incubate for 1 h.
16. For Akt aspirate blocking buffer and add 50-μL primary antibody solution (*see* **Note 10**). Incubate overnight at 4°C.
17. Aspirate primary antibody solution and wash three times with 100 μL PBS.
18. For Akt aspirate PBS and add 100 μL of PBS 1% H_2O_2. Incubate 30 min at room temperature. Wash three times with 100 μL of PBS.
19. For PKC aspirate the PBS and add 50 μL of staining solution. Incubate for 1 h (*see* **Note 11**).
20. For Akt aspirate the PBS and add 50 μL of staining solution (*see* **Note 12**). Incubate for 1 h.
21. Aspirate staining solution and wash three times with 100 μL PBS. For PKC go to 24.
22. For Akt aspirate the PBS and add 50 μL of tyramine staining solution (*see* **Note 3**). Incubate 10 min.
23. Aspirate the tyramine staining solution and wash three times with 100 μL PBS.
24. Aspirate PBS and add 200 μL of PBS. Seal plates and scan on ArrayScan HCS Reader using ×20 objective and cytoplasm to membrane Translocation bioapplication.
25. Store sealed plates in the dark at 4°C. Plates are stable for 48 h after preparation.

3.4. HCS Morphology Assays in Primary Screen for Therapeutic Proteins

Neurons assemble into functional networks by growing out axons and dendrites (collectively called neurites). Neuronal cell morphology, including neurite outgrowth, elongation, cell body hypertrophy, and growth cone behavior, is modulated by a variety of conditions such as trophic factors, electrical activity, synaptogenesis, and functional maturation and differentiation of neurons *(22,23)*. This method was used to screen for potential therapeutic proteins that activate the neurite outgrowth *(24)* of PC12 subclone NS-1. In cell culture, the cells are routinely split when they reach 70–80% confluency.

3.4.1. Day 1

1. The cells from flasks of 80% confluent cultures are detached by treatment with trypsin-EDTA.
2. The cells are resuspended in culture medium and counted (*see* **Note 1**).
3. The cell suspension is diluted with medium at 2×10^4 cells/mL for NS-1 cells 90 μL of cell suspension is seeded into a collagen I coated 96-well microplate containing 10 μL of controls with or without 2000 ng/mL of NGF or 10 μL of test proteins.
4. Incubate 3 d at 37°C in a humidified atmosphere of 5% CO_2.
5. Aspirate medium and add 100 μL PBS 3.7% formaldehyde, 1/2000 diluted Hoechst Dye Solution to each well. Incubate 20 min in fume hood at room temperature.
6. Aspirate fixation/Hoechst solution and wash three times with 100 μL 1X neurite outgrowth buffer.
7. Aspirate neurite outgrowth buffer and add 50 μL primary antibody solution (*see* **Note 14**). Incubate for 1 h.
8. Aspirate primary antibody solution and wash three times with 100 μL 1X neurite outgrowth buffer.
9. Aspirate neurite outgrowth buffer and add 100 μL secondary antibody solution (*see* **Note 15**). Incubate 1 h.
10. Aspirate secondary antibody solution and wash twice with 100 μL 1X neurite outgrowth buffer.
11. Aspirate neurite outgrowth buffer and wash twice with 100 μL PBS.
12. Add 200 μL of PBS and seal plate and run on ArrayScan HCS Reader using a ×5 or ×10 objective and the neurite outgrowth bioapplication.
13. Store sealed plates in the dark at 4°C.

3.5. HCS in Genotoxicity Screening

The in vitro micronucleus assay is a genetic toxicology assays, in which cultured cells are treated with compounds and scored for micronucleus induction. Micronucleus (are pieces of chromosomes or entire chromosomes that have failed to be included in daughter nuclei during cell division) formation can be because of clastogens, which cause chromosomal breaks, and/or aneugens, which affect the spindle apparatus. This method is used to screen compounds for there potential to form micronuclei *(25)*. In cell culture, the cells are routinely split when they reach 70–80% confluency.

3.5.1. Day 1

1. The cells from flasks of 80% confluent cultures are detached by treatment with trypsin-EDTA.
2. The cells are resuspended in culture medium and counted (*see* **Note 1**).
3. The cell suspension is diluted with medium at 4×10^4 cells/mL and is seeded into a Biocoat Collagen I 96-well plates.
4. The plate is incubated 12–18 h at 37°C in a humidified atmosphere of 5% CO_2.

3.5.2. Day 2

1. Remove medium and add 100 µL cellular dye solution, prepared as specified by the supplier (*see* **Note 16**). Incubate 1 h at 37°C in 5% CO_2.
2. In a 96-wells round bottom plate, prepare compounds dilutions in DMSO from a 10 m*M* stock. Add DMSO in control wells and MMC solution (33.3 ng/mL; 100 µ*M*) in positive control wells.
3. Transfer 2.5 µL into a new plate containing 247.5 µL of medium per well.
4. Wash 1X the cells with medium.
5. Transfer 100 µL of the compound plate on the cells.
6. Incubate 20h at 37°C in 5% CO_2.

3.5.3. Day 3

1. Remove medium from cells.
2. Wash once with medium and remove medium.
3. Add cytokinesis blocking agent, prepared as specified by the supplier (*see* **Note 16**).
4. Incubate 28h at 37°C in 5% CO_2.

3.5.4. Day 4

1. Add 50 µL permeability dye solution to the cells, prepared as specified by the supplier (*see* **Note 16**).
2. Incubate 30 min, 37°C, 5% CO_2.
3. Remove medium and wash once with medium.
4. Discard medium and add 100 µL fixation solution (PBS 3.7% formaldehyde).
5. Incubate 20 min.
6. Remove medium and wash twice with 100 µL PBS.
7. Add 200 µL PBS and seal plate and scan using the Cellomics Arrayscan reader using ×20 objective and the Micronucleus bioapplication.

4. Notes

1. Add 10 µL of cell suspension to 90 µL of 0.4% Trypan Blue and count living cells using a hemocytometer.
2. The amount and ratio between the standardizing plasmid pTK Renilla and the stimulation plasmid should be at least 1:1 or 2:1 and the total DNA content 6 µg for the FuGENE 6 ratio used.
3. Cell seeding density for nuclear translocation should be set so that cells are sufficiently separated and that at least 100 cells per field of view can be observe using a microscope with a ×10 objective. Seeding density can vary from cell type to another.
4. For some stimulation or cell types, one can obtain better pathway activation by serum starving the cells for 2 h. A dose–response for concentration of nuclear translocation inducer for each cell type and each translocating macromolecule should be tested.
5. Using the top dose determined in **Note 4** a time-course of nuclear translocation should be established for each cell type and each translocating macromolecule.
6. The time and the concentration of permeabilization buffer can be adapted if the staining is note optimal. An alternative to using PBS 0.2% Triton X-100 incubated for 90 s can be PBS 0.1% Triton X-100 incubated for 15 min or any permutations of these conditions.

7. The primary antibody from the HitKits: c-Jun, NF-κB, and STAT2 are diluted in PBS 1:200, 1:100, and 1:200, respectively. For other cell type, if background is a problem other dilution can be tested, 1% BSA can be added to the diluted antibody solution, a preincubating the cells with blocking buffer (PBS 10% FCS, 1% BSA) for 1 h and also the Detergent Buffer can be adapted from PBS 0.01% Tween-20 to PBS 0.1% Tween-20.
8. The staining solution for the HitKits: c-Jun, NF-κB and STAT2 contains secondary antibody diluted in PBS 1/100 (antimouse Alexa Fluor 488), 1/100 (antirabbit Alexa Fluor 488) and 1/100 (antirabbit Alexa Fluor 488), respectively, with Hoechst Dye diluted at 1/2000. For other cell type, if background is a problem other dilution can be tested, 1% BSA can be added to the diluted antibody solution and also the detergent buffer can be adapted from PBS 0.01% Tween-20 to PBS 0.1% Tween-20.
9. The membrane stain is prepared by diluting in PBS the membrane marker (Alexa Fluor 488) 1/120, if the secondary antibody used is conjugated to Alexa Fluor 555. If the Secondary antibody used is conjugated to Alexa Fluor 488 or the TSA amplification is with Alexa Fluor 488 then the membrane marker used is Weat Germ Agglutinin TRITC conjugate 1 mg/mL stock solution diluted 1/170 in PBS.
10. The primary antibody for the HitKit PKC is diluted in PBS 1/100 and for Akt the antibody is diluted 1/250 in PBS 1% BSA. For other cell type and if background is a problem other dilution can be tested.
11. The Staining Solution for the HitKit PKC contains the secondary antibody diluted in PBS 1/200 (antimouse Alexa Fluor 555) and if the Membrane stain used is conjugated to TRITC, the secondary antibody is diluted in PBS 1/200 (antimouse Alexa Fluor 488), the solution contains also Hoechst Dye diluted at 1/2000. For other cell type and if background is a problem other dilution can be tested and 1% BSA can be added to the diluted antibody solution.
12. The staining solution for Akt contains the secondary antibody (HRP conjugated antirabbit) diluted in PBS 1% BSA 1/400; the solution contains also Hoechst Dye diluted at 1/2000.
13. The tyramide amplification solution is prepared by diluting the tyramide stock solution 1/150 in amplification buffer/0.0015(H2O2.
14. The primary antibody for the Neurite outgrowth HitKit is diluted in 1X Neurite outgrowth buffer 1/800.
15. The secondary antibody for the Neurite outgrowth HitKit is diluted in 1X Neurite outgrowth buffer 1/200.
16. The cellular dye is prepared by diluting 5.5 µL of cellular dye stock solution in 11 mL of cell culture medium. The cytokinesis blocking agent is prepared by diluting 6.6 µL of cytokinesis blocking agent stock solution in 11 mL of cell culture medium. The permeability dye is prepared by diluting 3.6 µL of permeability dye stock solution in 11 mL of cell culture medium.

Acknowledgments

The author would like to thank Karen Yeow, Paul Lang, and Alexander Scheer for fruitful discussions. The author thanks the members of the Serono Pharmaceutical Research Institute, especially the members of the Biomolecular Screening and Cellular Pharmacology department and especially François Duval for his technical help and discussions.

References

1. Gribbon, P. and Sewing, A. (2003) Fluorescence readouts in HTS: no gain without pain? *Drug Discov. Today* **8,** 1035–1043.
2. Johnston, P. A. and Johnston, P. A. (2002) Cellular platforms for HTS: three case studies. *Drug Discov. Today* **7,** 353–363.
3. Abraham, V. C., Taylor, D. L., and Haskins, J. R. (2004) High content screening applied to large scale cell biology. *Trends Biotechnol.* **22,** 15–22.
4. Alessi, D. R. and Downes, C. P. (1998). The role of PI 3-kinase in insulin action. *Biochim. Biophys. Acta* **1436,** 151–164.
5. Giuliano, K. A., Haskins, J. R., and Taylor, D. L. (2003) Advances in high content screening for drug discovery. *Assay. Drug Dev. Technol.* **1,** 565–577.
6. Johnston, P. (2002) Cellular assays in HTS. *Methods Mol. Biol.* **190,** 107–116.
7. Russello, S. V. (2004) Assessing cellular protein phosphorylation: high throughput drug discovery technologies. *Assay. Drug Dev. Technol.* **2,** 225–235.
8. Taylor, D. L., Woo, E. S., and Giuliano, K. A. (2001). Real-time molecular and cellular analysis: the new frontier of drug discovery. *Curr. Opin. Biotechnol.* **12,** 75–81.

9. Vogt, A., Cooley, K. A., Brisson, M., Tarpley, M. G., Wipf, P., and Lazo, J. S. (2003) Cell-active dual specificity phosphatase inhibitors identified by high content screening. *Chem. Biol.* **10,** 733–742.
10. Ramm, P. and Thomas, N. (2003) Image-based screening of signal transduction assays. *Sci. STKE.* **2003,** E14.
11. Gaillard, P., Jeanclaude-Etter, I., Ardissone, V., et al. (2005) Design and synthesis of the first generation of novel potent, selective, and in vivo active (benzothiazol-2-yl)acetonitrile inhibitors of the c-jun N-terminal kinase. *J. Med. Chem.* **48,** 4596–4607.
12. Manning, A. M. and Davis, R. J. (2003) Targeting JNK for therapeutic benefit: from junk to gold? *Nat. Rev. Drug Discov.* **2,** 554–565.
13. Ruckle, T., Biamonte, M., Grippi-Vallotton, T., et al. (2004) Design, synthesis, and biological activity of novel, potent, and selective (benzoylaminomethyl)thiophene sulfonamide inhibitors of c-Jun-N-terminal kinase. *J. Med. Chem.* **47,** 6921–6934.
14. Lin, A., Minden, A., Martinetto, H., et al. (1995). Identification of a dual specificity kinase that activates the Jun kinases and p38-Mpk2. *Science* **268,** 286–290.
15. Ding, G. J., Fischer, P. A., Boltz, R. C., et al. (1998) Characterization and quantitation of NF-kappaB nuclear translocation induced by interleukin-1 and tumor necrosis factor-alpha. Development and use of a high capacity fluorescence cytometric system. *J. Biol. Chem.* **273,** 28,897–28,905.
16. Vakkila, J., DeMarco, R. A., and Lotze, M. T. (2004) Imaging analysis of STAT1 and NF-kappaB translocation in dendritic cells at the single cell level. *J. Immunol. Methods* **294,** 123–134.
17. Karin, M., Yamamoto, Y., and Wang, Q. M. (2004) The IKK NF-kappa B system: a treasure trove for drug development. *Nat. Rev. Drug Discov.* **3,** 17–26.
18. Mielke, K., Damm, A., Yang, D. D., and Herdegen, T. (2000) Selective expression of JNK isoforms and stress-specific JNK activity in different neural cell lines. *Brain Res. Mol. Brain Res.* **75,** 128–137.
19. Buckman, T. D., Sutphin, M. S., and Mitrovic, B. (1993) Oxidative stress in a clonal cell line of neuronal origin: effects of antioxidant enzyme modulation. *J. Neurochem.* **60,** 2046–2058.
20. Hurley, J. H. and Meyer, T. (2001). Subcellular targeting by membrane lipids. *Curr. Opin. Cell Biol.* **13,** 146–152.
21. Lundholt, B. K., Linde, V., Loechel, F., et al. (2005) Identification of Akt pathway inhibitors using redistribution screening on the FLIPR and the IN Cell 3000 analyzer. *J. Biomol. Screen.* **10,** 20–29.
22. Fields, R. D. and Nelson, P. G. (1992) Activity-dependent development of the vertebrate nervous system. *Int. Rev. Neurobiol.* **34,** 133–214.
23. Thoenen, H. (1991) The changing scene of neurotrophic factors. *Trends Neurosci.* **14,** 165–170.
24. Ramm, P., Alexandrov, Y., Cholewinski, A., Cybuch, Y., Nadon, R., and Soltys, B. J. (2003) Automated screening of neurite outgrowth. *J. Biomol. Screen.* **8,** 7–18.
25. Kirsch-Volders, M., Sofuni, T., Aardema, M., et al. (2003) Report from the in vitro micronucleus assay working group. *Mutat. Res.* **540,** 153–163.

Discovery of Protein Kinase Phosphatase Inhibitors

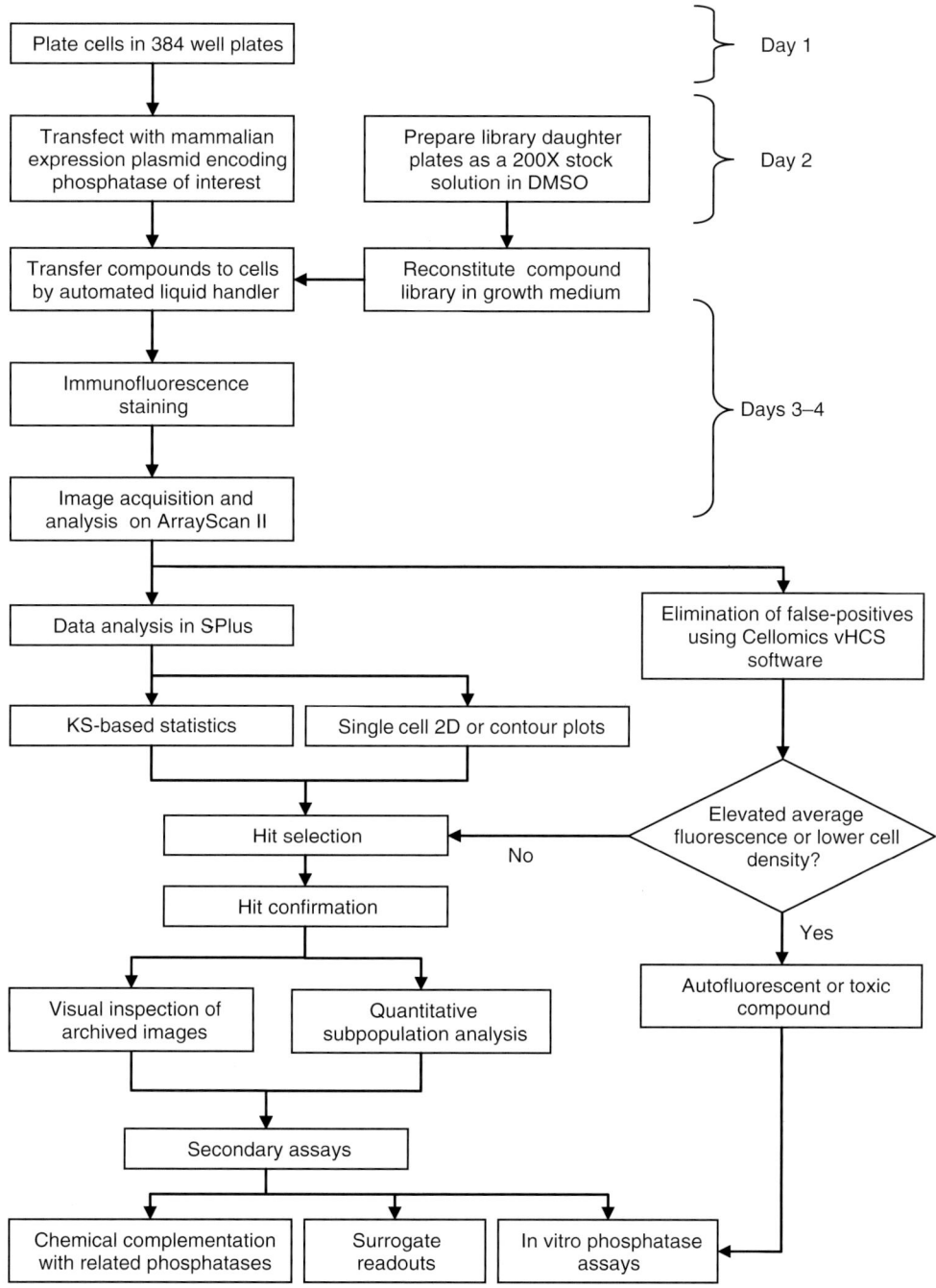

Fig. 3. Flow diagram of the chemical complementation procedure.

3.1.3. Day 3

1. Prepare 96-well treatment plates (clear polystyrene) containing 50 μL of compounds at the desired concentration in complete growth medium.
2. Treat cells with library compounds by transferring 30 μL of treatment solution from 96-well plates. A robotic liquid handler is necessary for this step.
3. Prepare 15 mL of 1.5 μg/μL TPA by adding 22.5 μL of 1 mg/mL DMSO stock of phorbol ester (TPA) in DMSO to 15 μL complete growth medium.

4. Dispense 15 mL TPA directly to cells from a single reservoir.
5. Return plates to incubator for 20 min.
6. Add 15 μL 16% formaldehyde directly to each well without removing medium and incubate at room temperature for 10 min.
7. Remove fixative and wash once with PBS.
8. Permeabilize for 5 min with Triton X-100.
9. Wash with PBS and incubate in blocking solution at room temperature for 1 h.
10. Remove blocking solution and incubate with primary antibody cocktail at room temperature for 1 h.
11. Remove primary antibody cocktail and incubate for 15 min with Tween-20.
12. Wash with PBS.
13. Incubate with secondary antibody cocktail for 1 h at room temperature in the dark.
14. Remove secondary antibody cocktail and incubate for 10 min with Tween-20.
15. Wash cells thrice with PBS, leaving last wash on cells.
16. Seal plate, and store at 4°C in the dark until analysis.
17. Add 60 μL PBS to each well, seal and barcode plate, and store at 4°C until needed for imaging.

3.2. Image Acquisition and Analysis

Immunostained plates were analyzed on the Cellomics, Inc., ArrayScan II using the Target Activation or Compartmental Analysis Bioapplication. Both algorithms are powerful image analysis tools that automatically set backgrounds, identify objects, and define regions of interest within the cells. Up to 100 readouts are provided for each cell, including fluorescence measurements, object counts, and nuclear morphology descriptors. Some of the features are ratiometric measurements.

For the chemical complementation screen, images were acquired in three independent fluorescence channels using an Omega XF93 filter set at excitation/emission wavelengths of 350/461 nm (Hoechst), 494/519 nm (AlexaFluor 488), 650/665 nm (AlexaFluor 647), respectively. The assay can be conducted using Texas red in place of Alexa 647, resulting in shorter exposure and screen times. Texas red compatible fluorophores require the use of an XF53 filter set. A nuclear mask was generated from Hoechst 33342-stained nuclei, and object identification thresholds and shape parameters were set such that the algorithm identified over 90% of the nuclei in each field. Objects that touched each other or the edge of the image were excluded from the analysis. The number of cells in each well was determined by enumerating objects in the Hoechst channel. At least 1000 individual cells were captured for each condition. For determination of MKP-1 and phospho-Erk expression levels, the nuclear mask was dilated by one pixel and AlexaFluor 488 and AlexaFluor 647 fluorescence intensities measured. The ArrayScan system in connection with Cellomics Store automatically archives images and plate data in a proprietary Structured Query Language (SQL) server database. Data and images are retrieved from the SQL database through a suite of HCS analysis algorithms termed the HCS toolbox.

3.3. HCS Data Set Evaluation

Average pixel intensities need to be evaluated for each individual cell to relate phospho-Erk levels to MKP-1 expression. Assuming that a minimum of 1000 cells are acquired in each well, a full 384-well plate will yield about 400,000 individual objects. This necessitates the use of non-Excel-based software, as Excel spreadsheets are limited to 65,000 objects. Our version of the Cellomics, Inc., software provides convenient well average readouts but is Excel-based and does not permit the export of all of the objects in the data set. Therefore, to perform the single-cell analysis, the individual cell-data was exported into the S-Plus statistical package (Insightful, Inc.) (*see* **Note 2**). Wells transfected with MKP-1 but not treated with compounds show low phospho-Erk signal in cells with high FITC signal. In the presence of a phosphatase inhibitor, many of the cells with high FITC signal become phospho-Erk positive. This increased density of phospho-Erk positive cells at higher FITC intensities appears as a "filling" of the individual two dimensional (2D) plots. Visualization can be improved by generating density contour plots such as is shown in **Fig. 6A**.

3.3.1. Subpopulation Analysis

One method of quantifying the HCS data is based on the differential measurement of phospho-Erk levels in MKP-1 expressing and nonexpressing cell populations *(9)*. Cell populations are gated into MKP-1 expressors and nonexpressors based on their average nuclear FITC intensity using S-plus scripts. The subsetting procedure can be automated by defining thresholds for FITC positive cells as the average FITC intensity of untransfected wells plus multiples of standard deviation. Cells are classified as "expressors" if their average FITC intensity exceeds this threshold. Alternatively, thresholds can be set manually (*see* **Note 3**). Once the data have been subsetted, phospho-Erk levels can be averaged in the two cell subpopulations as described *(9)*. This method of evaluating the data set has been shown to be quantitative, but it is sensitive to outliers and shows extensive data scatter when used for screening. **Figure 4A** shows the results of the 720 member MicroSource natural products library screen using phospho-Erk subpopulation differences. The signal-to-noise ratio is low and positive controls (indicated by a dotted line) are not well separated from background, making it difficult to set a cutoff for identification of positives. Despite this, the method is still useful to confirm the activity of positives identified from HCS screens (*see* **Fig. 3**).

3.3.2. Kolmogorov–Smirnov Statistics

Because the algorithm described in **Subheading 3.3.1** lacked discriminating power as a high-throughput data evaluation tool, we developed an alternative method to identify compounds with phosphatase inhibitory activities based a two-dimensional Kolmogorov–Smirnov (2D KS) analysis *(12)*. KS statistics are a method to quantitatively compare the distributions of individual cell populations *(13–16)*, and were shown to be a valuable tool for the quantification of HCS data *(17)*. The 2D KS method ranges over data in an (x,y) plane in search of a maximum cumulative difference between two 2D data distribution *(18,19)*. The resulting KS values can assume a value between 0 (identical distributions) and 1 (completely different distributions). To fully exploit the information contained in our multiparameter high-content data, we developed *(12)* an algorithm that combined the KS statistics with ratio analysis of means *(15)* of both Alexa 488 (MKP-1) and Alexa 647 (phospho-Erk) in a minimum of 1000 individual cells in 192 wells on a 384-well microplate. Two-color fluorescence data from 16 controls (MKP-1 transfected but not drug treated wells) were pooled for each fluorescence channel and used to generate a 2D reference data distribution. The 2D data distribution of each individual well was then compared to the reference data distribution, and a KS value was calculated for each well. Calculations were performed with the S-Plus statistical software package. **Figure 4B** shows the results of the 720 member MicroSource natural products library 2DKS algorithm compared to the earlier-described subpopulation analysis. The algorithm dramatically reduced background noise and reduced the percentage of "hits" from 3% to a more manageable 0.7%.

3.3.3. False-Positives

False-positives are common in cell-based screens with fluorescent labels. Most false-positives fall into four categories: Toxicity (including morphological changes), autofluorescence, artifacts (from reagents or through outside contamination), and analysis abnormalities (such as out of focus wells). With phenotypic assays, we have found that the false-positive rate can be as high as 60% *(9)*. An automated microscopy platform permits rapid and convenient identification of autofluorescence and toxicity through measurements of cell densities and well average fluorescence intensities. We have found that eliminating toxicity and fluorescence artifacts reduces the number of hits enough to permit evaluation of remaining positives by visual inspection of archived images. It should be noted that compounds should not be eliminated from consideration merely

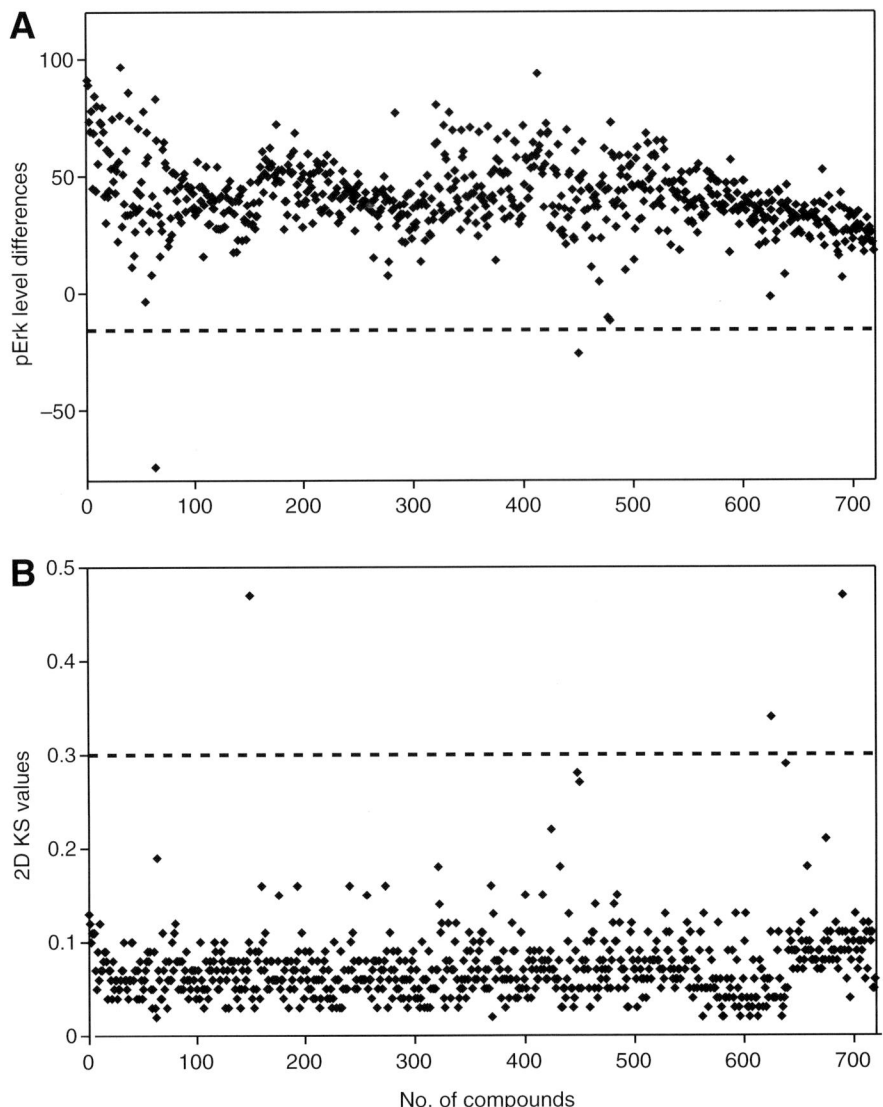

Fig. 4. Identifying hits in the chemical complementation assay by using subpopulation differences or 2D KS statistics. Differential expression of nuclear phospho-Erk was quantified in HeLa cells either transiently expressing or not expressing ectopic MKP-1. Cells were treated for 15 min with 0.5 µg/mL TPA in the presence or absence of compounds from a 720 member Microsource natural products library. (**A**) Represents the subpopulation difference in phospho-Erk levels. (**B**) Represents the 2D KS values for the entire cell population. Average pErk subpopulation differences or 2D KS values of wells treated with the positive control (PAO) are indicated by the dotted lines. Additional details can be found in **ref. 12**.

because they are autofluorescent, but instead be tested in secondary assays that do not require fluorescence measurements (*see* **Subheading 3.4.; Fig. 3**).

3.3.4. Visual Inspection of Archived Images

A major benefit of image-based high-content analysis is the fact that images are retained after the scan and can be downloaded and visually inspected for the high-phosphorylation/low-phosphorylation phenotype described in the introduction. **Figure 5** shows representative

Discovery of Protein Kinase Phosphatase Inhibitors

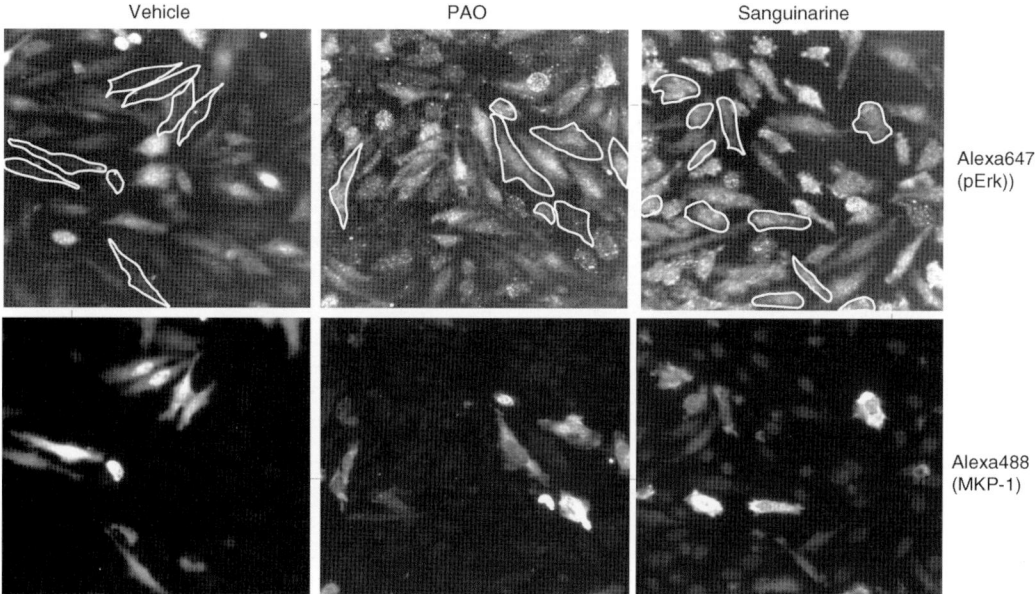

Fig. 5. Visual confirmation of a positive compound identified by chemical complementation assay with natural product library. Expression of nuclear phospho-Erk and MKP-1 were manually evaluated in archived immunofluorescence images acquired during the MKP-1 high content screen. Cells were identified by Hoechst 33342 staining. MKP-1 expression was determined on the basis of FITC (Alexa488) fluorescence (lower panels) and mapped onto phospho-Erk images acquired in the Cy5 (Alexa647) channel (top panels).

immunofluorescence images, in which MKP-1 expressing cells were identified on the basis of FITC (Alexa 488) fluorescence (lower panels) and mapped onto images acquired in the Cy5 (Alexa 647) channel (top panels). In the absence of small molecule compounds, MKP-1 expressing cells did not respond to TPA with increased Erk phosphorylation. Both the broad spectrum tyrosine phosphatase inhibitor, PAO, and one of the positive compounds in the MicroSource library, sanguinarine, restored Erk phosphorylation in the MKP-1 expressing cells, confirming the results from the 2D KS analysis. Visual confirmation of archived images also identifies image acquisition or analysis artifacts.

3.4. Secondary Assays

To illustrate the power of the chemical complementation assay for both primary and secondary assays, we compared the ability of sanguinarine to inhibit MKP-1 with a related phosphatase, MKP-3. HeLa cells were transfected with MKP-1 or MKP-3, treated, fixed, and stained, and single-cell contour plots were generated for each well as described in **Subheading 3.3.** The plate contained MKP-1 and MKP-3 transfected cells on the top half and the bottom half of the microplate. Wells A12:D12 were transfected with the phosphatase-inactive GFP, and wells E12:G12 and H12:L12 were transfected with MKP-1 or MKP-3, respectively, and treated with PAO. The graphical representation clearly shows high densities of phospho-Erk positive cells in GFP-expressing cell subpopulations and MKP-1 or MKP-3 expressing cell subpopulations that were treated with PAO. Wells that received sanguinarine show a concentration-dependent response in MKP-1 expressing cells but had little effect on MKP-3 expressing cells. The visual changes in cell population densities were then quantified using 2D KS statistics and produced concentration-response curves for inhibition of MKP-1 and MKP-3 by sanguinarine (**Fig. 6B**).

Other secondary assays are a Western blot implementation of the chemical complementation assay and in vitro assays using recombinant phosphatases, both of which have been described in

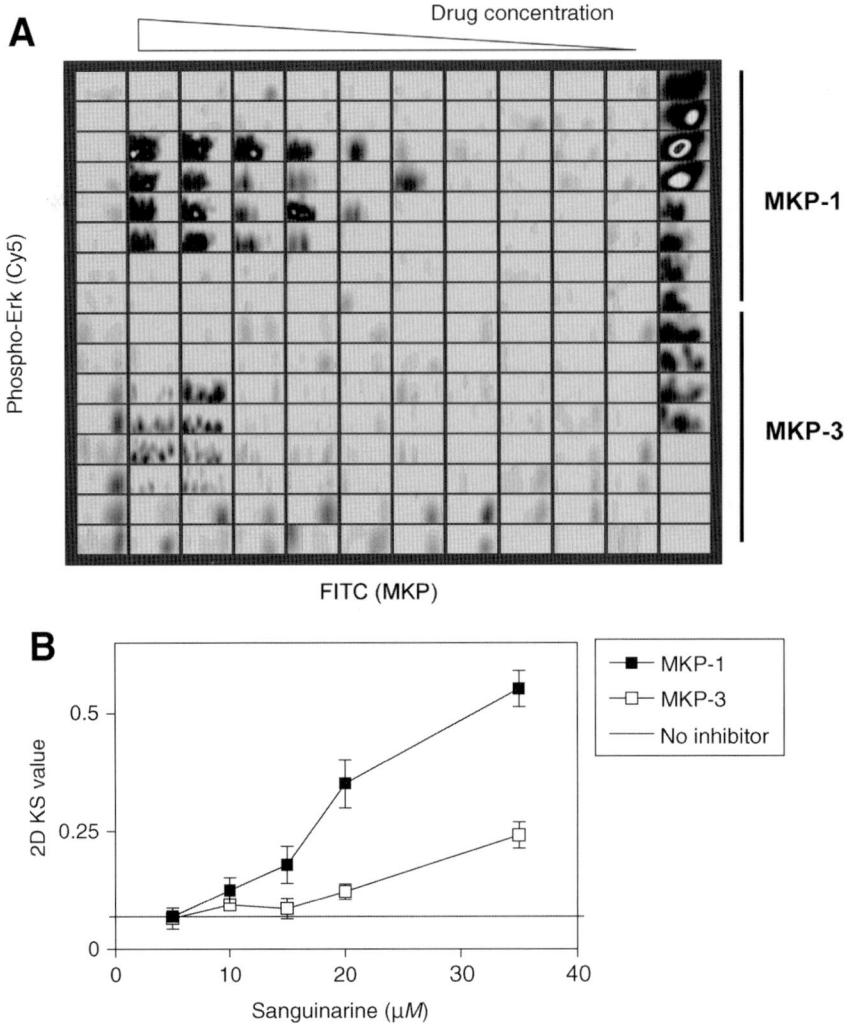

Fig. 6. Selective inhibition of MKP-1 but not MKP-3 by sanguinarine. (**A**) Single-cell density contour plots of phospho-Erk changes as a function of MKP-1 or MKP-3 expression in cells treated with different concentrations of sanguinarine. Cells transfected with MKP-1 are found in the upper portion of the panel, whereas those transfected with MKP-3 are shown in the lower portion of the panel. (**B**) Quantification of visual changes in cell population densities by 2D KS statistics.

detail *(10,12,20)*. Another option is surrogate readouts consistent with target inhibition. In the case of MKP inhibitors, this would encompass measurement of the phosphorylation status of MKP-1 target substrates, either by immunofluorescence or by Western blot analysis. This has been demonstrated in **ref. 12** in which compounds with MKP-1 inhibitory activity activated the MKP-1 substrates Erk and JNK/SAPK.

3.5. Conclusions and Prospectus

Chemical complementation provides a potentially powerful tool to interrogate small molecules for their ability to interfere with specific molecular targets in intact cells. The strategy is especially useful when in vitro, target-based assays are not readily available, or in cases in which cellular phenotypic readouts are influenced by a multitude of factors. The approach is most informative when the most proximal substrate of the target protein is used as an endpoint, as was illustrated with the Erk/MKP protein pair. Phenotypic readouts such as cell division, motility, shape, or DNA content, are subject to indirect effects.

The multiparametric version of chemical complementation should gain further utility if implemented in true high-throughput formats to screen diverse compound collections in intact mammalian cells. Preliminary results suggest that the assay can be robust enough for large-scale library screening *(12)*, however room for improvement remains.

For large-scale screening, it might be beneficial to develop cell lines that stably overexpress an intrinsically labeled form of the phosphatase of interest, perhaps with GFP, and use such constructs in coculture with the parental cell line. The assay in its current form requires a tightly timed incubation during stimulation with TPA. In principle, this step would not be required if cells could be identified that possess phospho-Erk levels high enough to be measured by immunofluorescence. To date, we have not been able to identify a cell line that shows constitutively high levels of nuclear phospho-Erk. Nonetheless, we believe the chemical complementation assay is the first definitive cell-based assay for phosphatase activity, and it appears reasonable to predict that in the near future a large-scale chemical complementation screen will identify potent, cell active inhibitors of phosphatase targets that are currently eluding drug discovery efforts.

4. Notes

1. We have found that Lipofectamine 2000 is best suited for high-throughput transfection followed by imaging. Other cationic lipids (Lipofectamine plus, Lipofectin) give fluorescent artifacts. Superfect (QIAGEN, Valencia, CA) also has low fluorescence artifacts but tended to give lower transfection efficiencies.
2. To perform data analysis on populations of single cells, we queried the Microsoft Access database produced by the BioApplication to extract single-cell average intensities in fluorescence channel 2 (FITC) and channel 3 (Texas Red or Cy5) by database query, followed by import into S-Plus for data manipulation and viewing. A useful means to quickly check the performance of an assay is the generation of a whole-plate trellis graph where Channel 2 and Channel 3 intensities are plotted on the x-axis and y-axis, respectively.
3. It is advisable to perform a visual check of transfection efficiency by manually counting the percentage of cells that appear green microscopically. The HCS platform allows the user to easily perform manual inspection of the images from which each set of cell features were extracted.

Acknowledgments

We thank Stephen Keyse and Nicholas Tonks for providing MKP-3 and MKP-1 expression plasmids, and Kenneth Giuliano for generating S-Plus scripts. This work was supported in part by NIH grants CA78039 and CA52995, and the Fiske Drug Discovery Fund.

References

1. Zolnierowicz, S. and Bollen, M. (2000) Protein phosphorylation and protein phosphatases. De Panne, Belgium, 19–24 September 1999. *EMBO J.* **19,** 483–488.
2. Tonks, N. K. and Neel, B. G. (2001) Combinatorial control of the specificity of protein tyrosine phosphatases. *Curr. Opin. Cell Biol.* **13,** 182–195.
3. Alonso, A., Sasin, J., Bottini, N., et al. (2004) Protein tyrosine phosphatases in the human genome. *Cell* **117,** 699–711.
4. Zou, X., Tsutsui, T., Ray, D., et al. (2001) The cell cycle-regulatory CDC25A phosphatase inhibits apoptosis signal-regulating kinase 1. *Mol. Cell Biol.* **21,** 4818–4828.
5. Chen, P., Hutter, D., Liu, P., and Liu, Y. (2002) A mammalian expression system for rapid production and purification of active MAP kinase phosphatases. *Protein Expr. Purif.* **24,** 481–488.
6. Slack, D. N., Seternes, O. M., Gabrielsen, M., and Keyse, S. M. (2001) Distinct binding determinants for ERK2/p38alpha and JNK map kinases mediate catalytic activation and substrate selectivity of map kinase phosphatase-1. *J. Biol. Chem.* **276,** 16,491–16,500.

7. Camps, M., Nichols, A., Gillieron, C., et al. (1998) Catalytic activation of the phosphatase MKP-3 by ERK2 mitogen-activated protein kinase. *Science* **280,** 1262–1265.
8. Balis, F. M. (2002) Evolution of anticancer drug discovery and the role of cell-based screening. *J. Natl Cancer Inst.* **94,** 78–79.
9. Vogt, A., Cooley, K. A., Brisson, M., Tarpley, M. G., Wipf, P., and Lazo, J. S. (2003) Cell-active dual specificity phosphatase inhibitors identified by high content screening. *Chem. Biol.* **10,** 733–742.
10. Vogt, A., Takahito, A., Ducruet, A. P., et al. (2001) Spatial analysis of key signaling proteins by high-content solid-phase cytometry in Hep3B cells treated with an inhibitor of Cdc25 dual-specificity phosphatases. *J. Biol. Chem.* **276,** 20,544–20,550.
11. Vogt, A. and Lazo, J. S. (2005) Chemical complementation: a definitive phenotypic strategy for identifying small molecule inhibitors of elusive cellular targets. *Pharmacol. Ther.* **107,** 212–215.
12. Vogt, A., Tamewitz, A., Skoko, J., Sikorski, R. P., Giuliano, K. A., and Lazo, J. S. (2005) The benzo (C) phenanthridine alkaloid, sanguinarine, is a selective, cell-active inhibitor of mitogen-activated protein kinase phosphatase-1. *J. Biol. Chem.* **280,** 19,078–19,086.
13. Young, I. T. (1977) Proof without prejudice: use of the Kolmogorov–Smirnov test for the analysis of histograms from flow systems and other sources. *J. Histochem. Cytochem.* **25,** 935–941.
14. Lampariello, F. (2000) On the use of the Kolmogorov–Smirnov statistical test for immunofluorescence histogram comparison. *Cytometry* **39,** 179–188.
15. Watson, J. V. (2001) Proof without prejudice revisited: immunofluorescence histogram analysis using cumulative frequency subtraction plus ratio analysis of means. *Cytometry* **43,** 55–68.
16. Cox, C., Reeder, J. E., Robinson, R. D., Suppes, S. B., and Wheeless, L. L. (1988) Comparison of frequency distributions in flow cytometry. *Cytometry* **9,** 291–298.
17. Giuliano, K. A., Chen, Y. T., and Taylor, D. L. (2004) High content screening with siRNA optimizes a cell biological approach to drug discovery: defining the role of P53 activation in the cellular response to anticancer drugs. *J. Biomol. Screen.* **9,** 557–568.
18. Peacock, J. A. (1983) Two-dimensional goodness-of-fit testing in astronomy. *Mon. Not. R. Astron. Soc.* **202,** 615–627.
19. Fasano, G. and Francescini, A. (1987) A multidimensional version of the Kolmogorov–Smirnov test. *Mon. Not. R. Astron. Soc.* **225,** 155–170.
20. Lazo, J. S., Aslan, D. C., Southwick, E. C., et al. (2001) Discovery and biological evaluation of a new family of potent inhibitors of the dual specificity protein phosphatase Cdc25. *J. Med. Chem.* **44,** 4042–4049.

29

High Content Translocation Assays for Pathway Profiling

Frosty Loechel, Sara Bjørn, Viggo Linde, Morten Præstegaard, and Len Pagliaro

Summary

This chapter describes the design and development of cell-based assays, in which quantitation of the intracellular translocation of a target protein—rather than binding or catalytic activity—provides the primary assay readout. These are inherently high content assays, and they provide feedback on cellular response at the systems level, rather than data on activities of individual, purified molecules. Multiple protein translocation assays can be used to profile cellular signaling pathways and they can play a key role in determination of mechanism of action for novel classes of compounds with therapeutic potential. This assay technology has developed from laboratory curiosity into main stream industrial research over the past decade, and its promise is beginning to be realized as data acquisition and analysis technology evolve to take advantage of the rich window into systems biology provided by translocation assays.

Key Words: Drug discovery; fluorescent proteins; high content screening; lead optimization; modes of action; multiple modes of action; off target effects; pathway profiling; pathway screening; protein translocation assays; redistribution assay.

1. Introduction

1.1. Cell Signaling and Protein Translocation

Cellular signaling pathways have long been studied, and their function dissected, with molecular and biochemical assays of various kinds. As powerful as these approaches have been, it is now clear that these nonsystems biological approaches give incomplete information about pathway biology. Because of this, assays performed at the cellular level have taken a predominant place in signaling studies, both in academia and in industry in recent years. Perhaps the most compelling reason to study pathways at the cellular level is the fact, now widely accepted, that all signal transduction occurs through transfer of mass within the cell (*1*). Because of this, it is virtually impossible to assess the net effect of a signaling process without performing cell-based studies. A convergence of science and technology over the past decade has made this possible.

1.2. The Advent of Fluorescent Proteins

A milestone which enabled widespread development of cell-based protein translocation assays was the advent of the use of fluorescent proteins, and particularly the development of cell lines expressing fusion constructs between a target protein and a fluorescent protein. The first of these, *Aequoria victoria* green fluorescent protein (GFP) was initially purified and characterized by Shimomura and colleagues in 1962 (*2,3*). The gene for GFP was cloned in 1992 (*4*) and expression of the fluorescent protein in other organisms was demonstrated in 1994 (*5*). This pioneering work led to the rapid development of many variants of cell-based translocation assays, incorporating a wide variety of parental cell lines, targets, and fluorescent proteins. The assays described in this chapter were all developed with *A. victoria* GFP, using an "enhanced" version of the protein conferring better folding, hence greater fluorescence intensity, at 37°C *(6)*; this approach can also be used with a variety of other fluorescent proteins that are now commercially available.

1.3. Development of High-Content Technology

In addition to the biological requirements for cell-based translocation assay systems, there was also a requirement for data acquisition and analysis capabilities beyond those found in the research laboratory of the mid1990s. This challenge has been met with the advent of integrated, image-based high content data acquisition and analysis systems, beginning with pioneering early models almost a decade ago *(7,8)* and progressing to more sophisticated models currently available *(see* **Subheading 2.2.1.**). During this early work, the term *high content* was coined *(8,9)* and it continues to describe complex assay systems, including cell-based protein translocation quantitation, well. This complex field is still undergoing rapid change and development, and it is dealt with in detail in **Subheading 2.** of this volume.

1.4. Pathway Profiling Enters Mainstream Drug Discovery

The concept of large scale industrial pathway profiling has long been intellectually attractive to basic scientists in the pharmaceutical and biotechnology industries. A number of academic groups have embraced this concept from a research perspective in recent years *(10–13)*, but industrial acceptance of the approach has been somewhat more reserved. One factor that has been missing for industry has been solid validation of the cost effectiveness of pathway profiling. This is because of several factors, including corporate confidentiality and the long cycle time for pharmaceutical development, resulting in a long wait for "real" market validation of the competitive advantages of integrating pathway profiling in pharmaceutical discovery and development. The promise of cell-based pathway profiling in industrial drug discovery is starting to be suggested by conference presentations, and evidence of its value is bound to appear in the literature over the next few years.

1.5. Examples of High Content Translocation Assays

Three examples of high content translocation assays are shown in **Fig. 1**, with a transcription factor, a glucose transporter, and a protein–protein interaction as the targets illustrated. Other similar examples have been described *(14–19)*.

The Forkhead (FKHR, FOXO1A) assay (**Fig. 1A**) monitors trafficking of the FKHR transcription factor between the cytoplasm and nucleus *(20)*. When the PI3K pathway is active, FKHR constantly becomes phosphorylated and is exported from the nucleus. When the pathway is inhibited, for example, by addition of the PI3K inhibitor wortmannin, FKHR accumulates in the nucleus. It is relatively simple to use an image analysis algorithm to quantify the translocation of FKHR between the nuclear and cytoplasmic compartments, as a measure of compound activity. Test compounds that bind to cells and fluoresce at the same wavelength as GFP give artifactual activity, and an internal fluorescence assay can easily be incorporated by mixing a small proportion of non-GFP expressing cells with the assay cell line, such that only 80% of the cells are green. If the algorithm is designed so that it reports the percentage of cells in the population that are green, then it is simple to deselect fluorescent compounds (those compounds that result in 100% of the cells in the population being green). Toxic test compounds are another source of artifact: when cells round up on the substrate, a substantial portion of the cytoplasm lies above (and perhaps below) the nucleus, making it appear as if the GFP-tagged target protein has

Fig. 1. *(Opposite page)* Three examples of high-content translocation assays used for pathway profiling. Images of a Forkhead nuclear translocation assay (**A**), a GLUT4 membrane tranlocation assay (**B**), and a p53-Hdm2 protein–protein interaction assay (**C**) are shown.

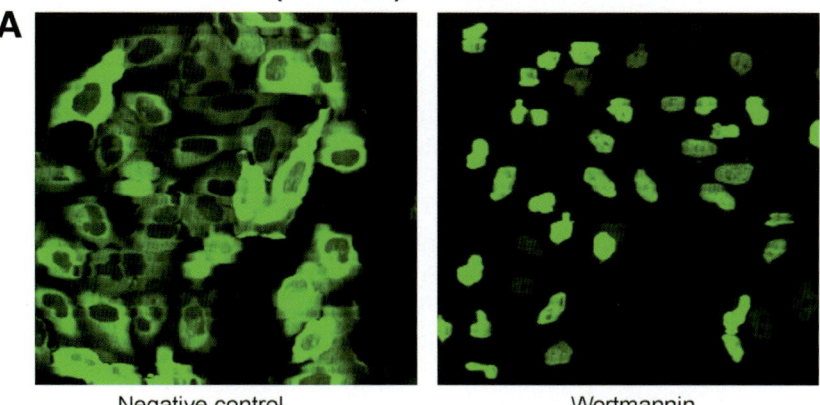
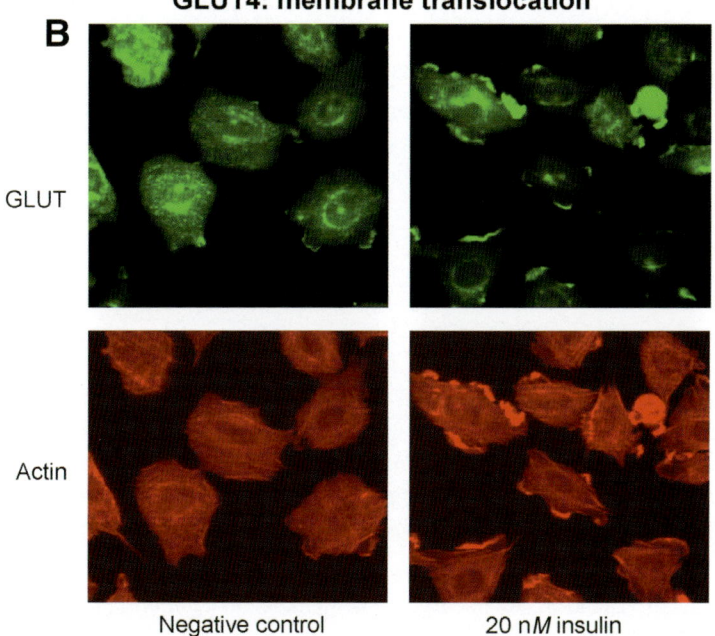
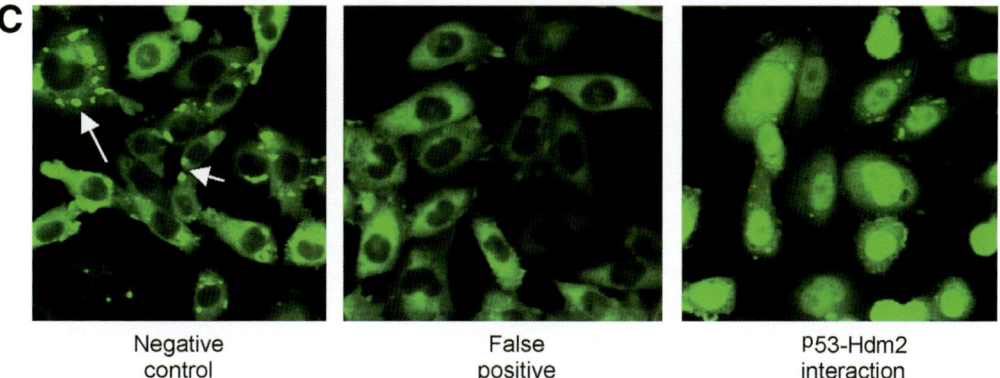

Fig. 1.

translocated to the nucleus. Algorithms can be designed to measure and report the cytoplasmic area for each cell, allowing such toxic compounds to be deselected.

The GLUT4 assay (**Fig. 1B**) measures translocation of the glucose transporter GLUT4 from intracellular vesicles to actin structures located near the plasma membrane, in response to insulin stimulation. Staining of the fixed cells with fluorescent phalloidin allows detection of the reorganized actin cytoskeleton, and gives the possibility of high content analysis of colocalization of GLUT4 with these actin structures.

The p53-Hdm2 assay illustrated in **Fig. 1C** is a high content screening assay based on a "bait and prey" technology, termed GRIP, for measuring protein–protein interactions (*see* **ref. 21** for an overview of this approach). The bait protein, in this case Hdm2, is fused to an inducible anchor, the phosphodiesterase PDE4A4, and is reversibly localized in compact cytoplasmic foci. The prey protein, GFP-p53, binds to the prey protein and is, therefore, localized to the same foci. Compounds that disrupt the binding between p53 and Hdm2 result in dissociation of GFP-p53 from the foci and translocation of fluorescence to the nucleus. False-positive test compounds that interfere with the anchor cause dissociation of the p53-Hdm2 complex from the foci, but the GFP-p53 remains in the cytoplasm, providing an internal deselection parameter.

2. Implementation of High Content Translocation Assays for Pathway Profiling

We describe how to design a cell-based translocation assay, generate an assay cell line, and optimize the assay; validation of the assay is an essential part of assay design and development. Finally, we give an example of how translocation assays can be used for pathway screening and hit deconvolution.

2.1. Assay Design
2.1.1. Cell Lines

The performance of cell-based assays is a combination of ease of use, reproducibility, and cell-line biology. The very best cell-based assays are derived from cells having the ideal biology for the assay and straightforward handling in terms of culture and automation; however, ease of use and biology often do not go hand in hand, and in such cases it might be necessary to make a compromise on the choice of cell line. Assay biology will of course vary from assay-to-assay, whereas conditions of cell handling by and large are known variables. For high content imaging assays we find that several parameters concerning the choice of cell line are important for the final assay quality (*see* **Note 1**):

1. Cell line morphology. The ideal cell line has flat morphology in a uniform monolayer, strong attachment to assay plates, and minimal tendency to form cell aggregates.
2. Culture conditions. Media, doubling time, and transfectability are important factors. Many primary cells require specialized media supplements or feeder layers, grow very slowly, are difficult to transfect by standard transfection agents, and senesce after relatively few doublings, whereas the opposite is true for many standard cell lines. This might not be a key issue for small experimental studies, but for high throughput assays in routine use, a rule of thumb is that simple culture conditions result in robust and reproducible assays.
3. Source of cells. Many laboratories, both in academia and industry, culture cell lines that have been around for years without knowing the original source of the cell lines. Some of these cell lines will have drifted significantly from the equivalent cell lines from a validated vendor. Therefore, only use cells or cell lines with known history and employ cell banking to maintain the cell passage number as close as possible to the source.
4. Assay alignment. Considering these cell culture guidelines along with the assay biology can be crucial for the final assay development process. Specialized biology often requires specialized cells, and in such cases it might be necessary to do significant optimization of cell handling to obtain acceptable assay quality. On the other hand, standard biology or ubiquitous signaling pathways typically can be measured in standard cell lines, and in such cases there is normally no reason to complicate assay

development by choosing an exotic cell line. However, a few additional considerations may be important for the final choice of cell line. First, the species of the cell line might be important for either biological or technical reasons. If an assay is intended for use with siRNAs, a human cell line will be required because most siRNA collections are directed against human mRNAs. Second, cell alignment with other assays is an important but often overlooked issue. If an assay is used for determining structure-activity relationships for compound series along with other cell based assays, for example, functional assays, it might be desirable to use the same parental cell line for all assays to reduce inter-assay variability, i.e., the structure-activity relationships will be based on real biological activity without interference from secondary effects coming from the use of different cell lines. Thoughtful planning before initiating cell line development, both in terms of operational and biological issues, is always a good investment.

2.1.2. Selection of Targets

If studies in the literature show convincing translocation of the protein of interest on stimulation or inhibition of the signal pathway, then there is a good chance that a robust translocation assay can be developed using a GFP fusion to the target protein. The images published in such articles (in which protein localization is visualized either by immunofluorescence or GFP-tagging) allow a reasonably good evaluation of how easy it will be to quantify the target protein's translocation in the final assay. Marginal translocation is unlikely to result in a high-quality assay.

In cases in which the target itself does not undergo translocation but exerts its cellular activity in static location (or submicroscopic movement), a good strategy is to select a pathway component downstream of the target that will translocate in response to activation or deactivation of the target. This indirect strategy opens up for assaying substances that have activity in the pathway upstream of the protein chosen for translocation, and can thus be used for identification of compound classes that inhibit the pathway by a variety of mechanisms. Deconvolution of compound mode of action should be performed subsequently. When using this strategy it is crucial to investigate the cellular activity of the pathway (*see* **Subheading 2.3.3.**).

2.1.3. Choice of Fluorescent Protein

The objective is to design and construct a genetic fusion between a fluorescent protein and the target protein, which can be expressed in the selected cell line, and in which the fusion protein will display the same translocation behavior as the target protein in response to a given stimulus. A number of fluorescent proteins with various properties are commercially available today. A very important consideration is that the fluorescent properties of the protein should be compatible with the detection equipment, i.e., it can be efficiently excited by the light source of the platform, and the emission wavelength can be detected. When the fluorescent protein is to be used as a marker of target protein translocation, it is important that the fluorescent protein does not itself harbor any signals that direct it to a cellular compartment or cause oligomerization. One should also choose a fluorescent protein that is as fluorescent as possible under the conditions tested, because the fewer molecules that needs to be expressed to do the job, the less intrusive it is on cellular functions.

In mammalian cells, EGFP is a good choice. It has been optimized to express well, and it has a long and excellent track record as a translocation marker. Its fluorescent properties are compatible with the most commonly used imaging platforms.

2.1.4. Choice of Expression Vector

The most important elements to consider in the expression vector are the promoter and selective marker. Both should be considered in relation to the host cell line that is to be transfected. The promoter should be one that expresses well in the host cell, also after many passages if the objective is to generate a stable cell line. Viral promoters such as pCMV and pSV40 can be used in most cell lines. These are considered to be strong promoters and will give rise to highly fluorescent cells. On the other hand, overexpression of the fusion protein might be deleterious to the normal cellular biology to be assayed, in which case a weaker promoter should be used; alternatively, cell clones should be selected for study not based on the maximal level of fluorescence but rather by their biology. The selective marker should be one that is effective in the host cell line, also during prolonged propagation. If the cell line already contains another DNA construct with a selective marker one must use a different selection for the fusion construct.

2.1.5. Choice of Orientation of the Fusion Protein

Some fluorescent proteins, for example, those based on *Aequorea* GFP such as EGFP, are amenable to fusion at both the N-terminus and the C-terminus, but one should consider whether there is a preferable orientation based on the fusion partner biology. For example, if either terminus of the protein of interest is known to harbor a localization signal or a phosphorylation site, it should probably not be chosen as a point of fusion to the fluorescent protein. Similarly, if the three-dimensional structure of the protein shows that one of the termini is buried, or if a terminal domain is known to interact with other proteins, it should probably be avoided as a point of fusion. If no such information is available, both orientations should be tested.

The length and composition of the linker between the two fusion partners might be of importance for proper function, but this parameter may be difficult to optimize for each construct. As a rule of thumb, a length of 10–20 amino acid residues is frequently used in the linker region, and ideally it should not contain any cryptic biologically active sites such as phosphorylation sites or protease recognition sites (*see* **Fig. 2**).

2.1.6. Stable vs Transient Transfection

Before going into the details of cell line development, it might be helpful to discuss differences between high content translocation assays based on stable cell lines expressing fluorescent probes vs assays based on transient transfection (*see* **Note 2**). Many researchers tend to think that transient assays are faster and easier than assays based on stable cell lines. This holds true for small-scale studies, or research experiments, in which an analysis can be based on few cells and the requirement for reproducibility is low. One advantage of transient assays is that the choice of cell line is not confined to single cell lines, but rather restricted by cell line transfection efficiencies. Often, viral transduction rather lipid-based transfection is required in order to obtain data of reasonable quality.

When assays are conducted routinely and the requirement for assay quality increases, it becomes complex and tedious to run assays based on transient transfections. First, there can be a considerable expenditure of time and transfection reagent used for each experiment. Second, the quality of a high content imaging assay will depend on the uniformity of a cell population in terms of expression level and biological response, both of which are very problematic to achieve with transient assays. The biological response often varies substantially among cells in a nonclonal population, from cells without any detectable response to cells with clear biological activity. There is no way around this other than development of clonal stable cell lines. Third, as assays based on transient transfection imply complex operational steps, it is complicated to maintain assay reproducibility and robustness, especially when assay protocols are transferred between laboratories or end users. Fourth, if viral transduction is necessary to obtain an acceptable percentage of expressing cells in transient assays, biosafety is an important concern, in particular when working with human oncogenes. The requirements for biological containment will be high for all steps involving living cultures and that might be operationally unachievable.

For these reasons, we prefer to develop high-content imaging assays as stable clonal cell lines expressing the protein of interest fused to a fluorescent protein, usually GFP. Such cell lines have uniform expression of the GFP-fusion protein and the biological response is normally aligned within the cell population in terms of timing and the level of response. Image analysis can be done in a robust and reproducible manner with high demands on assay quality. As a consequence these assay cell lines can be transferred between laboratories with very short start-up time for new users.

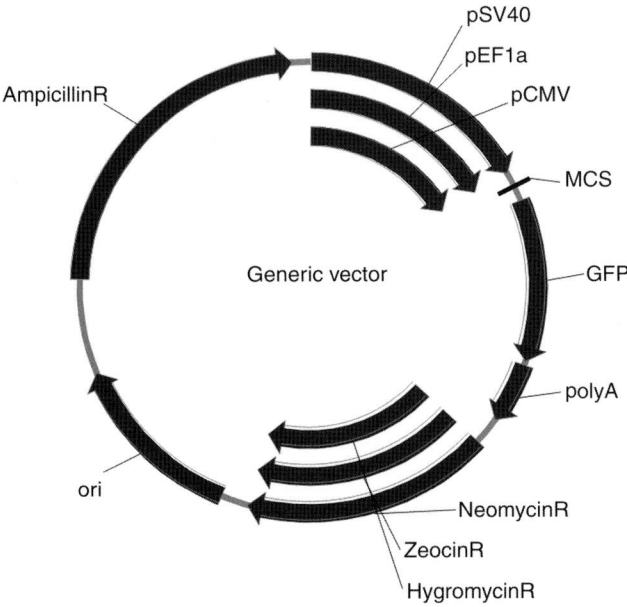

Fig. 2. A generic fusion construct vector map. A typical construct used to create fusions to GFP at the C-terminal of the protein of interest; the multiple cloning site is before GFP. To create fusions of the opposite orientation the MCS would be behind GFP. The fusion protein is expressed from a promoter, for example a viral promoter pCMV or pSV40 or a cellular promoter pEF1a. The plasmid is selected with for example g418, zeocin or hygromycin. In addition there are functions for propagation and selection in *Escherchia coli*.

2.2. Assay Development

2.2.1. Transfection

The first step in the development of stable cell lines for high content translocation assays is transfection of the expression plasmid into suitable host cell lines. Based on knowledge in the literature, complexity of the assay biology, and operational issues, we normally select three to eight candidate host cell lines that are developed in parallel as stable cell lines expressing the GFP-fusion protein of interest. The reason for developing several cell lines in parallel is that the attrition rate for high content assay cell line development is quite high—more than 50%. Many cell lines fail because of unworkable assay biology and some because of problems with protein expression. Among standard human cell lines that work well technically are U2OS, HeLa, and RKO, all of which can be easily transfected using standard transfection reagents such as FuGENE6 or Lipofectamine 2000.

2.2.2. Selection of Stable Cell Line

After transfection, cells are exposed to the relevant antibiotic selection agent, for which a resistance gene is carried on the expression plasmid. Several good selection agents are available including Geneticin (G418), blasticidin, and Zeocin. A resistant population of cells with stable integration of the expression plasmid will grow up after 2–4 wk of culture in selection medium. This cell population varies from cells with no expression of the GFP-fusion protein to cells showing high expression. A good estimate of the biological response of a cell population requires a high fraction of fluorescent cells. This fraction can be optimized by passing the cell population through a fluorescence-activated cell sorter that can be adjusted to collect cells having a defined level of fluorescence. Having a "green" fraction of the stable cell population, the biological activity profile of the cells can be assayed.

2.2.3. Testing the Stably Transfected Cell Line

A protocol for the initial translocation test should be prepared, based on current knowledge of the target protein's translocation behavior. The best available reference compounds (either agonist or antagonist) should be used. The compounds should be tested in concentration response, because cell lines can vary considerably in their sensitivity, and the potency of the same compound purchased from two different suppliers can also vary. Important assay parameters to consider include serum concentration and incubation time. If in doubt, test several different combinations of conditions and time.

Keep in mind that the cell line will be heterogeneous. Some of the cells will not contain detectable levels of the GFP-tagged target protein, and there can be a considerable range of expression levels. It is commonly observed that of the GFP-expressing cells, only a subset show the expected translocation behavior. In some cases it is possible to develop an assay on the multiclonal stably transfected cell line, but in general it is worth taking the time and effort to clone the cell line, in order to maximize assay quality and robustness.

2.2.4. Testing Clones

A simple procedure for obtaining a clonal cell line is to perform dilution cloning. The stably transfected cells are seeded in 96-well plates at a density of 0.5–1 cell per well. After 2–3 wk growth, the plates can be replicated, giving two plates for test of translocation and one mother plate for propagation. A clone should be as homogeneous as possible, both with respect to the expression level of the GFP-tagged target protein and the translocation response. Do not assume that the brightest cells will make the best assay cell line; overexpression can in some cases give aberrant subcellular localization or cellular toxicity. Cellular morphology and growth rate are also factors that can be used when selecting clones. Evaluating the translocation response can in some cases be difficult, because of the fact that the clones by definition are not tested using an optimized protocol. It is sometimes advantageous to perform a limited amount of optimization on the stably transfected nonclonal cell line, in order to develop a protocol that is sufficiently good to allow selection of the best clones.

2.2.5. Optimization

Assay optimization can be a complicated process, because of the fact that there is a large number of parameters to test, generally far too many to allow simultaneous testing in all possible combinations. A good strategy is to focus on two or three parameters at the beginning of the optimization process. Once the best conditions are established for these, move on to test other parameters. Return occasionally to the previously tested parameters if necessary, and retest them in the evolving assay protocol. This is an iterative process that can be repeated as desired until satisfactory assay quality is achieved.

When testing assay parameters, choices must constantly be made as to which conditions should be used in the final assay protocol. These choices can be made based on: (1) maximizing translocation response, evaluated both by visual examination of the images and by the output from the image analysis algorithm; (2) maximizing a key assay parameter such as Z'; (3) minimizing plate-to-plate and day-to-day variability. Important parameters to test include:

1. Serum concentration. Serum contains a myriad of growth factors that activate signaling pathways and affect the translocation of target proteins, often in unpredictable ways. Low serum concentration or serum-free conditions are preferable, because they allow better control of signal pathway activation through addition of defined agonists/antagonists and enhanced sensitivity for serum-binding test compounds, but a balance must be struck with cell viability and morphology. Some cell lines require high serum concentrations, particularly for extended incubations.
2. Assay time. Translocation events can occur on a time-scale of seconds, minutes, or even hours. Optimization is required for each target protein. Very long assay times (e.g., 18 h) can be problematic, because test compounds are often cytotoxic and cause cell rounding, making it difficult or impossible to measure translocation.

3. Cell density. Cell viability and signal pathway activation are sometimes improved when using confluent cells. On the other hand, certain image analysis algorithms work best with well-separated cells. A compromise is sometimes necessary.
4. Temperature. Temperature can be a consideration with assay times of less than an hour. Compound addition is generally performed at room temperature, and the cell plate is then returned to the 37° CO_2 incubator. Thus, the cells are exposed to temperature variations that can affect translocation, and these temperature effects can vary from plate-to-plate or from well-to-well. Strategies for avoiding these effects include the use of prewarmed assay buffer for compound addition and minimizing the amount of time that the cell plate is outside of the incubator *(22)*.
5. Plate coating. Good cell adhesion is crucial for translocation assays. Poorly adherent cells can be washed off the plate during compound addition or plate wash. In addition, if the cells round up during the assay because of insufficient adhesion to the substrate, it can be difficult to accurately quantify translocation with image analysis software. Choice of cell line is important (*see* **Subheading 2.1.1.**). Cell adhesion can often be improved by coating the plates with reagents such as fibronectin, collagen, or polylysine.
6. Fixation. Translocation assays can be performed either on live cells or on fixed cells. Live cell assays are sometimes useful as a prelude to assay optimization, but it is necessary to use a fixed cell format in order to obtain high throughput. A 5–10 min incubation with 4% formalin buffer is an easy and effective fixation method.
7. Nuclear staining. Image analysis algorithms typically require a nuclear marker in order to measure translocation. Hoechst 33258 is a very effective nuclear stain (added to the fixed cells as a 1 μM solution in PBS). DRAQ5 is another useful nuclear stain that has the advantage that it also stains RNA, though to a lesser extent than DNA. This allows it to be used as both a nuclear marker (intense staining) and a cytoplasmic marker (weaker staining), by using two different thresholds in the image analysis algorithm. DRAQ5 is used at a concentration of 0.3–1 μM, depending on cell density. For some cell lines, DRAQ5 might not give sufficient contrast between the nucleus and cytoplasm, necessitating the use of Hoechst 33258.
8. Cell line stability. Multiple vials of the assay cell line should be frozen. A good assay cell line can be continuously cultured for 1–2 mo with no reduction in assay quality. A cell line that is unstable with respect to translocation response or GFP expression levels is of little value.
9. Image based instrument platforms. A number of image-based instrument platforms for high content analysis are currently available, and the approach described in this chapter is platform-independent. The data in this chapter were acquired with the GE Healthcare (Piscataway, NJ) IN Cell 3000; however, these assays are currently running on a wide range of instruments, including the Cellomics ArrayScan, the Evotec Opera, the CompuCyte ICyte, the Molecular Devices Discovery 1, the BD Biosciences Atto Pathfinder HT, and others.
10. Image-analysis algorithms. Manufacturers of the major imaging platforms provide standard algorithms with the instruments. Alternatively, it is possible for users with programming expertise to generate custom algorithms using programs such as MATLAB.
11. Assay optimization example: FKHR (FOXO1A) assay. The U2OS cell line was selected for this assay, because it is a human cell line with good morphology and a functional PI3K signal pathway. It was transfected with a plasmid expressing FKHR-GFP and a stable clone was selected, giving the assay cell line U2OS PS1564 GS cl. 1A. The assay was optimized as a 1 h assay at 37°C in the presence of 0.7% serum, with the PI3K inhibitor wortmannin as the positive control compound. After fixation, the cells were stained with DRAQ5 and read on the INCell 3000 Analyzer, allowing calculation of test compound activity with respect to nuclear accumulation of FKHR, as well as calculation of cell rounding (toxicity) and fluorescence. The assay cell line is stable up to passage 25 (three to four cell doublings per passage).

2.3. Assay Validation

Validation of the assay is crucial before it is released for use. The most important factors to validate are target identity, regulation of target activity, and presence of an intact and functional signaling pathway (**Fig. 3**).

2.3.1. Target Identity

The most straightforward method is to perform Western blotting on a cell extract using a specific antibody against the target. Obtaining a specific band at the expected location on the blot (relative to a molecular weight marker) is good evidence that the target identity is correct. Include a cell extract of the parental cell line as a negative control. It is sometimes not possible to obtain antibodies of sufficiently high specificity for Western blotting. Instead an siRNA approach using siRNAs directed against the fusion protein can be applied. Target expression level can be monitored by Western blotting against GFP, or alternatively by fluorescence imaging of the siRNA treated cells. This latter approach provides the advantage of detecting potential cytotoxic effects of the siRNA protocol that could compromise the result. A significant down-regulation, without cytotoxic effects, with one or more of the siRNAs is a safe indicator that the identity of the fusion is correct. It is important to include appropriate negative and positive control siRNAs.

2.3.2. Regulation of Target Activity

After confirming the identity of the GFP fusion protein in the cell line, it should be confirmed that its activity is regulated properly by the relevant signaling pathway, if it is feasible to do so. This is relatively straightforward to test in cases in which the target translocates in a phosphorylation-dependent manner following stimulation of the cells, because activity can be monitored using phospho-specific antibodies. This can be done with Western blotting on cell lysates, which has the advantage that in many cell lines there is endogenous target protein, which will get modified simultaneously with the GFP fusion protein. The regulation of the endogenous protein can then be used as an internal reference. When performing such Western blots it is important to use a sensitive procedure to allow identification of even low levels of endogenous protein. Immunofluorescence with phospho-specific antibodies is another approach that provides information about subcellular localization and phosphorylation level of the target, though it cannot distinguish between the endogenous target protein and the GFP fusion protein. The fluorescence microscopy images can be analyzed using the same technique as in the translocation assay and can, thus, be used in a more quantitative manner.

2.3.3. Presence of an Intact and Functional Signaling Pathway

When generating a clonal assay cell line it is important to verify that the intracellular pathway of interest is functional and properly regulated. Validating all of the components of a pathway would in many cases be a long and difficult process, so instead a limited number of key components of the pathway are selected for analysis. When possible, proteins for which phospho-specific antibodies are available should be selected, which makes it simple to assess the state of activation using Western blotting or ELISA. If information about intracellular localization simultaneously with phosphorylation level is required, immunofluorescence will be a better choice (remember to use secondary antibodies conjugated with a fluorophore whose fluorescence spectrum does not overlap with that of GFP). Once the components are selected, a plan for which samples to prepare can be made. Translocation assays are as a standard run in 96-well plates. However, in order to prepare samples for Western blotting larger cell quantities are required, and cell extracts are prepared from cultures in six-well plates. When treating cells in a different format it is important to reproduce the assay conditions as closely as possible, because parameters such as media volume and cell density can influence the biology of the pathway.

In order to modulate the pathway, select a set of reference compounds that either activate or inhibit specific pathway components. These compounds should be as specific as possible and be well described in literature (*see* **Note 3**). The expected activity profile of the compounds is later used as a reference when the actual data have been generated. If there are discrepancies they should be investigated, details rechecked and reconsidered, and if necessary the experiments are repeated until unequivocal results are obtained. It is crucial at this point in assay development to be confident that the assay is giving the desired read-out before continuing with compound testing.

Discrepancies occasionally occur and there can be several explanations for this. Two of the major causes are that the reference compounds are less specific than described in the literature and have off-target effects, or that the signaling pathway in the assay cell line is not as described in the literature (*see* **Note 4**). Less than optimal specificity of reference compounds can sometimes be dealt with by using a panel of moderately specific compounds, rather than relying on a single reference compound. It is also possible to obtain specific modulation of the pathway using a siRNA approach; however, issues with off-target effects and toxicity must be addressed in the experimental design. A much more severe problem is a signaling pathway that does not respond as expected, because this indicates that the pathway biology is aberrant. This could be a clonality issue, and testing of other clones might lead to identification of a fully functional cell line. Overexpression of the GFP fusion protein could perturb the pathway either because of increased activity at its level in the pathway or a dominant negative effect. In either case the assay cannot be properly validated and should be redesigned, either by choosing a different cell line, reducing the expression level of the fusion protein, or modifying the fusion protein construct (e.g., by eliminating catalytic activity with a well-chosen mutation).

2.4. Pathway Screening

Figure 4 shows an example of how translocation assays can be used for pathway screening. The phosphoinositide-3-kinase (PI3K) pathway is activated when a ligand binds to a growth factor receptor (GFR) on the cell surface. PI3K generates PIP_3 in the plasma membrane, leading to translocation of Akt and other pH domain-containing proteins from the cytoplasm to the plasma membrane. Akt phosphorylates FKHR, resulting in export of the FKHR transcription factor from the nucleus to the cytoplasm. Both Akt and FKHR translocate, and we have developed high content translocation assays for these two targets. When the FKHR translocation assay is used to screen a library of compounds, hit compounds are obtained that inhibit the pathway at the level of FKHR translocation, as well as at all levels upstream of FKHR.

The mechanism of action of the hit compounds can be deconvolved using a panel of translocation assays and traditional assays. The Akt translocation assay *(20)* can be used to distinguish between compounds that inhibit PI3K, GFR, or Akt itself (positive in the Akt assay) from compounds that work downstream of Akt (negative in the Akt assay). The mechanism of action of upstream hits can be further defined using kinase inhibition assays (receptor tyrosine kinase inhibition assays or PI3K inhibition assays). Compounds that specifically inhibit nuclear export of FKHR can be distinguished from general nuclear export (Crm1) inhibitors by using the Rev translocation assay. Crm1 inhibitors will be positive in both the FKHR and Rev translocation assays and negative in the Akt translocation assay. Specific FKHR translocation inhibitors will be negative in the Rev translocation assay.

A validated functional assay is an important part of pathway screening. In the case of the PI3K pathway, which many cancer cells are dependent on for proliferation, a good functional assay is a matched pair of cancer cell lines, one of, which is dependent on the PI3K pathway for cell growth and one of which is not. Hit compounds should give clear differential killing of the two cancer cell lines. Hence the best-hit compounds from a pathway screen will give convincing activity in the downstream functional assay, with mechanism of action elucidated by the upstream assay panel.

3. Notes

1. A cell-based translocation assay is much more than just a cell line *(23)*. Generating a cell line for a translocation assay that stably expresses the protein of interest is a fairly straightforward task, given sufficient time and basic molecular biology and cell culture skills. However, an established cell line is only the starting point for generating a reproducible and robust translocation assay. A broad range of parameters can influence the cellular activity of the pathway and thereby biological relevance of substances tested in the assay. For example, cell density influences activity of certain pathways and can thereby affect the translocation of the target. Serum present in culture medium activates GFRs affecting multiple pathways and cell cycle. Another parameter is the dynamics of the translocation event. In some

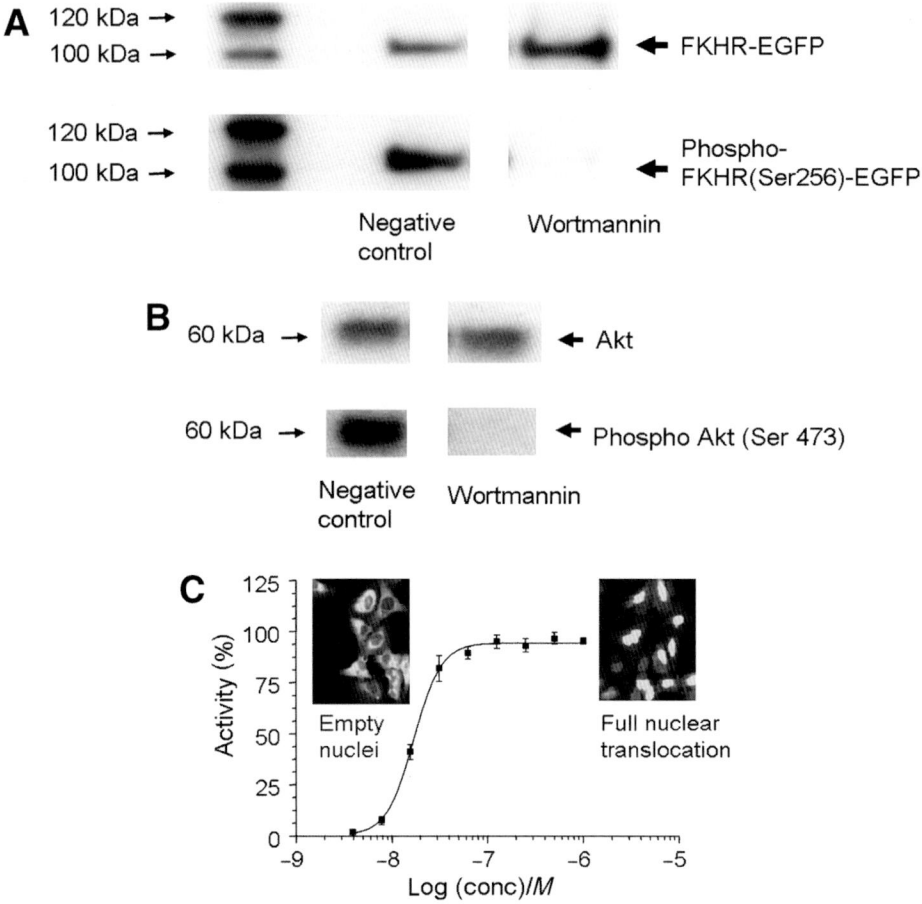

Fig. 3. Validation of FKHR-GFP translocation assay. Target regulation validated by phospho-FKHR specific antibody. In the negative control, FKHR-GFP is phosphorylated; wortmannin treatment leads to significant decrease in phosphorylation (**A**). Proper regulation of the pathway in the FKHR-GFP assay cell line validated using phospho-Akt specific antibody. Akt is phosphorylated and active in the negative control; wortmannin treatment eliminates Akt phosphorylation (**B**). In the negative control, FKHR-GFP is inactive and distributed in cytoplasm. Increasing concentrations of wortmannin leads to Akt inhibition, FKHR activation, and translocation of FKHR to the nucleus (**Insets C**).

cases it is a transient event requiring investigation of the optimal time-point for measurement. Many other parameters need investigation. The relationship between the parameters is impossible to predict and generation of a translocation assay from a cell line therefore includes a long optimization phase.
2. Translocation studies published in the literature are most often performed using transient transfections, because this is a quick and relatively easy way to generate the small number of images needed for a scientific article. For some targets, generation of a stable cell line is the major stumbling block in assay development. This problem can generally be resolved either by testing multiple cell lines, by reducing target protein expression levels (e.g., by using an inducible promoter), or by selective mutations in the target protein (e.g., elimination of kinase activity by mutation of a key amino acid in the catalytic site).
3. Good reference compounds are important for development of a good translocation assay. They are used during assay optimization, assay validation, and as controls in the final assay formatting. Unfortunately, perfect reference compounds are seldom if ever available, so one must make do with what has been described in the literature and is commercially available. Reference compounds are rarely completely specific, so it is important not to use too high a concentration for activation/inhibition of the relevant signaling pathway component, in order to avoid side effects (inhibition of off-pathway enzymes or cellular

High Content Translocation Assays for Pathway Profiling

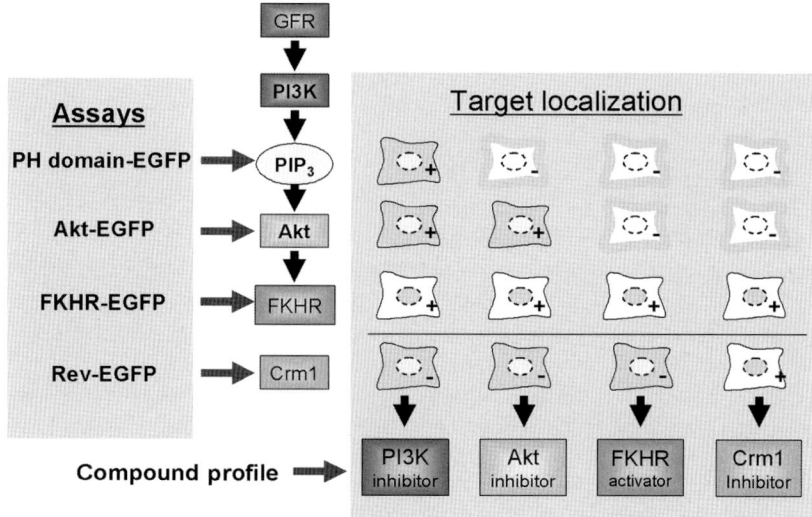

Fig. 4. Scheme for hit deconvolution/pathway screening. Four translocation assays targeting different levels of the PI3-kinase pathway are used for pathway screening and hit deconvolution by creating a family of different compound profiles. The target localization matrix generated by a series of assays at various levels in a signaling pathway enables rapid segmentation of the pathway space in which compounds are acting; subsequent chemo-proteomic analysis can lead to rapid mechanism of action identification.

toxicity). If a gold standard reference compound for a particular target is not available, using a panel of moderately specific compounds can often be useful in assay validation. Finally, RNAi reagents can sometimes be used as a substitute for reference compounds.

4. When performing pathway screening on a large compound library, it is not unusual to find off-pathway hits. The off-pathway mechanism is revealed by lack of activity in the upstream assays used for deconvolution of mechanism of action, or by a mismatch in potency in the screening assay and the downstream functional assay. Occasionally, off-pathway hits are found that are very potent in the functional assay. It can be worthwhile pursuing such hits: a compound with an unexpected mechanism of action that gives high potency in the functional assay can turn out to be extremely valuable as a drug candidate. The possibility of finding such compounds is one of the side benefits of performing a well-designed pathway screen.

References

1. Westwick, J. K. (2004) Drug discovery in the cellular context. *Eur. Pharm. Rev.* **1,** 79–84.
2. Shimomura, O., Johnson, F. H., and Saiga, Y. (1962) Extraction, purification and properties of aequorin, a bioluminescent protein from the luminous hydromedusan, Aequorea. *J. Cell. Comp. Physiol.* **59,** 223–239.
3. Shimomura, O. (2005) The discovery of aequorin and green fluorescent protein. *J. Microsc.* **217,** 1–15.
4. Prasher, D. C., Eckenrode, V. K., Ward, W. W., Prendergast, F. G., and Cormier, M. J. (1992) Primary structure of the Aequorea victoria green-fluorescent protein. *Gene* **111,** 229–233.
5. Chalfie, M., Tu, Y., Euskirchen, G., Ward, W. W., and Prasher, D. C. (1994) Green fluorescent protein as a marker for gene expression. *Science* **263,** 802–805.
6. Yang, T. -T., Cheng, L., and Kain, S. R. (1996) Optimized codon usage and chromophore mutations provide enhanced sensitivity with the green fluorescent protein. *Nucl. Acids Res.* **24,** 4592, 4593.
7. Kamentsky, L. A., Gershman, R. J., Kamentsky, L. D., Pomeroy, B. M., and Weissman, M. L. (1996) CompuCyte corporation. Pathfinder system: computerizing the microscope to improve cytology quality assurance. *Acta Cytol.* **40,** 31–36.
8. Giuliano, K. A. and Taylor, D. L. (1998) Fluorescent-protein biosensors: new tools for drug discovery. *Trends Biotechnol.* **16,** 135–140.

9. Ghosh, R. N., Chen, Y. T., DeBiasio, R., et al. (2000) Cell-based, high-content screen for receptor internalization, recycling and intracellular trafficking. *Biotechniques* **29,** 170–175.
10. Bailey, S. N., Wu, R. Z., and Sabatini, D. M. (2002) Applications of transfected cell microarrays in high-throughput drug discovery. *Drug Discov. Today* **7,** S113–S118.
11. Clemons, P. A. (2004) Complex phenotypic assays in high-throughput screening. *Curr. Opin. Chem. Biol.* **8,** 334–338.
12. Perlman, Z. E., Mitchison, T. J., and Mayer, T. U. (2005) High content screening and profiling of drug activity in an automated centrosome-duplication assay. *Chembiochemistry* **6,** 218.
13. O'Rourke, N. A., Meyer, T., and Chandy, G. (2005) Protein localization studies in the age of 'Omics.' *Curr. Opin. Chem. Biol.* **9,** 82–87.
14. Terry, R., Cheung, Y. F., Praestegaard, M., et al. (2003) Occupancy of the catalytic site of the PDE4A4 cyclic AMP phosphodiesterase by rolipram triggers the dynamic redistribution of this specific isoform in living cells through a cyclic AMP independent process. *Cell Signal* **15,** 955–971.
15. Almholt, D. L. C., Loechel, F., Nielsen, S. J., et al. (2004) Nuclear export inhibitors and kinase inhibitors identified using a MAPKAP kinase 2 Redistribution® screen. *Assay Drug Dev. Technol.* **2,** 7–20.
16. Almholt, K., Tullin, S., Skyggebjerg, O., Scudder, K., Thastrup, O., and Terry, R. (2004) Changes in intracellular cAMP reported by a Redistribution assay using a cAMP-dependent protein kinase-green fluorescent protein chimera. *Cell Signal* **16,** 907–920.
17. Pedersen, H. C. and Pagliaro, L. (2004) Redistribution approach for cellular analysis. *Genet. Eng. News* **24,** 36–38.
18. Grånäs, C., Lundholt, B. K., Heydorn, A., et al. (2005) High content screening for G protein coupled receptor targets using cell-based protein translocation assays. *Comb. Chem. High-Throughput Screen* **8,** 301–309.
19. Grånäs, C., Lundholt, B. K., Loechel, F., et al. (2006) Identification of RAS-mitogen activated protein kinase signaling pathway modulators in an ERF1 Redistribution® screen. *J. Biomol. Screen* **11,** in press.
20. Lundholt, B. K., Linde, V., Loechel, F., et al. (2005) Identification of Akt pathway inhibitors using redistribution screening on the FLIPR and the IN Cell 3000 analyzer. *J. Biomol. Screen* **10,** 20–29.
21. Pagliaro, L., Felding, J., Audouze, K., et al. (2004) Emerging classes of protein-protein interaction inhibitors and new tools for their development. *Curr. Opin. Chem. Biol.* **8,** 442–449.
22. Lundholt, B., Scudder, K., and Pagliaro, L. (2003) A simple technique for reducing edge effect in cell-based assays. *J. Biomol. Screen* **8,** 566–570.
23. Pagliaro, L. and Pra estegaard, M. (2001) Transfected cell lines as tools for high-throughput screening: a call for standards. *J. Biomol Screen* **6,** 133–136.

30

In Vitro Cytotoxicity Assessment

Peter O'Brien and Jeffrey R. Haskins

Summary

The most frequent reason cited for withdrawal of an approved drug is toxicity, yet no simple solution exists to adequately predict such adverse effects. Compound prioritization and optimization during in vitro screening cascades need to be based on confidence, not only in efficacy and bioavailability, but also in safety. A wider number and diversity of potential molecular and cellular effects of compound interactions might affect safety than might affect efficacy or bioavailability. Accordingly, cytotoxicity assessment is less specific, more multiparametric, and extrapolatable with less certainty, unless there are specific safety signals indicated by the chemical structure or by precedents. Cytotoxicity assessments have been limited by their inability to measure multiple, mechanistic parameters that capture a wide spectrum of potential cytopathological changes. Assays with multiple parameters for key, multiple, and different features, such as in high content screening (HCS), are more predictive because they cover a wider spectrum of effects. Assays need to be applied to a large set of marketed drugs that produce toxicity by numerous and different mechanisms for assessment of correlation with human toxicity. This will enable determination of the concordance between in vitro and in vivo results. Multiparametric, live cell, prelethal cytotoxic HCS assays for assessing the potential of compounds for causing human toxicity address some of the limitations of traditional in vitro methods. Assays of this class were used to screen a library of drugs with varying degrees of toxicity and it was found that the sensitivity of the assays was 87%, whereas assay specificity was more than 90%, thereby minimizing false positives.

Key Words: Autophagy; drug discovery; energy homeostasis; genotoxicity; hepatotoxicity; immune-mediation; oxidative stress.

1. Introduction

Early in drug discovery, in vitro cytotoxicity is becoming increasingly recognized as an effective indicator of human toxicity potential that must be addressed in order to maximize probability of successful progression of compounds into development *(1–6)*. Compound prioritization and optimization during in vitro screening cascades need to be based on confidence, not only in efficacy and bioavailability, but also in safety *(1)*. However, safety is more multifactorial, as it is dependent on homeostasis of virtually all cellular processes *(7)*. A wider number and diversity of potential molecular and cellular effects of compound interactions might affect safety than might affect efficacy or bioavailability. Accordingly, cytotoxicity assessment is less specific, more multiparametric, and extrapolatable with less certainty, unless there are specific safety signals indicated by the chemical structure or by precedents. Extrapolation needs a greater foundation of mechanistic understanding of both in vitro and in vivo pathogenesis of toxicities, as well as rigorous, empirical validation of models. Finally, it is important to recognize that cytotoxicity models are inevitably limited by their inability to account for toxicities arising from interactions between different cell types and structures and extracellular matrices, such as occur at the tissue, organ, and system level *(1)*. The previously mentioned complexity indicates that thorough and systematic consideration and evaluation of in vitro models should be undertaken before their implementation.

2. Mechanisms of Cytotoxicity

2.1. Nuclear Effects and Cell Proliferation

Cell proliferation is dependent on intact structure and function of all vital cell processes. Consequently it is affected early in all toxicities, even if only secondarily. Additionally, one of the numerous specific processes involved in replication might be affected primarily as the target of compound effect. Proliferation provides a "catch-all" screen for cytotoxic effects *(8,9)*. Characteristic morphological changes have long been noted to occur early with cell death, including nuclear condensation, shrinkage, and fragmentation. The opposite effect might occur with cell cycle inhibitors that cause nuclear swelling and nuclear-cytoplasmic asynchrony. Toxicity that is DNA-based and inheritable, that is genotoxicity, and occurs owing to mutagenicity or disruption of chromosomal separation during cell division is discussed at length elsewhere.

2.2. Cell Membrane Effects and Transport

The most well-known and assayed cytotoxic effects are cell membrane disruption causing leakage of cellular contents into the extracellular space or influx of extracellular dyes that stain cellular constituents *(1,6,8,9)*. These effects are typically nonspecific and occur when energy and ion homeostasis are compromised to the point in which the membrane barrier can no longer be maintained. However, they might occur more owing to specific interactions such as with membrane perturbing agents like detergents and volatile anesthetics, inhibitors of vital functions like ion transport or signal transduction, and with inhibitors of specific transporters of noxious substances such as in biliary and renal tubular cells. Cell membrane disruption might occur without necrosis if the effect is mild. For example, hepatic accumulation of lipid and carbohydrate in glycogenosis and steatosis might result in release of cellular enzymes without any evidence of necrosis. Release occurs by blebbing, in which cytoskeleton is disrupted and plasma membrane evaginates to form vesicles that bud off without loss of cell membrane continuity. Blebs contain cytoplasm complete with enzyme and even organelles. Alternatively, cellular constituents might be released by cell rupture, with more severe injury. Loss of the cell membrane's barrier function is frequently used in viability assessments to measure cellular influx of stains, such as Trypan blue, or DNA stains. Such stains might grade severity of injury on the basis of size and permeation properties of the dye.

2.3. Mitochondrial Effects and Energy Homeostasis

The third most well-studied and recognized mechanism of cytotoxicity is probably mitochondrial toxicity *(10–12)*. Mitochondria are ubiquitously involved secondarily to virtually all other cellular effects. Also, they are frequently a primary target of toxicities because of their complexity and numerous critical and diverse roles in energy and calcium homeostasis, biosynthesis, oxidative stress, and apoptosis. **Table 1** identifies and provides examples of the major mitochondrial functions that might be inhibited by drugs. Most of these mitochondrial dysfunctions are manifested as alterations in mitochondrial membrane potential or in reductive activity of the enzymatic oxidoreductases. Frequently, if cells are not overwhelmed by the toxic effects they make early compensatory adaptations, such as increasing mitochondrial biogenesis *(11,13)*.

2.4. Reactive Metabolites, Oxidative Stress, and Immune-Mediation

Many drugs have long been recognized to produce toxicity from oxidative stress by several mechanisms: (1) progressive reduction of oxygen into reactive oxygen species such as superoxide, peroxides, and hydroxyl radical and reactive nitrogen species such as nitric oxide and peroxynitrite

Table 1
Mechanisms of Mitochondrial Toxicity

1. *Oxidative phosphorylation*
    - Inhibition of complexes, for example, I by rotenone and fenofibrate, II by, IV by cyanide, V by oligomycin; depleters of coenzyme Q such as amitryptyline
    - Redox cyclers diverting electrons to form reactive oxygen and nitrogen species, for example, quinones
    - Uncouplers of electron transport from ATP synthesis, for example, protonophores, tolcapone, flutamide, cocaine, furosemide, fatty acids
2. *Fatty acid β-oxidation*
    - Inhibition by valproate, tetracyclines, nonsteroidal antiinflammatory drugs, antianginal cationic amphiphilic drugs, female sex hormones, CoA depleters such as valproate and salicylate
3. *Krebs cycle*
    - Inhibition of aconitase by superoxide and fluoroacetate, of succinate dehydrogenase by methamphetamine and malonate, of α-ketoglutarate dehydrogenase by salicylic acid
4. *Mitochondrial membrane transporters*
    - Inhibition of adenine nucleotide transporter by zidovudine
5. *Mitochondrial permeability transition pore*
    - Opening by reactive oxygen species, reactive nitrogen species, bile acids, thio crosslinkers, atractyloside, betuliniate, lonidamidem various anticancer drugs, to collapse mitochondrial membrane potential and activate mitochondrial apoptotic pathway
6. *Mitochondrial proliferation*
    - Inhibitor of DNA polymerase-γ for mitochondrial DNA synthesis by nucleoside reverse transcriptase inhibitors
    - Inhibition of mitochondrial protein synthesis by oxazolidinone antibiotics
    - Mitochondrial DNA mutation, for example, by oxidative injury by ethanol
7. *Mitochondrial oxidative stress*
    - Glutathione depletion by acetaminophen, bromobenzene, chloroform, allyl alcohol
    - Redox cyclers (*see* oxidative phosphorylation)
    - Reactive metabolites

(*see* **Fig. 1**); (2) depleting glutathione, (3) bioactivation to an electrophilic metabolite that forms adducts with cellular macromolecules *(14–17)*. As oxidative stress is a ubiquitous process found with normal intermediary metabolic activity and xenobiotic detoxification, cells have evolved complex antioxidant defensive systems that are readily upregulated, including signal transduction by transcription factor translocation (ap-1, nrf2), and numerous protective enzymes and free radical scavengers (*see* **Fig. 1**). Reactive metabolite formation has been associated with immune-mediated mechanisms of toxicity affecting skin, blood cells, and liver. The mechanism has not been defined but might relate to hapten formation and cell stress, which in the presence of an activate immune system and inflammation trigger an autoimmune response *(14)*.

2.5. Lysosomal Effects and Autophagy

One of the most commonly recognized toxicities that is specifically linked to chemical structure and primarily involves lysosomes is phospholipidosis *(1)*. This occurs because of interaction of cationic amphiphilic drugs with phospholipases or phospholipids such that their lysosomal catabolism is inhibited and they subsequently accumulate to excessive extent. Phospholipidotic drugs have a hydrophobic ring with hydrophilic side chain containing a charged cationic amino group and include amiodarone, perhexiline, chloroquine, clomipramine, imipramine, fluoxetine, norfluoxetine, gentamycin, propranolol, tamoxifen, and quinacrine. There are potential serious toxic effects of chronic administration. For example, amiodarone is an antiarrhythmic that produces serious pulmonary toxicity in 10% patients under chronic treatment and also liver injury.

A Oxidation by O₂ may from reactive oxygen species

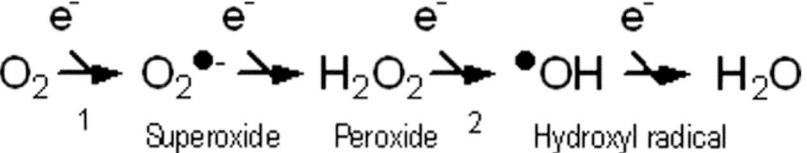

1 = oxidases (eg xanthine oxidase, monoamine oxidase, NADPH oxidase, cyclo oxygenase, lipooxygenase, amino acid oxidase), hemoglobin, cytochromes, mitochondria (NADH dehydrogenase, ubisemiquinone), NOS, redox cycling drugs
2 = Fe, Zn, Cu   **Antioxidants** include superoxide dismutase for superoxide, catalase and glutathione peroxidase for peroxide, and free radical scavengers such as tocopherol, glutathione, and ascorbate for hydroxyl radical

B Secondary reactive oxygen and nitrogen species

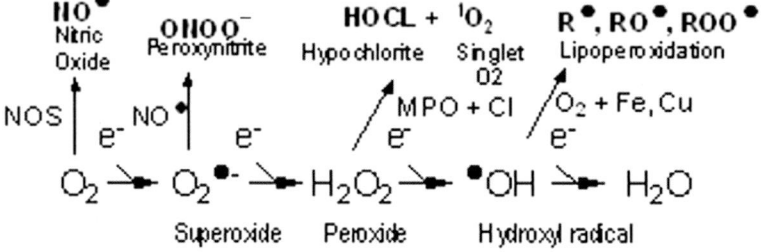

C Oxidative stress mechanism of paracetamol toxicity

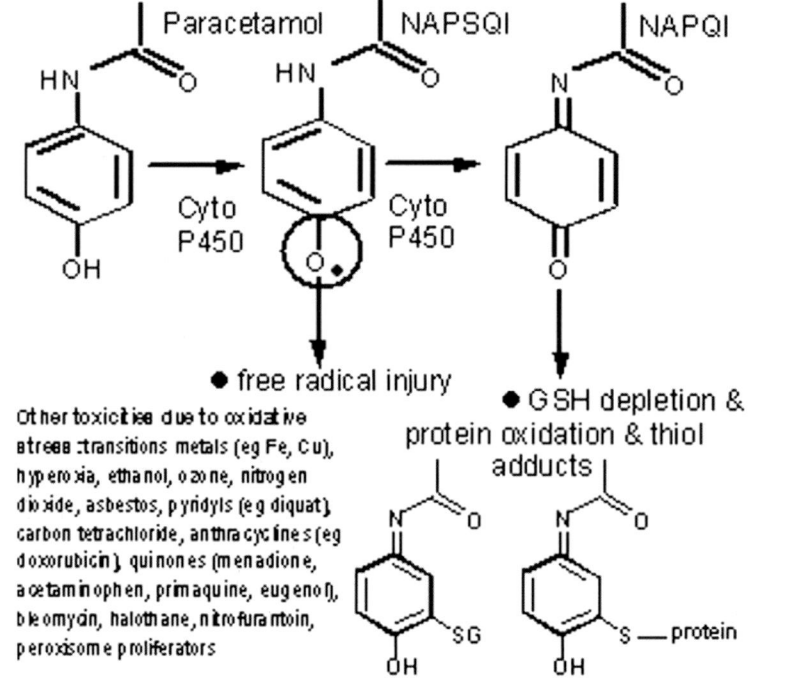

Other toxicities due to oxidative stress: transitions metals (eg Fe, Cu), hyperoxia, ethanol, ozone, nitrogen dioxide, asbestos, pyridyls (eg diquat), carbon tetrachloride, anthracyclines (eg doxorubicin), quinones (menadione, acetaminophen, primaquine, eugenol), bleomycin, halothane, nitrofurantoin, peroxisome proliferators

Note: Reduced glutathione is maintained by glutamyl-cysteine synthase, glucose-6-phosphate dehydrogenase and glutathione reductase

Fig. 1. Mechanisms of oxidative stress.

Perhexiline used for ischemic heart disease might produce severe liver injury in 1–2% of patients under chronic treatment.

A second and less commonly recognized primary lysosomal effect that is related drug structure has been referred to as vacuolization *(18,19)*. This also occurs because of lysosomal swelling, especially with tertiary amines that are charge neutral at physiological pH. In vitro it occurs at concentrations from 0.1 to 2.5 mM. Organic bases enter the acidic lysosome and are trapped by protonation at the lower pH that makes them positively charged, and prevent back diffusion. The drugs accumulate and increase intravacuolar osmotic activity. Consequently, there is water influx and swelling into large, clear vacuoles.

Lysosomes are also involved secondarily affected with other cytotoxicities. Their activity and mass might increase with apoptosis in which cells are undergoing programmed cell death, including autophagy as the cells involute. Their size and mass might decrease with general ill health of the cell.

2.6. Calcium Regulatory System

Calcium plays an integral role not only in regulating cellular functions such as secretion, contraction, metabolism, and gene transcription, but also in apoptosis and cell death. There is a 10,000-fold gradient of ionized calcium from the extracellular to intracellular space, that if not maintained and controlled, leads to dysfunction, activation of degenerative enzymes, and cell loss. Accordingly, calcium is an important prelethal signal of cell injury of all causes *(8,20,21)*. Calcium dyshomeostasis might also result from direct interaction of drugs or their toxic metabolites with Ca-regulatory channels and pumps, such as with fibrates, statins, anthraquinones such as doxorubicin, and thapsigargin.

3. Limitations of Traditional in Vitro Methods

Several limitations of traditional in vitro methods of assessment of cytotoxicity have resulted in their use being substantially restricted (*see* **Table 2**). Probably the most ineffective assays are those that have not allowed sufficient duration of exposure for cytotoxicity to be expressed *(6,8)*. An estimated three-quarters of drugs do not express their toxicity with acute exposure, but only after several days exposure. Some drugs, such as mitochondrial DNA and protein synthesis inhibitors might take up to a week to show cytotoxic effects in many assays. Dose–response curves are typically shifted to the left, toward increased sensitivity, with increasing duration of exposure from hours to days.

Another common cause of failure of some assays to detect in vivo toxicity potential is their dependence on measurement of events that occur only late in the life of the cell when severe cell injury has occurred, such as cell rupture with release of cellular constituents and uptake of extracellular substances. Such assays lack sensitivity in predicting in vivo toxicity primarily because cell death is not required for significant toxicity. Also, limited solubility of drugs might limit testing the drug to a sufficiently high concentration. Assay of total ATP content of cell populations is frequently used for cytotoxicity assessment, but has limited utility, especially when used alone as an indicator, because it changes quite late in pathogenesis *(22,23)*. The lateness of this change reflects the tight regulation and buffering of ATP concentration by interconversions with other high-energy phosphates, such as creatine phosphate, and by replenishment by metabolic conversion of substrate stores, such as glycogen. This tight regulation is required because ATP in turn regulates activities of numerous metabolic pathways, cell functions, and cell structure. Current assays for ATP are also significantly limited because of the need to apply them to cell populations rather than single cells (*see* below).

Cytotoxicity measurement of events that are too early in the pathogenesis of cytotoxicity might also be ineffective for assessment of in vivo toxicity potential. For example, measurement of concentration of a noxious substance, byproduct or metabolite, or of activation of a signal

Table 2
Major Criteria for an Effective Cytotoxicity Assay

1. *Prelethal: not too late or too early an indicator of adverse effect*
   - Prelethal effect as opposed to measurement of cell death is required
   - Measurement should not be made of noxious substance or a signal transduction event as opposed to measurement of adaptive and adverse effect
2. *Chronic: allows significant duration of exposure for toxicity expression*
   - Most toxicities are expressed only after multiple days of exposure
3. *Catches all: includes measurements of adverse effect in common to all toxicities*
   - There needs to be a "catch-all" measure of an activity that is affected in a final common pathway of cell injury
4. *Mulitplexing: makes multiple measurements of different end points for different processes*
   - A single end-point assay will not identify most cytotoxicities
5. *Measures morphological and functional characteristics*
   - Structural and functional measurement is complementary and additive
6. *Mechanistic: provides mechanistic information*
   - Signals generated in cytotoxicity assays need to indicate mechanisms to be followed up to derisk the compound
7. *Tracks individual cells*
   - Allows identification of hormesis and separation of compensatory adaptation from degenerative change
   - Allows more accurate identification of sequence of change in different cytotoxicity parameters as cells might not be synchronous or alike in their response
8. *Uses live cells under normophysiological conditions:* cell function is substantially affected by temperature, humidity, and oxygenation, pH and osmolality, as well as media growth factors and attachment substrate
9. *Predictive: high sensitivity and specificity and concordance with in vivo toxicity in species designed to be indicative of*
   - Inhibitor of DNA polymerize gamma for mitochondrial DNA synthesis by nucleoside reverse transcriptase inhibitors
10. *Precise: high level of within-run, across-day, and across-laboratory precision*
    - Assay must be reproducible
11. *Practical: sufficient operational performance: throughput and cost-effectiveness*
    - At least 100 to several hundred per week
    - Assays are available that can be run at significantly less than 100$ per compound including materials, staff, and instrumentation
12. *Dose–responsive: determines different response over range of concentrations*
13. *Effective cell model:*
    - Drug-metabolism competent
    - Same species as predicting for

transduction pathway is insufficient to conclude that a compound is toxic. Direct demonstration and measurement of the adverse effect, or of an adaptation to it, are required.

Cytotoxicity assessments have been limited by their inability to measure multiple, mechanistic parameters that capture a wide spectrum of potential cytopathological changes. The complexity of the biology behind a toxicological change requires several criteria for an effective cytotoxicity assay. First, there is no single parameter likely to be able to identify all toxicities, which immediately limits utility of uniparametric assays. Assays with multiple parameters for key, multiple and different features are more predictive because they cover a wider spectrum of effects. Additionally, those assay parameters that represent fundamental, cellular mechanisms of pathogenesis rather than being purely descriptive are more potent. Assays that combine morphology and functional assessments are more predictive as they measure more parameters by using more and independent analytic approaches, such as dimensional image analysis and fluorescence intensity measurements. Morphological assessments provide information about size and shape of cells and organelles, as well as on intracellular location, such as with transcription factor translocation or lysosomal sequestration. Thus assays measuring both morphology and function are making a more comprehensive evaluation over a wider spectrum of change.

Assays that measure end points for cell populations rather than multiple individual cells might produce contradictory findings (6). For example, mitochondrial reductive capacity will be decreased with decreased cell numbers but increased with cells that have been activated, such as lymphocytic immune-activation, or if the cells have adapted to the stress associated with the toxicity, such as with mitochondrial biogenesis. Thus, mitochondrial reductive capacity might be either increased or decreased with toxicity. Similar contradictory interpretations might occur with other cellular activities, for which there is a compensatory adaptive increase before their failure. This biphasic change is referred to as hormesis and occurs with not only reductive mitochondrial activity, but with mitochondrial number, cell number, mitochondrial membrane potential, antioxidant system activity, and numerous other activities. Finally, individual cell studies might be more accurate than cell population studies in which responses are variable over time or over different cells. Analysis of the sequence of changes in the different parameters might be important in elucidation of mechanisms and pathogenesis of toxicity.

One of the largest limitations of in vitro assays is their lack of full validation and determination of their sensitivity and specificity for prediction of human toxicity potential. Assays need to be applied to a large set of marketed drugs that produce toxicity by numerous and different mechanisms for assessment of correlation with human toxicity. This will enable determination of the concordance between in vitro and in vivo results. Typically, such assays show high specificity, in excess of 90% (6). When compounds react positively in the cytotoxicity evaluation, this is associated with in vivo toxicity. The major concern though, is the sensitivity with which the toxic potential is assessed. In a comparison of seven conventional cytotoxicity assays applied to 600 compounds with single end points measurement in an acute exposure experiment, only glutathione had significant sensitivity, 19% (6). Measures of mitochondrial reductive capacity and DNA synthesis were half as sensitive. Caspase induction, synthesis, protein synthesis, superoxide production, and membrane integrity were of negligible value.

4. Cytotoxicity Assessment in Drug Discovery
4.1. Screening for and Derisking Human Toxicity Potential

Acute toxicity is frequently assessed in drug discovery programs, in which cell-based efficacy assays are used and in which cytotoxicity could be a false-positive for efficacy. Although, efficacy is frequently expressed by cells in short-term studies of hours duration, toxicity is most often expressed only over a duration of several days. Furthermore, concentrations causing toxicity are frequently greater than those for efficacy, and so efficacy assays are limited for this reason alone in detecting in vivo toxicity potential. As discussed above, acute toxicity is typically effective at identifying in vivo toxicity potential only for overtly and potently cytotoxic compounds (**Table 3**).

Go or no go decision-making might frequently be made based on occurrence of marked and overt cytotoxicity at concentrations near the estimated efficacy concentration. However, chronic rather than acute cytotoxicity assessment is needed for prediction of human toxicity potential for most effective optimization and prioritization of compounds based on confidence in safety. Such early ranking of compounds for their progression is important for early initiation of potential hazard identification and for flagging up compounds needing follow-up safety assessment, and early development of risk management strategies.

Table 3
Cellular Functions Assessed in Cytotoxicity Assays

1. *Cell membrane and permeability barrier to extracellular environment*
   - Leakage of cell contents, for example, lactate dehydrogenase, chromium release
   - Entry of extracellular dyes, for example, Trypan blue, DNA stains
2. *Nucleus and cell proliferation*
   - Cell number
   - Nuclear size: contraction with apoptosis and swelling with cell cycle inhibition
   - Frequency distribution of nuclear DNA content of cell population
   - Protein synthesis, for example, ^{14}C-labelled methionine incorporation
   - DNA synthesis, for example, tritiated thymidine incorporation, DNA stains
   - Mass tracker dyes, for example, for lysosomes or mitochondria
3. *Mitochondria and energy homeostasis*
   - Dye oxidation, for example, tetrazolium salts (e.g., MTT, MTS), alamar blue
   - ATP concentration, for example, luciferase assay
   - Oxygen consumption, for example, oxygen electrodes, Luxcell oxygen sensor
   - Mitochondrial protein and nucleic acid synthesis
   - Mitochondrial mass, for example, mitotracker dyes
4. *Antioxidant system and oxidative stress management*
   - Oxidant production, for example, dihydroethidium, dichlorofluorescein
   - Antioxidant changes, for example, glutathione, antioxidant system enzyme, resistance to dye oxidation (e.g., total antioxidant status)
   - Macromolecular oxidation byproducts, for example, malondialdehyde, hydroxynonenal, 8-hydroxyguanosine
5. *Lysosomes and macromolecular disposal*
   - Phospholipidosis—Nile red, lysotracker dyes, electron microscopy of lysosomal multilamellar bodies
   - Vacuolization
   - Autophagy
6. *Cytoskeleton and cell shape*
   - Blebbing
7. *Ca regulatory system and Ca homeostasis*
   - Ionized and total calcium concentration
8. *Signal transduction and adaptive gene expression*
   - Cytoplasmic-nuclear translocation, for example, proportioning of immunocytochemical stain for NFKB with inflammation and AP-1 and NRF2 with oxidative stress
9. *Cell involution and apoptosis*
   - Nuclear condensation and lobulation, caspase activation, phosphatidyl serine externalization, annexin, immunocytochemistry, DNA fragmentation and labeled-dUTP incorporation by terminal deoxynucleotidyl transferase
10. *Intermediary metabolism and specific enzyme activities or metabolite concentrations*
    - Bench-top or automated chemistry analyzer assays of cell lysates for key enzymes of intermediary metabolism, antioxidant system, ion transport
11. *Cell-based efficacy targets in drug discovery*
    - Live-cell assays in which assessment of efficacy is based on a cellular function that might be inhibited by cytotoxicity

4.2. Strategies for Implementation of Cytotoxicity Assessment in Discovery

Cytotoxicity data cannot be used in isolation for decision making, excepting in which there are profound effects at concentrations similar to those for efficacy. Toxicity potential needs to be considered in the context of efficacy and bioavailability properties. Typically evaluation of cytotoxicity needs to be integrated with off-target pharmacological activity, drug metabolism and transport and distribution, chemical properties, in vivo animal study results. It is also important to consider differences in protein binding in vitro compared to in vivo, whether toxicity is mediated by a toxic metabolite, in which the drug is concentrated in vivo, relationship of in vivo exposure of Cmax needed for efficacy vs exposure or concentration causing toxicity in vitro.

A second tier of cytotoxicity testing based on the type of toxicity identified from the first, screening tier is usually warranted if the compound or series of compound progresses. This approach enables progressively more specific in vitro studies to define narrower dose–response relationships around the cytotoxic concentration, mechanisms of toxicity, and biomarkers for risk management.

4.3. Basic Assay Design

Optimal features for cytotoxicity assay design are outlined in **Table 2** and are based on considerations previously made. For adequate sensitivity, there must be multiple parameters incorporated into the assay and these must represent different relevant processes that occur early in cell injury, rather than before or after. Measurements must be precise and accurate. There must be sufficient duration of exposure for the toxicity to be expressed at relevant concentrations. For drugs that exert their toxicity through reactive metabolites, the cell model must have some metabolic competence for the relevant species. For assessment of the relevance of the cytotoxicity and designing follow-up studies, parameters must provide some mechanistic understanding. For practicality, assays must be cost-effective, have moderate throughput and be widely accessible for implementation.

5. Methods for In Vitro Assessment of General Cytotoxicity (Necrosis)
5.1. Conventional Methods

Conventional methods used for cytotoxicity assessment are typically adapted for use in multiwell plates for fluorescent plate readers and flow cytometers in order for relevant throughput to be available **(Table 3)**. Measures are typically based on fluorescence intensity from dyes that directly interact with intracellular constituents and reflect their concentration or activity, or from fluorescent labels attached to antibodies specific for cellular antigens. The latter immunocytochemical stains and the dyes for measurement of parameters that are static over the duration of the measurement, can be applied to dead or fixed cells or, if morphological information is not obtained, to their lysates. Such dyes would include those for DNA or lipid, cell or organelle mass, or ATP. However, measurement of dynamically changing parameters such as free ion concentrations, reactive oxygen species, or membrane potential need to be measured in live cells. Such measurements require continuous maintenance of normophysiological conditions of temperature, humidity, and oxygen tension.

Alternatives to fluorescence-based staining or using specific fluorescence probes have been used for cytotoxicity screening. Using reverse transcriptase polymerase chain reaction, small profiles of specific gene expression patterns might be applied effectively *(24)*. Assays for specific enzymes or metabolites of intermediary metabolism and oxidative stress have been adapted to automated chemistry analyzers and applied to cell-based systems for assessment of toxicity *(8)*.

There is a wide array of fluorescent probes currently available including for nuclear and mitochondrial DNA, lysosomal and mitochondrial mass. Immunocytochemistry can be used for morphological assessments, intracellular localization, quantitation of specific antigens, immunophenotyping, and transcription factor translocation.

5.2. High Content Methods

The recent development of capacity of fluorescence plate readers and, more recently, flow cytometers for multiple fluorescence measurements combined with image analysis has markedly enhanced their value in cytotoxicity assessment. These new capacities can now be used to simultaneously measure multiple parameters for both morphological and biochemical cell features *(25–27)*. Furthermore, they can be applied at the single cell level, as well as at the

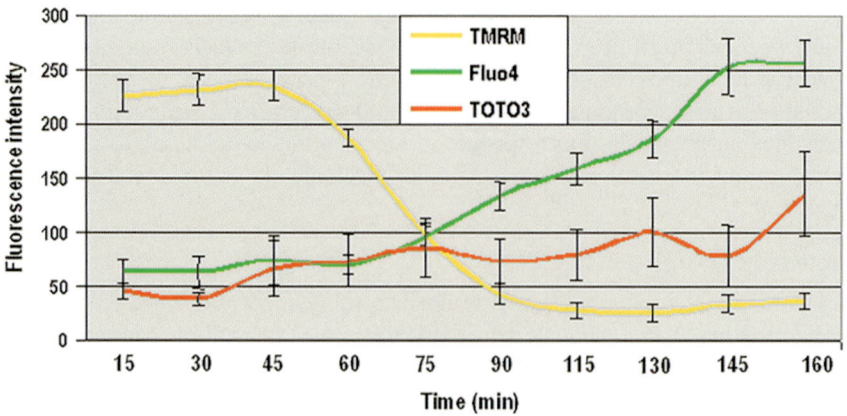

Fig. 2. Multiparametric, live cell, prelethal cytotoxicity assay.

well level and level of cell populations or subpopulations derived from thresholding for certain values of a specific parameter. Finally, they are able to identify intracellular location.

Separable fluorescence signals can be measured simultaneously for different cellular parameters by choice of dyes with nonoverlapping spectral properties. Dyes with overlapping fluorescence might also be used to discriminate different cellular targets simultaneously if they have nonoverlapping subcellular locations, such as nuclear and mitochondrial DNA. Also, dye replacement could allow use of dyes with similar fluorescence, such as application of an immunocytochemical stain following loss of a vital stain with cell death or fixation.

6. Assay Example: Quad-Probe Assay for Predictive Hepatotoxicity

A multiparametric, live cell, prelethal cytotoxicity assay for assessing the potential of compounds for causing human toxicity has been recently reported using high content screening (*1,6*; *see* **Fig. 2**). This was based on the "quad probe" assay reported by Haskins and colleagues (*25–27*). Human hepatocellular carcinoma cultured in 96-well plates and loaded with four fluorescent dyes, fluo-4 for intracellular calcium, tetramethylrhodamine methyl ester for mitochondrial transmembrane potential, Hoechst 33342 for DNA content to determine nuclear area and cell number, and TOTO-3 for plasma membrane permeability. In order to increase the assay sensitivity, the cells were continuously exposed to drugs for 3 d at a range of concentrations, up to at least 30 times the plasma concentration needed for an efficacious effect. Assay results were compared with those from seven conventional, in vitro cytotoxicity assays that were applied to 600 drugs and compounds and shown to have low sensitivity (<25%), although high specificity (approx 90%) for detection of toxic drugs. For 250 drugs with varying degrees of toxicity, the sensitivity of the novel multiparametric in vitro assays was 90%. Specificity of these in vitro assays was more than 95%, thereby minimizing false-positive results.

References

1. Xu, J. J., Diaz, D., and O'Brien, P. J. (2004) Applications of cytotoxicity assays and pre-lethal mechanistic assays for assessment of human hepatotoxicity potential. *Chem. Biol. Interact.* **150**, 115–128.
2. Perlman, A. E., Slack, M. D., Feng, Y., Mitchison, T. J., Wu, L. F., and Altschuler, S. J. (2004) Multidimensional drug profiling by automated microscopy. *Science* **306**, 1194–1198.
3. Dambach, D. M., Andrews, B. A., and Moulin, F. (2005) New technologies and screening strategies for hepatotoxicity: use of in vitro methods. *Toxicol. Pathol.* **33**, 17–26.
4. Schoonen, W. G. E. J., Westerink, W. M. A., de Roos, J. A. D. M., and Debiton, E. (2005) Cytotoxic effects of 100 reference compounds on Hep G2 and HeLa cells and of 60 compounds on ECC-1 and

CHO cells. I. Mechanistic assays on ROS, glutathione depletion and calcein uptake. *Toxicol. in Vitro* **19**, 505–516.
5. Schoonen, W. G. E. J., Westerink, W. M. A., de Roos, J. A. D. M., and Debiton, E. (2005) Cytotoxic effects of 110 reference compounds on Hep G2 and for 60 compounds on HeLa, ECC-1 and CHO cells. I. Mechanistic assays on NAD(P)H, ATP, and DNA contents. *Toxicol. In Vitro* **19**, 491–503.
6. O'Brien, P. J., Irwin, W., Diaz, D., et al. (2006) High concordance of drug-induced human hepatotoxicity with in vitro sublethal, live-cell cytotoxicity determined by high content screening. *Arch. Toxicol.* in press.
7. Jaeschke, H., Gores, G. J., Cederbaum, A. I., Hinson, J. A., Pessayre, D., and Lemasters, J. J. (2002) Mechanisms of hepatotoxicity. *Toxicol. Sci.* **65**, 166–176.
8. Slaughter, M. R., Thakkar, H., and O'Brien, P. J. (2002) Effect of diquat on the antioxidant system and cell growth in human neuroblastoma cells. *Toxicol. Appl. Pharmacol.* **178**, 63–70.
9. Sussman, J. L, Waltershied, M., Butler, T., Cali, J. J., Riss, T., and Kelly, J. H. (2002) The predictive nature of high-throughput toxicity screening using a human hepatocytes cell line, Vol.3: *Cell Notes*, Promega Corporation, Madison, Wisconsin, pp. 7–10.
10. Lewis, W., Day, B. J., and Copeland, W. C. (2003) Mitochondrial toxicity of NRTI antiviral drugs: an integrated cellular perspective. *Nat. Rev. Drug Discov.* **2**, 812–822.
11. Lee, H. -C., and Wei, Y. -H. (2005) Mitochondrial biogenesis and mitochondrial DNA maintenance of mammalian cells under oxidative stress. *Int. J. Biochem. Cell Biol.* **37**, 822–834.
12. Fariss, M. W., Chan, C. B., Patel, M., Van Houten, B., and Orrenius, S. (2005) Role of mitochondria in toxic oxidative stress. *Mol. Interv.* **5**, 94–111.
13. Kultz, D. (2005) Molecular and evolutionary basis of the cellular stress response. *Annu. Rev. Physiol.* **67**, 225–257.
14. Kaplowitz, N. (2005) Idiosyncratic drug hepatotoxicity. *Nat. Rev. Drug Discov.* **4**, 489–499.
15. Nassar, A. -E. F., DeMaio, W., Davis, M., and Talaat, R. E. (2004) Detecting and minimizing reactive intermediates in R&D. *Curr. Drug Discov.*, **July,** 20–25.
16. Kalgutkar, A. S., Gardner, I., Obach, R. S., et al. (2005) A comprehensive listing of bioactivation pathways of organic functional groups. *Curr. Drug Metab.* **6**, 1–65.
17. Park, B. K., Kitteringham, N. R., Maggs, J. L., Pirmohamed, M., and Williams, D. P. (2005) The role of metabolic activation in drug-induced hepatotoxicity. *Annu. Rev. Pharmacol. Toxicol.* **45**, 177–202.
18. Morissette, G., Moreau, M. C., -Gaudreault, R., and Marceau, F. (2004) Massive cell vacuolization induced by organic amines such as procainamide. *J. Pharmacol. Exp. Ther.* **310**, 395–406.
19. O'Brien, P. J., Kalow, B. I., Brown, B. D., Lumsden, J. H., and Jacobs, R. M. (1989) Porcine malignant hyperthermia susceptibility: halothane-induced increase in cytoplasmic free calcium of peripheral blood lymphocytes. *Am. J. Vet. Res.* **50**, 131–135.
20. Waring, P. (2005) Redox active calcium ion channels and cell death. *Arch. Biochem. Biophys.* **43**, 33–42.
21. Inoue, R., Tanabe, M., Kono, K., Maruyama, K., Takaaki, I., and Endo, M. (2003) Ca^{2+}-releasing effect of cerivastatin on the sarcoplasmic reticulum of mouse and rat skeletal muscle fibers. *J. Pharmacol. Sci.* **93**, 279–288.
22. Nerurkar, P. V., Pearson, L., Frank, J. E., Yanagihara, R., and Nerurkar, V. R. (2003) Highly active antiretroviral therapy (HAART)-associated lactic acidosis: in vitro effects of combination of nucleoside analogues and protease inhibitors on mitochondrial function and lactic acid production. *Cell Mol. Biol. (Noisy-le-grand)* **49**, 205–211.
23. Leandri, R. D., Dulious, E., Benbrik, E., Jouannet, P., and de Almeida, M. (2003) Deficit in cytochrome c oxidase activity induced in rat sperm mitochondria by in vivo exposure to zidovudine. *Int. J. Androl.* **26**, 305–309.
24. Bulera, S. J., Eddy, S. M., Ferguson, E., et al. (2001) RNA expression in the early characterization of hepatotoxicants in Wistar rats by high-density microassays. *Hepatology* **11**, 1239–1258.
25. Abraham, V. C., Taylor, L., and Haskins, J. R. (2004) High content screening applied to large-scale cell biology. *Trends Biotechnol.* **22**, 15–72.
26. Giuliano, K. A., Haskins, J. R., and Taylor, D. L. (2003) Advances in high content screening for drug discovery. *Assay Drug Dev. Technol.* **1**, 565–577.
27. Haskins, J. R., Rowse, P., Rahbari, R., and de La Iglesia, F. A. (2001) Thiazolidinedione toxicity to isolated hepatocytes revealed by coherent multiprobe fluorescence microscopy and correlated with multiparameter flow cytometry of peripheral leukocytes. *Arch.Toxicol.* **75**, 425–438.

31

Neurite Outgrowth in Retinal Ganglion Cell Culture

John B. Kerrison and Donald J. Zack

Summary

Retinal ganglion cells (RGC) are the projection neurons of the eye. The RGC is the primary cell type injured in a variety of diseases of the optic nerve, including glaucoma and optic neuritis. The most well-established extrinsic signal of RGC survival and axonal outgrowth is the neurotrophin brain-derived neurotrophic factor. An immunopurification system has been adapted in order to filter large enough quantities of RGCs from the mixed population of retinal neurons in order to perform high-throughput screening in a 96-well format. Using this assay, the screening of a combinatorial chemical library for compounds with a similar effect to brain-derived neurotrophic factor may be preformed. Follow-up validation studies are performed by evaluating for a dose–response relationship.

Key Words: Axon; BDNF; calcein AM; cellomics; chembridge; immunomagnetic; neuron; optic nerve; retinal ganglion cells.

1. Introduction

Retinal ganglion cells (RGCs) are the projection neurons of the eye, their axons extending along the inner retina to the optic nerve, coursing through the chiasm and optic tract, and synapsing on lateral geniculate neurons, which project to the visual cortex. The human optic nerve is made up of slightly over one million RGC axons, which are bundled by a meshwork of connective tissue septae containing small arterioles, venules, and capillaries. Other cellular components include astrocytes, oligodendrocytes provide myelination to retinal ganglion cell axons.

RGC axons might be injured along their course by a number of disease processes, which result in axonal degeneration, apoptosis of RGCs, and irreversible vision loss. RGCs are well suited for the study of axonal injury because of their accessible location outside the brain. From a clinical point of view, many RGC diseases have their initial insult at the axon level, including glaucoma, optic neuritis, anterior ischemic optic neuropathy, traumatic optic neuropathy, and compressive optic neuropathy. As such, the adoption of this model for the general study of axonal injury and regeneration has led to many insights, yet there has been no definitive impact on the clinical practices, which diagnose and manage optic nerve diseases: neuro-ophthalmology and glaucoma.

When the optic nerve is injured at the orbital apex, the entire length of individual axons degenerate simultaneously as early as 3 wk and as late as 6 wk after injury *(1)* and death of RGCs ensues. The mechanism by which RGCs die is programmed cell death or apoptosis, which is observed in both clinical *(2)* and experimental *(2,3)* optic neuropathies. Why do RGCs die following injury? Although the decisive event triggering RGC apoptosis after axotomy is not known, the downstream molecular signaling pathways involved in RGC apoptosis have been an area of considerable investigation. Once activated, the highly regulated cell death pathway is generally launched in an all or none fashion. In the final stages, a series of proteases of the caspase family (cysteine proteases that cut after an aspartate residue) undergo aggregation and activation in a self-amplifying cycle called the caspase cascade. Inhibition of the caspase cascade at the time of optic nerve transection is protective against RGC death *(4)*. The caspase cascade might be initiated by cell surface receptors, such as the Fas ligand binding to Fas receptor. Patients with Leber's hereditary optic neuropathy causing mitochondrial mutations might be particularly susceptible to Fas *(5)*. The caspase cascade might also be initiated internally, when mitochondria are induced to discharge the electron carrier protein cytochrome-*c* into the cytosol, where it binds and activates an adaptor protein called Apaf-1.

Central to the apoptosis pathway in general and to RGCs in particular are the Bcl-2 family of intracellular proteins. Bcl-2 and Bcl-xL are important antiapoptotic genes that function in part by blocking release of cytochrome-*c* from the mitochondria. Other members of the Bcl-2 family, including Bad, Bax, and Bak, are proapoptotic, either inactivating antiapoptotic factors or stimulating the release of cytochrome-*c* from mitochondria. In the retina, the antideath gene Bcl-xL is the predominant Bcl-2 family member, and its level decreases after optic nerve crush *(4,6)*. In retinal ganglion cell culture, expression of Bcl-2 reduces cell death *(7)*. With the ability to modulate the signaling pathways in apoptosis comes the ability to dissociate survival from axonal regeneration.

Following axotomy, RGC axons fail to regenerate past the injury site *(8)*. This is thought to be because of a developmental loss of the intrinsic capability of RGCs to regenerate and a prohibitive environment of the glial scar and myelin. The extent to which the signals are controlled in the axonal outgrowth during development and axonal regeneration in adulthood overlap is not yet known. The loss of the intrinsic ability of RGCs to grow neurites occurs early in postnatal development is irreversible, and extrinsically signaled by amacrine cells *(9)*. Another important developmental change is the switch from axonal extension to the elaboration of the dendritic arbor *(10)*. This event is mediated by a calcium responsive transactivator called CREST (for calcium-responsive transactivator) *(11)* and NeuroD *(12)*.

In another development, it was reported that inhibition of the ubiquitin ligase anaphase-promoting complex (APC) in primary neurons specifically enhanced axonal growth in the developing rat cerebellum *(13)*. The APC is essential to the coordination of cell cycle transitions, particularly exit from the cell cycle, and is highly expressed in postmitotic neurons. In this study, APC was inhibited by means of small hairpin RNA knockdown Cdh-1, a gene that activates APC. This suggests a mechanism by which axonal outgrowth might be under the developmental influence of cell cycle exit.

The most well-established extrinsic signal of RGC axonal outgrow is the neurotrophin brain-derived neurotrophic factor (BDNF). BDNF interacts with the Trk B receptor and signals through the ras-raf-MAP kinase and the phosphatidylinositol-3-kinase pathways *(14–16)*. With the development of the ability to dissociate cell survival from neurite outgrowth signaling by introducing Bcl-2 into RGCs, it was shown that BDNF, as well as electrical activity, specifically promote axonal outgrowth *(7)*. How well will a small molecule modulate RGC survival? BDNF and forskolin synergistically promote RGCs in culture. Forskolin increases intracellular cAMP by stimulating adenylate cyclase and subsequent activation of protein kinase A. The role of cAMP in synergistically modulating response to BDNF signaling has been shown to occur by translocation of TrkB receptors from the intracellular compartment to the membrane surface, thus making cells more responsive to BDNF *(17)*. The use of a high-throughput screen to identify potentially therapeutic compounds has been successfully used in neuronal cell cultures, and more specifically in retinal cell cultures. In a search for molecules that might be effective in the treatment of Huntington's disease, high-throughput screen of a small molecule library successfully identified a novel molecule that supported Huntington's disease neurons in cell culture, inhibits polyglutamine aggregation, and suppress neurodegeneration in vivo *(18)*.

In retinal cell culture, Leveillard et al. *(19)* demonstrated the power of this approach, when they screened a retinal expression library in chicken retinal cultures for a cone survival factor that might be effective in the treatment of retinal degeneration. For this study, an expression library from wild-type mouse neural retina was constructed in pcDNA-3. The library was collected into pools containing approx 100 clones each and introduced into COS-1 cells for a total of 2100 pools of 100 clones each. After successfully screening those 2100 pools for the cone survival phenotype, positive pools were diluted into subpools of 10 clones and finally tested as individual clones. The novel survival factor was named rod-dependant cone viability factor-1 (RdCVF1). RdCVF1 rescues cones in chicken cultures as well as cones from a mouse with retinitis pigmentosa (rd1 mouse). Expression of RdCVF1 is restricted to the retina. It is expressed by rods and its expression is markedly reduced following retinal degeneration. Restoration of RdCVF1 following rod degeneration might provide a novel strategy for the treatment of retinitis pigmentosa and/or other retinal degenerations. Because of that time, the library has been sequenced in collaboration with Novartis and is presently being organized into a set of individual clones that should make it more versatile.

Following identification of a drug that promotes growth of RGCs in culture, several hypothesis driven questions follow. One might hypothesize that the factor influences the competence of the cells to respond to neurite promoting cues or alternatively stimulate the cell directly, either by influencing known signaling pathways or electrical activity. This might be examined by culturing cells from animals at different ages in the presence of the factor and determining the percent of outgrowth neurons (*see* **Note 1**). One might hypothesize that it influences survival to a greater degree neurite extension and examine whether the factor promotes outgrowth in RGCs that are overexpressing Bcl-2, which allows cells to survive in culture without promoting axonal outgrowth *(7)*. One might hypothesize that the factor might differentially modulate axonal vs dendrite promoting factor and assess this with immunostaining. One might hypothesize that the factor modulates a known signaling pathway, particularly if it is synergistic with known RGC growth factors. This can be tested in vitro with known inhibitors of MAPK kinase, phosphoinositide 3-kinase, or RhoA. In order to identify the drug target, one might biotinylate the compound followed by hybridization it to a proteome array *(20)*. These are some of the lines of inquiry that might be pursued following drug identification.

A convergence of technologies in robotics, cell culture techniques, use of fluorescence in cell culture, digital imaging, image analysis, and combinatorial chemistry has made the outlined strategy possible. As with many of the recent technology driven advances in science, which have allowed broader questions to be asked, this represents a development of scale and proportion rather than a completely novel paradigm. That is to say, the underlying principle of identifying specific molecules in cell culture that promote cell survival that led to the discovery of BDNF and ciliary neurotrophic factor (CNTF) is no different from the strategy being employed in this proposal.

The screening of a combinatorial chemical library is highlighted by the NIH Roadmap and offers a distinct translational benefit. The development of such libraries is fundamental to the emerging field of chemical genomics. Chemical compounds used by biological systems represent a exceedingly small fraction of the total possible number of small carbon-based compounds with molecular masses in the same range as those of living systems *(21)*. In terms of numbers of compounds, "biologically relevant chemical space" is only a tiny fraction of complete "chemical space" *(21)*. The goal of chemical genomics is to identify small molecules that might be used as analogs of genetic mutations for studying mammalian systems.

Depending on the goals of the study one might wish to screen target oriented vs diversity oriented libraries. The diversity of a library is a quantitative description of how different compounds are from each other. A target-oriented library is made up of a collection of small molecules that has a specific attribute thought to be of biological significance often targeting a specific protein. Screening a library of FDA approved drugs led the observation that β-lactams promote expression of a specific target, the glutamate transporter, and were neuroprotective in an animal model of amyotrophic lateral sclerosis *(22)*. This has prompted the development of a collection of small molecules based on the β-lactam structure, that might have more potent activity. Diversity-oriented libraries are not targeted to any specific protein class and are often used in broad screens in which the target proteins are not known. Such libraries are made up of an assorted collection of small molecules based on a computational approaches used to sift through much larger numbers of more varied compounds thought to be biologically active. Although employing the Chembridge Diverset, a diversity oriented compound library, this assay can be applied to nonsmall molecule libraries. The retinal expression library described in the preceding section is of particular interest for vision science. Another library of significant interest is lentivirus-based RNA interference (RNAi) library that is being developed by the Viral shRNA Library Consortium, a joint effort of several research groups at the Whitehead Institute, the Broad Institute, the Dana Farber Cancer Institute, and the Harvard Institute of Proteomics *(23)*.

2. Materials

1. DNAse (0.4%; 12,400 U/mL).
2. Poly-D-lysine (dilute 1/100 leads to a concentration of 0.01 mg/mL) (Sigma, cat. no. P-6407).
3. L-cysteine.
4. Papain (Worthington Biochemical Corporation, Lakewood, NJ).
5. Ovomucoid solution (10X; 40 mL).
    a. To 40 mL DPBS, add 600 mg of BSA (Sigma, St. Louis, MO; cat. no. A8806), 600 mg of Trypsin.
    b. Inhibitor (Roche Basil, Switzerland; cat. no. 109878), and adjust pH to 7.4 (add approx 400 µL of N NaOH).
6. Laminin (Sigma, cat. no. L-2020, maintain initially at –80°C).
7. Tris-buffer, 50 mM, (Sigma, cat. no. T1503).
8. Rat Serum: (Jackson Immunoresearch West Grove, PA; cat. no. 012000-120).
9. Mouse antirat Thy1 (Chemicon International, Temecula, CA; cat. no. MAB1406).
10. Mouse antirat cd11 b/c (BD Biosciences, Pharmingen, Bedford, MA; cat. no. 554859).
11. Calcein AM (Molecular Probes, Eugene, OR).
12. Hocest (Molecular Probes).
13. BDNF.
14. Forskolin.
15. Neurobasal medium.
16. Glutamine.
17. Penicillin/streptomycin.
18. Dynal microbeads.

3. Methods

3.1. Methods Overview

Our culture system, based on the work of Barres et al. *(24)* employs an efficient two-step immunomagnetic purification of the mixed cell population of cells from the postnatal rat retina to yield a highly enriched population of RGCs for culture. The first step consists of immunomagnetic depletion of Thy1 bearing non-RCGS using a cd11b/c antibody followed by a second step consisting of immunomagnetic selection of the RGCs with an antibody that binds Thy1 on the RGC cell surface. In the final step, the cells are released from the magnetic beads by digesting the DNA linker with a low concentration of DNAase. Cells are grown on poly-D-lysine/laminin coated plastic plates in a basic medium consisting of neurobasal media, B27 supplement, glutamine, and penicillin/streptomycin.

These cultures are highly enriched for RGCs, are morphologically typical of RGCs in cell culture, and reproducibly respond to BDNF and forskolin as expected for RGCs. To assess the efficiency of our selection technique, RGCs were retrograde labeled by bilateral transcranial injection of DiI into the superior colliculus 3 d before sacrifice. With this technique, 94.8 ± 1.3%

of the cells in culture were labeled. After 5 d in culture, the cells have developed a robust arbor of neurites typical of what has been reported for RGCs. Finally, the addition of BDNF and forskolin, factors known to promote RGC survival, to our basic culture medium results in increased overall cell survival and more significantly in the survival of neurite bearing cells.

We typically harvest between 18 and 20 retinas per experiment in a procedure which takes less than 5 h. Based on an estimate of 100,000 RGCs per retina, our yield is approx 50%. Thus, we typically harvest 1 million cells per experiment. We estimate our initial postseeding viability by staining with Hoechst (a blue fluorescent nuclear dye taken up by all cells) and TOTO3 (a far red fluorescent nuclear dye retained by dead cells). Our viability ranges from 90 to 95% using this technique.

3.2. Cell Source and Purification

Retinas are dissected from postnatal d 3–5 Sprague Dawley rats, digested with activated papain (Worthington) and DNase (Sigma) before dissociation by trituration. Immunomagnetic depletion of macrophages is performed by incubating the retinal cell suspension with magnetic microbeads (Dynal Biotech) conjugated to mouse antirat cd11 b/c antibody (BD Biosciences, Pharmingen) and removal by magnetic separation. This is followed by immunomagnetic selection of Thy1 antigen bearing RGCs by incubating the cell suspension with magnetic microbeads conjugated to mouse antirat Thy1 antibodies (Chemicon) and selection by magnetic separation. In a final step, the magnetic beads are cleaved from the cells by incubation in DNAse, which cuts the DNA linker between the magnetic bead and the antibody.

3.3. Culture Conditions and Controls

The purified Thy-1 immunoreactive cells are resuspended in growth media (neurobasal media, B27 supplement, glutamine [2 mM], and penicillin/streptomycin [1 U/mL]) and quantified on a hematocytometer. Cells are seeded in 100 µL at a density of 4000 cells/well (150/mm^2) in 96-well culture dishes (Falcon BD Biosciences, Bedford, MA) that had been sequentially coated with poly-D-lysine (0.01 mg/mL) and laminin (0.01 µg/µL) (Sigma). Each well either contains 25 µL of media alone or 5X drug or 5X BDNF and forskolin. Final concentrations are: DMSO (1%), drug (20–30 µM), BDNF (50 ng/mL) and forskolin (10 µM). Culture dishes are placed in an incubator maintained at 37°C and 5% CO_2. One hour after seeding, a postseeding viability is determined.

The Diverset (Chembridge, Inc. San Diego, CA) is a diversity-oriented library of synthetic small molecules picked from the Chembridge collection of 300,000 chemicals. The organizing criteria for the collection are diversity, heterocyclic, predicted bioactivity, low molecular weight, production through multistep large-scale manual synthesis, and over 95% purity. The entire Diverset collection consists of five groups of 10,000. The compounds are diluted to 3 mM in DMSO and individually arrayed in columns 2–11 on a 96-well plate, leaving columns 1 and 12 for controls. Negative controls are placed in columns 1 and 2, wells A–E. Positive controls, which consist of media supplemented with BDNF and forskolin are placed in columns 1 and 2, wells F–H. The plates are stored at –20°C until ready for use. The final concentration of drug is approx 20–30 µM.

3.4. Staining and Imaging

After 120 h of incubation, cells are stained with Hoechst (5 µM, Molecular Probes) and Calcein AM (10 µM) (live cell stain; Molecular Probes). Following 30 min of incubation, each well is rinsed three times with PBS and imaged. In earlier studies, a third dye was used to stain dead cell nuclei (TOTO3). It was observed that it was a rare cell that could not be distinguished as "live" without calcein alone; rare cells stain robustly with both Calcein AM and TOTO3. Thus, in order to reduce scan time and simplify our protocol, TOTO3 was eliminated. However, TOTO3 is still used to determine viability because of low background without rinsing.

In a standardized fashion that is identical for each well, twenty stereotypic fields are autofocus imaged on the Cellomics KineticScan High Content Scan Reader using epifluorescence and filter sets corresponding to each of the two dyes. Scanning the entire plate takes just over 3 h and collects 3840 images.

3.5. Image Analysis

Images are processed using the Cellomics, Inc., Extended Neurite Outgrowth program. In the initial image processing, cells are identified on the Hoechst channel and selected based on size such that clusters of cells might be excluded from analysis. In addition, each nuclei on the Hoechst channel is matched to a calcein stained cell and counted as a "valid neuron," which might be regarded as a live cell. Neurons can be further selected based on size in the calcein channel. The Extended Neurite Outgrowth program is used to trace neurites extending from these selected neurons. The mean neurite length, mean neurite count, mean number of branch points, and mean number of crosspoints are determined. In addition, the percent of cells with neurite features (i.e., neurite length, neurite count) surpassing a threshold of one standard deviation more than the mean neurite characteristic of a set of reference wells containing growth medium without additional growth factors is determined. The "neurite outgrowth" is based on the combined mean neurite length and mean neurite count. Cells surpassing a threshold of one standard deviation more than the mean neurite outgrowth index for control wells are regarded as "outgrowth neurons." The number of outgrowth neurons is the most reliable indicator of the presence of growth factors in our system and the primary end point of this assay, yet several other parameters are assessed as well.

3.6. Statistical Analysis

The outgrowth count for each well is normalized to counts from all wells except for positive controls. As studies are done in quadruplicate, a mean, and standard deviation is calculated for each compound and a z-statistic as an indicator or reproducibility calculated:

$$z = [1 - (3 \times \text{standard deviation of controls}) + (3 \times \text{standard deviation of drug})] / (\text{mean of drug} - \text{mean of controls}).$$

Those drugs with the highest z-scores are regarded as the most reproducible.

3.7. Validation of Hits

Standard rationale for demonstrating a causal relationship between two factors is to modulate the level of function of one factor and correlate it with the desired outcome. The standard paradigm for studying gene functions is gain-of-function and loss-of-function manipulations. For small compound studies, the standard approach is a dose–response analysis. As such, one approach to the validation of hits is to demonstrate a dose–response relationship (*see* **Note 2**). This will demonstrate the underlying biological response. In addition, it will help to determine the potency of drugs so that the most potent drugs are identified. Although the developed technique for performing cell culture is effective and objective, it is desirable to assess ones outcome using a different technique. As such, the standard technique of culturing cells on poly-D-lysine and laminin-coated cover slips in 24-well culture dishes under various culture conditions followed by immunostaining with antineurofilament antibody might be performed followed by quantitative analysis.

As with genes in which precise functions can be further localized to specific domains, the essential bioactive element of small compounds can be determined. As with genes in which this involves site-directed mutagenesis and truncation, a combinatorial chemist can alter specific parts of a compound for testing in order to enhance the potency. Combinatorial chemistry is a group of methodologies developed by researchers in the pharmaceutical, agrochemical, and biotechnology fields to facilitate the production of effective new drugs. Combinatorial chemistry is used to create large populations of molecules that can be screened. By producing larger, more diverse compound libraries, the probability is increased that one might a find novel compounds of significant therapeutic value. The field represents a convergence of chemistry and biology, made feasible by fundamental advances in miniaturization, robotics, and receptor development.

For dose–response analysis, RGC harvest, and immunopurification is performed with the same technique as previously discussed. Drugs to be validated will be diluted to a final medium concentration of 3, 10, 15, 30, 60, 90, 120, and 150 μM while controlling DMSO at 1% in all wells including media only and BDNF/Forskolin controls. Drugs and growth factor supplements are added, 25 μL to each well in a 5X concentration, before the seeding cells in a volume of 100 μL giving a final volume of 125 μL. Cells are incubated, stained, and imaged as stated earlier. Experiments are performed in duplicate. Following normalization to media only well, the dose–response is plotted using SPSS statistical software in order to perform curve fitting and analysis.

For each drug, resupply in quantities from 1 to 5 mg is available through online ordering from Chembridge, Inc. If larger quantities are needed, further drug can be synthesized. For validation studies, 1 mg of desiccated drug is resuspended in 22.2 μL of DMSO to a final stock solution of 150 mM. If the drug does not go into solution, the stock is further diluted until it is in solution. Afterwards, the drug is stored in the dark at –20°C until ready for use.

4. Notes

1. Phenotypic vs target end point. The use of a phenotypic end point, in this case "outgrowth neuron," as opposed to a target-based screening, in high-throughput analysis has the advantage of having a functional end point yet has the challenge of more difficult assay development and interpretation as well as downstream challenge of target identification.
2. What is a hit? The criteria of what constitutes a hit are important. The criteria should be broad enough to minimize type II error (failing to reject the null hypothesis when it is false), in which we might fail to select a particular compound for follow-up studies. Compounds selected for follow-up are those with an outgrowth count and z-score comparable to BDNF and forskolin as well as compounds that are the most robust in comparison with media alone. This might range from 0 to 3 per compound plate. Some of these are "false-positives" representing either an outlier or the tip in the distribution of outgrowth counts for a given plate. Others of course represent a "true-positives." Sorting out the false hits and the true hits are the reason for validation studies.

Acknowledgments

This work was supported by grants from the National Eye Institute, the Glaucoma Foundation, the Glaucoma Research Foundation, and the American Health Assistance Foundation, Research to Preven Blindness, Inc., the MICC, the Marriott Foundation, the William and Ella Owens Medical Research Foundation, and by a generous gift from Mr. and Mrs. Robert Smith.

References

1. Quigley, H. A., Davis, E. B., and Anderson, D. R. (1997) Descending optic nerve degeneration in primates. *Invest. Ophthalmol. Vis. Sci.* **16(9)**, 841–849.
2. Kerrigan, L. A., Zack, D. J., Quigley, H. A., Smith, S. D., and Pease, M. E. (1997) TUNEL-positive ganglion cells in human primary open-angle glaucoma. *Arch. Ophthalmol.* **115(8)**, 1031–1035.
3. Quigley, H. A., Nickells, R. W., Kerrigan, L. A., et al. (1995) Retinal ganglion cell death in experimental glaucoma and after axotomy occurs by apoptosis. *Invest. Ophthalmol. Vis. Sci.* **36(5)**, 774–786.
4. Chaudhary, P., Ahmed, F., Quebada, P., and Sharma, S. C. (1999) Caspase inhibitors block the retinal ganglion cell death following optic nerve transection. *Brain Res. Mol. Brain Res.* **67(1)**, 36–45.
5. Danielson, S. R., Wong, A., Carelli, V., et al. (2002) Cells bearing mutations causing Leber's hereditary optic neuropathy are sensitized to Fas-Induced apoptosis. *J. Biol. Chem.* **277(8)**, 5810–5815.
6. Levin, L. A., Schlamp, C. L., Spieldoch, R. L., Geszvain, K. M., and Nickells, R. W. (1997) Identification of the bcl-2 family of genes in the rat retina. *Invest. Ophthalmol. Vis. Sci.* **38(12)**, 2545–2553.
7. Goldberg, J. L., Espinosa, J. S., Xu, Y., et al. (2002) Retinal ganglion cells do not extend axons by default: promotion by neurotrophic signaling and electrical activity. *Neuron* **33(5)**, 689–702.
8. Ramon, Y. and Cajal, S. (1928) Traumatic degeneration and regeneration in the optic nerve and retina, in *Degeneration and Regeneration of the Nervous System* (May, ed.), Hafner, New York.
9. Goldberg, J. L., Klassen, M. P., Hua, Y., and Barres, B. A. (2002) Amacrine-signaled loss of intrinsic axon growth ability by retinal ganglion cells. *Science* **296(5574)**, 1860–1864.

10. Goldberg, J. L. (2004) Intrinsic neuronal regulation of axon and dendrite growth. *Curr. Opin. Neurobiol.* **14(5),** 551–557.
11. Aizawa, H., Hu, S. C., Bobb, K., et al. (2004) Dendrite development regulated by CREST, a calcium-regulated transcriptional activator. *Science* **303(5655),** 197–202.
12. Gaudilliere, B., Konishi, Y., de la Iglesia, N., Yao, G., and Bonni, A. (2004) A CaMKII-NeuroD signaling pathway specifies dendritic morphogenesis. *Neuron* **41(2),** 229–241.
13. Konishi, Y., Stegmuller, J., Matsuda, T., Bonni, S., and Bonni, A. (2004) Cdh1-APC controls axonal growth and patterning in the mammalian brain. *Science* **303(5660),** 1026–1030.
14. Klesse, L. J. and Parada, L. F. (1999) Trks: signal transduction and intracellular pathways. *Microsc. Res. Tech.* **45(4–5),** 210–216.
15. Klocker, N., Kermer, P., Weishaupt, J. H., et al. (2000) Brain-derived neurotrophic factor-mediated neuroprotection of adult rat retinal ganglion cells in vivo does not exclusively depend on phosphatidyl-inositol-3′-kinase/protein kinase B signaling. *J. Neurosci.* **20(18),** 6962–6967.
16. Kaplan, D. R. and Miller, F. D. (2000) Neurotrophin signal transduction in the nervous system. *Curr. Opin. Neurobiol.* **10(3),** 381–391.
17. Meyer-Franke, A., Wilkinson, G. A., Kruttgen, A., et al. (1998) Depolarization and cAMP elevation rapidly recruit TrkB to the plasma membrane of CNS neurons. *Neuron* **21(4),** 681–693.
18. Zhang, X., Smith, D. L., Meriin, A. B., et al. (2004) A potent small molecule inhibits polyglutamine aggregation in Huntington's disease neurons and suppresses neurodegeneration in vivo. *PNAS* **102(3),** 892–897.
19. Leveillard, T., Mohand-Saed, S., Lorentz, O., et al. (2004) Rod-dependant cone viability factor-1, a secreted factor promoting cone viability. *Nat. Genetics* **36,** 755–759.
20. Zhu, H., Bilgin, M., Bangham, R., et al. (2001) Global analysis of protein activities using proteome chips. *Science* **293(5537),** 2101–2105.
21. Stockwell, B. R. (2004) Exploring biology with small organic molecules. *Nature* **432,** 846–854.
22. Rothstein, J. D., Patel, S., Regan, M. R., et al. (2005) Beta-lactam antibiotics offer neuroprotection by increasing glutamate transporter expression. *Nature* **433(7021),** 73–77.
23. http://web.wi.mit.edu/sabatini/pub/siRNA_consortium.html. Accessed August 1, 2005.
24. Barres, B. A., Silverstein, B. E., Corey, D. P., and Chun, L. L. (1988) Immunological, morphological, and electrophysiological variation among retinal ganglion cells purified by panning. *Neuron* **1(9),** 791–803.

Index

A

Adipocyte, *see* Differentiating cell systems
AGT, *see* Alkylguanine-DNA alkyltransferase
Akt
   high content screening assays for target validation
      cell line selection, 368, 373
      immunoassays of phosphorylated proteins, 372, 373–375
      materials, 368
      multiplexed apoptosis and proliferation assay, 373–375
      reagent identification, 368–370, 373, 374
   pathway profiling with translocation analysis, 411
   signaling, 221
   translocation inhibitor discovery with high content screening, 380, 383–386
Alkylguanine-DNA alkyltransferase (AGT), SNAP-tag, 181
Anaphase-promoting complex (APC), axonal growth role, 428
Angiogenesis, high content screening, 26–28
APC, *see* Anaphase-promoting complex
Apoptosis
   Akt signaling pathway high content screening assay, 373–375
   fluorescent probes, 239, 240
   high content screening cancer studies of RNA interference knockdown effects on apoptosis and proliferation
      cells
         seeding on plates, 355, 363
         selection of cell lines, 354, 363
      data analysis
         apoptosis analysis, 362
         proliferation analysis, 362, 363
      image processing
         apoptosis assay, 360–362, 364
         proliferation assay, 362
      imaging of plates, 356, 357, 359, 360
      materials, 353, 354, 363
      overview, 353
      staining
         bromodeoxyuridine, 357–359, 364
         Hoechst 33342 counterstaining, 356
         YO-PrO-1, 356, 363
      transfection, 355, 356
   retinal ganglion cells following optic nerve injury, 427, 428
Aptamers, fluorescent fixed end-point high content screening assays, 147
ArrayScan(r) VTI, features, 42, 52, 54
β-Arrestins, protein-fragment complementation assays for dynamics studies, 228–230
Artificial neural networks, *see* Machine learning
Axiovision, image analysis, 64

B

BacMam, cell engineering, 30
BDNF, *see* Brain-derived neurotrophic factor
Biocarta, pathway database, 321
Bioinformatics, *see* Informatics, high content screening
Brain-derived neurotrophic factor (BDNF), axonal growth role, 428
Bromodeoxyuridine, high content screening cancer studies of RNA interference knockdown effects on apoptosis and proliferation
   cells
      seeding on plates, 355, 363
      selection of cell lines, 354, 363

data analysis
  apoptosis analysis, 362
  proliferation analysis, 362, 363
image processing
  apoptosis assay, 360–362, 364
  proliferation assay, 362
imaging of plates, 356, 357, 359, 360
materials, 353, 354, 363
overview, 353
staining
  bromodeoxyuridine, 357–359, 364
  Hoechst 33342 counterstaining, 356
  YO-PrO-1, 356, 363
transfection, 355, 356

C

Caged compounds
  advantages, 253, 254
  batch transfection, 256
  cage requirements, 253
  delivery to cells, 254, 256
  examples and properties, 254–256
  small interfering RNA and gene knockdown experiments
    glyceraldehyde-3-phosphate dehydrogenase knockdown, 258, 259
    light–dosage working curve generation, 258, 261
    materials, 257
    photoactivation, 259–261
    rationale, 256
    reagent preparation, 257, 258, 260, 261
    transfection, 258, 261
    transfection, 258, 261
  UCOM instrumentation for high content screening, 254
Calcium flux
  fluorescent probes
    protein sensors, 236, 237, 235, 236
    small molecule probes, 235, 236
  mechanisms, 235
CARD, *see* Catalyzed reporter enzyme deposition

Catalyzed reporter enzyme deposition (CARD), fixed end-point high content screening assays, 147
CellCard™
  antiproliferative compound assessment in several cell types simultaneously
  carrier mixing and microtiter plate dispensing, 131, 132
  CellCard preparation, 130, 131, 137
  compound addition, 132
  examples, 133, 134, 136, 137
  experimental design, 130, 131
  image analysis and data visualization, 133
  materials, 130
  scanning, 132, 133
  staining, 132, 137
  tissue culture, 131
  overview of system, 129, 130
Cell counting, fluorescence assays, 240, 241
Cell Lab IC 100, features, 42, 52
Cellome, definition, 3
Cellomics
  definition, 4
  overview, 5, 6
Cell plating, automated
  materials, 109
  plating, 111, 117
  single cell suspension generation, 111, 117
CellProfiler, image processing and analysis, 37
CellwoRx, features, 42, 54
Chemical complementation assay, *see* Protein phosphatases
Chemotaxis
  automated assay
    chemokinesis assay, 111, 112, 118
    materials, 110, 111
    single cell kinetic and immunocytochemical assay, 114–119
  directed algorithm for analysis, 76, 77
Coimmunoprecipitation, protein–protein interactions, 322

Collagen thin film
  alkanethiol-coated support preparation, 102, 103
  automated quantitative microscopy of cells, 99, 104–106
  cells
    culture and specimen preparation, 103
    fixation and staining
      morphology analysis, 104
      green fluorescent protein quantification, 104, 106
    film preparation and characterization, 103, 105, 106
    materials for quantitative measurements of cells, 101, 102
    reference extracellular matrix application, 96–99
    variability within cell populations, 99, 100
Computer vision
  image analysis workflow, 84, 85
  limitations of conventional imaging systems, 85–87
Cytotoxicity
  assay limitations and effective criteria, 419–421
  cellular functions in assessment, 421, 422
  drug discovery
    assay design, 423
    implementation of cytotoxicity assessment, 422, 423
    importance of cytotoxicity assessment, 415, 416
  genotoxicity, high content screening assays, 381, 384–386
  hepatotoxicity quad probe assay, 424
  mechanisms
    calcium dyshomeostasis, 419
    cell membrane effects and transport, 416
    lysosomal effects and autophagy, 417
    mitochondrial effects and energy homeostasis, 416, 417
    nuclear effects and cell proliferation, 416
    oxidative stress, 416–418, 419
  necrosis
    conventional assays, 423
    high content screening, 423, 424

D

Data mining, *see* Informatics, high content screening
Data visualization
  combining data, 307, 308
  data sources, 302, 303
  data transformation, 303–307
  filtering, 311
  interactive visualization, 311
  overview, 301, 302, 312
  prospects, 311, 312
  related data visualization, 310
  spatial and temporal views, 309
  tools, 275, 276
  well- and plate-level information, 309, 310
Differentiating cell systems
  adipocytes
    preadipocyte culture, 122, 123, 126, 127
    triglyceride accumulation assay, 123, 124, 127
  applications, 121, 122
  materials, 122, 125, 126
  osteoclasts
    precursor culture, 124, 127
    resorption assay, 125, 127
Directed algorithms
  associated target identification and measurement, 70, 71
  automated directed algorithm analysis steps, 66
  BioApplications, 64
  developer tools versus specific algorithms, 65, 66
  high content screening problems
    categorization of problems
      cell cycle, 78
      cell health and toxicity, 79
      cell movement, 76, 77
      cell size and shape changes, 75, 76
      colocalization, 74, 75
      counting objects, 73, 74
      interconnected tubular object analysis, 76
      intracellular intensity changes, 72, 73
      neurite outgrowth, 78
      overview, 72, 73

receptor internalization and translocation, 77, 78
spot analysis, 74
translocation, 77
practical utilization of directed algorithms, 79, 80
raw measurement analysis, 71, 72
primary objects
identification, 68
property measurements, 68
Discovery-1, features, 42, 54, 55
Dual-specificity phosphatases, *see* Protein phosphatases

E–F

Extracellular matrix protein, *see* Collagen thin film
FlAsH
applications, 210, 211, 218, 219
protein conformation studies, 210, 218
protein labeling in live cells
background staining reduction/suppression, 216–218
expression, 214, 215, 219
FlAsH loading and staining, 215, 216, 219
fluorescence microscopy, 218, 219
materials, 211, 212
tetracysteine-tagged protein generation, 212, 213, 218
protein tagging, 180, 181
tetracysteine binding and fluorescence induction, 210
Flash photolysis, *see* Caged compounds
Flow cytometry
historical perspective, 6–8
standards, 241
Fluorescence microscopy
autofocus and system performance, 48, 49
automation, 11, 64
cameras, 49
confocal versus wide-field systems, 53, 54
FlAsH in living cells, 218, 219
historical perspective, 6–11
immunofluorescence of small interfering RNA treated cells, 249–251
lifetime imaging, 59
light sources, 45, 46
multiwavelength imaging, 45, 59
numerical aperture of objectives, 47, 48
optical performance parameters, 46, 47
standards, 241
Fluorescence resonance energy transfer (FRET), green fluorescent protein applications, 150
Fluorescent probes, *see also* specific probes
brightness, 144
cellular manipulation combination with high content screening, 155
cellular probes, 238–241
classification, 143, 144
collection modalities and options, 234, 235
fixed end-point high content screening reagents
aptamers, 147
catalyzed reporter enzyme deposition, 147
immunoreagents, 145, 146
molecular beacons, 147
quantum dots, 146, 147
four-color multiplexed immunoassay design
cell plating and incubation, 190, 191
materials, 190, 191
overview, 189, 190
plate reading and interpretation, 191
staining, 190–192
ion indicators, 235–237, 239
live cell and kinetic high content screening reagents
fluorescent analog cytochemistry applications, 149
cell loading, 149, 150
green fluorescent protein fusion proteins, 150
living cell probe capture, 150
overview, 147, 148
fluorescent protein biosensors
engineering, 152, 153

Index

fluorescence resonance energy
transfer, 153, 154
pH indicators, 154
prospects, 154
translocation studies, 154
physiological indicator dyes,
147, 148
positional biosensors, 150–152
live cell experiment design, 234, 235
nonspecific binding, 144
perturbing reactions, 145
photobleaching, 144
phototoxicity, 144
prospects, 155, 156
stability, 144
standards, 241, 242
vendors, 233, 234
voltage-sensing dyes, 237
Forkhead assay, principles, 402,
404, 411
FRET, *see* Fluorescence resonance energy
transfer

G

GAPDH, *see* Glyceraldehyde-3-phosphate
dehydrogenase
Gene MicroArray Pathway Profiler,
pathway database, 321
Genotoxicity, *see* Cytotoxicity
GFP, *see* Green fluorescent protein
Glyceraldehyde-3-phosphate dehydrogenase
(GAPDH), caged small interfering
RNA knockdown, 258, 259
GPCRs, *see* G protein-coupled receptors
G protein-coupled receptors (GPCRs)
activation, 20
high content screening of orphan
receptors, 21, 22, 30
protein-fragment complementation
assays for dynamics studies, 228–
230
Green fluorescent protein (GFP)
biosensors
engineering, 152, 153
fluorescence resonance energy
transfer, 153, 154
pH indicators, 154
prospects, 154
translocation studies, 154, 402

cell engineering for fusion protein
expression
cell line selection, 171
cell selection for expression, 177, 178
design elements, 168, 170
fluorescent protein variant selection,
172
stable transfectant validation, 178
target protein selection, 168,
170, 171
vector design, 172, 174, 176
collagen thin film cell fixation and
staining for automated microscopy,
104, 106
expression effects on cell cycle, 178–180
fluorescent analog cytochemistry, 150
high content screening application
overview, 167, 168
history of research use, 65
limitations, 209
prospects, 180, 181
variants and properties, 165, 166

H

HaloTag™
advantages, 204, 205
cell culture, transfection, and labeling
with ligands, 201, 205
cell-based applications, 197
expression vectors, 198–201
fixation and immunocytochemistry, 201
fluorescence detection, 203, 204
hydrolase reporter, 195, 196,
205, 206
ligand structures, 195, 196
materials, 198
protein tagging overview, 181
Western blot analysis, 204
HCS, *see* High content screening
High content screening (HCS)
components, 11–13, 44
data characteristics, 270, 271,
279, 280
definition, 4
drug discovery, 380
fixed end point versus live cell assays,
13, 14, 54, 55
historical perspective, 6–11, 41, 143

informatics, *see* Informatics, high content screening
installation and operation considerations, 56, 57
market and commercial suppliers, 43
overview, 4, 5, 379
prospects, 14–16, 31
spatial resolution, 34, 44
throughput, 43, 56
time-resolved assays, 35, 36
whole-tissue assays, 36
HPRD, *see* Human Protein Reference Database
Human Protein Reference Database (HPRD), protein–protein interactions, 323

I

Image analysis, *see also* Computer vision; Machine learning; Software
data mining, 52, 53
data storage and management, 53
ImageExpress 5000A, features, 42
ImageExpress Micro, features, 42, 49, 54
Imaging, *see* Fluorescence microscopy; Image analysis; Software
ImageJ, image processing and analysis, 37
ImagePro, image analysis, 50, 64
InCell 1000, features, 42, 54, 55
InCell 3000, features, 42, 45, 55, 56
Informatics, high content screening
data characteristics, 270, 271, 279, 280
data integration with other systems
application/software integration, 278, 279
database integration/federation, 289
data-level integration, 278
overview, 276
data management, 275
data mining, 276
database structure, 272
functional workflow, 294, 295
hardware and network considerations, 273–275
high throughput screening, 285–289, 291

implementation decisions, 285
information explosion, 293, 294
information retrieval challenges, 294
metadata structure, 272
organizational structure
championing research needs, 283
core activities, 283, 284
noncore activities, 284
partnering, 282
results of research organization/information technology partnering, 284
trust building, 283
overview, 269, 270
prospects, 279, 291
research needs for computing, 281, 282
system architecture, 272, 273
Velocity for Life Sciences™, *see* Velocity for Life Sciences™
visualization, *see* Data visualization
volume of data, 270, 271
iPath, pathway database, 322

J–M

JNK, *see* Jun N-terminal kinase
Jun N-terminal kinase (JNK), inhibitor discovery with high content screening, 380–382
KEGG, pathway database, 321
KineticScan[(r)], features, 42
Kolmogorov–Smirnov statistics, high content single-cell chemical complementation assay analysis, 395
Lead candidates, high content screening, 22, 24–26
Machine learning
custom solutions, 92–94
imaging systems
classification, 87–89
image analysis workflow, 87
segmentation, 89–91
limitations of conventional imaging systems, 85–87
overview, 85
training techniques, 91–93
MetaCore
architecture

Index

block, 327, 328
component, 327
effect, 327
overview, 326
transformation, 327
content, 326
experimental data mapping on networks, 330
genomics data mapping
  signature networks for radiosensitive cervical cancer patients, 340
  Tat-upregulated genes at G1/S phase, 336, 339
high content screening data mapping on networks, 336
metabonomics data mapping, 336
network algorithm and filters
  analyze networks algorithm, 329
  analyze transcriptional regulation algorithm, 329
  auto expand algorithm, 329
  direct interactions algorithm, 329
  expand by one interaction algorithm, 330
  overview, 328, 329
  self regulations algorithm, 329
  shortest paths algorithm, 329
network statistical analysis, 330, 332
overview, 345, 346
pathway database, 321, 322
proteomics data mapping, 340–342, 345
p-value and statistical significance evaluation, 332, 333
MetaDrug
  applications, 334–336
  overview of features, 334, 345, 346
  prospects, 334
MetaMorph, image analysis, 50
Mitogen-activated kinase phosphatases, *see* Protein phosphatases
Molecular beacons, fixed end-point high content screening assays, 147
Morphology profiling, subpopulation dynamics, 35

N

Natural language processing (NLP), data mining, 322

Necrosis, *see* Cytotoxicity
Networks
  definition, 320
  information flow in cells, 320, 321
  MetaCore, *see* MetaCore
  theory and tools for analysis, 323–326
Neural networks, *see* Machine learning
Neurite outgrowth
  activator discovery with high content screening, 381, 384, 386
  automated assay
    addition of analytes, 112, 118
    fixation and labeling, 113, 114, 118
    imaging of plates, 114
    materials, 110
  directed algorithm for analysis, 78
  retinal ganglion cell growth promoter screening
    cell purification and culture, 431
    image analysis, 432
    materials, 430
    principles, 429–431
    staining and imaging, 431, 432
    statistical analysis, 432
    validation of hits, 432, 433
NLP, *see* Natural language processing
N-tier architecture, informatics systems, 273
Nuclear translocation, *see* Protein translocation

O–P

Opera, features, 42, 45, 54, 56
Organelles, fluorescent probes, 239
Osteoclast, *see* Differentiating cell systems
p53, Hdm2 interaction assay, 404
PathArt, features, 334
Pathway
  databases, 320–323
  definition, 320
  high content translocation assays, *see* Protein translocation
  information flow in cells, 320, 321
  profiling in drug discovery, 402
PathwayAnalysis, features, 333
Pathway HT, features, 42, 54–56
Pathway Studio, features, 334

PCA, *see* Protein-fragment
        complementation assay
Phosphatases, *see* Protein phosphatases
Phosphatidylinositol-3-kinase (PI3K),
        translocation analysis of pathway,
        411
PI3K, *see* Phosphatidylinositol-3-kinase
PKA, *see* Protein kinase A
PKC, *see* Protein kinase C
Plating, *see* Cell plating, automated
Potassium, fluorescent probes, 239
Protein kinase A (PKA), fluorescent cyclic
        AMP biosensors,
        150–152
Protein kinase C (PKC), translocation
        inhibitor discovery with high
        content screening, 380, 383, 384,
        386
Protein Lounge, pathway database, 321
Protein phosphatases
    dual-specificity phosphatases, 389
    high content single-cell chemical
        complementation assay
        cell transfection, treatment, and
            processing, 392–394, 399
        data set evaluation
            archived image inspection, 396,
                397
            false-positives, 395, 396
            Kolmogorov–Smirnov statistics,
                395
            software, 394, 395, 399
            subpopulation analysis,
                395, 399
        image acquisition and
            analysis, 394
        materials, 391, 392
        principles, 390
        prospects, 398, 399
        secondary assays, 397, 398
    mitogen-activated kinase phosphatases,
        389
Protein translocation
    directed algorithms, 77
    fluorescent proteins, 154, 401, 402
    high content screening
        nuclear translocation
            Forkhead assay, 402, 404, 411
            materials, 380

transcription factor activation, 382,
    383, 385, 386
pathway profiling
    Akt pathway, 411
    cell lines, 404, 405, 411, 412
    clone testing, 408
    expression vectors, 405, 406
    fluorescent protein selection and
        orientation, 405, 406
    optimization of assay, 408, 409
    overview, 401, 402
    pathway function assessment, 410–
        413
    phosphatidylinositol-3-kinase
        pathway, 411
    target selection, 405
    target validation, 409, 410
    transfectant selection and testing,
        407, 408
    transfection, 406, 407
plasma membrane translocation
    GLUT4 assay, 404
    materials, 380
    p53-Hdm2 assay, 404
    protein kinase C/Akt assay, 383–
        386
Protein-fragment complementation assay
    (PCA)
    drug discovery advantages, 226
    drug effect studies, 230
    pharmacological profiling, 230, 231
    principles, 223–225
    protein complexes
        drug interactions, 221–223
        dynamics studies, 228–230
        localization and quantification in
            cells, 223–225
    subcellular localization of protein
        complexes, 226, 228
p-value, network statistical significance
    evaluation, 332, 333

Q–R

Quantum dots, fixed end-point high content
    screening assays,
    146, 147
Raf, protein-fragment complementation
    assays for dynamics studies, 229,
    230

Ras, protein-fragment complementation assays for dynamics studies, 229, 230
ReAsH, principles, 210
Reference standards, high content screening, 57, 58
Retinal ganglion cell (RGC)
  apoptosis following optic nerve injury, 427, 428
  axonal growth factors and identification, 428, 429, 433
  neurite outgrowth promoter screening
    cell purification and culture, 431
    image analysis, 432
    materials, 430
    principles, 429–431
    staining and imaging, 431, 432
    statistical analysis, 432
    validation of hits, 432, 433
  pathology, 427
RGC, see Retinal ganglion cell
RNA interference, see also Small interfering RNA
  genomic screens for gene target identification, 245
  high content screening cancer studies of gene knockdown effects on apoptosis and proliferation
    cells
      seeding on plates, 355, 363
      selection of cell lines, 354, 363
    data analysis
      apoptosis analysis, 362
      proliferation analysis, 362, 363
    image processing
      apoptosis assay, 360–362, 364
      proliferation assay, 362
    imaging of plates, 356, 357, 359, 360
    materials, 353, 354, 363
    overview, 353
    staining
      bromodeoxyuridine, 357–359, 364
      Hoechst 33342 counterstaining, 356
      YO-PrO-1, 356, 363
    transfection, 355, 356
  immunofluorescence study validation
    cell plating for transfection, 246, 247
    immunofluorescence of small interfering RNA treated cells, 249–251
    materials, 246, 251
    overview, 245, 246
    siIMPORTER™ transfection reagent, 247, 251
    small interfering RNA expression plasmids
      oligonucleotide design and cloning, 248
      transfection, 248
      transfection complex, 248, 251
    transfection with small interfering RNA duplexes, 247, 248
  physiological functions, 245
  sample manipulation, 36
  target validation, 409, 410
siRNA, see Small interfering RNA
Small interfering RNA (siRNA), see also RNA interference
  caged small interfering RNA and gene knockdown experiments
    glyceraldehyde-3-phosphate dehydrogenase knockdown, 258, 259
    light–dosage working curve generation, 258, 261
    materials, 257
    photoactivation, 259–261
    rationale, 256
    reagent preparation, 257, 258, 260, 261
    transfection, 258, 261
    transfection, 258, 261
  profiling, 28
SNAP-tag, protein tagging, 181
Sodium, fluorescent probes, 239
Software
  commercial sources by application, 50–52
  customization, 28–30, 92–94
  directed algorithms, see Directed algorithms
  imaging, see also Computer vision
    acquisition and control, 50
    analysis, 37, 50–52

integration between systems, 278, 279
machine learning, *see* Machine learning
prospects for automated assay development, 59
statistical analysis, 394, 399
Systems cell biology
  definition, 4
  levels, 5

T

Target validation
  Akt signaling pathway high content screening assays
    cell line selection, 368, 373
    immunoassays of phosphorylated proteins, 372, 373–375
    materials, 368
    multiplexed apoptosis and proliferation assay, 373–375
    reagent identification, 368–370, 373, 374
  directed algorithms, 70, 71
  overview of process, 367

Tetracysteine, *see* FlAsH
Transcription factors, high content screening, 26, 402, 404, 411

V–Y

Velocity for Life Sciences™ (VVLS)
  advantages, 300
  content integration, 296, 297
  document clustering, 296
  intelligent query routing, 296–300

Visualization, *see* Data visualization
Voltage-sensing dyes, fluorescent probes, 236, 237
VVLS, *see* Velocity for Life Sciences™
Western blot
  HaloTag protein fusions, 204
  target validation, 409, 410
Yeast, high content screening applications, 33, 34
Yeast two-hybrid system, protein–protein interactions, 322